The Ultimate Guide To Alabama Fishing

Mike Bolton

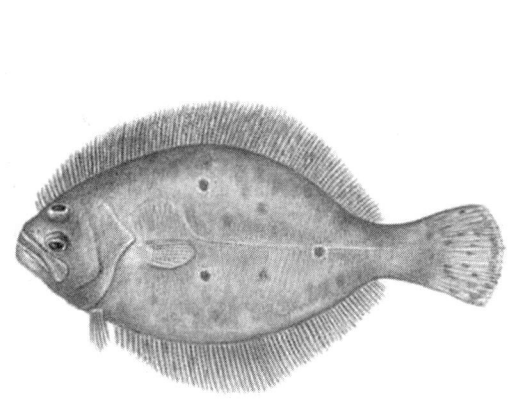
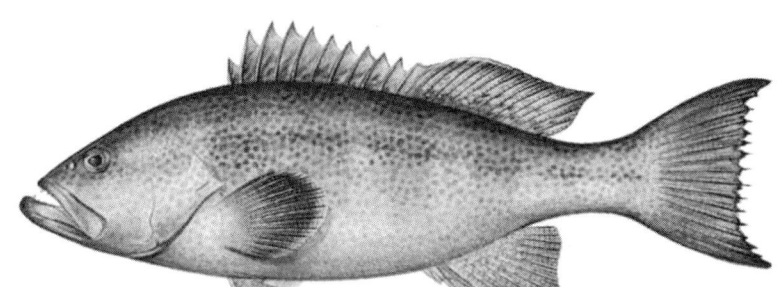

B Direct LLC
Birmingham, Alabama

The Ultimate Guide to Alabama Fishing

Dedication

To my No. 1 fishing and hunting buddy, my daughter Lauren.

Published by B Direct LLC
221 Red Oak Road
Birmingham, Alabama 35206

Copyright © 2013 Mike Bolton
All rights reserved.
Reviewers and writers of magazine and newspaper articles are free to quote passages of this book as needed for their work. Otherwise, no part of this book may be reproduced or transmitted in any form or by any meahns, electronic or mechanical, including photocopying, recording or by any information storage and retrieval system, without the written permission of the publisher.

Library of Congress Control Number: 2012954474

ISBN 978-1-59421-094-5

Manufactured in the United States of America by
Seacoast Publishing, Inc.
1149 Mountain Oaks Drive
Birmingham, Alabama 35226

To obtain copies of this book, please contact:
B Direct LLC
221 Red Oak Road
Birmingham, Alabama 35206
205.987.0354
www.gofishingal.com

Contents

Introduction 5

Fishing Alabama, Region by Region 7
North Alabama.................................8
Central Alabama 11
West Alabama 15
East-Central Alabama 18
Southeast Alabama 22
Southwest Alabama 27

Tennessee River 31
Guntersville Lake 32
Wheeler Lake 40
Wilson Lake 50
Pickwick Lake 57

Coosa River 66
Weiss Lake 67
Lake Neely Henry 78
Lake Logan Martin 88
Lay Lake 99
Lake Mitchell 110
Lake Jordan 119

Warrior River................................ 129
Smith Lake 130
Bankhead Reservoir 139

North River 146
Lake Tuscaloosa 147

Chattahoochee River 154
West Point Lake 155
Lake Eufaula 168

Tallapoosa River 178
Lake Harris 179
Lake Martin 186

Alabama River 197
Jones Bluff Reservoir 198
Miller's Ferry 210

Tombigbee River 220
Aliceville Lake 221
Gainesville Lake 229
Demopolis Lake 239

Mobile Delta 249

The Gulf Coast 260
Gulf of Mexico 261
Mobile Bay 275
Perdido Bay 279

More Fishing Opportunities
Alabama's State Lakes.................. 283
Pond Fishing 287

Fish Identification
Freshwater Fish 289
Saltwater Fish 294

Tackle & Bait
Rods and Reels 302
Terminal Tackle 306
Lines and Knots 309
Live Bait 311
Other Tackle............................... 314

Catching Fish 317

The Ultimate Guide to Alabama Fishing

About the author

    Mike Bolton has served as an outdoors writer for *The Birmingham News* more than 30 years and has spent thousands upon thousands of hours on Alabama waterways with not only the state's best tournament and weekend anglers but pro anglers as well. It was during those hundreds of fishing trips that he was able to observe the fishing locations and techniques used by Alabama's top fishermen. Many of those top anglers were gracious enough to mark structure on maps and discuss at length their successful techniques on tape. All told, more than 80 fishermen contributed to this book.

    Bolton attended the University of Alabama and worked in the Sports Information Department of the U of A Athletic Department from 1976 until 1980. He served as editor of *'Bama Magazine* and has worked for newspapers in Tuscaloosa, Montgomery, Enterprise and Birmingham. In Birmingham he won eight AP Sweepstakes awards for the best sports story of the year in Alabama.

    He has been married to the former Elizabeth Rae Shryock of Tampa, Fla. for more than 28 years and they have two children, Cory, and Lauren. They live on an eight-acre lake in Springville.

Alabama: America's best fishing?

Get a map of the United States and spread it out before you.

Other than Alabama, count the states that you imagine have the potential to give up 16-pound bass, 4-pound bream, 8-pound spotted bass, 10-pound smallmouth bass, 4-pound crappie, 40-pound striped bass and 80-pound catfish, not to mention 60-pound grouper, 100-pound tarpon, 75-pound amberjack, 35-pound red snapper, 8-pound speckled trout and 55-pound king mackerel.

Make a list of states with fishing diversified enough that you can catch a rainbow trout on a flyrod and hours later catch a 30-pound striped bass (at Smith Lake), or drop a minnow and catch largemouth bass, smallmouth bass, five species of catfish, sauger, white crappie, black crappie, drum, skipjack herring, five species of sunfish, striped bass, white bass and hybrid bass (at Wheeler Lake) or catch a 5-pound bass on one cast and a 10-pound redfish on the next (in the Mobile Delta).

Can the overall fishing in any state compare with Alabama's? I honestly believe not. California has given up enormous largemouth bass, but the state has never seen a 4-pound bream or a 4-pound crappie. Texas has given up its share of big bass in recent years, but smallmouth bass don't exist there. Nor has the Lone Star state ever given up a 4-pound bream or a 4-pound crappie.

Georgia? It gave up the world-record largemouth in 1932 but hasn't come close since. Almost every state fishing record in Alabama is larger than Georgia's. Florida you say? It has lots of big bass, but not the first 4-pound crappie. Smallmouth bass and spotted bass can't be found in that state.

Check out the state record bass in each of the 50 states. The state record for 44 of those states wouldn't make the list of the 20 largest bass ever caught in Alabama. More than 75 percent of the states have never given up a bass larger than 9 pounds.

Study the nation's 60 largest metropolitan areas. Can anywhere compare with 55th-ranked Birmingham? The state's largest city is less than 100 miles from Weiss Lake, the crappie capital of the world; Lake Guntersville, the lake that pros call one of the nation's best bass lakes; Smith Lake, home of 40-pound plus striped bass; Lake Martin, which has given up a 51-pound striped bass; Lake Pickwick, which gave up a IGFA line-class world record smallmouth bass; Aliceville Lake, which gave up one 16-pound bass and two 14-pound bass in less than a one-month period and Wheeler Lake, which gave up a 70-pound-plus and an 80-pound-plus blue catfish in 19 days.

That doesn't include other public water in a two-hour drive—water such as Wilson Lake, the Warrior River, Demopolis Lake, Holt Reservoir, Inland Lake, Lake Jordan, Lay Lake, Lake Logan Martin, Lake Mitchell, Lake Neely Henry, Oliver Reservoir, Lake Harris, Lake Thurlow, Lake Tuscaloosa, the Warrior Reservoir in Eutaw and Yates Reservoir.

That adds up to 26 bodies of water and more than a quarter-million acres of excellent fishing anglers in most states would die for.

The lifelong Alabama resident fishermen may take that for granted, but the transplanted angler doesn't. One has to only look at nearby Southern cities such as Louisville, where an angler's best bet is the nearby filthy Ohio River, or Atlanta, where the angler's best close bet is overcrowded Lake Lanier, to see what Alabama has to offer.

While serving as outdoor writer for *The Birmingham News* for the past 30 years, I was fortunate enough to wade the Cahaba River with my flyrod and catch a 2-pound bream, catch and release more than 60 largemouth bass caught on topwater in one day on Lake Guntersville, catch a 5-pound smallmouth in Wheeler Lake, catch a 25-pound striped bass from Lake Martin, catch and release 123 hybrid bass in one day below Wilson Dam, catch a limit of slab crappie on Weiss Lake, catch two 10-pound bass in one day on a private lake and land a 55-pound grouper and 60-pound amberjack—all in a 67-day period.

That's the kind of fishing Alabama has to offer.

I hope this updated version of *The Ultimate Guide to Alabama Fishing* will give Alabama's anglers—from the beginning perch-jerker to the accomplished bass angler—a better insight into fishing in Alabama and greatly increase his fishing chances in a state where his fishing chances are already tremendous.

To call putting together *The Ultimate Guide to Alabama Fishing* work is stretching the truth a bit. Being able to fish every lake in Alabama with more than 100 of the state's best fishermen was nothing but pure joy. The long nights at the computer transferring those tape-recorded conversations into the written word was as close to work as it got, but those long nights also served to rekindle the wonderful memories of Vienna sausage dinners in rain-filled boats, hours on the water with wonderful people and the stories of the big one that didn't get away.

Good fishing,
Mike Bolton

Mike Bolton

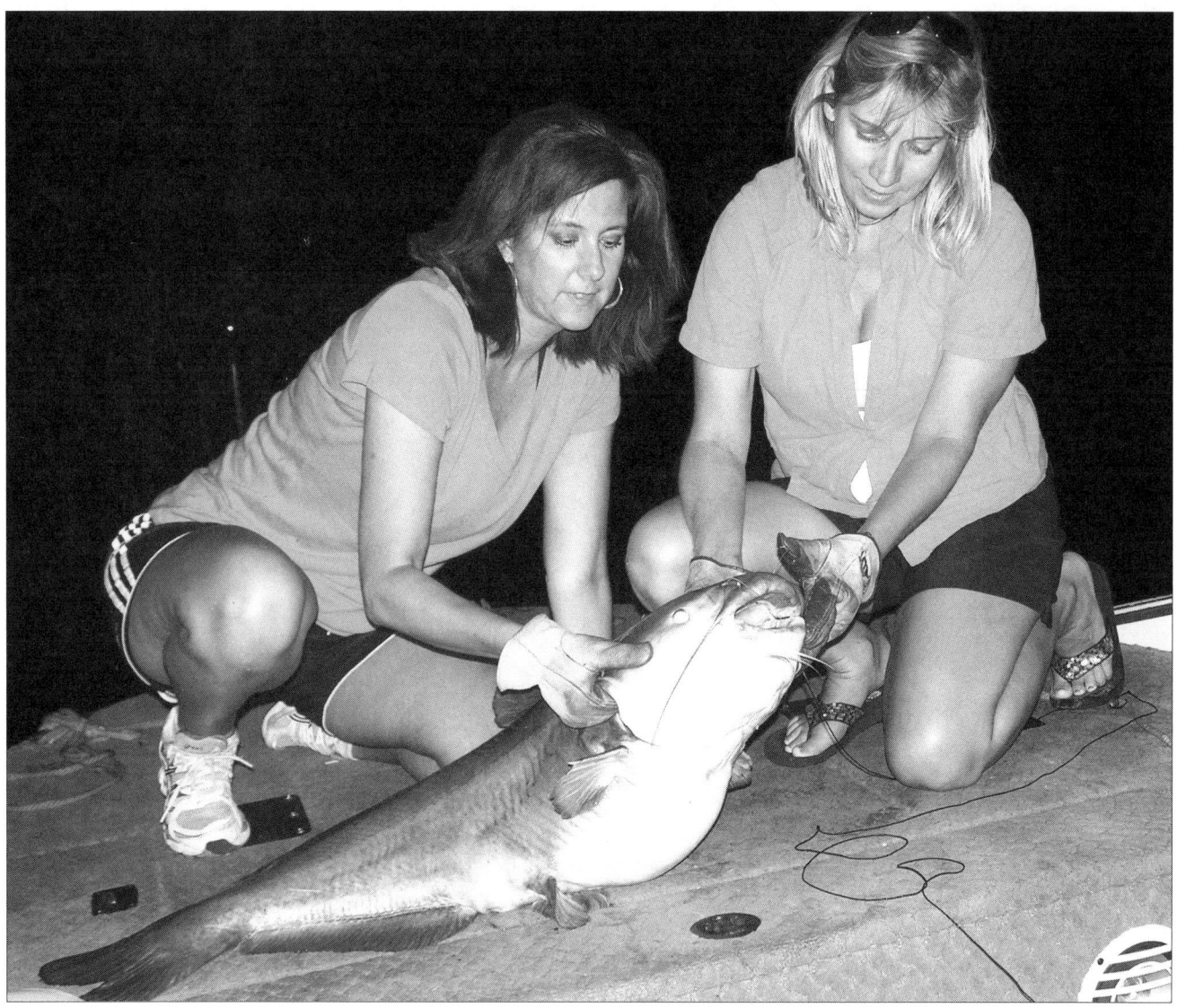

Fishing in North Alabama

Twelve counties in North Alabama provide fishermen with thousands of acres of prime fishing waters ranging from major lakes to small streams and creeks.

The four major lakes on the Tennessee River provide more than 182,000 acres of water. Smaller reservoirs, such as those found in the Bear Creek watershed, provide the opportunity to fish in lakes which are not quite as large.

Several streams offer the solitude of wading and float fishing. Farm ponds and fee-charging fishing lakes, including two state-operated lakes, round out the opportunities for angling enjoyment.

Reservoirs

The mainstream reservoirs of the Tennessee River —Guntersville, Wheeler, Wilson and Pickwick—provide excellent fishing for thousands of anglers each year. Crowds of fishermen can be found on these lakes in early spring when crappie are concentrated around submerged treetops and stumps. Many fishermen use live minnows for bait on cane poles as did their fathers and grandfathers. Other fishermen prefer rods, reels and fish small plastic jigs.

Bream fishing is most productive on all these impoundments during the spring when the bream are bedding in shallow water and during summer mayfly hatches. Fishing under trees heavily laden with mayflies, or willow flies as they are called locally, can result in large catches of bream as well as bass and catfish.

The best fishing for channel catfish usually occurs in the late spring when catfish are spawning. Catfish like

North Alabama Fishing Opportunities

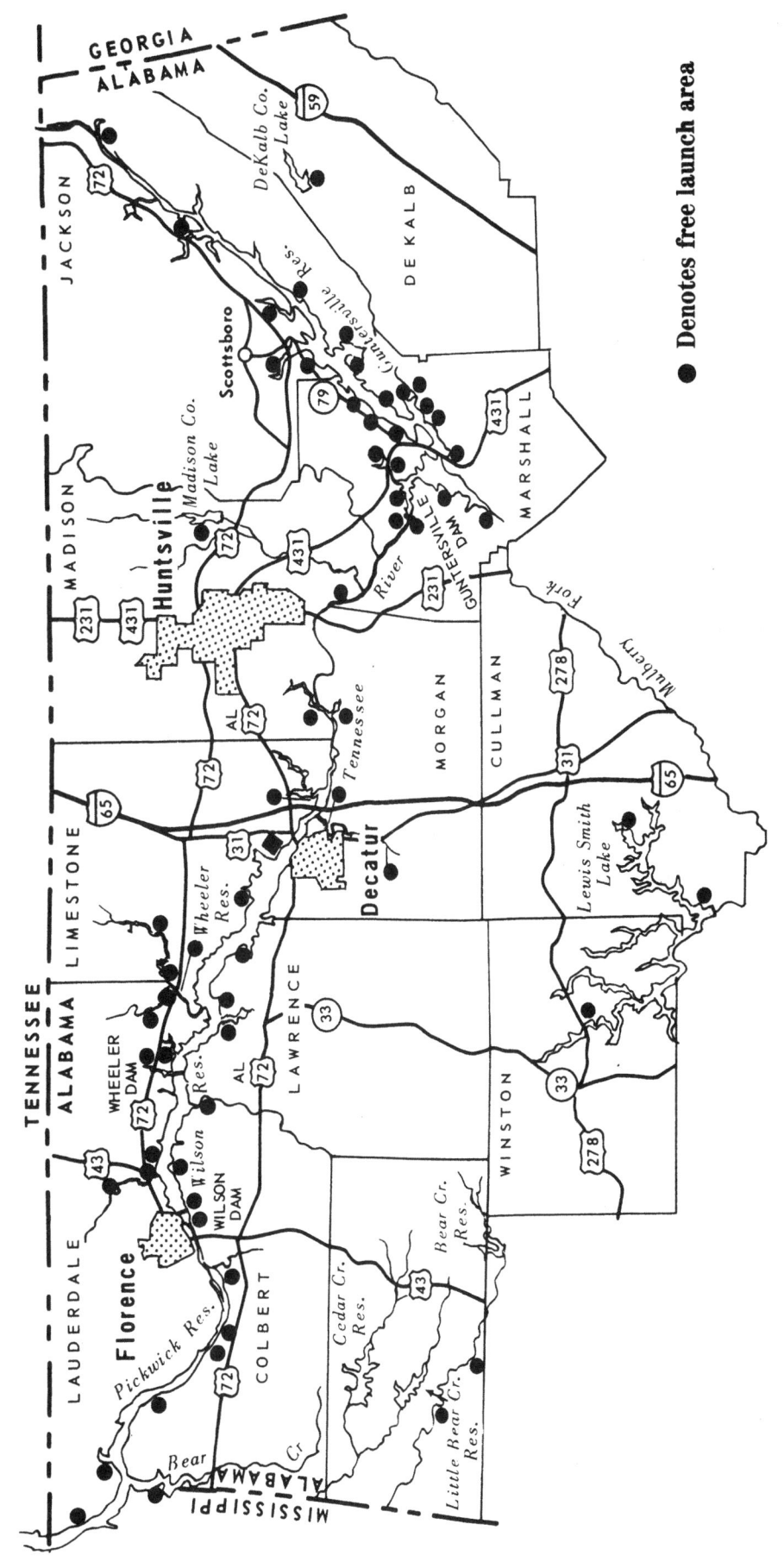

to spawn in holes under rocks, so fishermen frequently fish along steep rocky banks or riprap. They fish baits such as gobs of live worms or chicken liver along the edges of the rocks. Although most of the catfish caught from these reservoirs weigh less than 10 pounds, some large ones are present. A 106-pound blue catfish was caught by a commercial fisherman in Lake Guntersville. The state rod-and-reel record fell three times in 19 days below Lake Guntersville Dam in 1988. That record eventually ended up at 82 pounds. Flathead catfish larger than 50 pounds are caught occasionally.

Saltwater striped bass and hybrid striped bass have been stocked by WFF to establish an additional sport fishery for the fishermen of Alabama. More than 3 million saltwater and hybrid striped bass have been stocked in Guntersville, Wheeler and Wilson reservoirs. Catches in the reservoirs are sporadic, but the tailwaters have produced some excellent catches during the spring and fall.

White bass are popular game fish in these reservoirs. Fishermen harvest them most frequently by jump fishing during the summer when white bass break water chasing shad. Trolling or fishing under lights at night is also an effective technique.

Largemouth and smallmouth bass are present in the Tennessee River reservoirs. Largemouth bass fishing is good in all four reservoirs. Thanks to incredible catches in Bassmaster events in recent years, Lake Guntersville has become well-known nationwide as one of the nation's top bass fishing lakes.

Smallmouth bass fishing is confined primarily to the lower end of Wheeler Lake and in Wilson and Pickwick lakes. Some of the most popular smallmouth fishing occurs below dams in November and at night during the summer. Fishermen drifting live shad below the dams catch October and November fish while anglers casting spinnerbaits or jigs often have excellent catches at night.

Tailwaters

The tailwater areas below Guntersville, Wheeler and Wilson Dams offer some of the most exciting fishing in the Tennessee Valley. These dams serve as barriers to upstream migrations of fish. Consequently, large concentrations of fish seasonally congregate below the dams. Because of these concentrations, catches are usually excellent. TVA records over a seven-year period indicate that fishing trips in these tailwater areas averaged 198,426 fishing trips per year.

These areas provide the greatest diversity of fishing in the state. Although the majority of tailwater fishermen pursue largemouth bass, smallmouth bass, sauger, catfish and saltwater striped bass, they can expect to catch any of the approximately 20 species of game fish present.

The type of fishing found in tailwater areas is seasonal. During the winter, sauger are abundant and are sought by many fishermen braving the elements. Walleye are also present but not as abundant as sauger.

In the spring after the spawning runs of sauger has tapered off, crappie, white and yellow bass are caught in large numbers. Saltwater striped bass and hybrid striped bass start hitting in late spring and continue until fall. Largemouth bass also are often caught below the dams in the spring. The smallmouth bass fishing below Wheeler and Wilson Dams is often referred to as some of the best in the South. Spring and fall produce the best smallmouth fishing.

Catfish are readily caught below these dams. Live or cut baits fished on bottom catch the most fish, including blue, channel and flathead catfish. June is usually the month of choice for catfishing.

One should always exercise caution and basic common sense in tailwater areas. The Alabama Marine Police Division requires that each boater within 800 feet of hydroelectric dams and/or navigation locks must wear a Coast Guard approved life preserver.

Small impoundments and streams

Four small reservoirs, constructed by TVA in the Bear Creek watershed, provide an opportunity for fishermen in the Franklin County area to have good fishing close to home.

The four lakes, ranging from 670 acres to 4,200, are controlled by the Bear Creek Authority. A BCA user permit must be obtained prior to swimming, fishing or boating. Camping areas are available at most of the lakes, and restaurants and motels are present in Russellville and Haleyville.

The Sipsey River, in the Bankhead National Forest, and Little River, above the fall at Highway 35 near Fort Payne, are excellent float fishing streams. Tributary streams to the Paint Rock River offer scenery and serenity to the wade fisherman. All these streams provide beautiful scenery and super fishing for small bass and bream. Canoes and small flat-bottom boats are excellent for floating the streams and fishing. Light spinning tackle with a four- or six-pound test line is ideal.

Light spinners are preferred and will catch both bream and bass. Late spring is the most pleasant and productive season to float fish.

Mike Bolton

Fishing Central Alabama

Fishing in Central Alabama provides anglers with many diverse fishing experiences. In this 11-county region are major lakes, reservoirs, small streams, swamps, ponds and the longest stretch of undammed river east of the Mississippi.

Reservoirs

Twelve major lakes are in this area and each exceeds 500 acres. Lake Jordan, a 6,800-acre reservoir, is the number one choice for many area fishermen. This Alabama Power Company reservoir is on the Coosa River a few miles north of Wetumpka. Largemouth and spotted bass fishing is best from March through May. This lake has relatively little shallow water so most bass are caught along the shoreline from the mouths of creeks to the upper end of sloughs.

From February to April, crappie fishing is in full swing with fish commonly being caught in the three-quarter pound range. Preferred habitat for spawning crappie is submerged tree tops along steep banks.

After bass and crappie fishing slows down, catfishing begins to pick up with the summer months being the peak season. Trotlines are used to harvest most catfish. The bait is suspended to 20-foot depths.

Striped and hybrid striped bass have been stocked heavily in Lake Jordan. These fish have established a popular fishery with catches of stripers in the 5- to 10-pound range. These fighters are caught in the mouths of creeks during late spring, early summer and winter. Autumn brings on jump fishing where fishermen scan the lake with binoculars for the splash of white bass and stripers feeding on schooling shad, then rush to the feeding frenzy as quickly as possible casting spoons, spinners and medium-size jigs.

A unique feature of Lake Jordan is that it is impounded by two separate dams, Jordan Dam and Walter Bouldin Dam. Bouldin Dam created a 1,000-acre embayment which is connected to Lake Jordan by a canal. The upstream end of this canal is a top fishing spot on the lake.

For the fisherman wishing to catch a trophy largemouth, Lake Harding, also known as Bartlett's Ferry, is a good lake to fish. This area has produced several bass in the 13-pound range, with eight-pounders caught quite regularly.

Lake Harding is primarily a river-run impoundment of the Chattahoochee River. The major backwater area is the Halawakee Creek arm and most fishing occurs there. Besides the noted largemouth bass fishing in Lake Harding, excellent catches of crappie are reported each spring. Hybrid striped bass have been stocked and are commonly caught from most areas of the lake.

Two unique small reservoirs on the Tallapoosa River are Yates and Thurlow. Both reservoirs are famous for their shellcracker populations, with many of those caught being in the one-pound range. Few public waters have trophy bream so these lakes are highly recommended to avid ultra-light tackle and fly fishermen. Live worms or small spinners fished along the shoreline during the summer months bring best results.

Jones Bluff Reservoir is between Montgomery and Selma on the Alabama River. It has become a popular fishing spot in recent years with bass fishing clubs. Only a small amount of surface water extends beyond the original river banks in this river-run impoundment.

Fishing this lake can be a real challenge because there is normally current flowing along the flooded river channel. A few creeks have been flooded along both banks and provide some backwater fishing. Largemouth bass can be caught along sandbars, bluffs and in the creeks.

A popular nighttime fishing experience during summer is jug fishing for catfish. Plastic bottles of all sizes are rigged at varying depths with shad pieces, minnows and freshwater clams. These jugs are scattered around the boat, and then the fishermen casually drift downstream and occasionally check the jugs with a flashlight. An actively moving jug means a fish is on the hook. Often a large catfish can take the jug completely under the surface, resurfacing several yards away, and adding to the excitement of this unique nighttime method of fishing.

Tailwaters

The most popular tailwater fishing for boaters in central Alabama is definitely below Mitchell Dam. In February, fishing begins with occasional catches of walleye and drum. March and April bring on fishing in earnest with spotted and largemouth bass, drum, catfish, white bass, and hybrid striped bass filling the creels.

The largemouth bass are generally caught along the rock-strewn river banks with live shad or artificial baits. Your best bet to catch the other species of fish is to anchor near the dam and fish with live or cut shad which can usually be dipped below the dam. Another popular live bait is the mayfly larvae, commonly called willow grubs. Be sure to take a variety of spoons, spinners and jigs in case the shad are not running.

The natural rocky habitat below Jordan Dam has long been famous for spotted bass. This area is good for catfish, drum, white bass and hybrid striped bass. Prime fishing time is from March through June. Small boats can be put in at a primitive ramp a few hundred yards below the dam on the east side. However, bank fishing is easy and there is plenty of parking.

The tailwater below Walter Bouldin Dam is a large canal dug in clay material with only a few large rocks. Species caught in the prime fishing season are the same as in the Jordan Dam tailwater. Bank access on the west side is good with adequate parking. Larger boats can navigate this tailrace area. However, boat ramps are several miles away at Fort Toulouse and in Wetumpka.

For fishermen who want to catch a large striped bass, the tailwaters below Martin, Yates, and Thurlow dams on the Tallapoosa River are prime striper areas. All of these tailraces have yielded fish in the 20- to 30-pound range. The state record of 55 pounds, in fact, was taken from this area and nearly broken in 1983. May, June and July are the best fishing months, and the most popular baits are live shad, one-half to 1-ounce bucktail jigs and large floating lures such as the Redfin.

Although the three areas can be reached by boat, most of the fish are caught by bank fishermen. The reported catches of smallmouth bass below Martin Dam and walleye, frequently caught below Thurlow Dam, further contribute to the uniqueness of this fishing area.

Fishermen in east-central Alabama also have excellent tailwater fishing. The area below Bartlett's Ferry Dam has developed into an excellent hybrid striper fishery in addition to the usual largemouth bass and catfish catches. Fishing techniques here are similar to those of other areas; however, catfish are commonly caught on freshwater clams collected from shallow water areas along the banks and sandbars downstream.

Bank fishing access below Bartlett's Ferry Dam is nonexistent but boat access is excellent. A launch ramp is immediately below the dam on the Alabama side.

Lucky are the fishermen who live in Phenix City. They are often able to catch their limit of hybrid stripers during an hour lunch break.

A small dam, the Eagle-Phenix Dam, in downtown Phenix City has helped create some of the best urban fishing in the southeastern United States.

Hybrids congregate in large numbers and provide some exciting fishing from March to early June. Only occasional high water will deter fishermen from lining the west bank in hopes of out-fishing friends next to them. Jigs, floating plugs and live shad are favorite lures.

The availability of excellent tailwater fishing in central Alabama means fishermen do not need an expensive boat and electronic equipment. A short drive in the family car can often place one in an area with the possibility of acquiring a full stringer of fish.

Rivers

Individuals who enjoy the solitude of float fishing a free-flowing stream or river should try the scenic Cahaba River. The tail end of this small river boasts a 185-mile stretch of river without dams.

Early in the spring several areas along the Cahaba produce excellent walleye fishing. During most of the year bass and catfish are the primary species sought. Access is relatively good. There are many bridge crossings but there are no ramps so you will want to use a canoe or jon boat which you can slide up and down the banks. The beauty of the high bluffs and overhanging trees makes any trip on this river worthwhile.

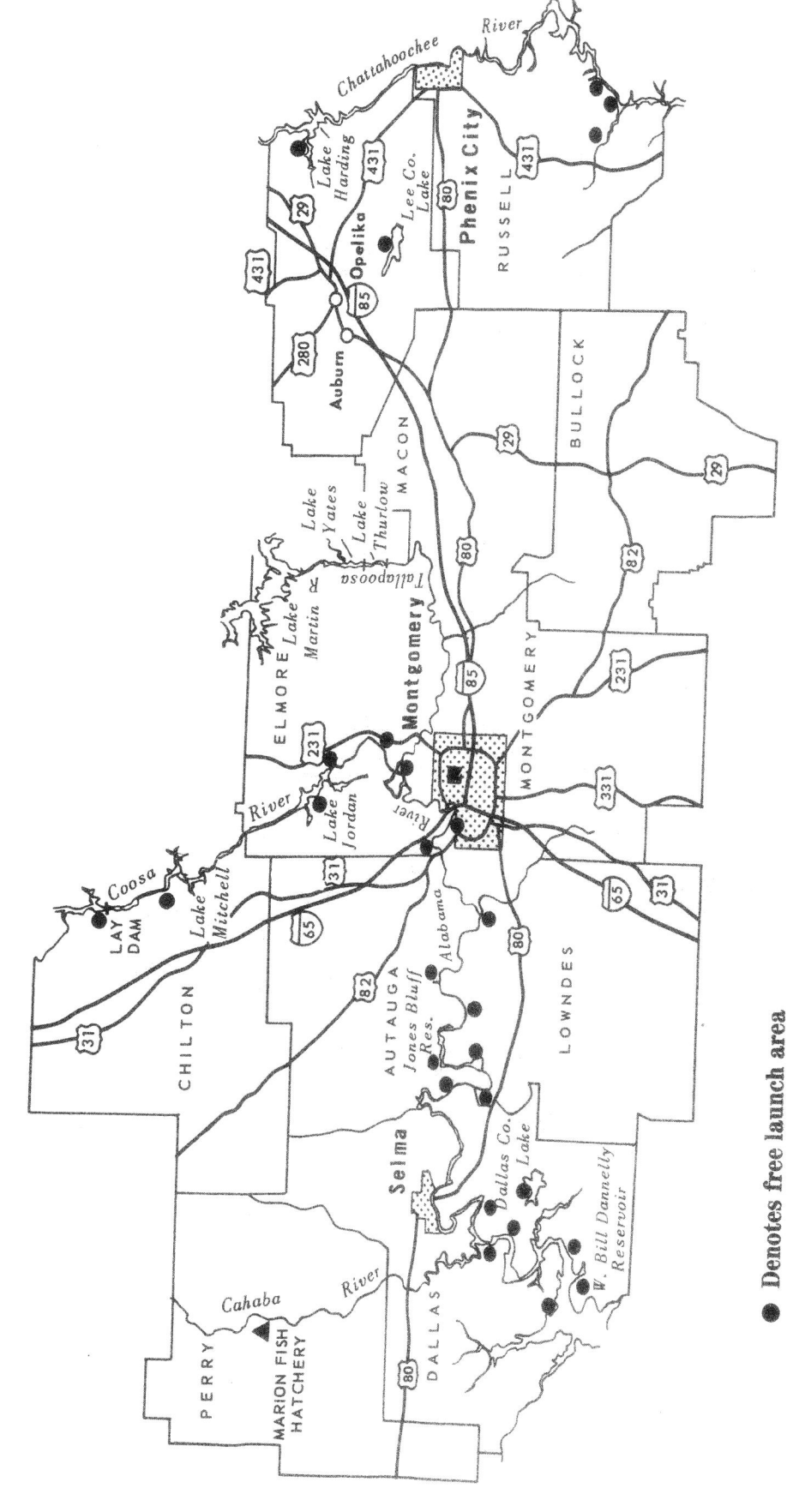

Small Streams

Central Alabama contains hundreds of small streams from six to more than 100 feet wide. Phenix City, Auburn, Opelika, Tallassee, Eclectic, Wetumpka, and Clanton, are located on or near the fall line, a physiographic feature that sharply limits the range of certain species of fish. Above the fall line, many streams contain cobbles, boulders and waterfalls. Below the fall line, streams typically have a sandy bottom, few rocks, and numerous large sandbars.

Fish in small streams are generally smaller than those in lakes and rivers thus ultra-light spinning tackle is ideal. Small lures are more productive and will take a wider assortment of species than larger lures. A small Mepp's spinner (1/32 to 1/16 ounce) is probably the best lure for creek fishing and small diameter line (1- to 4-pound test) is needed to cast small lures effectively.

Species commonly caught from small streams include shellcracker, bluegills, redbreast sunfish, spotted bass, largemouth bass and redeye bass.

In the spring, walleye, white bass, striped bass and hybrid striped bass often enter small streams to spawn. Other species caught in creeks include black crappie, white crappie, green sunfish, rock bass, channel catfish, bullheads, suckers and carp.

Halawakee, Osanippa, Wacoohee, Uchee and Little Uchee Creeks hold the distinction of being the only creeks in the area where shoal bass occur.

A few of the more impressive creeks in the area are Autauga Creek (Autauga County), Bogue Chitto and Barron Creek (Dallas County), Sofkahatchee and Channahatchee Creek (Elmore County), Chewaca, Saugahatchee, and Little Uche Creek (Lee County), Big Swamp Creek (Lowndes County), Cubahatchee and Thobolocco Creek (Macon County), Catoma Creek (Montgomery County), Oakmulgee and Coffee Creek (Perry County), Watermelon and North Fork of Cowikee Creek (Russell County).

The lower portion of some of the larger creeks in the area can be float fished from a canoe or flat-bottom boat. Boat launch sites on small streams are few, crude and generally on private property so the best way to gain access to a floatable creek is to slide a canoe or light boat in at a bridge crossing on a public road.

Swamps

There are numerous swamps in Central Alabama, particularly in Macon, Bullock and Lowndes Counties. Even the Pea and Conecuh Rivers in Bullock County are mostly swamp habitat. Although swamps are probably the most overlooked fishing habitat, the fishing in them can be fantastic, especially when the water is low and the fish are concentrated.

Bluegill, bass, warmouth, bowfin, shellcracker, chain pickerel and redfin pickerel are common in swamps and oxbows. People fishing in them should take along a can of mosquito repellant and exercise extreme caution to avoid being bitten by a cottonmouth.

Wading a small stream or swamp is a delightful way to fish. One of the best features of fishing a creek or swamp is the solitude. Rarely do you encounter another fisherman in a day's fishing. To avoid trespassing on private property, ask yourself, "Can this stream be floated by a regular canoe under average flow conditions?" If the answer is "no," then the adjacent landowner's permission is required before wading the stream. If "yes," then the stream can be waded without the landowner's permission. You would, however, need the landowner's permission to leave the stream and walk along the bank.

Ponds

Fishermen surveys indicate that 20 to 33 percent of all Alabama fishing trips are to private ponds. This represents about three million fisherman-trips annually.

There are many reasons why pond fishing is so popular with people of all ages. Ponds are often only a short distance from home.

They are convenient for people who like to fish after work, and they are perfect for rural youngsters who are too young to drive.

Most trophy largemouth bass and bream come from properly managed ponds. Montgomery County alone has more than 3,000 ponds, so there are plenty of pond fishing opportunities. The vast majority of ponds are privately owned, so the owner's permission must be obtained prior to fishing. Also, many ponds are open to the public for a small fee. These ponds usually can be found by asking at country stores or by contacting county extension agents.

Most pond owners will not allow the use of minnows as bait, so be prepared to bass fish with artificial lures. Help prevent overharvesting the pond by keeping only a small number of bass. To insure a successful trip, take along some live crickets. This is a favorite bluegill food. Red worms and mealworms are popular for both bluegill and shellcracker.

Mike Bolton

Fishing West Alabama

If you are a fisherman who enjoys varying angling challenges then the fishing in west-central Alabama will delight you.

This 12-county region is blessed with an abundance of resources including large reservoirs, major rivers, scenic streams and creeks, swamp and backwater areas, and small impoundments such as watershed lakes and farm ponds. Each of these areas offers a unique type of fishing.

Perhaps the best known and the oldest impoundments in west Alabama are the waters of the Black Warrior River Basin. The second largest river system in Alabama, the Black Warrior contains more than 52,000 surface acres of water. Tributaries of the Warrior system include the Sipsey Fork, Mulberry Fork and Locust Fork. The most widely sought fish is the largemouth bass. However, striped bass, hybrid bass, catfish, spotted bass, bluegill and crappie are popular favorites as well.

The Warrior River system has approximately 25 access areas along its length. Numerous private ramps are available to fishermen for a nominal launching fee. This river system includes the Bankhead, Holt, Oliver and Warrior reservoirs as well as part of Demopolis Reservoir.

The first signs of mild weather in February usually bring the fisherman out of winter hibernation, with March and April the most productive bass and crappie months. During this time, fish are found near heavy cover such as log jams and weed beds. Spinnerbaits, crankbaits and plastic worms are the best producers.

Smith Lake is a scenic reservoir on the Sipsey Fork of the Black Warrior drainage system. This reservoir is appealing with its green waters and its beautifully wooded and steep rocky shoreline.

Fishermen from all over the world have come here seeking the elusive and hard fighting Alabama spotted bass. Five previous consecutive world records were caught there years ago but it is no longer a world-record type producer. Largemouth and spotted bass fishing is best during February, March, April and May. Plastic worms and deep-diving plugs are the most popular baits.

White bass, bluegill, and crappie are other popular species commonly caught in Smith Lake.

Because of its deep, clear water, Smith Lake's late spring and summer bass fishing is best for night owls. Many fishermen catch their limits of bass and crappie at night fishing around lighted piers and near underwater structures such as bridges or pilings.

Smith Lake has also developed into one of the nation's best lakes for saltwater striped bass. The stripe were introduced in Smith Lake as 3-inch fingerlings in 1983 and by the summer of 1989, fish in the 32-pound range were being caught. The lake has given up several stripes larger than 40 pounds and 20-pounders are common.

In addition, a tailwater fishery has been established for rainbow trout through the efforts of WFF.

Lake Tuscaloosa, a 5,885-acre reservoir just north of Tuscaloosa, was created by the impounding of the North River in 1969. Largemouth bass, spotted bass, crappie and bream are the most highly-sought sportfish. Trophy bass in the 5- to 10-pound category were regularly caught by bass fishermen when the lake was new but the quality of those fish declined as the lake aged. Lake Tuscaloosa has received large numbers of Florida bass fingerlings as part of a stocking plan. More than 900,000 striped bass and hybrid striped bass have also been stocked into Lake Tuscaloosa to establish a fishery for these species. Hybrid stripers are commonly caught in the tail waters below the dam. Live shad and swimbaits are the most popular baits there.

West-central Alabama is the site of Gainesville Res-

The Ultimate Guide to Alabama Fishing

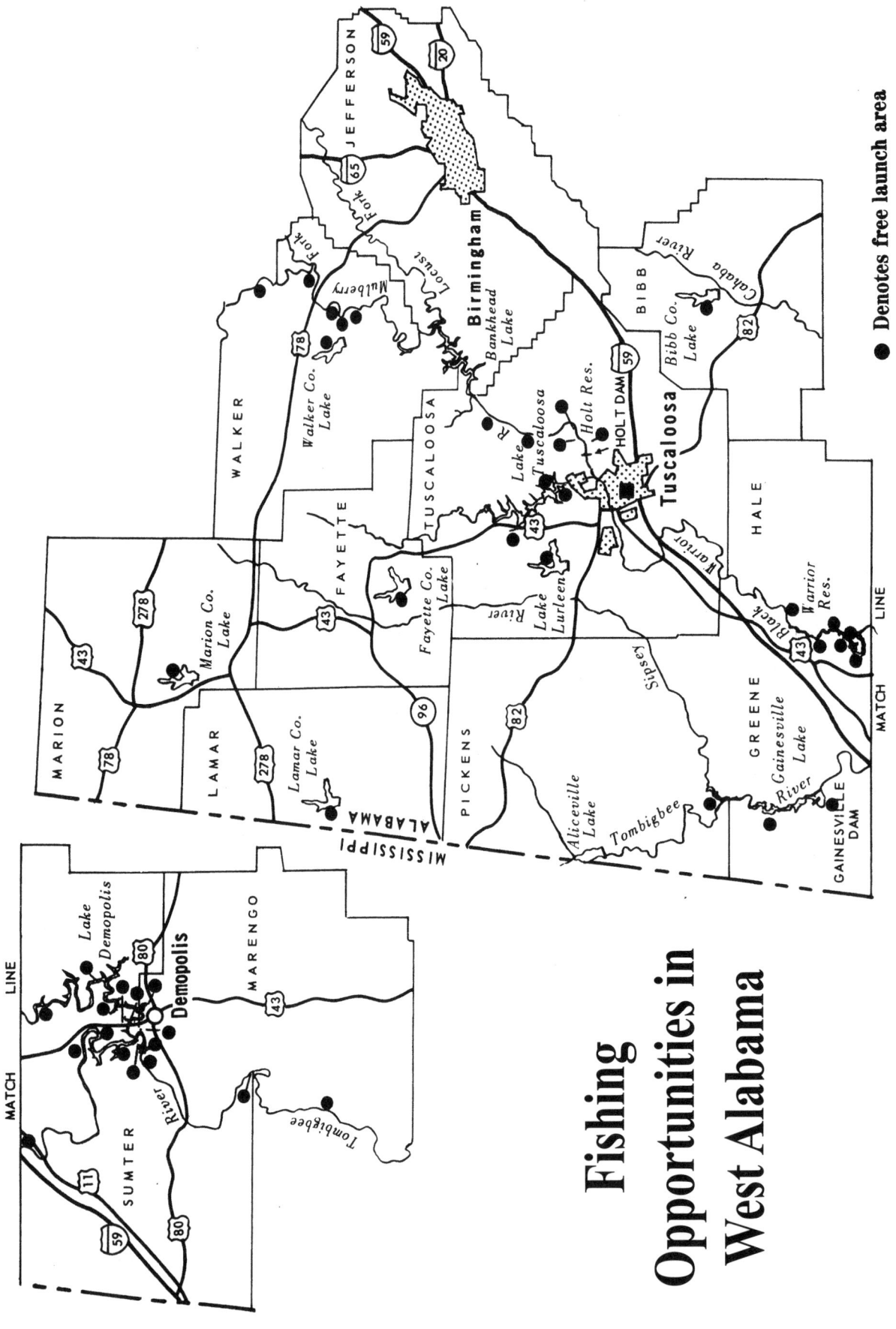

ervoir (7,200 acres) and Aliceville Reservoir (8,300 acres). Those lock-and-dam reservoirs on the Tombigbee River link the Tennessee and Tombigbee rivers to provide a shortcut for river traffic between the Tennessee Valley and the Gulf of Mexico.

Both reservoirs were built with fisherman in mind. They contain large areas of standing timber which provide excellent habitat for many game fish species. Cut-offs were constructed at several locations to keep bend ways open and avoid disturbance by barge traffic. Fourteen public boat ramps are available. Fishing on both reservoirs is good for largemouth bass, bream, crappie and catfish. The Fisheries Section has stocked both reservoirs with Florida largemouth bass and trophy bass fishing is excellent.

The Gainesville Reservoir shoreline is completely within Alabama. Aliceville Reservoir lies on the Alabama-Mississippi border.

Both lakes shook up Alabama fishermen early in 1989 when they produced two 14-pounders and one 16-pound largemouth. Ten-pound-plus largemouths were common then too. The lake saw the normal upsurge of big fish when it was new and has seen the normal stabilization of more moderate-sized bass in recent years.

Excellent access is provided for shore fishermen to the tailraces of both reservoirs. Tailraces yield many catfish and concentrate striped bass and hybrid bass.

In the 12 counties, more than 19,000 acres of private ponds have been stocked by WWF. Ponds are good producers of bass, bream and catfish. Many can be fished merely by obtaining permission from the owner.

State-owned lakes

There are five state-owned public fishing lakes in west-central Alabama which are intensively managed by WFF.

They are Bibb County Lake, Fayette County Lake, Lamar County Lake, Marion County Lake and Walker County Lake. In addition, Lake Lurleen in Tuscaloosa County is managed by the State Parks Division with assistance from Fisheries Section personnel. All of these lakes provide excellent fishing for bass, bream and catfish.

Fishing East-Central Alabama

The fisherman doesn't have to search long to find a good place to wet his hook within the 13 counties of east-central Alabama.

From the mainstream impoundments of the Coosa and Tallapoosa Rivers to the cascading streams of the lower Appalachian Mountains, a wide spectrum of fishing is available which offers something for everyone.

Whether you're a high-tech bass fisherman with all the latest electronic fishing gadgets, or a stream fisherman equipped with an assortment of flies and spinners, you can find desirable fishing in this region.

Reservoir fishing

Much of the fishing in east central Alabama is provided by the five reservoirs on the Coosa River ranging from sprawling Weiss Lake on the Alabama-Georgia border to Lake Mitchell in Coosa County. Between Weiss Lake and Lake Mitchell are Neely Henry, Logan Martin and Lay Lake reservoirs.

Weiss Lake is a 30,200-acre reservoir which has received national publicity for decades for its outstanding crappie fishing. Many out-of-state anglers flock to the lake to catch crappie. Spring crappie season begins in February with the peak of the season arriving between March 15 and April 15, depending on seasonal trends.

The crappie are found in creek channels and deep holes during the late winter and early spring. As the water temperature begins to warm, the crappie begin to congregate along the banks around whatever cover is available. In the spring, crappie can be caught on 1/8-ounce grubs and later in the summer fishermen use minnows when fishing from lighted piers.

Spring bass fishing usually begins in early April. Swimming jigs fished in the weed beds and spinnerbaits fished near weed beds are recommended lures.

Crankbaits, thrown around rocks, road beds and creek channels produce healthy stringers of largemouth bass. As water temperatures warm, most bass fishermen change to plastic worms and crankbaits fished around deep points and creek channels. Another good tactic is flipping around boat docks and piers in the early morning and late afternoon.

A good depth-finder is important when fishing Weiss Lake because shallow water, stumps and rock outcroppings are common.

Lay Lake is one of the favorite Coosa River Reservoirs among fishermen. It is within easy driving distance from Birmingham. Typical of many of the Coosa impoundments the most sought after species in Lay Lake are crappie and bass. The state record crappie for several years was a 4-pound 5-ounce giant taken from Lay by

Frank Gilliam of Birmingham. This lake has many areas with standing timber which provide excellent crappie habitat. The crappie season extends from late winter into May.

Standard crappie bait includes small jigs and minnows. Depth is the critical factor for putting fish in the boat. Early in the year fish deep and then try shallower water as the fish move near shore to spawn.

Bass fishermen on Lay Lake can expect a mixture of largemouth and spotted bass in their catches. In the spring, swimming jigs is a favorite tactic in the grass and weed beds that line much of the main body of the lake.

Weed beds are also productive in the spring as the bass move into shallow water during spawning season.

In the winter, bass can be caught in the vicinity of the Gaston Steam Plant due to the heated discharge warming the river.

The other Coosa River Reservoirs of Neely Henry, Logan Martin and Mitchell offer a variety of fishing. In addition to crappie and largemouth bass, white bass fishing can be excellent in the spring as the white bass begin moving up the feeder creeks. Fishing the jumps is popular on Logan Martin in the summer months when the white bass surface in large schools and feed on shad.

West Point Reservoir is the only impoundment of the Chattahoochee River within the boundaries of east central Alabama. Approximately 3,285 acres are in Alabama and the remaining 22,708 acres are in Georgia. Crappie and hybrid fishing is excellent on the lake

Tallapoosa River reservoirs in the area include 10,660-acre Lake Harris in Randolph County and Lake Martin (39,000 acres) near Alexander City.

Lake Harris Reservoir was impounded in October 1982, and good populations of largemouth bass, spotted bass and crappie are present. One 13-pound Florida largemouth has been caught from the lake.

Lake Harris also received a stocking of hybrid striped bass. Catfishing is also good.

Lake Martin is one of the most scenic lakes in Alabama and a fishing trip to its clear waters, set amid rolling pine-covered hills, can be a memorable experience. Sightings of bald eagles there are relatively common.

The lake is noted for its abundance of spotted bass. Much of Lake Martin is excellent spotted bass habitat with its rocky points and steeply sloping banks.

Crappie compose about 25 percent of the total catch in pounds of fish per year from Lake Martin. The prime months are February through May but there are many crappie caught during the summer months by night fishing. A lantern or a floating crappie light, a bucket of minnows and a rod and reel are all you need to take advantage of this technique. Fishing near creek mouths in 15 to 25 feet of water is often productive.

Lake Martin is currently Alabama's top lake for big striped bass. The most productive area for stripers has been the headwaters of the lake in the vicinity of Irwin Shoals. The prime time has been late March and April when the rockfish move upstream and attempt to spawn. Many 30-pound-plus fish have been caught in this area. Live shad are recommended as bait.

Large stripers also have been caught by fishing shad in 20 to 40 feet of water just above the dam. The Kowaliga Bridge area continues to be a good spot for striped bass in summer months. Techniques include night fishing with big shad near the old bridge pilings or trolling through the area with large diving plugs.

Lake Purdy (bait shop phone 205-991-9107) is a 1,050-acre water storage reservoir southeast of Birmingham just off Alabama 119. Crappie fishing begins the last of February and continues through April. Bream fishing picks up in mid-April and hefty stringers of bluegills and shellcrackers are common. Bass fishing is best in the spring and in the fall when the draw-down begins.

No special permit is required to fish from the bank of Lake Purdy although a state fishing license is required. No private boats are allowed and only outboards of 10 horsepower or less or trolling motors can be used on rental boats.

Tailwater fishing

East Central Alabama has some of the finest tailwater fishing in the state. Perhaps the best known area is the tailwater below Logan Martin Dam. This area has something to offer everyone, from the bank fishermen to the saltwater striped bass and hybrid striped bass angler.

Depending on the season, everything from catfish and drum to crappie, white bass and spotted bass can be caught there. The hybrids are caught generally within several hundred yards of the turbine discharges. Most anglers anchor and cast Sassy Shads or grub jigs in white, chartreuse or yellow colors.

Be prepared to quick release your anchor because a large hybrid can make several long powerful runs and you may need to move downriver with him to wear him down. The best months for hybrids are March through early May when they are concentrated in the waters below the dam.

The Logan Martin tailwater has also yielded some trophy saltwater striped bass of 20 pounds or larger. Catches have been reported of up to 40 pounds.

Techniques are similar to hybrid striped bass fish-

The Ultimate Guide to Alabama Fishing

Fishing Opportunities in East-Central Alabama

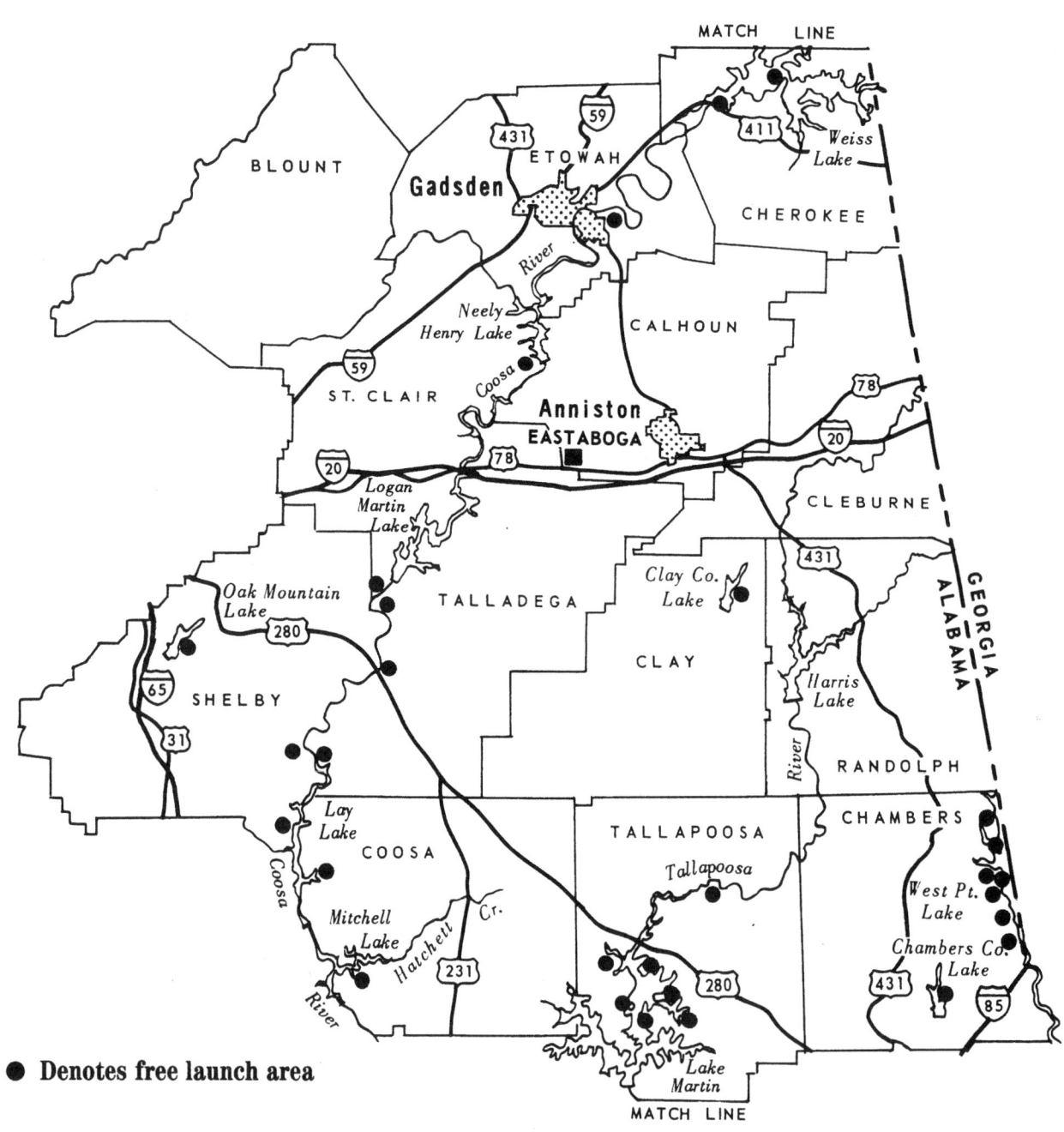

ing although striped bass seem to congregate farther below the dam. The area near the first island below the dam and the gas-line crossings can be productive sites.

White bass and crappie action usually begins in late winter and early spring while catfish activity peaks in May and June. Largemouth and spotted bass are caught by anglers floating downriver and casting among the trees lining both banks.

The tailwater below Neely Henry Dam is a favorite fishing hole for many fishermen. It is similar in many respects to the Logan Martin tailwater. Fishermen can be observed below Neely Henry with hefty stringers of white bass and crappie in the spring and excellent catfish harvests in late spring and early summer.

Large concentrations of saltwater striped bass occur in April and May, although fishing pressure is still light on them. The area on the east bank across from the old lock approach, and on the west bank, below the lock approach, are good areas.

Fishing rivers and streams

For the angler desiring to float a scenic river or wade a remote stream, numerous opportunities are available in east-central Alabama. The Tallapoosa River is floatable at times even though the Harris Reservoir Dam has produced erratic river levels and downstream flows. Good catches of spotted bass are made in the river but float fishing conditions are unpredictable and depend on the power generation schedule at the dam.

The stretch of the river below Horseshoe Bend to the headwaters of Lake Martin has continued to yield some nice stringers of spotted bass, but be prepared to do some boat dragging if the water is low at Harris Dam.

Hatchet Creek should not be overlooked when you are planning a float fishing trip. This scenic creek in Coosa County is aesthetically appealing and has a healthy population of spotted bass. The creek can be floated below the U.S. 231 bridge to Lake Mitchell, and there are several bridge crossings that provide access sites. The stretch from the new U.S. 280 bridge to U.S. 231 bridge is floatable in the spring and is quite remote.

Fishing can be good in this section, but you should allow two days to make the trip. White bass fishing is outstanding from the middle of March to early April when the bass make a spawning run up the creek. The best section to fish is in the vicinity of the Highway 29 bridge in Coosa County.

For the wading fisherman, there is a wealth of small streams from which to choose. Many contain the feisty redeye bass. The Talladega National Forest has many excellent streams where a person can fish all day without seeing another fisherman.

Equipped with ultra-light spinning gear or a light fly rod, the angler can get into some fast action on these redeye bass streams. Shoal and Little Shoal creeks near Heflin are good streams and are easily accessible.

Farther south near Cheaha State Park, Salt, Hillabee and Horse Creeks are teeming with redeye bass.

Fishing in Southeast Alabama

The Wiregrass area, named for the wiry, stiff grass that grows there, includes Barbour, Butler, Coffee, Covington, Crenshaw, Dale, Geneva, Henry, Houston and Pike Counties.

The area is unusual in Alabama in that except for having what many consider one of the state's top lakes (Lake Eufaula), it lacks the big reservoirs found in other parts of the state.

River and stream fishing

Major streams in Southeast Alabama are the Chattahoochee River, Choctawhatchee River, Pea River, Conecuh River and the upper reaches of the Yellow River. Actually, the area of the Chattahoochee River from Columbia Lock and Dam to the Florida line is the only portion of the river in this part of the state that is not impounded.

Except for the Chattahoochee River, streams are relatively shallow with mostly sandy bottoms. Bank fishing is limited and fishing is generally confined to highway rights-of-way at bridge crossings.

One area on the Pea River that produces good bank fishing is below the power dam at Elba (access can be found on the Kinston highway). Another good spot is below Point A Dam on the Conecuh River near Andalusia. The principal species taken below these dams are bluegills and catfish. Fishing is best in the spring.

Float fishing is one of the more enjoyable and successful ways to fish these rivers. A small boat can be launched either from access areas or from the rights-of-way at bridge crossings. Before starting your trip, check county road maps to determine the distance between put-in and take-out points. Allow an hour for each mile of stream traveled. As mentioned before, the streams are relatively shallow and during dry months it is impossible to use an outboard motor to travel upstream.

For this reason, it is best to use two vehicles for a float fishing trip—one at the put-in point and one at the take-out point.

The headwaters of these rivers should be avoided because they are narrow and contain many fallen trees. Fishing is best in the spring and fall although good fishing can be found the year round dependent upon river conditions. Bluegill and shellcracker fishermen usually have the best luck fishing in slow-moving eddy water with the usual assortment of worms, crickets and small spinners.

Bass fishermen experience the most success when fishing crankbaits or spinner baits around tops of submerged trees, stumps or log jams. Jigs and small lures can be effective in the pools below rapids and logjams. Bass fishing is best in the early morning and late afternoon along the shady side of the river. Some areas of these rivers are swift in the spring, making it impossible to fish them effectively while floating.

To overcome this, most fishermen use a small outboard motor (an electric trolling motor is not strong enough) to hold position in the current long enough to fish each area before moving downstream. Normally two people will fish together taking turns operating the motor. Good strings of largemouth bass are taken out of sunken tree tops and from around submerged logs by fishing this method. Diving lures such as medium-lipped crankbaits and Rat-L-Traps are commonly used.

Many people try their luck at catfishing and are often successful on these rivers. Set lines and bush hooks, as well as trotlines, usually are the best methods. Cut bait, liver and small salamanders take the most catfish.

There are several beaver ponds near the Choctawhatchee River and many of them are accessible. These areas produce good bream and bass fishing in the spring, especially after they have been flooded by the river. Wading or using small aluminum boats is best for fishing. One important thing to remember is that landowner permission is required to fish beaver ponds on private property.

In addition to the fishing in the Wiregrass area you will enjoy scenery seldom seen elsewhere. You are likely to see many of nature's creatures such as deer, turkeys, ducks, owls, hawks, river otters, beavers and alligators.

Reservoir fishing

Southeast Alabama has several moderately large impoundments in addition to Lake Eufaula.

Lake Eufaula with its 45,000 acres of diversified fishing water is nationally known for its largemouth bass fishing. It is on the Chattahoochee River. The dam is approximately 11 miles east of Abbeville. The lake extends 85 miles upstream to Phenix City.

Bass fishing is generally good all year, but the better catches are made during late winter and spring months when there are large numbers of bass along the banks and are easy to find.

The most popular lures at this time are spinnerbaits, buzz baits and deep-diving crankbaits. As the water begins to warm, plastic worms become more effective.

After most of the spawning activity has subsided, the bass tend to migrate to the creek and river channels. The better areas appear to be outside bends which have water from 10 to 15 feet deep along the ledges dropping off into the deeper channels. Plastic worms are the favorite lures but many people also like jig-and-pig fishing.

Bass make a short migration back to the shoreline and up the creeks during the fall when the water begins to cool. Topwater fishing is probably at its best during this period. As the water cools further, the bass migrate to the deeper water of the river channel. They remain in the deeper water until the latter part of January, when they begin moving into the many flats found on the lake.

The flats are areas generally four to five feet deep containing a fair amount of cover. Yellow/white or chartreuse/white spinnerbaits are productive in these areas. From the flats the bass move back to the shoreline and up the creeks looking for bedding areas.

In addition to the bass fishing, Lake Eufaula generally has good to excellent crappie fishing. The best time to catch these fish is from late fall through the spring. Most fishermen prefer cane poles and small minnows. Good catches can also be made by using small white jigs.

Most of the crappie are taken near submerged tree tops and other cover. If you are unfamiliar with the lake, ask one of the area bait dealers who will gladly direct you to a likely spot.

WFF, in cooperation with the U.S. Army Corps of Engineers and the Georgia Game and Fish Commission, placed 56 fish attractors in Lake Eufaula. These attractors were made by placing cedar trees at various locations in the lake. These areas are marked with buoys.

There is virtually unlimited access to the waters of Lake Eufaula with boat ramps in every well-known area.

Columbia Reservoir is directly downstream from Lake Eufaula. A river-run impoundment, it contains approximately 1,600 surface acres of water. Access to this reservoir is limited. However, there is a good ramp approximately one mile below Walter F. George Lock and Dam and one at Omusee Creek near Columbia.

Fishing Opportunities in Southeast Alabama

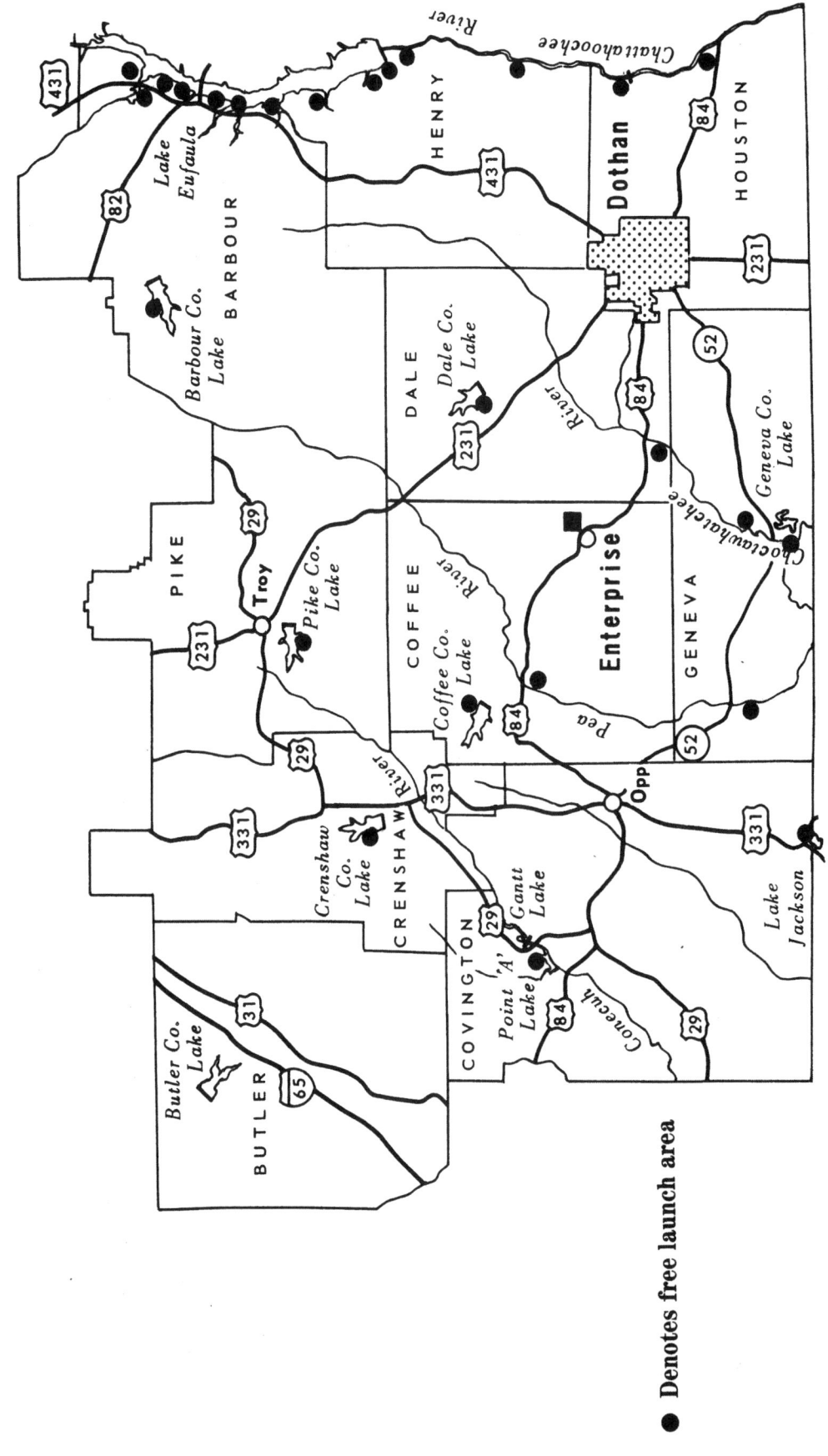

Good catches of largemouth bass, hybrid striped bass, crappie, white bass, bream and channel catfish are made each spring. The largemouth bass fishing generally is good most of the year, but it takes several trips to locate the most productive spots.

Gantt Lake, a lesser known reservoir of the Wiregrass, is a 1,600-acre impoundment on the Conecuh River near Andalusia. The lake offers fair-to-good fishing for largemouth bass and crappie, but the favorite species of fishermen is the shellcracker. Good strings of these panfish are taken during March and April when the shellcrackers are bedding. Most of the fish are caught in the shallow, upper end of the lake around the many cypress trees, stick-ups and aquatic weeds. Earthworms are used to take ice chests full of shellcrackers which commonly average half a pound.

The spring months are the best times to fish the lake for bass and crappie. The grass beds and stump rows in the sloughs near the dam provide the best bass fishing. Six-inch plastic worms or long, floating or diving minnow jerkbaits such as Rapalas are the most effective lures.

Chartreuse crankbaits also will produce well. During the summer the bass go deep and usually are caught around the old river channel with plastic worms. Some are taken by trolling with deep-diving crankbaits.

Crappie fishing is best in Cottle Creek or along the steep banks of the main lake where the old river channel approaches the shoreline. Night fishing around lighted piers is productive.

Use of live minnows is by far the best way to fill a stringer. Boat ramps are available at Dunn's Bridge off County Road 43 or at Clearview Campground on Highway 29 North. Several privately owned fee ramps exist and directions can be obtained from local citizens.

Point A is an impoundment of the Conecuh River just below Gantt Lake. As at Gantt, the most sought species in this 600-acre reservoir are largemouth bass, crappie and shellcracker.

The best fishing is generally found in March, April and May. The sloughs around Covington County Park, the FOP Lodge and the Mallette Creek area provide the best bass fishing. The same lures and methods which produce on Gantt also will work here.

Good catches of crappie are taken on minnows in the Patsaliga arm and along the deep banks of the lake proper. Shellcrackers can be caught on earthworms in most of the shallow sloughs by fishing the stickups and grass beds.

A good boat ramp is available at Covington County Park, on the northwestern side of the lake.

Tailwaters

Hybrid striped bass fishing is good during the spring and fall. The best area to fish is the tailwater immediately below Walter F. George Lock and Dam. The best time to fish for the hybrids is when the hydroelectric turbines are operating. Large white jigs, swim baits and live shad are the best baits.

Another well-known area for hybrid striped bass fishing is the tailwaters of Columbia Lock and Dam. During the early spring, hundreds of fishermen flock to this area to catch hybrid stripers, one of the best fighting fish in the Chattahoochee River.

There are two main areas to fish. One is from the bank on the Alabama side immediately below the dam. This area is lined with riprap and the fishermen stand on these rocks while casting into the swift, swirling water which flows over a rough bottom that is lined with bigger rocks. White and yellow jigs are the best lures. The jig should be retrieved as slowly as possible, but kept just off the bottom to prevent hang ups.

The other area is on the Georgia side of the river. However, one needs a boat. The boat should be anchored at the edge of the swift water between the end of the lock wall and the boat ramp. The favorite method of fishing here is to use live shad, the bigger the better. The shad should be hooked from the bottom of the jaw up through the hard area of the nostrils. Fish the shad as close to the bottom as possible.

Catches of eight and nine pound hybrids are common here and you usually experience little trouble catching the limit, once you have mastered the technique used by the local fishermen.

Lake Jackson

Lake Jackson is on the Alabama-Florida line in Florala. This is a shallow, bowl-shaped natural lake filled with aquatic vegetation. The most sought after species is the largemouth bass. The water is clear and fishing is difficult for most anglers.

Those anglers willing to put in the time and effort are often rewarded with that once-in-a-lifetime trophy, however. Ten-pound-plus fish are taken each year.

Most of the larger bass are caught in and around the mats of aquatic weeds. The best time to land a bragging-size bass is during the spring spawning season. The spawning beds and the large fish near them can be spotted easily in the clear water. A live shiner is usually cast into the bed and allowed to stay until the fish takes it.

Heavy tackle is usually in order since the bass will most assuredly become entangled in the vegetation.

During summer, most anglers prefer to fish the lake at night. This is almost a necessity because of the clarity of the water and the heavy use of the lake by skiers. The lure of choice is a l0-inch black plastic worm rigged weedless.

Lake Jackson also offers good fishing for chain pickerel and crappie. Most of the pickerel are caught around the weed beds by casting brightly colored crankbaits. Live minnows are the preferred bait for crappie. These panfish are taken around the weed beds during the spring and in the deeper holes in the summer.

Private lakes and ponds

Because this area is primarily agriculture-oriented, literally thousands of private lakes and ponds dot the landscape. Many of these private lakes are open for a fee. To locate these lakes, contact the local county agent's office.

Fishing in Southwest Alabama

Southwest Alabama is fisherman's paradise. The vast areas of freshwater and saltwater waterways in this region provide many opportunities to enjoy fishing experience. This area includes Baldwin, Choctaw, Clarke, Conecuh, Escambia, Mobile, Monroe, Washington and Wilcox Counties.

The Alabama and Tombigbee Rivers, the main tributaries of the Mobile Bay drainage system, flow through this area and provide the foundation for some of the finest fishing in Alabama.

The Mobile Bay system drains 43,560 square miles of land and includes portions of Mississippi and Georgia as well as a major portion of Alabama. This system empties into the Gulf of Mexico by way of the Mobile Delta, a major fishing area in Southwest Alabama.

In addition to those waters associated with the Mobile Bay drainage, there are waters in the Escatawpa River system, the Perdido River system, and the Conecuh River system which also provide good fishing opportunities.

Alabama River system

The Alabama River system, one of the two major tributaries of the Mobile Bay Drainage, has two large reservoirs and many miles of river in this area. The reservoirs are Miller's Ferry and Lake Claiborne.

Miller's Ferry is a 22,000-acre reservoir built by the U.S. Army Corps of Engineers in Wilcox and Dallas counties. This reservoir provides excellent fishing.

Noted as both a haven for big crappie and bass, this lake has 516 miles of shoreline and an abundance of structure, deep channels, shallow flats, sand bars and flooded standing timber. One of the outstanding features of this reservoir is that fish can be caught during the spring flooding season. Normal spring lake level fluctuations do not flood boat ramps or vast areas in the flood plain.

Lake Claiborne is a 5,850-acre reservoir below Miller's Ferry and is the last reservoir in the Alabama chain. Claiborne is smaller and more of a river-run impoundment than Miller's Ferry.

In addition to excellent fishing in these reservoirs, the tailwaters below both provide fishing excitement. Many species of fish tend to congregate below dams. There are also a number of oxbow lakes on the lower Alabama which offer good fishing.

The major sport fish in the Alabama River system are largemouth bass, crappie, bream and catfish. Spotted bass occur here and there is a decent striped bass fishery.

Overall, spring and fall are the most productive fishing seasons on the Alabama system. Largemouth bass fishing begins in later winter and continues to be good until early summer. After a midsummer lull, bass fishing picks up again in the fall. Plastic worms and deep diving crankbaits are effective.

Spring crappie fishing peaks at spawning time, which occurs when the water temperature is approximately 60 degrees. During spawning, crappie can be found close to the bank in four to eight feet of water. Otherwise, they are most likely found associated with submerged brush and trees.

Striped bass will congregate below dams in the spring. In the summer, stripers can be found in channels near the mouths of streams where cool water empties into the river. They also can be found in schools suspended in mid-water in open river and reservoir areas. Shad are a favored food item for stripers. Fishing around schools of shad is often successful. Striper baits include live shad and bluegills, jigs and large shad-like lures.

Tombigbee River system

The Tombigbee River, the other major tributary of the Mobile Bay drainage system, enters this area at the north end of Choctaw County. Coffeeville Lake, a river-run reservoir of 8,800 acres, is on this river. This lake provides good fishing for largemouth bass, especially in the spring and fall. Populations of crappie, bream and catfish inhabit this reservoir and provide a good sport fishery.

The tailwater below Coffeeville Lock and Dam provides a good fishery for crappie, bass, bream and an excellent fishery for catfish.

As you move downstream, oxbow lakes and tributaries provide excellent fishing for crappie, bream and largemouth bass. Some of the more popular lakes are Three Rivers Lake, Fishing Lake and Bates Lake, all near McIntosh.

Mobile Delta

The Mobile Delta is formed by the confluence of the Alabama and Tombigbee Rivers in Clarke County. This river delta, the second largest delta system in the United States, is approximately 40 miles long and 10 miles wide. It consists of an intricate network of rivers, creeks, bayous, lakes, marsh and bays, which eventually empty into Mobile Bay. The Mobile Delta is the upper end of an estuarine system which is the basis for important sport and commercial fisheries.

The lower quarter of the delta consists of shallow bays, bayous and river channels bordered by wetland plants and few trees. The upper three quarters of the delta consists of deeper lakes and rivers bordered by typical bottomland hardwoods. Most of the delta is undeveloped and in a near-pristine state. The Mobile Delta is one of the most important ecological areas in Alabama.

By definition, a river delta is rich in natural fertility. As the river slows, the suspended and dissolved nutrients acquired from the 43,560 square miles of drainage area are assimilated by the delta system before being passed on to the marine environment.

Fish production in the Mobile Delta is high. In addition to the common freshwater species, the lower delta serves as a nursery for numerous marine fish and shellfish. A seasonal influx of desirable adult marine sport fish such as speckled trout, sheepshead and redfish also occurs. This invasion of marine species enhances an already good sport fishery in the Mobile Delta.

Sportfish in the delta include largemouth bass, bluegill, shellcracker, crappie, warmouth, catfish, pickerel, yellow bass, white bass, striped bass, alligator gar and mullet. The major fishing seasons are spring and fall.

Largemouth bass are abundant in the Mobile Delta. There are not as many large bass caught in the delta as are harvested in upstate impoundments. However, no other area in Alabama can match the delta in number of bass per acres of water. Bass fishing in the upper delta is best in deep lakes and dead river runs. These areas have plenty of cover for bass, such as submerged logs, brush tops and drop-offs. The shallow bays and bayous of the lower delta provide good bass structure including grass beds, pilings, drainage cuts and old duck blinds. All the popular bass baits are effective. Live shrimp are an excellent bass bait in the lower delta.

Bream fishing is also good in the Mobile Delta.

The shallow bays of the lower delta and the deeper oxbows of the upper delta produce heavy stringers of shellcrackers, bluegill, and warmouth (goggle eye).

Crappie are more numerous in the deeper lakes of

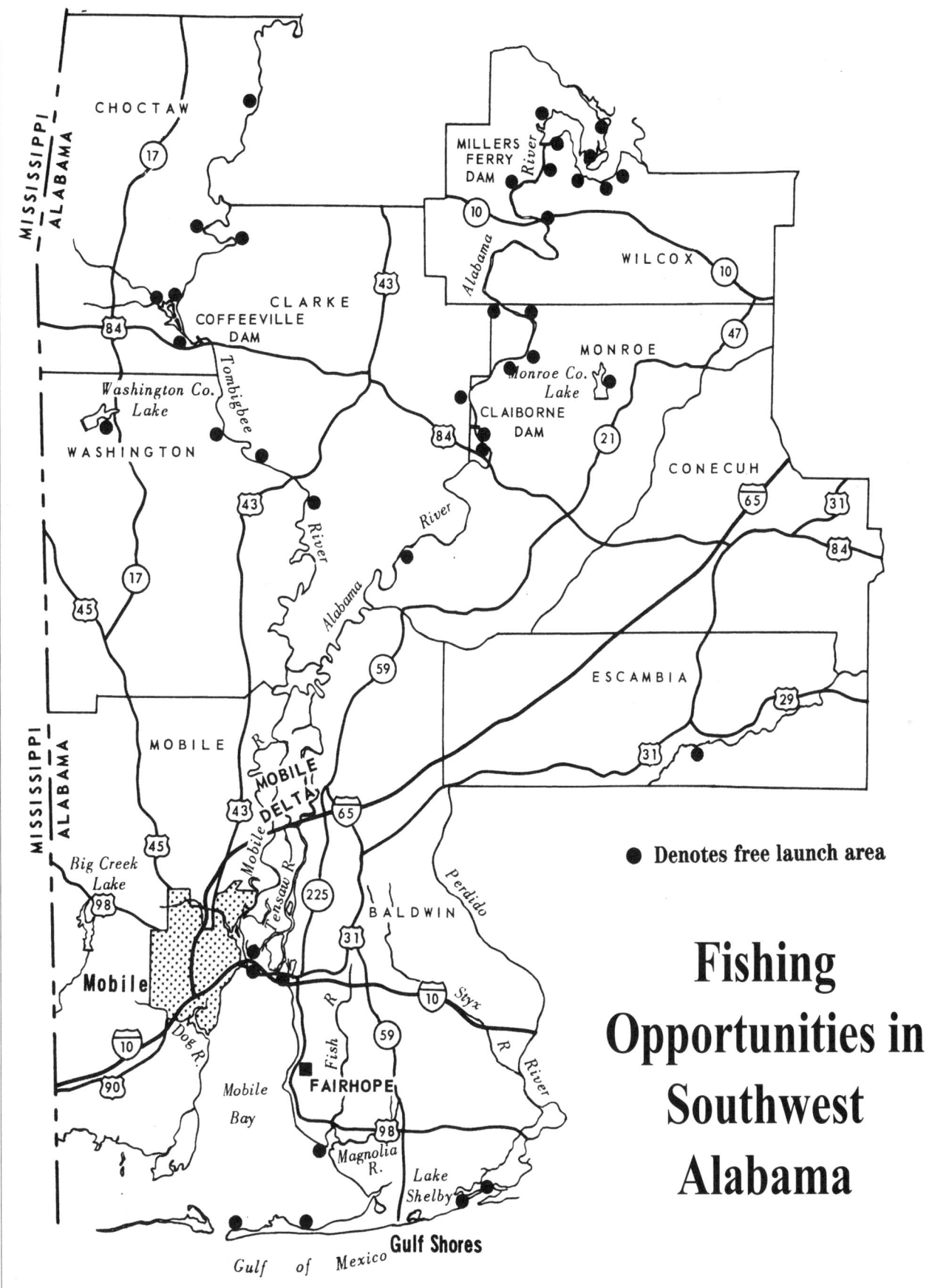

the upper delta than in the shallow bays of the lower delta. Favorite crappie fishing areas include Miflin Lake, Negro Basin, Tensaw Lake, Fishing Lake and David Lake. Many other areas in the delta also provide good crappie fishing.

The lower Mobile Delta and other tidal streams in Mobile and Baldwin counties offer an extra bonus in the fall. From September through January speckled trout, redfish, and flounder enter the mouths of rivers and streams and add to the excellent freshwater fishing.

Big Creek Lake

Big Creek Lake is a 3,600-acre public water supply reservoir for the City of Mobile. This reservoir is on the Escatawpa River System which flows into Mississippi. Big Creek Lake, in west Mobile County, has great fishing, especially for largemouth bass, crappie and hybrid striped bass. The water in the lake is generally clear because of the wooded drainage area and low fertility. Water as deep as 90 feet can be found here which is unusual in south Alabama.

Threadfin shad were stocked into the lake by WFF and have benefitted the crappie, largemouth bass and hybrid striped bass populations by providing more forage fish. Shad school near the surface during early morning and late afternoon hours of summer when the wind is calm. Excellent catches of hybrid stripes and largemouth bass can be made near shad schools.

Shelby lakes

The Shelby Lakes are natural freshwater lakes in Gulf State Park in Gulf Shores. The unusual situation of freshwater lakes within view of the Gulf of Mexico provides some unusual fishing opportunities for park visitors.

Lake Shelby is the largest in a chain of three contiguous lakes which cover a total of 829 acres. The water in these lakes has tannic acid stain typical of many aquatic systems in south Alabama. These lakes are relatively shallow with water rarely deeper than eight feet. There are a few holes that drop off to 20 feet. Although the bottom contour is rather constant, a depth finder will enable the angler to find some desirable bottom structure.

The sportfish available in the Shelby system include Florida largemouth bass, channel and blue catfish, bluegill, shellcracker, crappie and redfish. Since Florida bass were first introduced into Lake Shelby, the lake has gained a reputation of being a trophy bass lake. Although large numbers of bass were killed in this lake as a result of salt water intrusion from hurricanes, more Florida bass have been stocked and a healthy population of largemouth are again present in the lake.

Channel and blue catfish are periodically stocked into Lake Shelby and are an important facet of the sport fishery. These fish seem to thrive in the slightly saline water, and their taste is unsurpassed by river or farm-raised catfish. A good fishing method for catfish is drifting and bottom fishing with dead shrimp. Other common catfish baits are also productive.

Bream fishing is best in the canals which connect the three lakes, especially near submerged brush and grass and reed beds that line much of the shoreline. Shellcracker fishing is particularly good in Lake Shelby.

Redfish, or red drum, were introduced naturally into this system by Hurricane Frederic. These naturally-stocked fish grew so fast and created such a desirable fishery, that more fingerlings have been stocked by the Game and Fish Division. This marine species which often occurs in the mouths and lower reaches of rivers is an aggressive fighter which can be taken on live, cut or artificial bait.

Spring and fall are the best fishing season for the Shelby system; however, summertime angling can also be productive.

Small streams and ponds

Some of the better small streams are the Escatawpa River in Mobile County, the Styx River and the Blackwater River in Baldwin County, Murder Creek and Sepulga River in Conecuh and Escambia counties, and Pine Barren Creek in Wilcox County. Most of these slow-moving and sand-bottom streams support populations of largemouth bass, spotted bass, bluegill, warmouth, chain pickerel and catfish.

Stream fishing is generally best during the summer and early fall. Many private ponds are present in southwest Alabama. These ponds are stocked with shellcracker, bluegill, catfish and largemouth bass and provide many fishing trips for local fishermen. In many cases, permission of the pond owner is all that is required for an enjoyable fishing trip.

Mike Bolton

The Tennessee River

Guntersville Lake

Wheeler Lake

Wilson Lake

Pickwick Lake

Lake Guntersville

Weeds, grass, moss, milfoil, hydrilla. Call it what you like, but the aquatic vegetation in Guntersville Lake has made the Tennessee Valley Authority reservoir into one of the nation's top bass fishing lakes.

The vegetation scattered from one end of this lake to the other is an excellent hatchery for bass and bream. It offers a safe haven for the smaller fry of both species to grow and become good fish. It does the same for minnows and shad. The result is an excellent food chain and terrific numbers of good fish.

The amount of grass in the lake either complicates or uncomplicates fishing here, depending on how you look at it. It uncomplicates things for many because it is the dominant structure in the lake and limits the number of patterns, or methods, needed to fish it well. Unlike a lake which may have a river-like upper end full of rocks and a ledge-filled lower end full of timber, Guntersville Lake offers a consistent makeup throughout its length. Wide water with grass is standard from one end to the other. A pattern that is working upriver is most likely working downriver, too.

For the competitive angler, the amount of grass tends to complicate things. It is possible to catch so many fish in so many places, it's difficult to force yourself to stop catching good fish and move elsewhere in search of bigger fish.

The incredible catches in Bassmaster tournaments have drawn nationwide attention to Guntersville Lake. Only the lake's lower end is featured in this section. It is here that the lake's better fishermen concentrate. It is recommended that a Guntersville Lake map such as the one produced by Carto-Craft Map Co. be purchased and studied when reviewing this lake. The thousands of sandbars and underwater islands are easily pinpointed on these maps.

Spring bass fishing

Guntersville Dam to 431 bridge

Bass on this lake relate to aquatic vegetation. That's no one's secret. There are a number of fish on this lake that bury themselves in the weeds no deeper than 11 feet and stay there all summer and winter. The angler who knows the location of grass-ridden deep humps and old creek channels in this end of the lake can start his year off right and put some big fish in the boat before most anglers get on the lake for the first time of the year. A jig-and-pig fished in the dead grass on these humps and along these channels usually will produce quality bass starting in late February. A 1/2-ounce orange/brown jig with a black pork frog trailer makes an excellent slow bait for these fish.

A depthfinder is needed to locate these areas. First, find areas you want to try by locating them on maps in this book. Then, ride the area in your boat and pinpoint them with a depthfinder. Use marker buoys to mark the spot. A swim jig worked slowly through the dormant grass is the key to catches of big fish.

Another pattern also works well during this time. In late February and early March, the water temperature is still low, but the fish are able to detect patches of warmer water and will move to them. This lake is full of riprap around bridges, culverts and banks and the large rocks used to prevent erosion absorb heat and radiate warmth into the nearby water. The water surface temperature next to this riprap will sometimes be three or four degrees warmer than surrounding water and that makes a big difference this time of the year.

A jig-and-pig fished in this riprap early in the spring often results in not only numbers of fish, but quality fish as well.

When mid-March arrives and there are three or four days of continuous sunshine, get ready for some good bass fishing. Those three or four days of sunshine will warm the water into the mid- to high-50s and that's warm enough to get out the Rat-L-Traps and chatterbaits and start fishing the milfoil.

The best color of Rat-L-Trap this time of the year is a matter of debate. It takes an awful lot of rain to stain this lake and it is normally in pretty good shape in the spring. A chartreuse combination may be needed in a Rat-L-Trap on the rare occasions the lake is muddy, but a gold, white, chrome or a chrome with a blue or black back will work under normal conditions.

The key to fishing this bait, or any bait around the milfoil on the lake, is occasionally allowing it to drop and tick the weed tops on the retrieve. The reason for that is simple. Fish can tell this bait is coming toward them, but the weeds tend to disorient them to a small degree. Ticking the weed tops seems to help them find the bait much easier.

Bass are beginning to think spawn this time of year and they will be in the milfoil in shallow water in the back of pockets on the main river or the back of creeks. This shallow (3 to 5 feet) water may be only two degrees warmer than water 30 yards away, but it makes a major difference.

April means bass are thinking hard about the spawn and topwater fishing becomes feasible. It's often still chilly this time of the year and the bass are like you and me in that they enjoy the feel of the sun on their backs. They will be close to the water surface. Stumps and the fringes of milfoil will hold bass this time of the year and buzzbaits and rats and frogs result in vicious strikes.

This is also the time of the year for those who like to fish spinnerbaits, but spinnerbaits usually take a backseat to Rat-L-Traps on this lake. The Rat-L-Traps cover more water quickly. Spinnerbaits are an excellent bait on this lake, nevertheless.

There are several key areas to concentrate on this end of the lake in the spring. Brown's Creek is a tremendous spring area with large expanses of stump-filled flat water, humps and a winding creek channel.

Another key area is the what locals call the Islands, the series of islands northeast of the U.S. 431 bridge. The area is full of small islands, humps, stumps and channels that concentrate bass.

Honeycomb Creek produces numbers of spring fish, but is an area most of the better anglers have marked off of their maps because the number of fish caught there doesn't justify the amount of time needed to catch them.

U.S. 431 bridge to bridge to South Sauty Creek

The spring pattern above U.S. 431 bridge is basically the same pattern the lake's better anglers use below the bridge. Shallow, grass-filled pockets are normally the areas of choice for springtime fishing, while humps on the main river come a close second. Anywhere 12 to 15 feet of water suddenly rises to 4 to 5 feet will hold fish this time of the year. The grass will be dormant in the early spring and this is the time of year that experience on the lake really comes into play. The pockets and humps that held fish the spring before will hold fish this time of year, even though the grass they want may not yet be growing or visible.

There are several areas to concentrate on and several areas to avoid during this period. Town Creek is one of those areas you need to consider first. The stumpy flats, winding creek channel, humps and milfoil make for excellent spring habitat.

The very back of Town Creek, back where it narrows, offers excellent fishing with jig-and-pigs in February. The Claysville area, known to some fishermen as Siebold Branch, has everything you need for spring fishing, including stumpy flats, islands, winding creek channels and grass. It, too, is a key area.

The flats on both sides of the river from Town Creek to South Sauty Creek also offer good spring bass fishing in spots. Pay special attention to the sloughs in the back of these flats. These also are key areas for bass during the spring.

South Sauty Creek is another key area. The creek is filled with stump patches, a winding creek channel and plenty of flats covered in milfoil. It should be noted that just because parts of this creek don't have milfoil, that doesn't mean it shouldn't be fished. Most areas without milfoil have underwater stumps that hold bass.

One would think that due to the amount of fish turned loose in Spring Creek following bass tournaments that it would be an excellent place to fish. It just doesn't work that way, however.

The creek just doesn't hold fish that well for some reason. The only exception to that being the far back area past the bridge. This area has some shallow water and boathouses and plays home to some spawning fish, but it is by no means an area to concentrate.

Hot-weather bass fishing

Guntersville Dam to South Sauty Creek

Like on all Alabama lakes, summer bass fishing on Lake Guntersville means the angler must move to deeper water.

Top anglers have a successful pattern of fishing crankbaits on the river drop-offs during the summer months. It isn't difficult to fish this pattern. The angler should parallel his boat to the bank in 25 feet of water on a ledge or creek channel that has 10 or 12 feet of water, as well as hydrilla or stumps. He should then cast his lure as close to the ledge as he can without becoming entangled in the milfoil, then crank it down at a speed that allows him to slightly bump the stumps or hydrilla.

The entire area should be worked in all directions to locate fish. These fish will usually be holding in tight around the structure, with groups rarely being in a stretch longer than 20 yards. Once the fish are located, the angler should back his boat off and position it in deeper water where the fish-holding area can be fished vertically, rather than horizontally.

The bass will be in different water depths depending on the time of summer, so the type of crankbait may need to be adjusted. A pearl or white deep-diving crankbait with a black or blue back are both effective when the fish are deep. When the fish are closer to the bank, medium-running crankbait or chatterbait is needed.

Another pattern is strong in summer months when the fish are buried deep in the grass. Lake Guntersville's milfoil normally only grows into water 10 feet deep or less; but there, hydrilla has begun to take over in many places. Unlike milfoil, which is stringy and can entangle hooks easily, hydrilla is crisper, much like lettuce. Lures that hang in hydrilla often pull free easily.

Hydrilla is different in another respect, too. It has a long, thin stem that sometimes grows 12-feet-tall in summer months. It also has a broad leaf that forms a canopy. This is perfect habitat for summer bass not only because it is in deep water, but also because the large leaves block out much of the sunlight. As you probably know, bass have no eyelids, nor do their pupils constrict. Their only method of controlling the light in their eyes is moving into shade.

The canopy formed by hydrilla often holds numbers of bass and good bass, but it would seem tough to get to them under these leaves. Not so. A 1-ounce leadhead bass jig with no trailer can be flipped into these areas and the heavy jig will punch its way through the hydrilla. It should be allowed to hit bottom then jigged.

Another strong pattern during this time is flipping the piers. Permanent piers with wood pilings in deep water are best. These piers offer shade and cooler water bass need during this time and the bass gather under them in numbers. A 1/2-ounce black bass jig on 12-pound line is an excellent bait for flipping these piers.

It should be noted that the piers are non-productive during overcast days. When there's cloud cover, the bass leave the piers and roam.

Many smaller bass stay in the milfoil throughout the summer, but they are tough to get to during most hours. Rats and frogs can be used in the milfoil during summer months, but only in the early morning and late afternoon when bright sunlight is not present. On overcast days, they can be used all day.

These patterns work up and down the river during the summer months. Although there are some summer fish in the creeks with deeper water, the better anglers concentrate on the main river. This is the area that receives the current and the water flow makes both the shad and bass active.

Cool and cold weather fishing

Guntersville Dam to South Sauty Creek

The are four seasons in the year, but Guntersville Lake really has only two fishing seasons. The spring and fall fishing are very much alike, as is the summer and winter fishing.

The cooling temperatures of October once again send the bait fish seeking shallow water and the bass follow. The major difference in fishing the lake in fall is the milfoil will be out of the water good and it will still be alive.

Buzzbaits become an excellent bait at this time. A black buzzbait thrown into the thinner, outer edges of the grass will evoke some exciting strikes. Rats and frogs will do the same.

You'll find bass in the grass on the main river channel, in the pockets and in the creeks during this time. Brown's Creek, Spring Creek, Town Creek and South Sauty Creek are excellent fall fishing spots.

When winter arrives, the bass once again drop into the deeper water, much like they did in summer. Jig-and-pigs and worms become the bait of choice at that point.

Other types of fishing

Bream Fishing

For the same reasons this lake offers tremendous bass fishing, it also offers excellent bream fishing. The thousands of acres of aquatic vegetation and stumpy flats are excellent habitat for bream to not only spawn, but to hide and become big fish.

Bream fishing is tremendous during the spring. The larger bream head for the weedless pockets, sloughs and flats that have gravel or sand bottoms and lay their eggs. The areas are often covered with submerged stumps and the angler must plan on losing some tackle, but the gains are worth the losses. Guntersville Lake's spawning bream are usually good size, ranging from 1/2 to 3/4 of a pound.

Tackle needed for this type of fishing is basic. A spincast reel allows longer casts which decreases the chances of spooking the fish. A small foam float, 4- or 6-pound test line, a single bb shot, a long-shanked bream hook and a cricket make a perfect combination.

A cane pole or fiberglass pole also will work at this time, but because of the opportunity to scare fish while pole fishing and the low cost of a spincast outfit, the spincast tackle is the way to go.

The first full moon in May usually will find bream hard on the bed in Guntersville Lake, and some species will spawn on the full moon every month throughout the summer.

Guntersville Lake's mayfly hatches are unpredictable, but they are regular throughout the summer months. The hatches send bream to the bank. Anything that hangs over the water from piers to bushes will concentrate bream during this time. The bream gather waiting for a bug to hit the water so it can become a tasty meal.

The same rig used for spring bream fishing can be used for this type of fishing. An especially fun way to fish for bream at this time is a flyrod. A mayfly imitation will sometimes produce 200 or more bream a day around boathouses and piers.

Bream fishing is much better on this lake during the summer than it is on most lakes in Alabama. The reason is the bream stay relatively shallow under the milfoil. A milfoil-covered hump that has an opening in the grass patch will produce large numbers of fish.

A spincast may be used at this time, but the opportunity for hanging the hook in the grass is great. A fiberglass pole or cane pole is much better at this time because it allows a worm or cricket to be gingerly placed on target.

Bream will stay in the milfoil until late fall when they drop into deeper water and become almost impossible to locate and catch.

The Ultimate Guide to Alabama Fishing

Lake Guntersville's best fishing spots

1. Underwater hump.
2. Mouth of Honeycomb Creek. Good summer fishing.
3. Underwater hump.
4. Sandbars and humps. Covered with milfoil. Next to deep water. Good spring and fall fishing.
5. The Islands. Stumps and milfoil. Spring and fall fishing.
6. Underwater hump in Brown's Creek.
7. Winding creek channel in Brown's Creek.
8. Good milfoil patch. East of 69 bridge.
9. Underwater roadbed. Good in spring and fall.
10. Underwater depression. Good in spring. Surrounded to the north by stumps.
11. Stumpy flats all the way to back of the creek. Good spring fishing.
12. Riprap around U.S. 431 bridge. Good in early spring.
13. Mouth of Spring Creek. Good year-round. Good drop-offs.
14. Humps with milfoil.
15. Winding creek channel. Good interstate area to back of creek in spring.
16. Riprap around bridge. Good in spring.
17. Good spawning area. Boathouses and piers.
18. Henry Island. Good underwater island covered with milfoil.
19. Sandbars with stumps and milfoil.
20. Mouth of Short Creek. Good summer fishing area.
21. Claysville area. Lots of humps, stumps, islands, sandbars. Key area in spring and fall.
22. Town Creek. Lots of milfoil and hydrilla-covered humps adjacent to deep water.
23. Sharp bend in creek channel in Town Creek. Known as the Red Bank area. Good spring area.
24. Stump and milfoil-covered humps in Town Creek.
25. Stump patches in back of Town Creek. Spring and fall.
26. Bridge riprap. Good in spring.
27. Excellent flipping area in back of Town Creek in January and February.
28. Good humps with stumps and mil foil.
29. Same as 28.
30. Mouth of Mill Creek. Good hump adjacent to shallow water.
31. Bridge pilings and riprap. Good in spring.
32. Stump patches with milfoil. Good in spring.
33. Murphy Hill area. Winding creek channel leading out of main river. Excellent year-round fishing.
34. Pine Island. Underwater island splits main river channel. Milfoil one end to other. Excellent year-round fishing.
35. Stump-filled slough. Good spring and fall.
36. Stump-covered sandbars.
37. Mouth of Preston Creek. Stumps and milfoil.
38. Highway 79 bridge pilings. Good spring fishing around riprap.
39. West finger of Preston Creek. Stumps in shallow water. Good in spring.
40. Islands. Deep water on north side; shallow water on south side. Good year-round.
41. Long, underwater island. Lots of milfoil. Good year-round.
42. Underwater humps. Lots of milfoil and stumps.
43. Bridge pilings and riprap.
44. Sharp drop as creek goes under bridge. Good spring and fall fishing area.
45. Winding creek channel. Locate with depthfinder. Excellent travel area in spring and fall.
46. Milfoil patches.
47. Shallow slough filled with stumps and milfoil.

Lake Guntersville

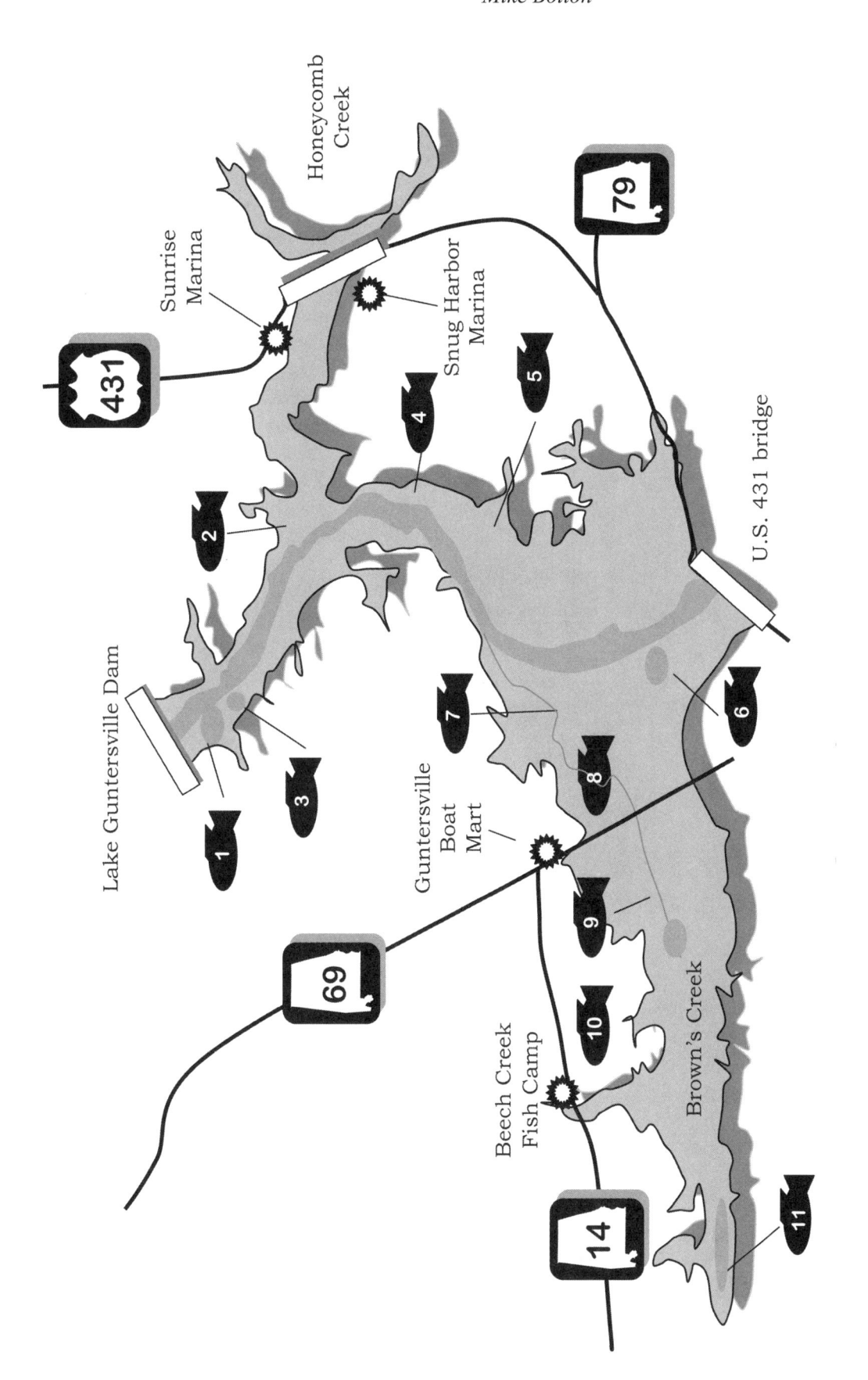

Lake Guntersville Dam to U.S. 431 bridge

Lake Guntersville

U.S. 431 bridge to Siebold Branch

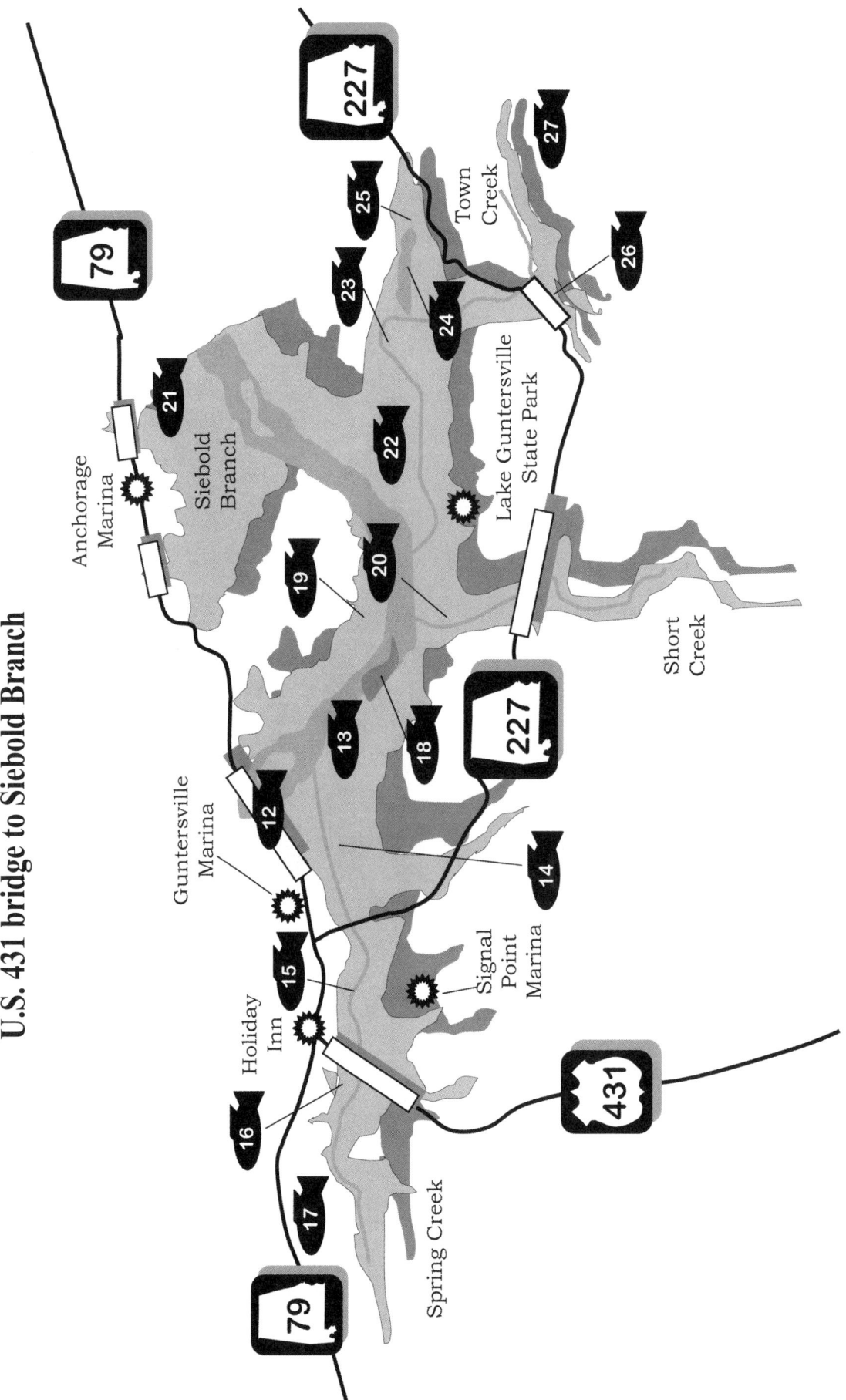

38

Lake Guntersville

Siebold Branch to Mink Creek

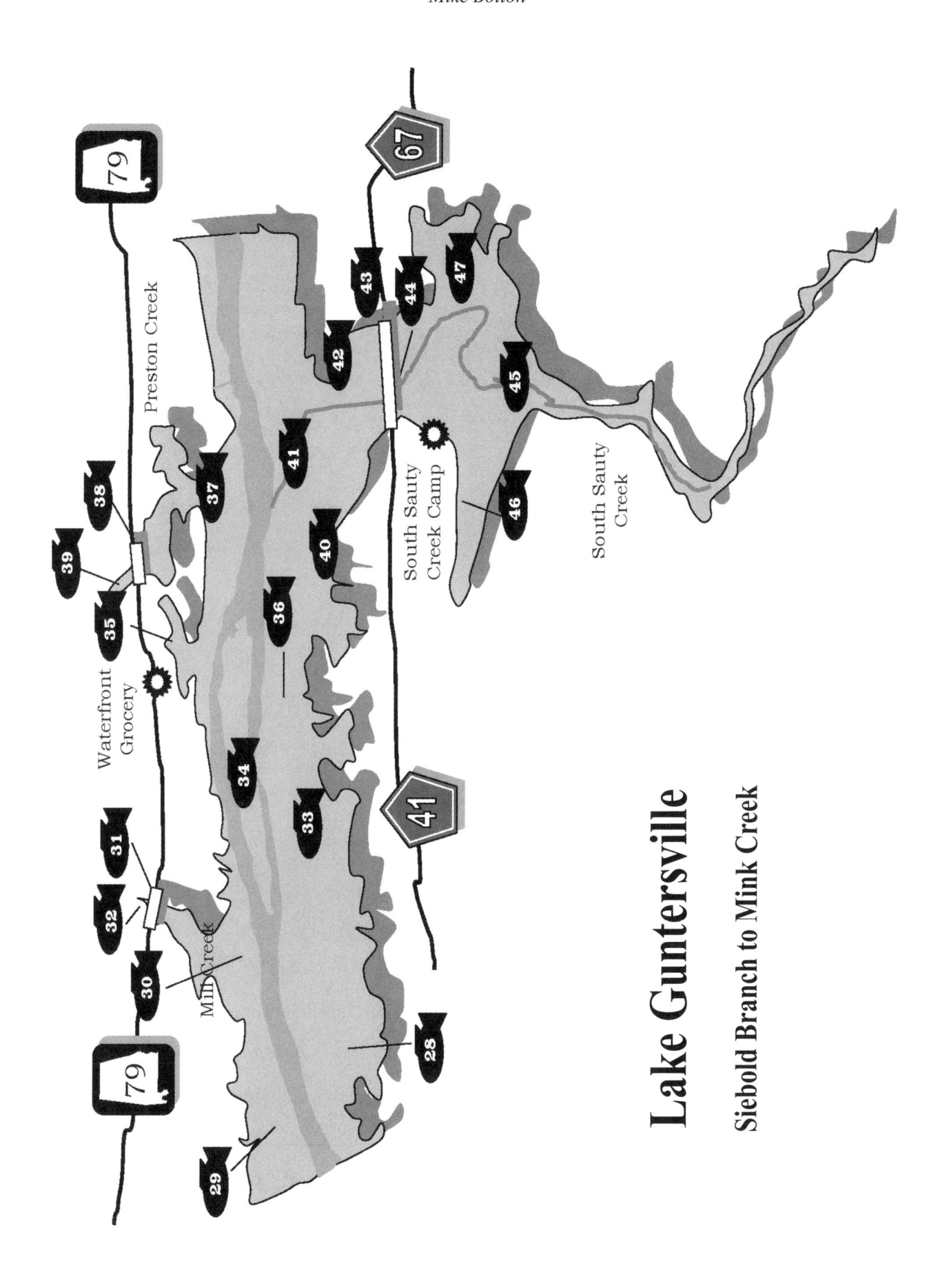

Wheeler Lake

The second-largest lake in Alabama, Wheeler Lake offers north Alabama some of the most diversified trophy fishing of any lake in North America, This lake has given up 10-pound smallmouth bass, 12-pound largemouth bass, 3-pound crappie, an 84-pound blue catfish and a 38-pound striped bass.

Even if record-pushing fish aren't your bag, this massive lake is bound to have some type of fishing that you will enjoy. Largemouth and smallmouth bass, catfish, crappie and sauger have long been staples of this lake, and the addition of striped bass and hybrid bass has enhanced the fishery even more.

It's not impossible for an angler fishing a curly-tail grub to catch as many as seven different species in a day's time.

Spring bass fishing

Wheeler Dam to Fox Creek

You will find little debate among anglers that the western end of Wheeler Lake offers the best spring fishing. The location of Second Creek, First Creek, Elk River, grass-filled Spring Creek and the numbers of tiny, shallow, structure-filled cuts on this end of the lake make it hard to compete with early in the year when spawning is on the mind of bass.

Both First Creek and Second Creek have the type of structure that make for excellent spring fishing. Sandbars, cuts with brush, rocky banks and points and flats dot the length of both creeks. Each creek also has some grass in its upper end. Second Creek has two roadbeds that produce bass in spring, and year round as well. The cuts on the main river are havens for big springtime bass.

There is just a smattering of grass in these cuts, but they are filled with blow-down timber, man-made brush piles and other wood structure. Many also have sand and pea gravel bottoms which bass love for spawning.

The Elk River also offers good spring fishing. The hundreds of tiny cuts sit adjacent to a stump-lined channel. Wash-in brush, blow-down timber and logjams can be found up and down this river.

Spring Creek somehow got the grass that much of the rest of this end of the lake missed. It also has a series of ditches, ledges, islands and other structure conducive to spring fishing.

Mallard Creek and Fox Creek are two more creeks

on the lower end of the lake, although both are upriver a bit from the concentration of good spring fishing locations mentioned before.

Many anglers start fishing on the lower end and, because of such good spring fishing there, never get this far upriver. Fishing both creeks, and especially Fox Creek, will not be a mistake if you decide to give fishing up that way a try.

During the pre-spawn period, which most anglers agree begins approximately Feb. 15, anglers should keep in mind that the bass find themselves in a predicament. They are ready to start feeding in anticipation of the spawn, but the water temperature still has their metabolism low and they aren't close to being the voracious feeders they'll be two months from then.

In other words, these fish can't catch a spinnerbait or crankbait ripped by them. The angler should fish accordingly.

A 1/2-ounce brown jig-and-pig with a chartreuse pork-frog trailer makes for a good, slow, visible bait in water that is almost always stained this time of the year. This bait is heavy enough to fall rapidly following the cast and it drops from ledges and other structure quickly into deeper water. That is important this time of the year when the bass are coming out of deeper water.

Another bait that works well is the spinnerbait because its speed can be easily controlled. A white or chartreuse/white spinnerbait with large nickel blades is preferred because it can be easily seen and because, by design, it must be fished slow.

It should be tipped with a white or chartreuse pork frog. The pork frog serves several purposes. First, it gives the lure added weight so it can be cast a great distance. Two, it gives a natural feel and taste when the fish hits the hook. Some fishermen also argue it makes the spinnerbait more weedless.

A 1/2-ounce or 3/4-ounce spinnerbait is best because it drops into deeper water quickly.

Once the water temperature rises and the fish move shallow, switch to a 1/4-ounce or 3/8-ounce spinnerbait. You want a lighter bait then that won't fall quickly and hang up in structure.

Some anglers also use a worm at this time, but most anglers agree jig-and-pigs and spinnerbaits catch more and better fish this time of year.

Once the water warms and the fish begin their forays onto the ledges and into the mouths of creeks in search of food, the faster baits can be used. A 1/4-ounce spinnerbait is still tough to beat, but a medium-running crankbait is also good.

Once the fish are in water three feet or less and thinking spawn, or chasing shad in shallow water, buzzbaits, floating lizards, stick baits, jerk baits and shallow-running crankbaits are key.

There are not many reasons to do a lot of traveling on this lake during the spring. An angler can decide on fishing First Creek, Second Creek, the cuts on the main river channel, etc., and simply set up shop there for the day. If he fishes that area properly, he'll catch enough fish that moving won't enter his mind.

This author has set on the first secondary point on the right bank entering First Creek in April and caught and released a limit of bass in less than an hour.

It's also possible to sit in Fox Creek all day with a spinnerbait and buzzbait anytime during the spring.

Fox Creek to U.S. 31 bridge

When you move into this section of the lake, you'll find a much different lake than you found on the lower end. This area is made up of wide flats with water depth that varies from one to six feet outside of the main channel. What anglers call drop-offs in this section may only be a 1-foot ledge, but this 1-foot ledge will still hold fish.

This area is full of islands, as well as flats with sprinklings of milfoil, and to a lesser degree, hydrilla.

Two spring baits are a must in this area—spinnerbaits and Rat-L-Traps. Just as on Wheeler's sister lake, Lake Guntersville, the milfoil will be submerged this time of the year and ripping a Rat-L-Trap across the top of it can be deadly.

It makes no sense to pinpoint just one area in this portion of the lake that offers good fishing. The presence of the milfoil and hydrilla make it possible for any spot to hold fish. Anglers should use common sense, however, and fish the deeper weeds early and progress to the shallow water as spawning time approaches.

Once the fish move shallow and the growth of the weeds increases, it will become necessary to switch from a Rat-L-Trap to a spinnerbait. Leave the pork trailer off the bait in this instance because you need the lightest bait possible. Go with a 1/4-ounce chartreuse spinnerbait and concentrate on the shallow water closest to the islands that have grass.

Unfortunately, that's like telling someone in Alabama to deer hunt near pine trees. It doesn't narrow it down much, but that's where you'll find fish.

U.S 31 bridge to U.S. 231

Once the angler goes east (upriver) of the U.S. 31 bridge, he'll encounter a brief stretch of flats similar to those in the middle section, then he'll find a lake that narrows and is river-like the rest of the way to Guntersville Dam. This well-defined river channel has small ledges on both sides throughout its length.

Flint Creek, Piney Creek, Limestone Creek, Beaverdam Creek, Cotaco Creek, Barren Fork Creek, Indian Creek and Huntsville Spring Branch offer decent spring fishing.

These creeks have adequate water depth and enough stumps for good spawning grounds. These creeks have some grass, but not enough to be much of a factor in patterning fish to grass.

Due to the fact some of these creeks are the only shallow water in the area, they are the only places capable of confining spring baitfish. Bass are forced to go into the areas.

Another factor that make these areas worth trying is the fact most of the creeks have small mouths which are easy to cover, especially before and after the spawn. Bass traveling to and from the creeks to deep water are funneled into a tiny area. An angler can fish these areas thoroughly and he can't help catching fish.

During the pre-spawn period, when bass are leaving the channel and coming into the creeks, jig-and-pigs, big-bladed spinnerbaits and Carolina-rigged lizards can be effective in the creek mouths.

Following the spawn, when the bass are forced back to deep water by the water temperature, faster baits such as Sassy Shads, Rat-L-Traps and medium-running crankbaits will do the trick. When the bass are up in the creeks, the traditional spring baits—buzzbaits, spinnerbaits, lizards and jerk baits—will work.

Hot weather bass fishing

Wheeler Dam to Fox Creek

Once the spring spawn is over, the bass will resume the heavy feeding pattern they abandoned during the spawn and slowly move to deeper water. When the water surface temperature reaches the mid 70s, it's safe to assume you'll no longer find quality fish in water less than 8 feet deep.

Between the time the bass quit spawning and the time they head for deeper water, they are making up for lost time in the eating department. The mouths of the creeks are good places to find these transition fish. A chrome or white Rat-L-Trap, or even a worm or spinnerbait, fished on the primary and secondary points can result in good catches.

Once you don't find the quality fish shallow anymore, it's a safe bet they have gone off the ledges to find a summer home on deep bars, underwater humps and islands, dropoffs, deep rocky banks and long, deep points. It's not uncommon for these fish to be 30 feet deep before the summer is over.

Fishing during this time is tough, especially for the angler who lacks patience. He's probably not going to catch numbers of fish as he did during the spring, and instead of moving regularly in hopes of finding great fishing, he needs to fish key areas slowly and thoroughly.

Several baits will work during this period. The angler with an arm strong enough to hack it may find success with a pearl-colored 20-Plus crankbait, but few anglers have the gusto to battle the sun and a deep-diving crankbait all day.

One way around that, if you're not tournament fishing, is to troll a 30-Plus crankbait over the humps and bars in this part of the lake. One of the lake's better anglers does this with a three-rod setup and keeps a marker buoy close by his side. When there is a hookup, he throws the buoy to the spot, reels in the fish, then switches rods.

He first throws a jigging spoon he already had rigged to the spot, and if that doesn't work, he throws a pre-rigged Carolina-rig. One of those rigs usually results in another fish or two and they are usually good fish. It's not unusual for bass to school in tight schools this time of the year and multiple catches from one spot aren't uncommon.

Another pattern that works well is locating bass on the deep, rocky, stumpy points with a depthfinder, then attacking them with a jigging spoon. Not many anglers stop to think about it, but unlike most other lures, a spoon can mimic an injured baitfish and do it over and over again in the same spot. This often triggers an angry strike from a fish that isn't interested in hitting a bait, or even a fish that is leery of what he's seeing and knows something isn't right. The spoon will catch fish this time of the year that worms, crankbaits and spinnerbaits won't catch.

It takes little more than your eyes and a depthfinder to locate summer fish on this end of the lake. Any long, sloping point that goes into deep water is likely to be good, as is any sudden change in the bottom. The underwater islands in the main river channel between the Elk River and Spring Creek is an excellent place to start.

Fox Creek to U.S. 31 bridge

Unlike on the lake's lower and upper ends, the shade and cover provided by the acres of milfoil and hydrilla in this area delays the movement of bass into the deep river channel during the summer. While you may find bass on deep structure in early summer on the lake's upper and lower sections, it's possible you may catch them on buzzbaits and other topwater lures here on the same day.

The grass provides shade which holds the water temperature lower longer, and provides cover for the baitfish. This growing grass also produces oxygen during the critical time of the year when the water needs oxygen.

Getting the bass to leave is akin to making somebody leave the air-conditioned den during the summer to go sit in the backyard.

The baits used here during the spring will catch fish into the summer as long as the bass are holding in the grass. As the water temperature rises, the bass will eventually be forced to go to deep water at least for part of the day. That exit from the grass will begin earlier and earlier each day during the summer. However—except during an unusually hot July or August—the bass will come back to the grass sometime during the evening or night.

Almost any summer morning at daybreak, or summer evening at dusk, it's possible to throw a torpedo lure, buzzbait or jerk bait on the fringe of this grass facing the channel and catch bass. On rainy, overcast days, it's sometimes possible to catch fish that way all day long.

The ability to catch bass shallow throughout the summer on this lake will be in direct contrast to some other lakes in the state, such as grass-filled Lay Lake. The reason for that is Wheeler Lake's slightly lower water surface temperature.

There are two reasons for that minor water temperature difference. First, Lay Lake and some other grass-filled lakes are further south and are a degree or so warmer because of it. Second, Wheeler Lake is a navigable waterway and Lay Lake isn't. The dams and locks on the Tennessee River move much more water during the summer than does the Coosa River. Moving water is more difficult to heat than still water.

Back to Wheeler Lake. During dog days, once the sun clears the tree line, the better anglers switch to plastic worms and Carolina-rigged lizards and fish the ledges adjacent to the grass. To do this, the boat should be positioned in the deep water at least 25 yards into the main channel and the cast made onto the ledge. The retrieve should bring the worm across the ledge and eventually allow it to fall in the channel.

Many strikes come when the worm "falls off the table." Why? Because the schooling bass will often chase shad onto the ledge, then retreat to deep water and suspend themselves a few feet below the ledge out of view. The shad will swim back toward the deep water and once they exit the ledge, they have their last surprise.

U.S. 31 bridge to U.S. 231 bridge

Because of the well-defined, deep river channel, this part of the lake offers good summer fishing for the angler who understands river and ledge patterns.

The ledges are scattered with blow-downs and other debris, and summer ledge fishing can be excellent.

The bass are forced to suspend in the deep water in this area during the summer. The baitfish attempt to get as close as possible to the deep water without giving up their life-saving structure, meaning they usually end up on stump rows, brush piles and other such debris on the edge of the ledge.

There are two ways to entice these fish to bite. The most tiring and difficult is to spend the day fishing parallel to the ledges with deep-diving crankbaits. This is a tough pattern to fish during the summer heat, but it can be effective and can produce big fish. The angler should first locate the fish with a depthfinder, then drop a marker buoy. He should then make a long cast well past the schooling fish and, using his judgment, make the retrieve fast enough so the lure runs right over the top of the fish.

Another pattern that works well is locating the stumps and other structure on the edge of the ledge and fishing for the bass running up on the ledge to feed. Locating this structure can be time consuming, but running the boat along the ledge will show this structure on the depthfinder. Another way to locate the structure is by positioning your boat in open water and watching for the tell-tale signs of bass knocking shad out of the water. It's a safe bet that wherever there's a sign of feeding activity, some type of structure can be found nearby on the ledge.

Once this structure is located, several baits can be used. Two of the best are a 3-inch pearl-colored Sassy Shad or a white bucktail jig fished on 6-pound test line with a spinning outfit. These shad imitations should be cast onto the ledge and hopped back to the boat. Once they fall off the ledge, allow either to fall and make the hops, or jigging motion, more pronounced to allow them to cover both deep and shallow water.

Plastic worms and small crankbaits also will work.

This is also the area on Wheeler Lake in which flipping can be productive during the summer. Pig-and-jigs or lizards flipped into structure on the deeper ledges and onto stumpy points will also produce bass.

Cool and cold weather bass fishing

Wheeler Dam to Fox Creek

The bass you found in summer won't be far away from that same location during the fall and winter. The bass will move into shallower water during the fall in search for food on which to fatten up for the winter, but not the shallow water in which you found them during the spring.

Bass holding on long, deep points during the summer will use that point as a roadway on which to move back and forth to less deep water. It's no secret that the baitfish use this pathway, too. For the bass who find summer homes on underwater islands or sandbars, the fish will move to the uppermost point of this structure in search of food. If food isn't available, they'll move onto ledges or in the mouths of creeks.

Several patterns can be used to catch these fish. A Carolina-rigged worm is one bait that works well, especially when the water temperature drops below 72 degrees. A jig-and-pig also works well during this time, and on into the cold of the winter. A semi-jigged Rat-L-Trap, its speed of retrieve determined by the water temperature, also works well, especially coming down the points.

Early in the fall, a 3-inch pearl curly tail grub fished on stumpy points will produce bass if you can keep the crappie and hybrids off of it.

The better anglers stay out of the creeks during this time, even though it's possible to catch some bass on the ledges of the creeks that have deep channels.

Keep in mind that bass are finicky during this time. A change in color of a lure, or adding a shorter worm to a Carolina rig, can bring strikes in areas where you couldn't buy a strike before

Why? Who knows. That's why they call it fishing instead of catching.

Fox Creek to U.S. 31 bridge

Because this area is so full of shallow grass, it isn't a very good place to fish for bass during the fall and winter, especially during the colder months. The dying grass robs the water of oxygen and fish can't survive for long in the area.

The better anglers elect to fish the upper or lower ends of the lake where the water is deep enough to thwart grass.

U.S. 31 bridge to U.S. 231 bridge

During the fall and winter, the patterns for catching bass don't change much from those used during the summer.

The fish still make the deep-water channel home, but during the fall will use the ledges more to feed on, especially when the winter feeding frenzy starts. Fishing on the deep points also gets a little better during the fall.

During the winter, fishing is tough in this area. Bass suspended in a channel during the winter are almost impossible to get to except with jigging spoons and lipless crankbaits. Go hunting instead.

Other types of fishing

Spring smallmouth fishing

This lake takes a backseat to Pickwick Lake when it comes to smallmouth fishing. But Wheeler Lake can offer some mighty good smallmouth fishing in the spring.

Few anglers will argue that the lower end of the lake is not the best area for smallmouth fishing, or that the spring is the best season.

The deep, sloping rocky banks next to deep water in First Creek, Second Creek and the mouth of Elk River offer the best spring smallie fishing, with a Jumping Minnow the primary bait.

Wheeler Lake's best fishing spots

1. Mouth of Second Creek. Good drop-off from 5 feet to 45 feet. Good largemouth and smallmouth area year round.
2. Excellent point. Good spring in pre-spawn.
3. Good fishing around pilings and riprap on U.S. 72 bridge.
4. Good point for spring bass fishing.
5. Underwater ditch. Good spring fishing.
6. Underwater road bed.
7. Rocky cuts. Good for spring fishing.
8. Mouth of First Creek. Good year-round fishing.
9. Good bank for spring and summer fishing. Full of treetops.
10. Good spring bass.
11. Island in First Creek. Good spring fishing.
12. Good spring point.
13. Good crankbait area in spring.
14. Excellent points on south bank for spring fishing.
15. Excellent point for year-round fishing.
16. Island in mouth of Elk River. Good spring fishing.
17. Excellent bank for year-round fishing. Cuts, logjams, blow-downs, etc.
18. U.S. 72 bridge pilings and riprap. Good spring and summer. Good crappie fishing at night in summer.
19. Good cut. Good spring fishing
20. Bank filled with logjams. Good buzzbait area in spring. Good flippin' area in summer.
21. Bend in channel. Good ledge fishing in spring, summer and fall.
22. Good spinnerbait and shallow-running crankbait in spring and early summer.
23. Underwater islands. Good summer fishing for all species.
24. Oxbow-shaped underwater island. Good summer fishing area for all species.
25. Mouth of Spring Creek. Good schooling area for largemouth in summer and fall.
26. Island in Spring Creek. In shallow water next to deep water. Excellent spring fishing.
27. Islands in Spring Creek. Good spring fishing.
28. Good buzzbait fishing in back of Spring Creek.
29. Good cuts. Good spring fishing.
30. Underwater island in main river channel. Good summer area, especially when current is moving.
31. Excellent spring cuts.
32. Coxey Creek. Good spring fishing.
33. Point in mouth of Mallard Creek. Good year-round fishing.
34. Island in Mallard Creek. Good spring fishing.
35. Bridge pilings and riprap around bridge. Good crappie fishing in spring and summer.
36. Mouth of Fox Creek. Good year-round fishing.
37. Shallow island in back of Fox Creek. Good spring fishing with spinnerbaits.
38. Excellent point. Good year-round fishing up and down point.
39. Good spring flats. Milfoil and hydrilla.
40. Island. Good spring fishing.
41. Series of small islands. Good spring fishing.
42. Island. Good spring fishing.
43. Islands. Good spring fishing.
44. Sandbars with stumps. Tremendous spring fishing for largemouth and smallmouth.
45. Underwater island that splits main river channel. Good summer and winter fishing.
46. Underwater islands in mouth of Baker's Creek. Good year-round fishing.
47. Mouth of Baker's Creek. Good year-round fishing.
48. Long underwater ditch leading from bank to main river channel. Tremendous for bass year round. Bass travel up and down from shallow to deep water in this ditch.
49. Keller Bridge. Good crappie fishing in summer.
50. Mouth of Flint Creek. Good year-round fishing.
51. Bridge pilings in Flint Creek. Good in spring for bass. Summer crappie fishing at night.
52. 1-65 bridge pilings. Summer fishing at night for crappie.
53. Narrow mouth of Piney Creek. Good funnel area for bass pre- and post-spawn.
54. Good bank for spinnerbaits in spring.
55. The Bayou. Good spring spinnerbait and buzzbait.
56. Lily Pond. Good spring area with floating lizard.

Wheeler Lake

Wheeler Dam to Fox Creek

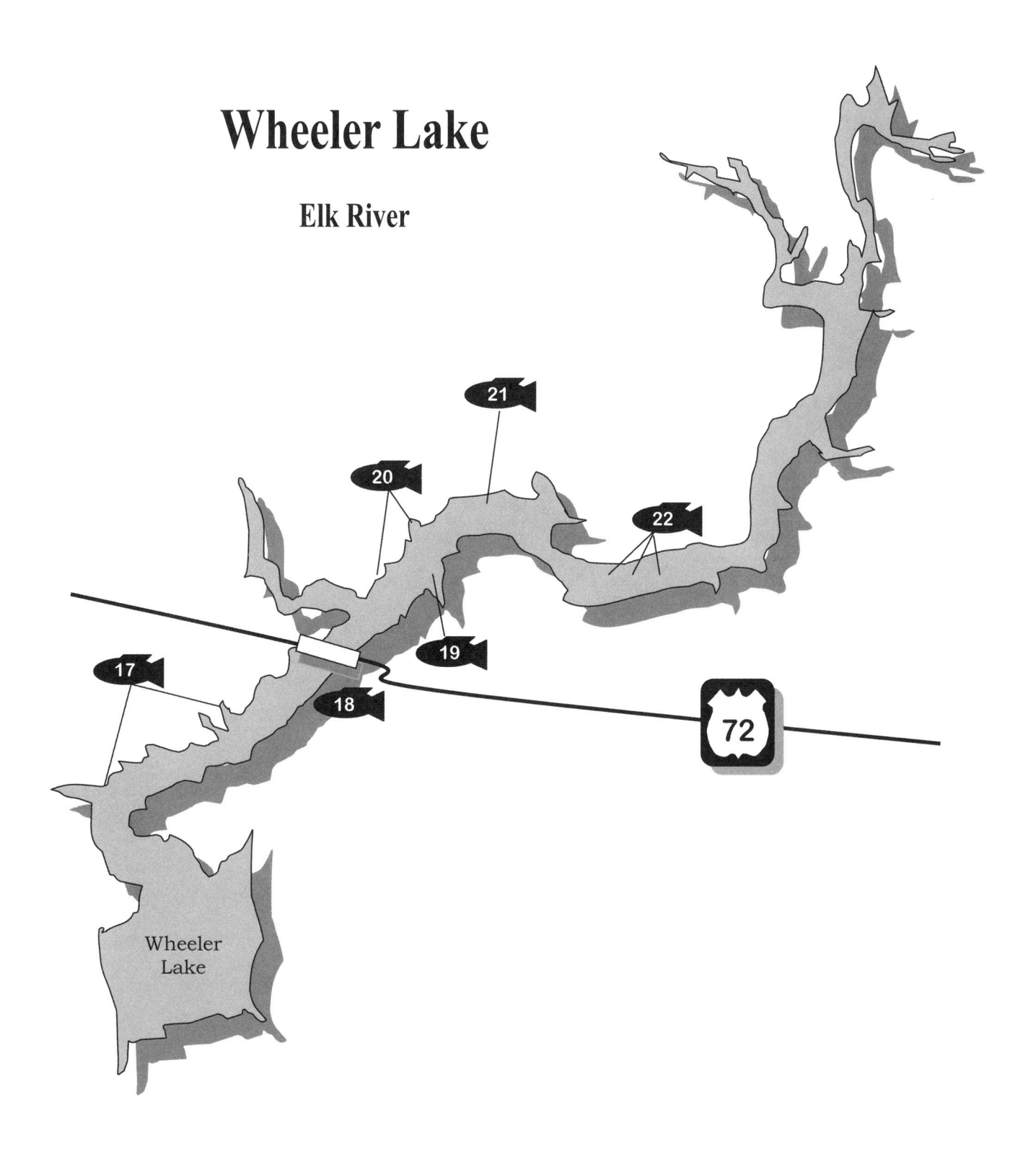

Wheeler Lake

Elk River

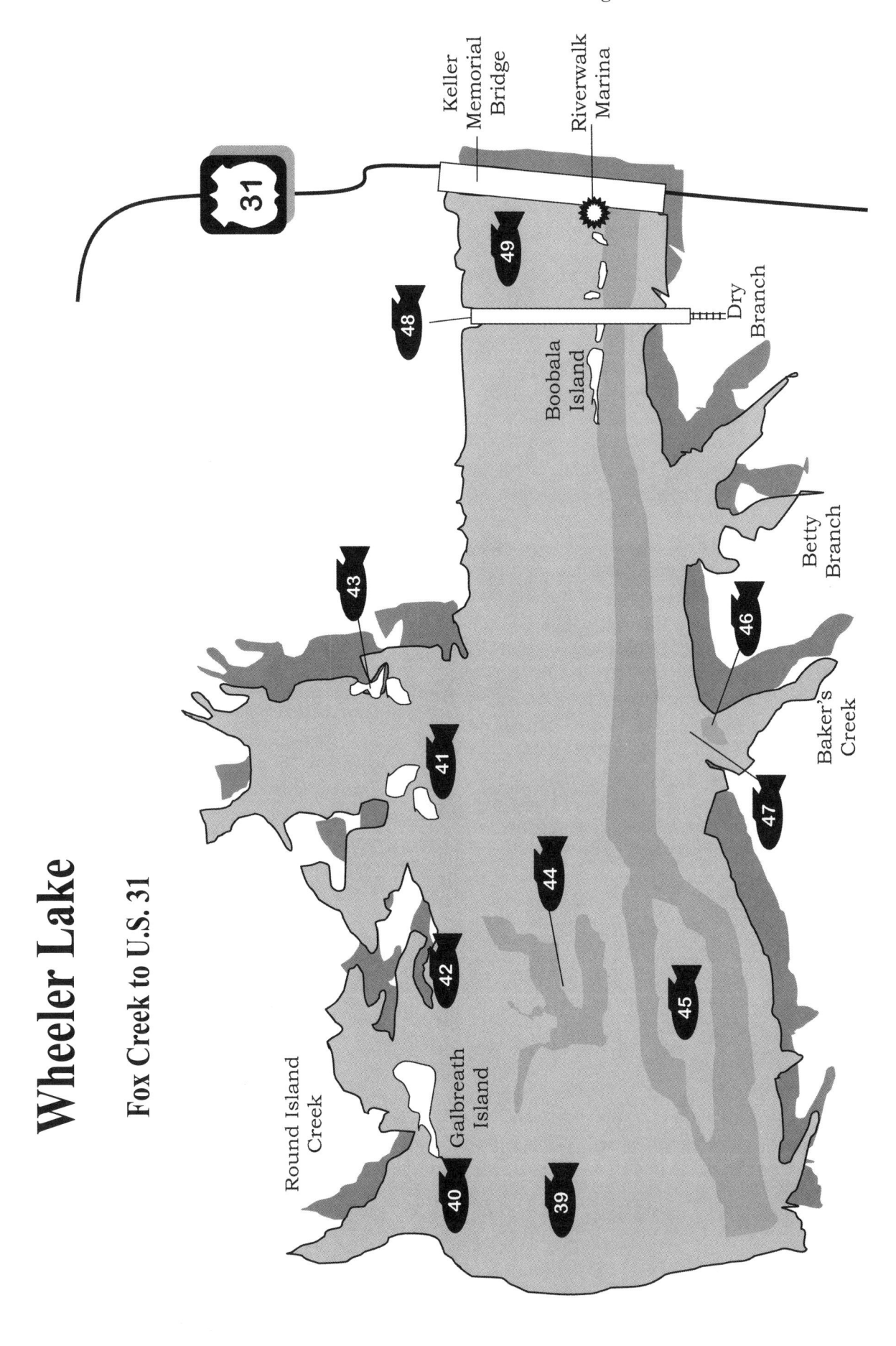

Wheeler Lake
Fox Creek to U.S. 31

The Ultimate Guide to Alabama Fishing

Wheeler Lake

U.S. 31 bridge to Lily Pond

49

Wilson Lake

Because Wilson Lake is a 15,930 acre lake wedged between 68,300-acre Wheeler Lake and 47,500-acre Pickwick Lake, "tiny" Wilson Lake doesn't always get the respect it deserves.

This tremendous fish producer may be old, but it's by no means feeble. It's probably the most varied fishery in Alabama and it annually produces excellent catches of largemouth bass, smallmouth bass and spotted bass, not to mention crappie, catfish, stripe, hybrid, white bass, sauger and rough species.

Two of the current state records have come from this body of water, both former world records. One of those catches, the l0-pound, 8-ounce smallmouth caught in 1950, was the world record for several decades.

Spring bass fishing

As is the case in all of the Tennessee River impoundments in Alabama, Wilson Lake has its share of aquatic vegetation. The presence of non-native species of weeds such as Eurasion milfoil and hydrilla, coupled with a variety of native species of grass, makes Wilson Lake a first-class spring bass fishery.

The lake lacks the numbers of large wide creeks that its Tennessee River neighbors Wheeler Lake, Pickwick Lake and Lake Guntersville enjoy, but the dozens of small, grass-filled pockets off the main river channel offer spring spawning grounds for bass that are actually easier to fish than creeks.

If you're a spring creek fisherman, don't fret, however. Wilson Lake has a few creeks that can keep you busy. Town Creek, Blue Water Creek, McCernan Creek and Shoal Creek offer excellent spring fishing.

The basics of the spring ritual for bass are no different here than on any other lake in Alabama. The bass begin leaving their deep-water winter homes at the first sign of warming water in late February and begin feeding heavily in anticipation for the spawning period, a time in which many of them do not eat.

This spawning period can be characterized as a human walking up a flight of stairs and coming back down. The bass begin in the deep water and step themselves up into more shallow water as the water temperature allows.

Once the preferred water temperature is available, usually around 68 degrees, the bass will spawn in shallow water, then begin their journey back down the steps. By summer, the bass will be back at the bottom of the steps where they were in winter.

Most Wilson Lake bass will be found in the deep water of the main channel in early February, but soon start moving toward the banks, pausing briefly at different levels on different types of structure.

Two key areas during this time are the underwater islands in the mouth of Big Nance Creek and in the mouth of Town Creek. These islands allow the bass to move up and down as the weather does its early spring flip-flops. A Carolina-rigged lizard or plastic worm, a 5/8-ounce brown jig with a black frog pork trailer or a shad-colored crankbait will take the fish off these islands in early spring.

Once the water temperature is right, the fish will move off the islands and up into Big Nance Creek and Town Creek. These creeks are perfect for spring bass because they have both deep and shallow water, as well as humps, logjams, piers, brush and grass-filled cuts. Several baits will work during this time. A 1/2-ounce chartreuse/white spinnerbait with a chartreuse pork frog trailer is probably the most effective bait because of its ability to cover so much water.

As said before, the hundreds of cuts and sloughs along the main river channel hold the majority of the spring bass, however. The bass are able to stage themselves out of deep water and onto the ledges before moving into the pockets.

A jig-and-pig or deep-diving crankbait is needed when the bass are still suspended on the ledges. Once the fish move up on the ledges and begin to make their move into the pockets, a spinnerbait is the key. Use a 1/2-ounce bait early and allow it to fall into the channel as part of the retrieve. Once the bass are up on the ledges to stay, switch to a lighter 1/4-ounce bait.

Once the bass are in the pockets, the topwater angler may switch to his favorite kind of baits. Buzzbaits, jerk baits and floating lizards will work on the grass fringes, as well as up in the grass.

These same patterns will work in Shoal Creek and McCernan Creek. Both have grass-filled cuts full of spring bass.

Hot weather bass fishing

Once the spawn is over, the bass will hang in the grass in the cuts and creeks until the increasingly warm water temperature (usually 80-plus degrees) forces them to escape to deeper, cooler water. This exit to deeper water will not be a sudden movement, for the bass will fight to stay in the grass as long as possible. They like it there. The fish will be in the grass at daybreak early in the summer and may stay there until 10 a.m., but the increasing summer heat will force the bass to leave the weeds earlier and earlier each day until, eventually, they'll be gone.

That's only in theory, of course. It actually may be possible to find some bass in this grass all summer, but these will almost never be quality bass. By mid-June, the water temperature will have forced most large bass to deeper water.

During this period that the fish are in the grass, several patterns will work. A buzzbait will usually work throughout the period the bass are in the grass, but the last few weeks that the bass are in the weeds, the grass may be too thick for a buzzbait. Either flipping or a swimming jig will be needed then. A 3/4-ounce brown jig with a black pork frog will be strong enough to poke through most of the grass that you'll find on Wilson Lake.

While some of the smaller bass will be fooling with the grass during the summer, the true quality bass will be deep on structure in the main river channel. The main river channel area is full of underwater islands and sandbars, many covered with brush piles and stumps.

Some of those areas that anglers should key on are the constantly changing humps and sandbars below Wheeler Dam, the series of underwater ridges and ledges near the mouth of Blue Water Creek, the major underwater island out from the mouth of Town Creek, the series of large islands and sandbars in the middle of the lake, the sandbar out from the mouth of McCernan Creek and the two large humps above Wilson Dam. It is from these deep underwater areas that better anglers catch most of the best summertime bass.

Several patterns will catch these fish. If you're fishing for fun and not in a tournament, deep-diving crankbaits trolled over these humps will catch fish. This is a tremendous pattern for big fish when the water is being pulled through both dams. It will also work at night.

Other lures that will work include Carolina-rigged plastic worms, jig-and-pigs with pork frog trailers and last, but surely not least, jigging spoons.

The jigging spoon is a well-kept secret of many of the lake's top tournament anglers who fish in summer. Once a school of summer-time bass is located with a depthfinder, a jigging spoon should be dropped directly on top of the fish. The spoon should then be lifted 18 inches in a medium-speed motion, then allowed to fall with a fluttering motion. Most strikes will take place when the lure is on the fall. Be forearmed. The strikes are violent. Have the rod steady in your hands and the drag set properly.

Quality fish are rarely found in the creeks during the hot summer months, so the angler looking for big bass should stay in the deep water in the main river channel.

If you're just looking for fish large enough to claim you caught some, plenty of small bass can be found in the creeks, however. The angler looking to just catch a fish or two of any size can find some success by fishing worms in the brush around piers and on the ledges of the deeper creeks, especially at night.

The only exception to avoiding the creeks during the summer to catch big fish is the cool-water spring in Blue Water Creek. This spring, which emits cool water in the summer and warm water during the winter, will hold quality fish—of all species—year round.

This spring can easily be located with the use of a surface temperature gauge.

Cool and cold weather bass fishing

Because of the unusual situation on this lake of some water being deeper on some banks than it is 50 yards out from the banks, the cuts and the mouth of cuts adjacent to that deep water offers good fishing during the fall.

Grass-filled cuts next to deep water means the bass have easy access to both deep and shallow water as the fall weather does its inevitable flip-flops. Keep in mind that the bass are schooling and feeding heavily at this time to store fat for the winter and that shad are their No.1 food.

Two of the key baits at this time are shad look-a-likes—swim baits and Rat-L-Traps. Early during the fall, when the bass are on the ledges and running up on the ledges in the mouths of the cuts to feed, either of the baits can take numbers of quality fish. A pearl 3-inch swim bait fished on spinning tackle and 8-pound test line should be against the law this time of the year. Some anglers catch 50 and 60 bass a day with this pattern when they hit it right. A chrome or pearl Rat-L-Trap will give similar results.

Several other baits will work well during this period, too. A white spinnerbait fished on the fringes of the grass and in the mouth of the cuts is a prime producer after the fish exit the channel and begin staying on the ledges. This bait covers a lot of territory quickly and tends to produce more quality fish than other lures.

The more the water temperature cools, the slower a bait must be. Many of the better spinnerbait anglers force themselves to slow down by using larger blades, while other anglers switch to naturally slower baits such as jig-and-pigs and worms. These worms and jig-and-pigs fished on the fringes of the grass and in ditches and creek channels in the cuts, as well as the mouth of the cuts, will sometimes produce fish through the late fall and winter, especially during mild winters.

During normal winter patterns, however, the falling water temperature will force the quality bass into warmer, deeper water. The sandbars, islands, humps and other structure over 12 feet deep will concentrate these winter fish and a good rule of thumb is to look for them in the same place you found them in July and August. Jig-and-pigs and worms are the only baits that give you much chance at these bass during the winter and even then, it's not that good a chance.

A few other types of fishing

Tailrace fishing

It may not be an exaggeration to say Wilson Lake's upper end, the tailrace below Wheeler Dam, may offer the best tailrace fishing in the U.S.

Nowhere else in Alabama can an angler have a more varied catch of large fish. It was in this tailrace in September one year that this author and Birmingham's Nolen Shivers had a day of fishing like many anglers only dream of. By lunchtime, the two of us caught and released 23 smallmouth between four and six pounds. Eleven of those fish topped five pounds, including the largest, a 6-pound, 2-ounce beauty. Included in that day's catch were a 6-pound largemouth, a 16- pound catfish, numerous smaller cats, some 10-pound-plus drum and several crappie.

It's almost impossible to find a negative in fishing this tailrace. Unlike many tailraces in Alabama, this one doesn't have a single underwater obstacle that poses a boating hazard. There is uniform depth of 12 feet for at least a half-mile or more below it. And because this is a navigable waterway, there will be periodic water flow through the dam even during the worst of droughts.

Fishing is always best when this water is moving.

This tailrace produces fish year round, too. During some summers anglers catch nice stringers of smallmouth throughout the hottest months there. You don't go there fishing for just one species, either. It's possible to spend a day fishing and end up with a catch of largemouth bass, smallmouth bass, Kentucky spotted bass, white bass, yellow bass, hybrid bass, striped bass, white crappie, black crappie, sauger, five species of catfish, skipjack herring, drum and on and on.

There is easy access to the tailrace from several areas, most notably John Hill's Fisherman's Resort in Big

Nance Creek.

If this area does have a negative, it's that bad weather can make the area treacherous. A good storm and a west wind can make it extremely dangerous. Keep that in mind when you see weather approaching.

An angler has a wide array of ways to catch these tailrace fish. Most anglers opt for live-bait fishing, choosing to scoop their bait free of charge with a dip net or cast net along the walls of the dam and powerhouse. Threadfin shad, gizzard shad and yellowtail minnows are the favorites.

Other anglers choose to fish with plastic grubs, swim baits such live worms, minnows (for crappie) or cut baits.

An angler may anchor in the swift current or run his boat up to the head of the dam, cut the motor and drift fish.

The fall run of smallmouth to the dam is something an angler can't afford to miss. Starting in mid-September, the larger smallmouth make a run at the dam to feed on the millions of minnows and shad that gather around the concrete walls. The fish are in a feeding frenzy, stuffing themselves in anticipation of the coming winter. The good news is that at no other time and on no other lake will you see so many quality smallmouth schooling in such a confined space.

Although the angler will try to convince himself it can't possibly matter, light line is a must here. Eight-pound test on a spinning reel is needed, as well as a No. 2 or No. 4 (depending on the size of the shad being used) pattern 84 Eagle Claw hook. A No. 5 or 6 split shot should be placed 18 inches above the bait.

The angler runs his boat within 50 yards of the turbines (water must be moving), cuts the motor and fishes as he drifts. The shad should be cast gently and reeled just fast enough to keep the hook from grabbing bottom.

A hookup in this swift water, coupled with the fact smallmouth will often make dazzling 5-foot jumps out of the water, is nice.

These fish begin close to the dam in early October and will slowly move further and further downriver as winter approaches. By the end of December, they may be as much as a mile below the dam.

Smallmouth fishing

Any time an angler is bass fishing on this lake, he's a threat to tear into a smallmouth bass. Smallmouth often share the same water (rocky points and banks, gravel bottom) as the Kentucky spots that are found on this lake. An angler can do things to increase his chances of catching smallmouth, however.

The smallmouth have a strong liking for smaller baits, most notably white grubs and 4-inch worms. The areas around and including Four-Mile Creek and Six-Mile Creek and the Wheeler Dam tailrace are prime smallmouth locations and novice smallmouth anglers should direct their attention to those places.

During the spring, the angler should locate rocky points and banks and fish them with grubs. Any cuts that have red clay banks will likely hold some smallmouth, too.

During the summer, the smallmouth gather on the deep islands, particularly the underwater humps closest to the dam. Slider worms, grubs and an occasional small crankbait will take the smallmouth then, especially when water is being pulled.

Some anglers prefer to fish for smallmouth at night during summer and numbers of fish, rather than large fish, can be caught then. A small, white spinnerbait fished along the rocky points and banks will produce.

During the fall, there isn't a reason to fish for smallmouth anywhere except below Wheeler Dam. This is when the quality fish come forth.

Wilson Lake's best fishing spots

1. Grass-filled cuts on right bank above Wilson Dam. Good fishing area in spring.

2. Underwater islands. Good fishing in summer and winter for all species. Especially good in summer for bass when water is being pulled.

3. Mouth of McCernan Creek. Good year-round fishing. Next to deep water and underwater island.

4. Back of McCernan Creek. Excellent spring spawning area. Lots of grass-filled cuts.

5. Excellents cuts on left bank of dam. Good spring and fall.

6. Excellent slough. Grass, points and piers. Good spring and fall.

7. Good spring cuts.

8. Mouth of Shoal Creek. Good year-round fishing. Shallow water adjacent to deep water.

9. Cuts up right bank of Shoal Creek. Good except in winter.

10. U.S. 43 bridge pilings. Good for spring bass and crappie. Good summer crappie fishing at night. Good riprap banks.

11. Edgewater Beach slough. Shallow slough. Full of grass. Good spring fishing for topwater.

12. Good deepwater slough. Good spring and summer.

13. Good cuts in upper end of Shoal Creeks. Excellent spring fishing with jig-and-pig.

14. to 20. Underwater Island.

21. Excellent, grass-filled pockets.

22. Six-Mile Creek. Good spring fishing.

23. Key point. Pay attention here. One of the best fish producers on the lake. Good year-round.

24. Four-Mile Creek. Good spring and fall fishing.

25. Another key point. Good year-round fishing.

26. Excellent grass-filled cuts. Good spring and fall.

27. Huge underwater island in mouth of Town Creek. Good year-round fishing. Lots of brushpiles and stumps.

28. Excellent grass-filled cuts.

29. Mouth of Town Creek. Good year-round fishing.

30. Foster's Mill Bridge pilings. Good spring fishing area.

31. Steep-cut bank. Good fishing in spring.

32. Mouth of Blue Water Creek. Good year-round fishing.

33. Location of underwater spring. Excellent spot for all species in summer and winter.

34. Logjams. Good flipping area.

35. Hog Island. Next to deep water on all sides. Good stopover point before and after spring spawn. Backside holds good bass when water is moving through dam.

36. Tailrace. Excellent year-round fishing.

37. Good rock piles. Good topwater area in spring. Backside holds fish when water is being pulled.

Mike Bolton

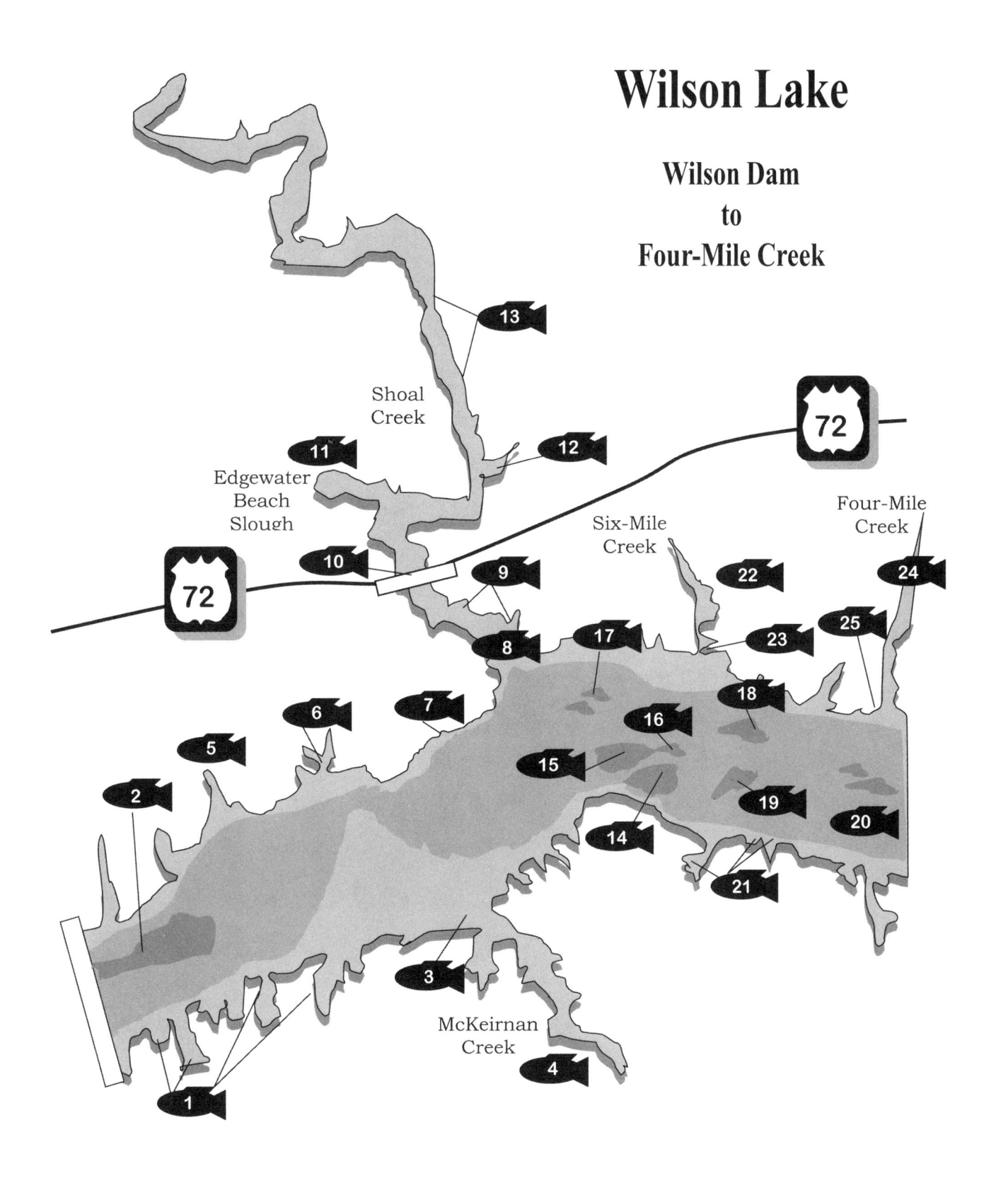

Wilson Lake

Wilson Dam to Four-Mile Creek

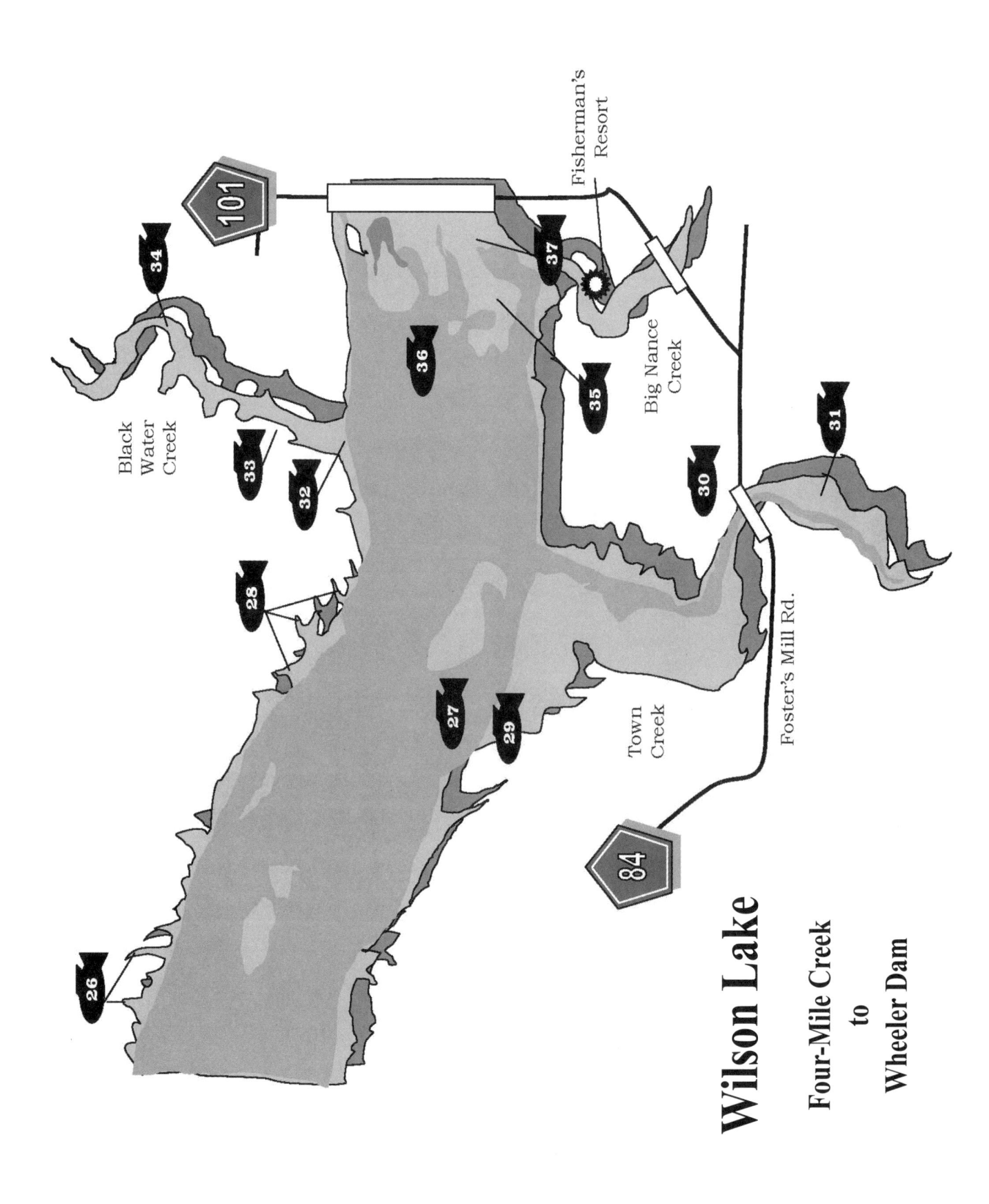

The Ultimate Guide to Alabama Fishing

Wilson Lake
Four-Mile Creek to Wheeler Dam

Mike Bolton

Pickwick Lake

If trophy smallmouth fishing is your bag, nowhere else in Alabama is better than Pickwick Lake. This 47,500-acre Tennessee Valley Authority impoundment is unquestionably trophy smallmouth headquarters in Alabama.

Fishing for largemouth bass is pretty danged good, too.

Never doubt this lake is full of other trophy fish, too. No less than three state records—sauger, muskellunge and paddlefish—came from this reservoir.

Spring bass fishing

Above the Natchez Trace Bridge

The deep points on the main river begin to release their bass in late February as the water temperature begins a slight rise. These cold-blooded bass are sluggish and a long way away from chasing a swift bait. Gravel bars, rocks and sandy points are staging areas during this time. These areas, usually in 12 to 15 feet of water on both sides of the main river channel, are easily located with a depthfinder. Those with stumps, brush and logjams are especially productive.

A 1/2-ounce black/brown jig-and-pig with a black pork frog trailer makes an excellent early season bait because it drops quickly and can be retrieved slowly. Small grub-type baits also are excellent at this time. A small white or smoke-colored grub fished slowly on light line is especially strong. Feathered crappie jigs and horse-hair jigs are also effective.

As the water warms, quicker baits such as 1/4-ounce spinnerbaits become more productive because they cover more water. The warming water allows the bass more movement and they will begin to chase faster baits. A chartreuse/white spinnerbait tipped with a white curly-tail jig body works well.

The warm waters of late March and April eventually send the bass into the shoals looking for spawning areas. This area has few creeks and most of the bass are forced to look for stump and brush-filled pockets off the main river channel.

Shallow-running crankbaits, plastic worms and Carolina-rigged lizards become the best baits in the shallow water. All handle brush relatively well and produce fish.

The Little River side of Seven Mile Island is one of the first areas to warm in spring and it plays host to numbers of shallow-water fish. The area between Coffee Slough and Woodland Stumps offers spring fishing as good as most anglers will ever see. Coffee Slough has a well-defined creek channel, plenty of stumps and several springs that feed it. Woodland Stumps is across from the steam plant and receives some warm water. It is also full of stumps and brush.

This area does have a few small creeks that are worth a look at this time. Spring Creek and Sinking Creek offer relatively poor fishing, but Little Cypress Creek, Little Bear Creek, Dry Creek and Cane Creek boast some spring fish worth trying. Cane Creek offers the best spring fishing.

While the spinnerbaits don't work that well in the

shoals, they are effective in the creeks.

Below Natchez Trace Bridge

The lower section of Pickwick Lake has many more gravel bars, sandbars, drop-offs, rocks, rock ledges, ridges, ditches and—most noticeably—fewer creeks than the section above Natchez Trace Bridge. The bass making their move out of deep water have no trouble finding structure on which to stage during their route to shallow water. These irregularities in the bottom are easy to locate with a depthfinder and it's for that reason that most anglers find pre-spawn fishing much easier in this area of the lake.

The most simple method of fishing these pre-spawn bass is to ride with the depthfinder on and drop marker buoys whenever a piece of irregular bottom shows fish. The angler can then pull his boat in deeper water and cast to the buoy, thus lessening his chances of spooking the fish.

Water temperature dictates the bait. Spoons are the best shot at trophy fish. The lure should be cast well past the buoy and slowly hopped back to the boat. A jig-and-pig will work just as well as long as the angler has the patience to allow the bait to sink to the bottom before beginning his retrieve.

Once the water temperature moves up, faster baits such as spinnerbaits and deep-diving crankbaits are best. These baits cover much more territory and greatly increase the odds of a catch.

Once spawning time arrives, fish in this area are somewhat hard pressed to find suitable habitat for spawning and many anglers believe many are forced to spawn in deeper water than they'd prefer.

Cuts, pockets, sloughs and small branches hold most of the spawning fish, especially between Natchez Trace Bridge and Bear Creek. Once past Bear Creek, things get somewhat better. Bear Creek itself is a major body of water that plays host to much spawning activity, as does Second Creek, Indian Creek and Panther Creek. The cuts on both banks just inside the mouth of Yellow Creek also hold some spawning fish.

The flatter banks on the lower end allow something the upper end doesn't and that's lots of grass. This makes excellent spring fishing when the fish are shallow. Shallow-running crankbaits, grubs and topwater baits work well on the grass fringes, while spinnerbaits, floating worms, rats, frogs and buzzbaits work well in the grass. Most of the grass is on the north banks.

Hot weather bass fishing

Above the Natchez Trace bridge

Following the spawn, the bass slip out of the shallow water and gather to rest on the backside of points in an attempt to escape the current. Spawning is hard on the fish and most will be in bad shape for up to three weeks. These fish are non-aggressive at this time.

Once they recover, their appetite returns and they become voracious feeders. Plastic worms and medium-running crankbaits thrown on the backside of points, especially when the water is running, result in some of the most vicious strikes that you'll encounter.

The glaring June sun eventually will drive these fish deep on the points where they will stay until fall. These deep fish are much more difficult to locate and catch than they were a month before. Worms and deep-diving crankbaits become the choice lures at this time, with worms producing the most fish and crankbaits the larger fish.

A deep-running crankbait is the choice of the better anglers. Crawfish-colored baits are best in clear water, with the Tennessee Shad color running a close second. Worms are a matter of choice. Pumpkinseed and appleseed are popular.

The drop-offs and points below Seven Mile Island to Natchez Trace bridge will all hold fish during the summer months. Fishing is better in the morning and afternoon hours.

Below the Natchez Trace bridge

The multitude of irregular bottom structure below Natchez Trace bridge makes this the area to fish during summer months. This area was full of islands before the Tennessee River was impounded, but time has taken its toll. The islands have eroded leaving rough underwater humps, bars and rock outcroppings. Many of these islands are covered with mussels.

These areas are on the north side of the main channel and easily located with the depthfinder. All make for perfect bass habitat during the summer months. These mounds can be found in 8 to 20 feet of water, so fishing can be good no matter the temperature or weather. The mounds in the shallow water will hold fish in early summer, overcast days and often during the morning and evening hours on hot days. The deeper mounds hold fish during July and August.

A good topographic map of the area before the lake was impounded is beneficial. It's sometimes possible to pull up on one of these spots and ambush schooling fish, catching 20 or more without moving the boat. Carolina-rigged worms, Texas-rigged worms, grubs and spoons will work on these fish and light line is a must.

Cool and cold weather bass fishing

Above the Natchez Trace bridge

Bass anglers fishing this area during the cooler months usually experience an increase in catches of spotted and smallmouth bass regardless of the fact that isn't what they're fishing for.

The weather plays a major role in largemouth fishing here in the fall. If the weather stays warm the bass will hold tight to deep points. This enables them a good route in which they dash to shallow water to chase baitfish, or to deeper water when a front approaches.

Crankbaits and spinnerbaits are key in this situation. A crawfish or shad-colored crankbait is probably the best bet, followed by a 1/2-ounce or 3/4-ounce spinnerbait. The heavier spinnerbait is needed because of the deeper water. The spinnerbait, either chartreuse and white or solid white, should be outfitted with a No. 5 nickel blade and tipped with a white pork chunk.

Grubs are a good bait for those who enjoy fishing light spinning tackle. A white 1/4-ounce grub is hard to beat, especially when the fish are finicky and longing for smaller baits.

On the occasions of a cold fall and winter the pattern is much different. The bass move much deeper and are much less active. Slower baits such as jigging spoons, grubs, Carolina rigs and jig-and-pigs are needed.

Ledges and humps with structure and in 15 to 25 feet of water are prime locations to look for largemouth bass during the colder periods. The underwater island in the mouth of Dry Creek is one of the better spots. Buck Island, the underwater island near the mouth of Little Bear Creek, is another. The upriver point of Seven Mile Island will hold some winter fish, as will the north ledges next to the channel across from Mulberry Creek.

Below the Natchez Trace bridge

The same irregular areas that offered good summer fishing also offer good fall and winter fishing. The mounds and bars in 12 feet of water or less hold the fish during September, October and early November under normal conditions, as do the many ditches and roadbeds in the area. Although there is a limited number of incoming creeks, the mouths of these creeks will hold fall fish.

Grubs, worms and crankbaits are the baits of choice during the fall.

Areas to concentrate on include the road beds leading from the bank to the main channel (No. 5, No. 9, No. 18, No. 22 and No. 35 on the map), the series of underwater islands on the north bank below Panther Creek, the rough bars on the south bank east of the mouth of Bear Creek, and the numerous bars and humps on the north bank stretching from Second Creek to Natchez Trace bridge.

During the colder months of December and January, grubs fished on the deeper humps in the same areas will produce some of the lake's better bass.

Other Types of Fishing

Smallmouth fishing

It's possible that you'll catch smallmouth bass when bass fishing any time of the year on Pickwick Lake, but an angler can greatly increase his chances by purposely fishing for them. Fishing certain baits in certain areas will significantly increase those odds.

Smallmouth need clean, fast-moving water and rocks to thrive and the river channel drop-offs, rocky ledges, deep rocky banks and gravel-covered bars on Pickwick Lake offer that. The rocky ledges on the river side of the area close to the head of Seven Mile Island is an area to concentrate on for smallmouth year round because it offers suitable habitat in both deep and shallow water. Other key areas are the rocky dropoffs along the north edge of the original river channel and the mussel-covered humps near the channel.

Cigar-shaped topwater baits and jerk baits are excellent for smallies in spring, but much of the rest of the year grubs are needed. A 3-inch smoke-colored grub with a 1/8 to 3/8 lead head will catch smallmouths most months of the year.

Since rocky dropffs are needed for the best summertime fishing and the rocks aren't visible, many inexperienced anglers have trouble with summer smallmouth fishing. Smallmouth will suspend on these rocky areas in 10 to 40 feet of water, with the bigger fish being deeper. A good rule of thumb is that anywhere up to 18 feet of water falls off in the river channel is likely to be rocky, thus likely to be holding smallmouth.

The good summer smallmouth angler needs a lot of time to fish and a lot of trial and error before he can become good at summer fishing. Some of the better anglers have no trouble catching summer smallmouth, however.

Fall fishing for smallmouths can be excellent, especially in the downriver section. Smallmouth will move somewhat shallower during this time and situate on humps and rocks. Spoons become the top bait then, with grubs a close second.

Early winter is good as many smaller fish move to the eddy areas behind points. Live baits such as shiners and yellowtails are good at this time because it's the type baitfish they are feeding on. The live bait should be hooked through the lips with a 2/0 hook and a bb shot placed 12 inches above the bait. Spinning tackle should be used to cast the bait into the eddies.

Pickwick Lakes best fishing spots

1. Mouth of Yellow Creek. Summer time schooling area for bass. Smoke grubs and Sassy Shads.
2. Cuts in Yellow Creek. Excellent spring spawning area for largemouth, smallmouth and crappie.
3. Excellent point.
4. Shallow sandbars. Good in spring.
5. Underwater roadbed leading from shallow to deep water. Good year round.
6. Underwater island.
7. Panther Creek. Both shallow and deep water. Good year round.
8. Cedar Fork. Good in spring.
9. Underwater road bed leading from shallow water to deep water. Good year round.
10. Whetstone Branch. Cuts offer good spring fishing. Point good year round.
11. Excellent point leading into Short Creek.
12. Shallow sandbars. Excellent spring area for smallmouth.
13. Mouth of Indian Creek. Excellent year round.
14. Excellent cuts and points. Cuts in spring. Points year round.
15. Excellent crappie area in spring.
16. Good smallmouth fishing in spring. Grubs and Sassy Shad.
17. Underwater road bed leading from bank to channel.
18. Underwater road bed leading from bank to channel.
19. Underwater road bed.
20. Underwater island.
21. Underwater island.
22. Underwater road bed.
23. Underwater road bed.
24. Underwater island next to main channel. Good in summer.
25. Underwater ditch out from mouth of slough. Best in summer.
26. Underwater islands and drop-offs. Good staging area in spring. Holds good summer fish.
27. Underwater island. Good year round.
28. Underwater hump in middle of large flats.
29. Underwater road bed.
30. Man-made underwater brush piles. Someone has placed brushpiles in row, using telephone poles on the bank as markers. Each brush pile lines up with pole.
31. Underwater road bed.
32. Underwater road bed.
33. Well-defined creek channel. Key area for spring bass.
34. Tremendous crappie fishing in spring. Underwater islands.
35. Mouth of Second Creek. Good year round.
36. Points in Second Creek mouth. Good year round; best in spring.
37. Second Creek. Shallow grass. Good in spring for largemouth.
38. Underwater road bed.
39. Underwater island splits main channel. Good summer and winter.
40. Fantastic spring fishing for smallmouth and largemouth. Hundreds of sandbars with grass.
41. to 43. Underwater island.
44. Shallow, grass-covered sandbars.
45. Underwater island.
46. Good smallmouth area year round.
47. Colbert Creek. Good spinnerbait area spring, early summer.
48. Natchez Trace bridge pilings. Good year round fishing. Old road bed can be found upriver on right bank right above bridge.
49. Good rock rows. Runs parallel to bank for 50 yards. Tremendous area year round for largemouth and smallmouth.
50. Koger Island. Good spring and summer fishing.
51. Stumpy ledge next to main river channel. Excellent early summer spot.
52. Long sandbars parallel to bank. Good smallmouth area in spring and summer.
53. Woodland stumps. Underwater sandbars. Good spring area for largemouth and smallmouth. Good buzzbait area.

54. Cane Creek. Good warm water in winter, cool in summer. Excellent for largemouth, smallmouth, hybrids and stripes.

55. Excellent catfish area in summer. Twenty-pound-plus cats here common.

56. Pipeline area. Good in spring-summer around pipeline.

57. Good underwater island that splits flow of main river channel. Good summer fishing.

58. Little river side. Good year round. Lots of small islands.

59. Big river side. Good year round.

60. Buck Island. Good year round.

61. Underwater rock rows. Excellent spring fishing for smallmouth and largemouth.

62. Underwater rock row.

63. Little Bear Creek. Good spring and summer fishing with deep water. Lots of stumps.

64. Coffee Slough. Good spring fishing in flats, good year round fishing in channel.

65. Points at mouth of Spring Creek. Summer schooling area.

66. Back of Spring Creek. Good spring fishing.

67. Head of Seven Mile Island. Good year-round fishing with grubs and worms.

68. Cypress Creek mouth. Good year round.

69. Cypress Creek. Good to bridge. Lots of stumps. Good spring and summer.

70. Main River Bridge riprap. Good topwater area all spring, early mornings in summer.

71. Rocky gravel bank with stumps. Good smallmouth area.

72. Stump rows.

73. Jackson Island. Good year-round fishing. Good escape area from current behind island year round.

Pickwick Lake

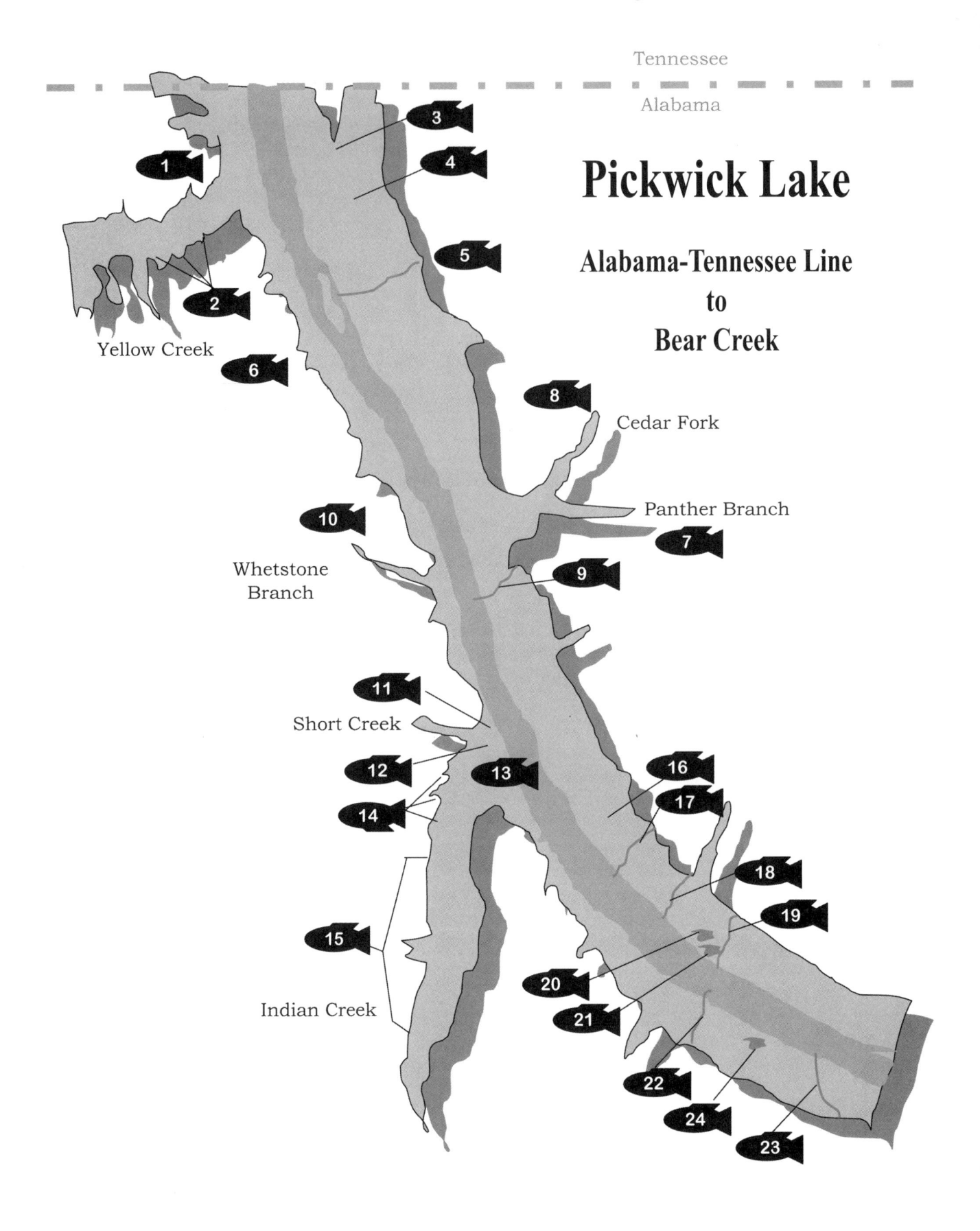

Alabama-Tennessee Line to Bear Creek

Yellow Creek

Cedar Fork

Panther Branch

Whetstone Branch

Short Creek

Indian Creek

62

Pickwick Lake

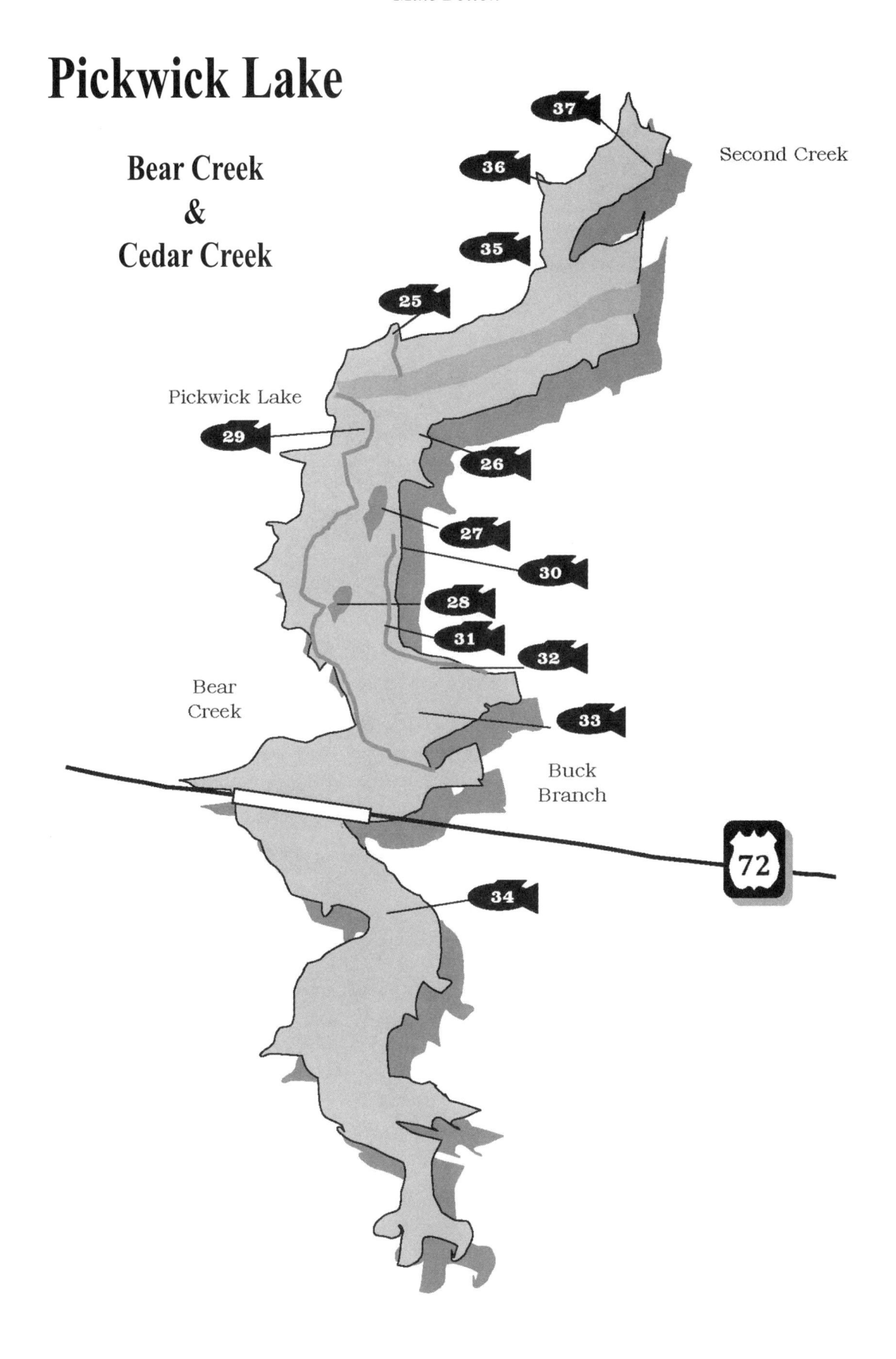

Bear Creek & Cedar Creek

The Ultimate Guide to Alabama Fishing

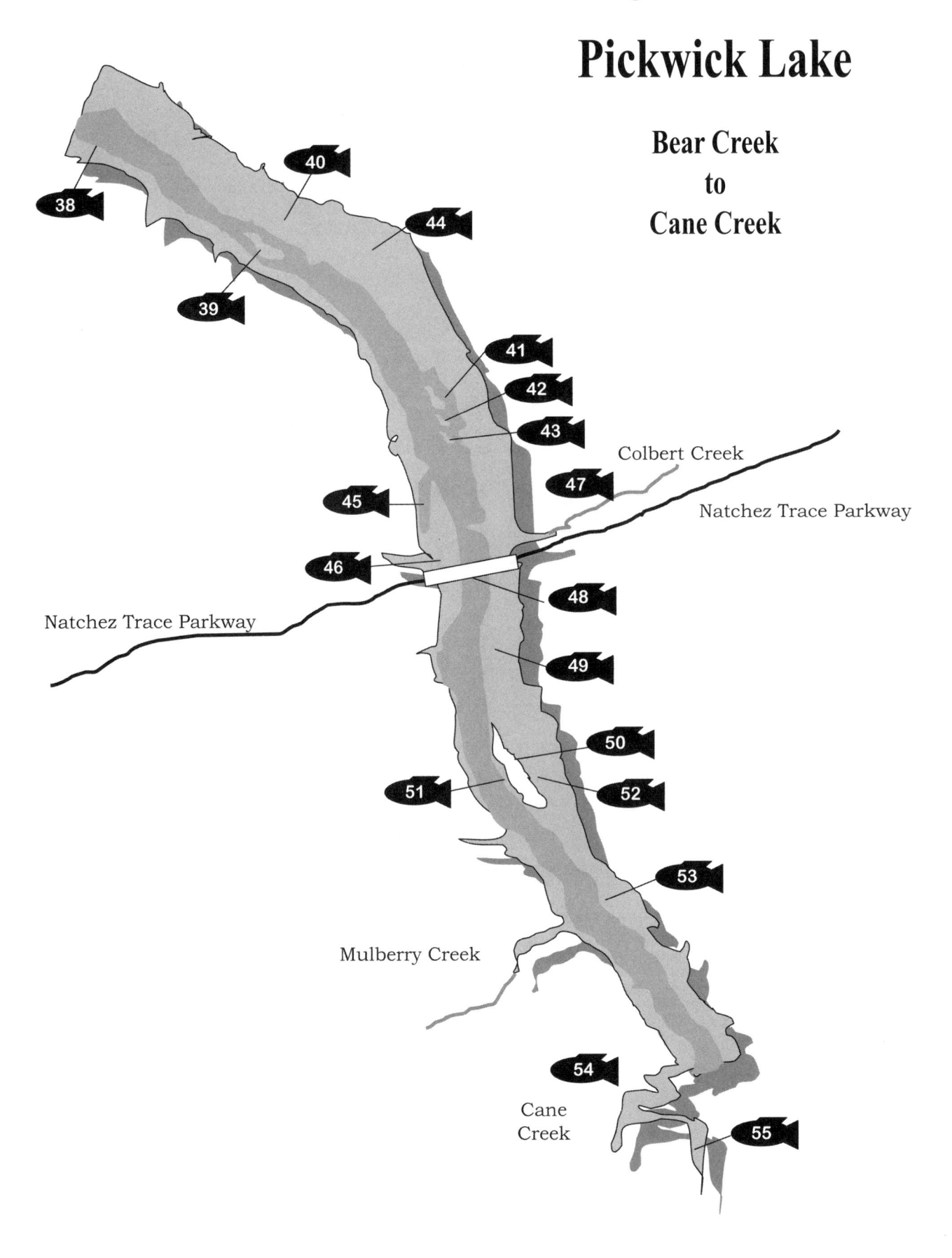

Pickwick Lake

Bear Creek to Cane Creek

Pickwick Lake

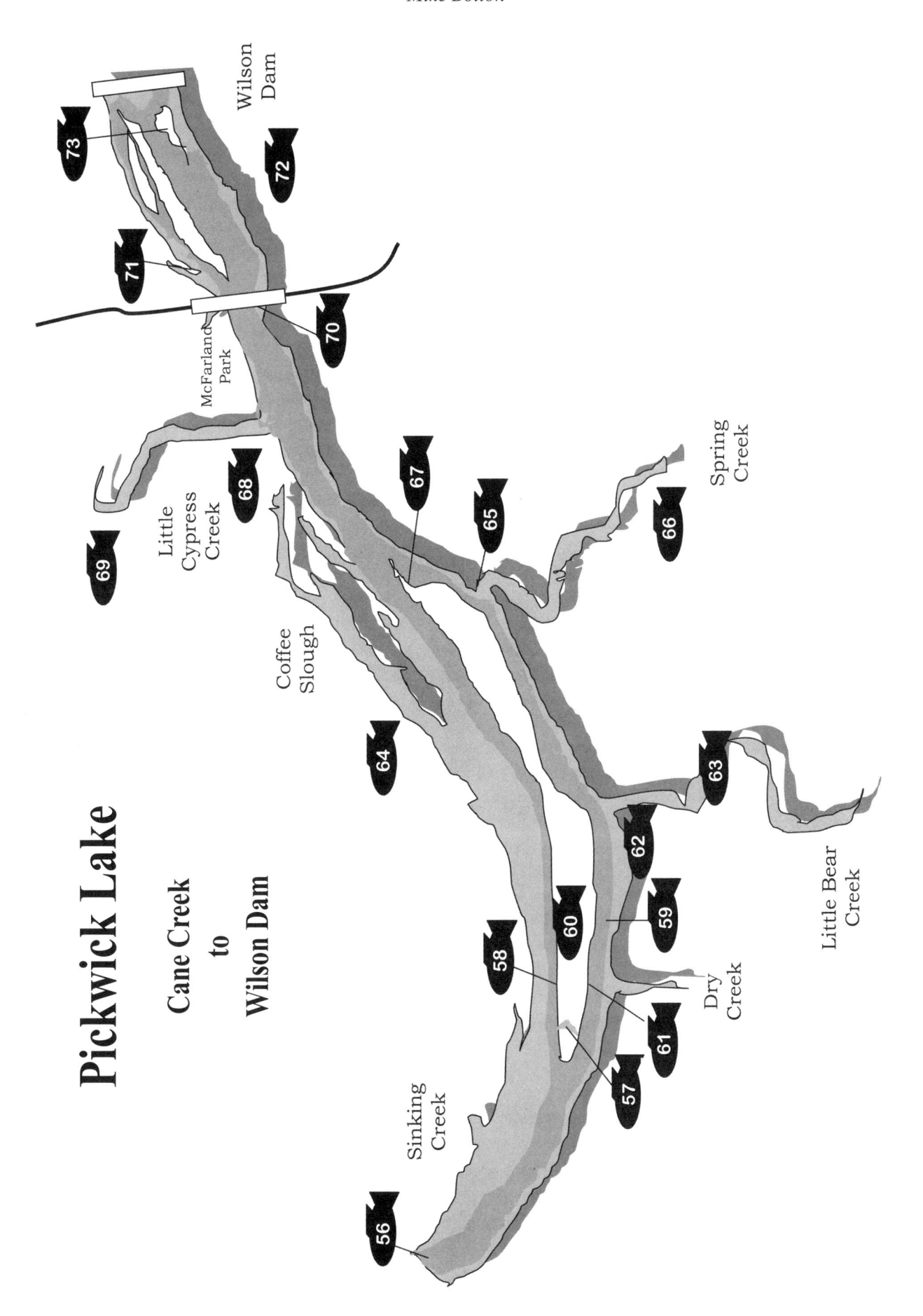

Cane Creek to Wilson Dam

The Ultimate Guide to Alabama Fishing

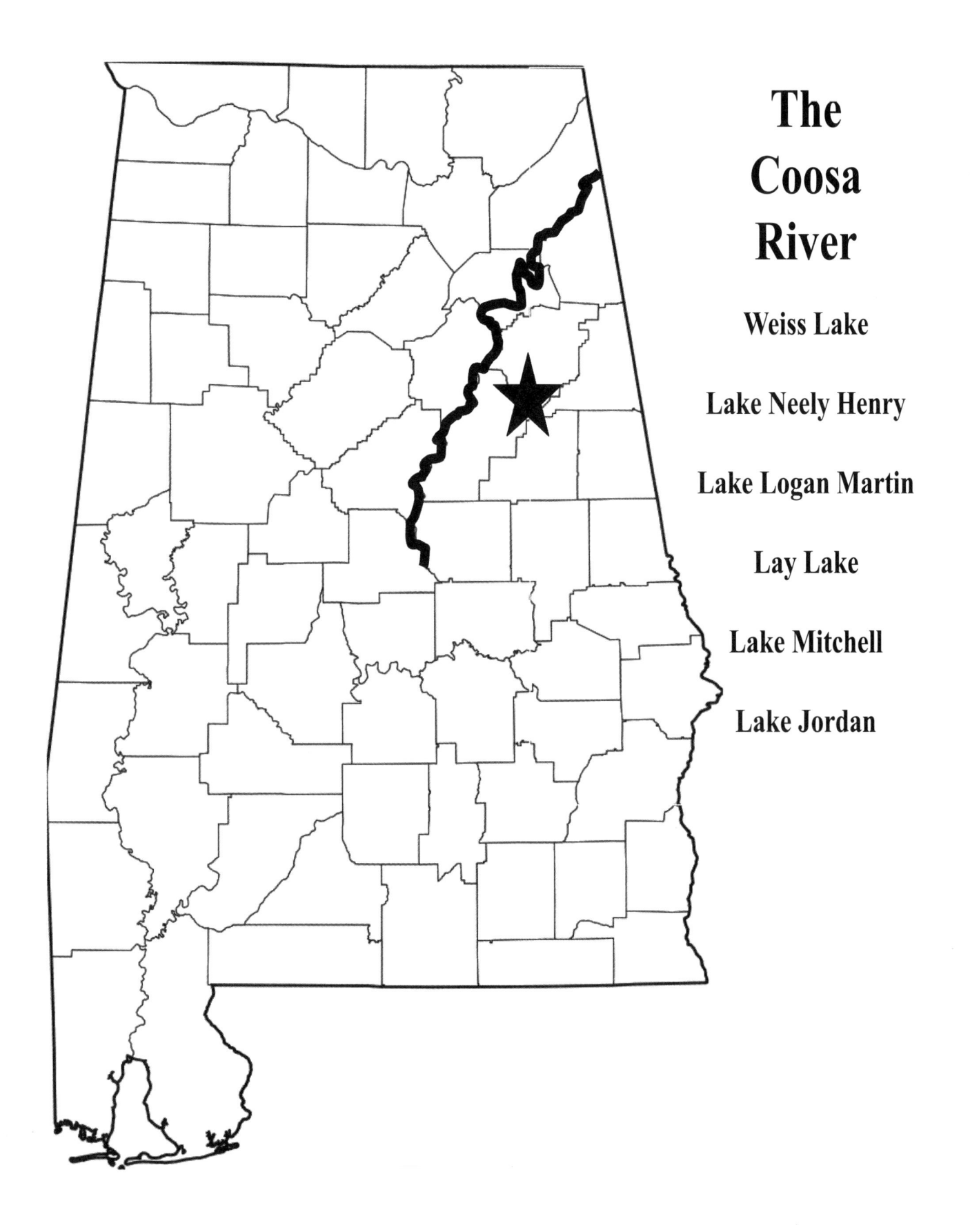

The Coosa River

Weiss Lake

Lake Neely Henry

Lake Logan Martin

Lay Lake

Lake Mitchell

Lake Jordan

Mike Bolton

Weiss Lake

"Crappie Capital of the World" is a self-appointed title, the result of the creative minds of former Bay Springs Marina owners Kenneth and Cathy Mackey, who spent many hours promoting Weiss Lake as the world's best crappie fishing lake.

Not many lakes in the world have cried foul over this lake's boastful claim, however. The northeast Alabama reservoir annually entertains more out-of-state fishermen than any lake in Alabama. It is not unusual to see marina parking lots full of vehicles with tags from Georgia, Tennessee, Kentucky and Illinois. Very few crappie fishermen have ever left disappointed. The lake is full of papermouths.

Unknown to many across the U.S., Weiss lake is without question one of Alabama's finest bass fishing reservoirs, too. The conditions that provide excellent crappie fishing are also conducive to an excellent bass fishery.

Here is a look at what makes Weiss Lake the "Crappie Capital of the World," and one of Alabama's top bass producers.

Spring Bass Fishing

Georgia State line to Cedar Bluff bridge

When the first warm rays of spring sunshine come to Weiss Lake, no section of water heats up faster or becomes more desirable to bass earlier than the area from Cedar Bluff Bridge to the Alabama-Georgia line.

This shallow, dingy water absorbs heat rather than reflecting and as a result it woos the bass out of the river channel earlier than other locations on the lake.

The area lacks the structure that can be found elsewhere on the lake, so shallow water pocket fishing becomes necessary. The northeast banks of coves, sloughs and islands receive the most sunlight in early spring, warm up the quickest and get fish first.

Some of the best areas for this early spring fishing is Cowan Creek and around Godfrey's Island. The main structure to look for to hold fish this time of the year is exposed brush. Due to minimum structure, locating exposed brush piles will just about guarantee fish. Another key fish-holding bit of structure is isolated stumps. Bass tend to gather around these isolated stumps rather than clumps of stumps. Gravel points and gravel bars are a key third area.

A jig-and-pig is the bait to use before the water reaches the 53- to 55-degree range. A brown/orange with a brown pork frog is best in the continuously dingy water.

This slow bait works equally well in the brush and stumps. A surprise bait the better anglers use once the water passes the 55-degree mark is a shallow-running lipped crankbait such as a Rapala. These work well with a slow retrieve over and around the stumps.

Of course, the spinnerbait is the lure most use effectively here this time of the year. A 1/2-ounce chartreuse or chartreuse and white spinnerbait with a larger blade works well in the stained water. When the water begins to clear, drop to a 3/8- or 1/4-ounce spinnerbait and a No. 5 or 6 blade. When the water gets a little warmer, a buzz bait also will work well.

Godfrey's Island, Cowan Creek, Spring Creek, James' Branch and Pruett's Island are they key spawning areas where the bass eventually end up at the height of the spawn.

Also pay close attention to boathouses, docks and piers in any of these creeks. Almost all have man-made brush piles around them. These brush piles will concentrate baitfish and will attract bass on the way to the spawning grounds.

Cedar Bluff to U.S. 411 bridge

During spring months, an estimated 75 percent of the lake's bass fishing takes place between the U.S. 411 bridge and the Cedar Bluff bridge and for good reason—lots of bass.

The Chattooga River, Little River, Big Nose Creek and Little Nose Creek all feed this portion of the lake and provide excellent spawning areas.

The cutoff area between Yellow Creek and Little River is the prime bass area this time of year. This area offers approximately 1,000 to 1,500 acres of shallow water full of stumps, rocks, road beds and humps. This water offers excellent structure, as well as the ability to warm fast. Those are always two key ingredients in good springtime bass fishing on any reservoir.

Little River, the back of Big Nose Creek and the back of Little Nose Creeks are key areas in the spring.

Avoid Yellow Creek. It's much too deep to provide consistently good spring fishing.

Big brush piles and individual stumps are the structure you're looking for. Sandy banks and pea gravel banks are key fishing areas, too. Avoid muddy, silty bottoms and banks because bass avoid them.

The bass will stage themselves this time of the year, moving from the main river channel, to the structure adjacent to it, to the structure in the creek mouths and then into the creek flats to spawn. Although crankbaits and, later, spinnerbaits are key baits, never underestimate the ability of a chatterbait in these conditions

The Powerhouse Lake

Although the powerhouse lake is technically a portion of Weiss Lake, it is literally a separate reservoir from Weiss Lake connected by a 1 1/4-mile man-made canal.

The powerhouse lake is part of an unusual situation where the dam and power generating facilities are separate units more than 1 3/4 miles apart.

Anglers may choose to launch their boats at several ramps on the powerhouse lake, or they may travel from the main portion of Weiss Lake by boat under the U.S. 411 bridge taking the man-made canal until it dumps into the smaller reservoir. This canal linking the two reservoirs is deep and safe to navigate.

The shallow lake, like the rest of Weiss Lake, is full of structure including road beds, stumps, islands and old home sites. The lake is dishpan in configuration and lacks steep drop-offs on its banks, but deep water can be found in the original river channel.

With the water normally being down in the early spring, the key to good fishing here is positioning your boat in five feet of water and casting into 18 inches to 3 feet of water with jig-and-pigs, spinnerbaits or shallow-running crankbaits. All three baits can be ultra successful catching bass from this structure this time of the year.

The makeup of this small reservoir provides an excellent spawning area. The area near the power house offers deep water adjacent to shallow water ledges full of stump rows. This is an excellent spawning area.

Although the powerhouse lake offers good spring fishing, much better spring fishing can be found on other portions of Weiss Lake. The powerhouse lake offers its best fishing in the summer and fall.

Hot weather bass fishing

Georgia state line to Cedar Bluff bridge

If you wish to be a good summer bass fisherman on this end of Weiss Lake, it wouldn't be a bad idea for you to spend the winter fishing here. The low water level at winter pool reveals a lake that many summer-only anglers don't know exists. Stumps, logjams and other such structure can be found all along the original river channel and it's in this structure that you'll find bass during the hot-weather months of May, June, July and August.

Pay close attention to the map of this area that accompanies this book. It shows the location of a great many the stumps, timber, roadbeds and other bass-holding structure adjacent to the river channel. These are the areas that you need to fish during the summer.

The stump rows above Pruett's Island, the sharp bend in the river between Dead Boy Branch and Water House Creek and the sharp bend in the river between Poole's Ferry and Ballpark Creek are among the best areas.

This structure on ledges should be fished as you would any other ledges on the Coosa Chain. For those who don't know the location of the original river channel well, use your depthfinder to locate it and drop marker buoys. Position your boat in the river channel and cast to the structure on the ledges. Stick with medium running crankbaits, spinnerbaits with a pork frog or a jig-and-pig with a pork frog until the water temperature reaches 70 degrees. At that point, switch to plastic worms, jig-and-pigs with a plastic worm trailer, or a swimbait.

Other areas to look for are the causeways of any incoming creek, such as the riprap-laden entrance to Cowan Creek. Fish these areas with a crankbait first, and if you're sure there's fish there, come back with a plastic worm or a jig-and-pig. A buzzbait will often work in these areas at daylight and dark.

Cedar Bluff bridge to U.S 411 bridge

The middle section of Weiss Lake provides hot weather ledge fishing much like you'll find on most reservoirs in Alabama. The bass exit the creeks following the spawn and are forced to seek water temperatures much more tolerable than the shallow water offers. The deep water in the original river channel is just what the doctor ordered.

Although it might not be evident to the casual observer, the bass don't gather in random locations along the channel. There's rhyme and reason to their madness. They must continue to eat and they'll concentrate in areas with structure that holds shad.

Finding these aren't that difficult. The major bends in the river tend to grab floating logs and form a jam, creating excellent structure immediately adjacent deep water like the bass like. Look at the map and you'll see these major bends.

Another area with good structure piled on it is the deep water mouths of inlets such as Big Nose Creek, Little Nose Creek, Yellow Creek, Little River and the entrance to Bay Springs.

Texas-rigged worms, small crankbaits, spoons and small swim baits all seem to work well.

Texas-rigged plastic worms are the No. 1 bait most anglers uses this time of the year in this structure, but it's not the No. 1 bait the better fishermen use. The worm's major drawback is that it's not a good bait to "hunt" fish with. It's time-consuming to use. If you're a fisherman out there for leisure, the worm's okay to locate fish. If you're trying to catch numbers of fish, it isn't.

Use a crankbait first. You'll either locate bass or structure with it. Once the bass or the structure has been located, switch to a worm.

The Powerhouse Lake

With so many open flats to cover in the powerhouse lake, the best bass anglers on the lake often choose to attack the area in summer months with a Carolina-rig. The worm hook is attached to a 3- to 5-foot leader, and then a swivel is added. A 3/4-ounce to l-ounce bullet worm lead is then added to the line coming from the fishing rod, a bead is added and this is then attached to the swivel. This results in a weight that will drag bottom, but the worm is allowed to float. The hook in the worm is left exposed. The rig works well in the open, deep water found in many areas of the powerhouse lake. Many of these open areas have small humps which hold fish and the Carolina-rigged worm is tailor-made for these areas.

Other summer areas to watch are the shallow flats next to the main river channel above the powerhouse lake. Bass suspend in the deep water and rush to the ledge to chase bait fish. Texas-rigged plastic worms, Prissy Shads and small, shad-colored crankbaits work well in these areas.

Cool and cold weather bass fishing

Georgia state line to Cedar Bluff bridge

Due to the vast expanses of shallow water, this end of Weiss Lake isn't your best area to find cool- and cold-weather bass. The bass will move out of the main river channel back into the structure nearby when fall arrives, but as it gets colder, the bass become harder and harder to locate.

One reason for this difficulty is apparently the bass' movement into the headwaters of many creeks. During the dead of winter, many crappie anglers show surprise at catching bass in the headwaters of Cowan Creek and Ballpark Creek.

These bass gather around the areas with incoming waters. Whether they gather there to chase sluggish shad or whether they find water a few degrees warmer isn't known.

Other areas that hold cold-water bass are the bridge pilings in Spring Creek and Cowan Creek. Bass gather around these piling and can be taken by crankbaits, jig-and-pigs or worms.

Cedar Bluff bridge to U.S. 411 bridge

From approximately Sept. 15 until the water surface temperature reaches 55 degrees, small, vibrating, shallow-running, lipped crankbaits are bass slayers in the cut-off area between Yellow Creek and Little River.

Bass seek the security that wood offers when the water turns cool. As mentioned before, this shallow-water area is full of stump rows, roadbeds and humps. For those who don't like fishing shallow-running crankbaits, spinnerbaits, worms and jig-and-pigs also work.

The Powerhouse Lake

You'll have to look long and hard to find better fall fishing on Weiss Lake than you'll find on the powerhouse lake. This reservoir is where the water comes from to pass through the generators in the powerhouse and the result is many hours of much-needed moving current. That translates into superb fishing.

Bass sometimes school by the dozens on brush-laden roadbeds and humps, ends of roadbeds and steep ledges in this area when the current is moving. Carolina rigs and crankbaits are the primary baits.

When winter hits and the water surface temperature drops below 48 degrees, Texas-rigged worms and jig-and-pigs will take bass from the same roadbeds and humps.

Other types of fishing

Crappie fishing

Weiss Lake is known as the Crappie Capital of the World and for good reason. The tremendous amount of crappie in the lake draws large numbers of crappie fishermen from not only Alabama, but Georgia, Tennessee, Kentucky, Indiana and Ohio as well. A liberal limit per angler and a very real chance of catching that many fish team up to make Weiss Lake a crappie fisherman's delight.

Because of several recent years of drought and the tremendous amount of crappie fishing pressure on the lake, fishing isn't as good as it was in the lake's heyday in the early 1970s, but a beautiful weekend during the spring will always result in the lake being full of anglers chasing crappie.

It's no secret that spawning crappie are easy to catch as they move into shallow water, so the lake gets its most pressure then. The area around Godfrey's Island is the perfect spring location because it offers vast acres of perfect spawning territory, as well as water that warms much quicker than on other portions of the lake.

Any area with water from 18 inches to 5 feet deep that has stumps, trees, brush or piers is likely to hold crappie during the spring. A long cane pole or a graphite pole outfitted with light line, a small split shot, a small wire hook and a cork or bobber work well enough for most fishermen. The rig should be baited with a small chartreuse jig or live minnow.

The cut-off area, Little River and the Chattooga River also offer excellent spring crappie fishing using the same methods. The powerhouse lake generally has too much current to be conducive to excellent spring crappie fishing, but the area around Pine Cone Marina and Budweiser slough often produce good spring crappie.

A good rule of thumb is that any slough on the lake which has a sandy or rock bottom and is void of current is a threat to hold spring crappie.

Unfortunately for most anglers, the best time they have to fish is the summer and the summer isn't the best time to fish for crappie on Weiss Lake. The best method for summer crappie fishing is the tedious pattern of bumping live minnows on the bottom of the main river channel. This can be accomplished by using light line, a small wire hook and a 1 1/2-ounce weight.

If you fish areas with good structure such as stump rows and brush, you'll find fish, but you'll also lose a lot of rigging. One of the best rigs for this summer pattern is a light spinning outfit rigged with 6-pound test. At the end of the line, attach a three-way swivel. From the bottom eye, hang approximately 36 inches of 2-pound test line. A 1 1/2-ounce lead weight attached to the 2-pound test line works well, but a cheaper weight (consider that you'll lose a lot of weights) is a spark plug. Garages and service stations usually have plenty of old spark plugs they'll give away. The electrode can be pushed down to form a loop you can attach the line to.

From the remaining eye of the swivel, attach approximately 18 inches of six-pound line and a light wire hook. Bait the hook with a live minnow.

If the weight (the spark plug) hangs on brush while bumping bottom, the 2-pound test line breaks easily and you don't lose the entire rig. If the hook hangs up, the wire hook will usually straighten with a steady pull.

You can make several weight rigs from spark plugs ahead of time. Simply bend the electrodes, attach the 36-inch piece of 2-pound test, and add a snap swivel to the end. Wrap the line around the spark plug and secure it with a piece of tape for easy storage.

Another summer pattern that works well and allows the angler to escape the summertime heat is fishing at night with lights. Many anglers choose to use a Coleman lantern, but a sealed-beam headlight attached to a foam block is a better rig because it's easier and safer.

Anchor your boat near or atop bridge pilings, road beds with brush, humps with brush, treetops or brush piles and be patient with this type of fishing. Let the light draw in the baitfish and the crappie will follow.

The area between Cedar Bluff bridge to the Leesburg bridge offers the best summer fishing. Little River, Chattooga River and Yellow River are all full of the type structure needed for good summer crappie fishing.

When fall and winter arrives, it's time to move back to the ledges in 10-17 feet of water with the bottom bumping rig. The Little River area offers the best fall and winter fishing.

When bumping bottom in both the summer and winter, outline a stretch of the river channel by dropping marker buoys. This is critical. Being five feet off of the channel can put you out of the picture.

Weiss Lake's best fishing spots

1. Excellent bass and crappie sloughs January and February. Good fishing when water is up.
2. Standing trees in back of slough. Good for crappie and bass in spring when water is up in slough. Use topwater or spinnerbaits for bass.
3. Ballpark Creek. Accessible when water is low elsewhere in area. Good spring crappie fishing. Bass use the mouth to escape current in the spring.
4. Standing bushes. Good when water is up in spring.
5. Mud Creek. Excellent bass and crappie area when this creek has water in spring.
6. Standing trees in back of mud creek. Buzz topwater bass in spring and summer.
7. Standing trees in shallow water.
8. Underwater roadbed. Old Pooles Ferry Road.
9. Small islands near Poole's Ferry. Topwater at night in summer; spinnerbaits in spring for bass.
10. Stump patch off of northeast end of Godfrey's Island. Holds late spring crappie and bass.
11. Godfrey's Island. Excellent cover and brush. March to May bass and crappie.
12. Standing timber. Good flipping.
13. Sandbar where main river channel splits. Good logjams and brush on top. Excellent when water is off at night in summer.
14. Stump rows.
15. Dead Boy Branch. Expanses of shallow water, rocks, stumps and man-made brush piles. Excellent fishing in spring for bass, crappie and white bass.
16. Riverside Slough. Closest access to Godfrey Island. Fish escape current in mouth in spring when boat traffic is slow.
17. Underwater road bed.
18. Stump rows.
19. Underwater road bed.
20. Extended stump row adjacent main river channel. Excellent for schooling summer bass.
21. Underwater road bed.
22. Spring Creek. Lots of shallow water in spring. Good crappie and bass fishing above and below bridge. Mouth an excellent spot for bass in late fall and winter.
23. County bridge in Spring Creek. Excellent dropoffs both side of bridge. Both dropoffs full of brush and stumps.
24. Back of Spring Creek. Acres of shallow water stumps. Excellent bass fishing in spring with buzzbaits and other topwater lures. Also good spawning area for crappie.
25. Cowan Creek. Excellent spring fishing for crappie and bass. Susceptible to high winds in March.
26. Bridge in Cowan Creek. Good bass fishing around pilings in spring and fall. Good crappie fishing at night in late spring and summer.
27. Yancey's Bend. Excellent area early spring and winter. Bass bunch up on stumps, brush and logjams in river bend.
28. Dead End Bridge Slough. River channel makes its first bend directly out of the mouth of slough and good fishing begins there. In the slough itself, shallow water slough behind restaurant offers good bass and crappie fishing in spring. Boathouses produce numbers of big fish.
29. Old home foundations. Use depthfinder to locate.
30. Underwater road bed.
31. Stump patch.
32. and 33. Old home foundations.
34. Cedar Bluff Bridge. Points offer schooling June to September.
35. to 37. Underwater road bed.
38. Island in mouth of Little River. Very good drop on north side of island. Stumps and cedar brush tops placed here by Weiss Lake crappie guides. Good for crappie in early winter, bass in fall and summer. Many Weiss Lake bass tournaments have been won here.
39. Underwater road bed.
40. Old home foundations. Locate with depthfinder. Defined drop in Little River. Good bass fishing in this drop during the summer.

41. Good bend in original river channel.

42. Meeting of the Chattooga River and Little River. Island offers deep water on three sides. Schooling bass in summer and fall; crappie in late fall and early winter.

43. Sailboat Club. Well-defined river ledge; stumps and brush.

44. The Hogback. Stretch of riverbank that shallows up to 3 feet. Spring crappie on minnows or chartreuse jigs; summer bass on worms or crankbait.

45. Gaylesville bridge. Excellent summer schooling for bass. Good dropoffs, good concrete pilings where old Gaylesville bridge was located.

46. Cornwall's Furnace. Historical landmark. Made cannonballs here during war between the Americans and Yankees. Shallow stumps directly across from furnace. Excellent for early spring crappie fishing.

47. Chattooga River. Once you cross silted place where No. 47 is located, you can go all the way to Triana, Ga.

48. J.R.'s Marina slough. Store has good sandwiches if you aren't catching fish.

49. Levee area, 600 yards long. Sandbags were placed here before river was impounded to keep river in the channel. Sandbag wall still remains. During the summer and fall, cast deep running crankbaits until schooling fish are located.

50. Back of North Spring Creek. Excellent creek channel usually holds first bass in winter. Jig-and-pigs, worms and shallow- and medium-running crankbaits are good here in winter. During the summer, grass produces good fish.

51. Wolf Creek meets Little River. Excellent spotted bass area in early spring. Clear water here means light line.

52. Stump patch in Black Mallard Cove. Good topwater for bass at night in summer. Good spinnerbait area in spring. Some spring crappie also caught here.

53. Intersections of underwater roadbeds.

54. Old home sites due south of Hog Island.

55. Stump patch.

56. Stump patch directly south of Hog Island.

57. Underwater road bed.

58. Stump patch.

59. Underwater road bed. Lots of brush tops placed on top. Summer and fall for bass.

60. Underwater road bed.

61. Stump patch. Danger area.

62. The Hat. Little River meets the Coosa River. Big sandbar offers excellent crappie fishing in spring, excellent schooling bass in summer.

63. Old home foundations.

64. Intersection of underwater road beds.

65. Stump patch.

66. Little Nose Creek.

67. Stump patches.

68. Big Nose Creek.

69. Old railroad trestle collapsed in mouth of Yellow Creek. Good schooling area for bass in summer. Excellent crappie fishing at night during summer.

70. Good sandbar in Yellow Creek runs up to 12 feet deep. Good for bass with worm or crankbait. Topwater bait good early in morning during summer. Look for seagulls diving into shad.

71. Yellow Creek falls dumps into lake here. Good early spring fishing for bass. Nearby bridge offers some of best crappie fishing at night during summer.

72. The tower area. Good schooling fish in area during summer. Texas-rigged red or blue worm best.

73. Stump patch.

74. Mouth of Crazy Woman Slough. Point on downstream side full of brush and stumps. Excellent summer, fall and winter. Excellent for bass and crappie.

75. Underwater road bed in back of Crazy Woman Slough.

76. Stump patch in back of Crazy Woman Slough.

77. Bay Springs. Most bass and crappie tournaments are held here, so area gets restocked about twice a week.

78. Pockets with brush and shallow-water stumps. Rat-L-Trap and spinnerbait year round for bass.

79. O'Neal's Point. Bass in spring, summer and fall.

80. The Canal. Connects Powerhouse Lake with Weiss Lake. Can be productive for spots.

81. Mouth of canal. Schooling hole for bass in summer and fall.

82. Shallow water with stumpy banks. Spring fishing for bass or crappie. Good cane pole area for crappie. Rat-L-Trap for bass in spring. Spinnerbait for bass in fall.

83. Underwater road bed.

84. Underwater hump. Underwater road beds intersect here. Also old home foundations on hump. Medium-running crankbait or worm is best in summer for schooling fish.

85. Underwater road bed.

86. Underwater hump. Underwater road beds intersect; old home foundations on top of hump. Good for schooling bass in summer.

87. Powerhouse area. Wing walls hold big bass.

88. Underwater road bed.

89. Underwater hump adjacent Pine Cone Marina. Schooling spring bass.

90. Stump-lined underwater island. Plastic worms for summer and winter bass.

91. Stumps in back of Cedar Point Slough. Crappie, bass in spring.

Mike Bolton

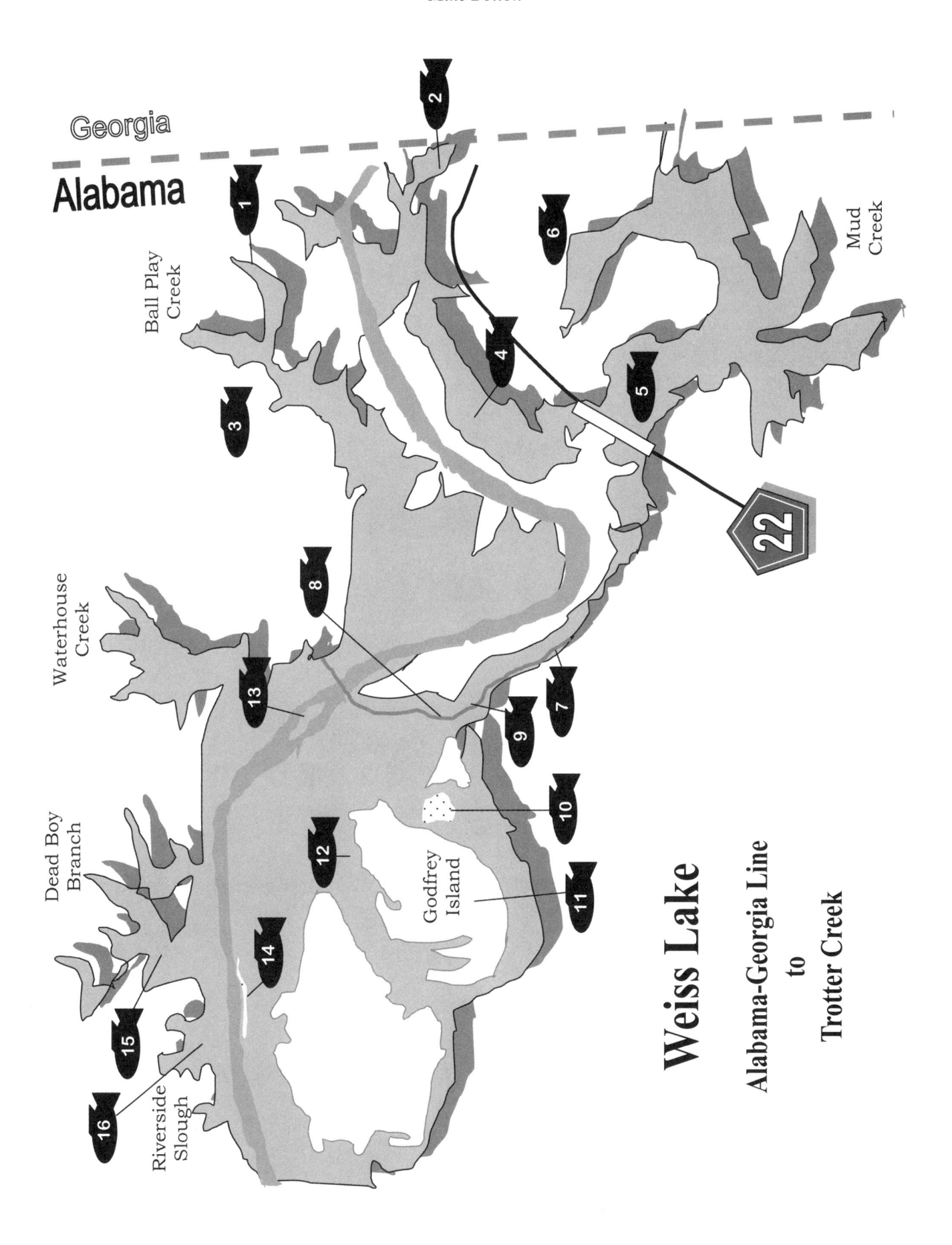

Weiss Lake

Alabama-Georgia Line

to

Trotter Creek

73

The Ultimate Guide to Alabama Fishing

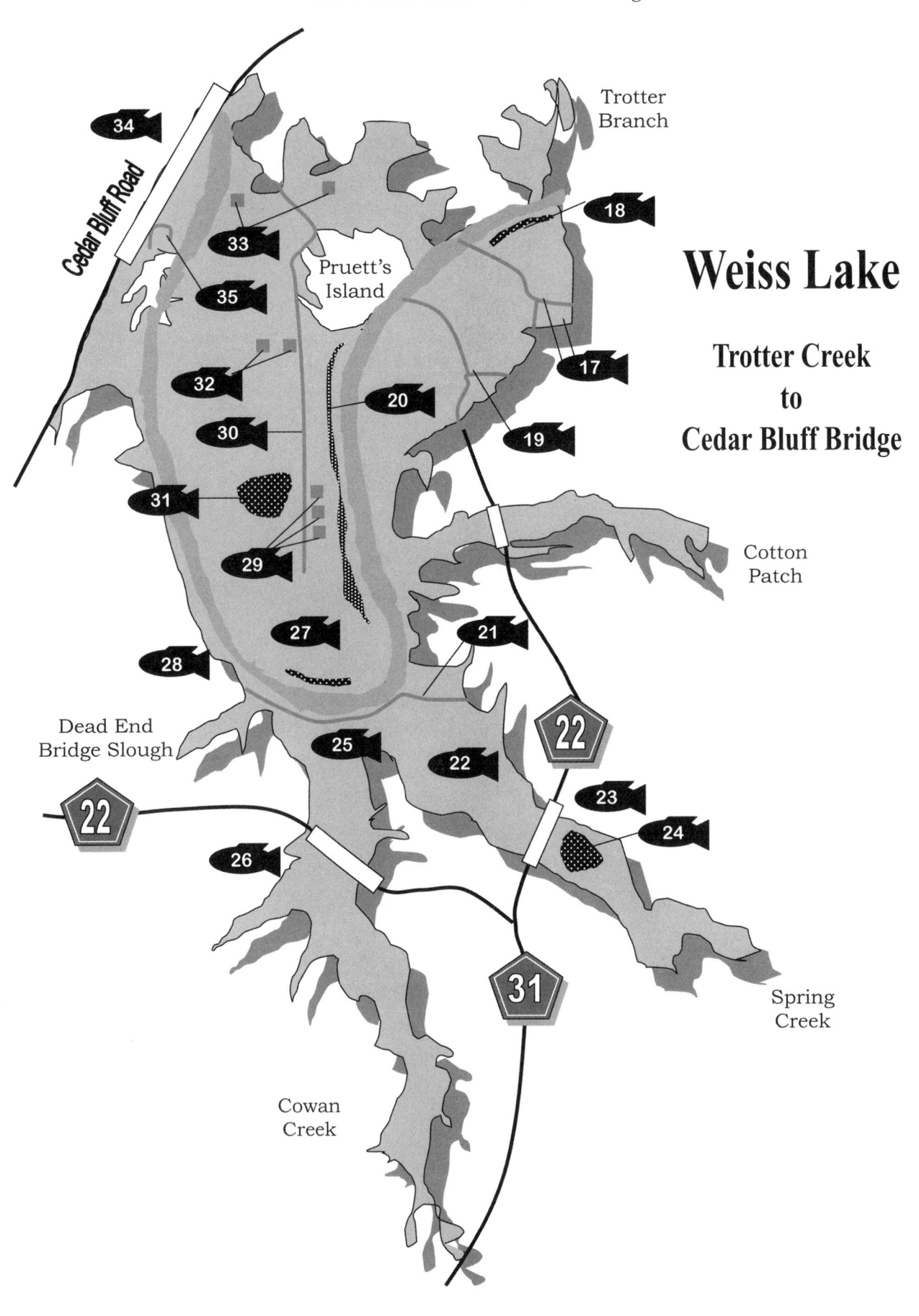

Weiss Lake

Trotter Creek to Cedar Bluff Bridge

Trotter Branch

Cedar Bluff Road

Pruett's Island

Cotton Patch

Dead End Bridge Slough

Cowan Creek

Spring Creek

74

Weiss Lake

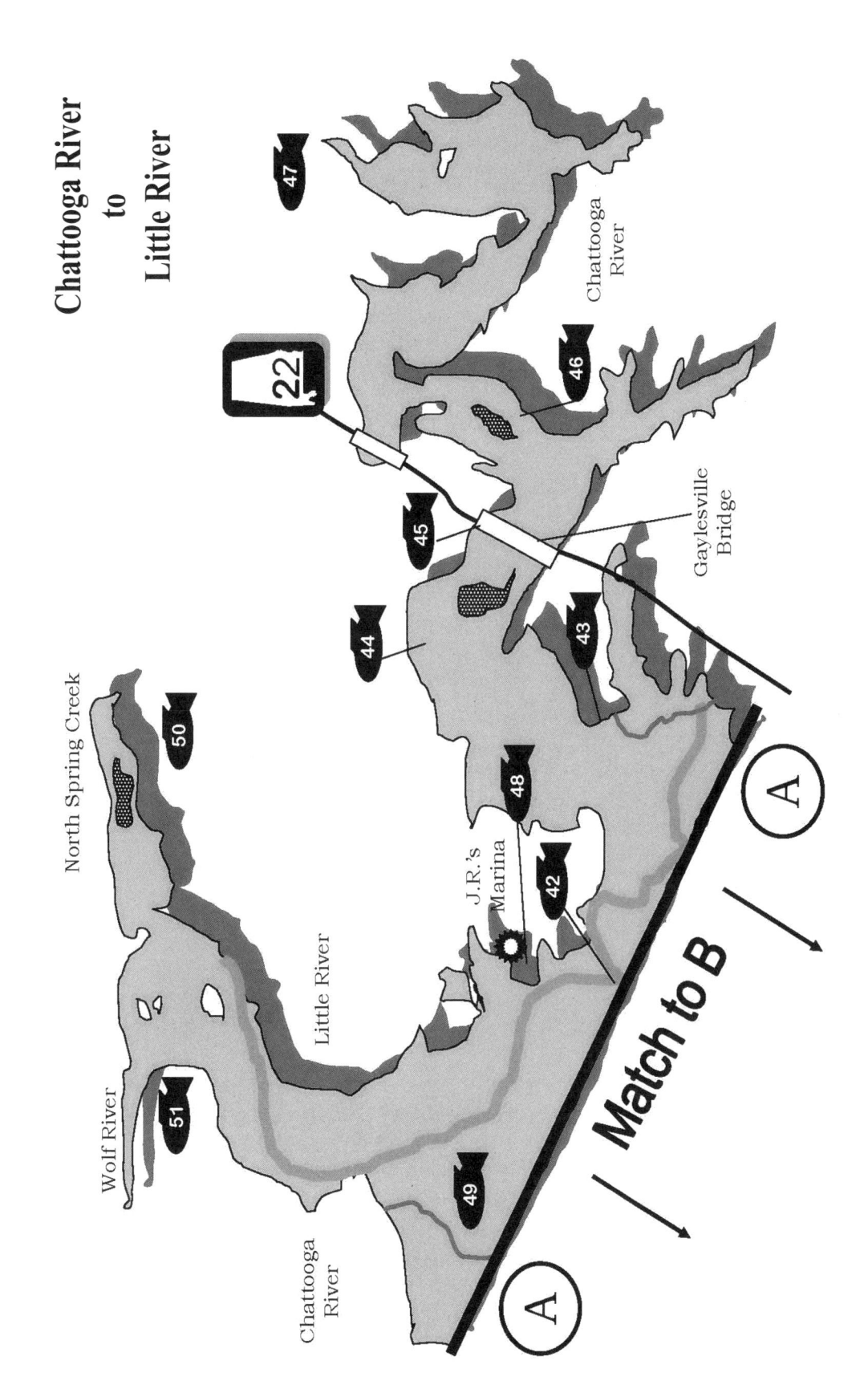

Chattooga River to Little River

The Ultimate Guide to Alabama Fishing

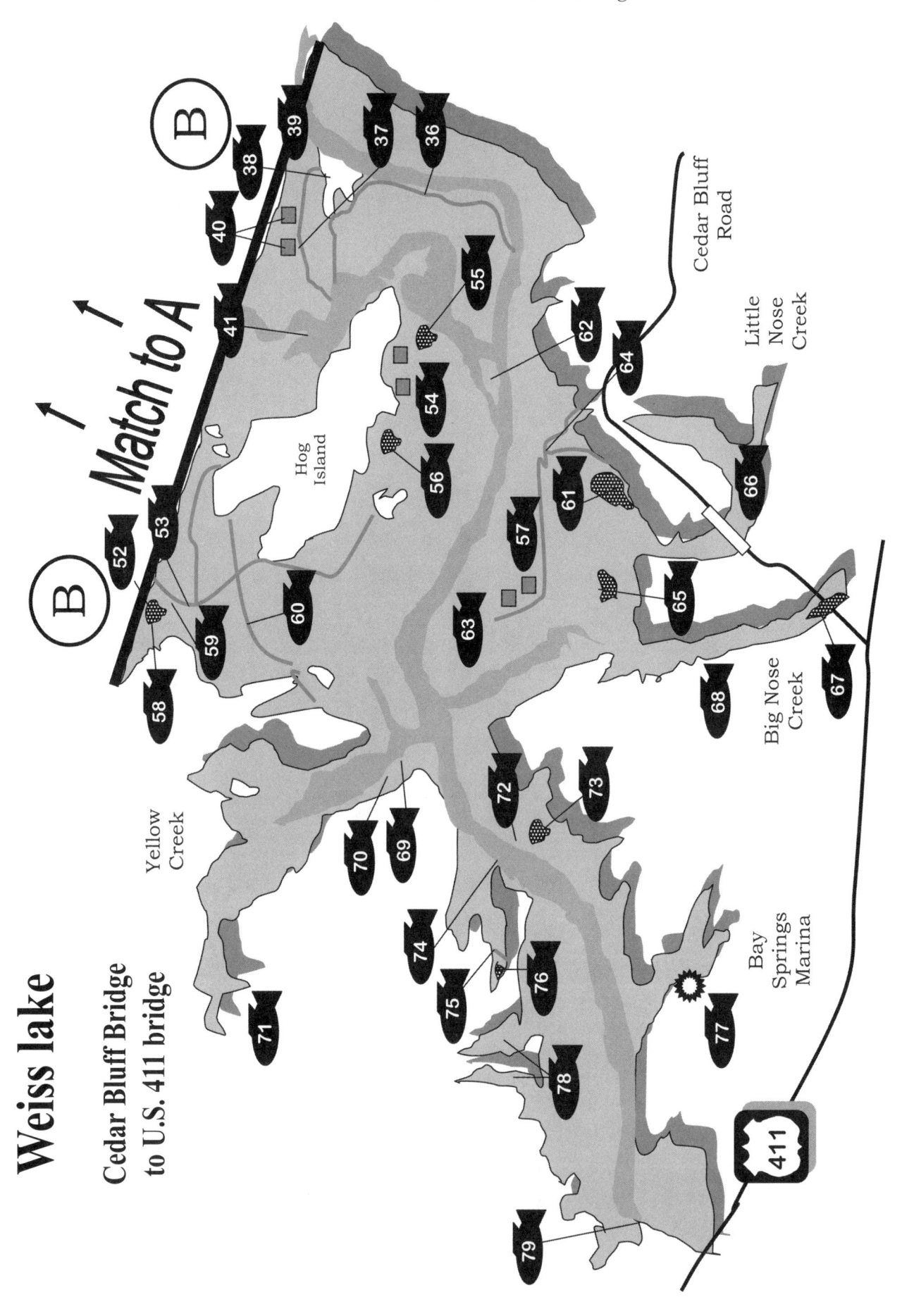

Weiss Lake
Cedar Bluff Bridge to U.S. 411 bridge

Weiss Lake

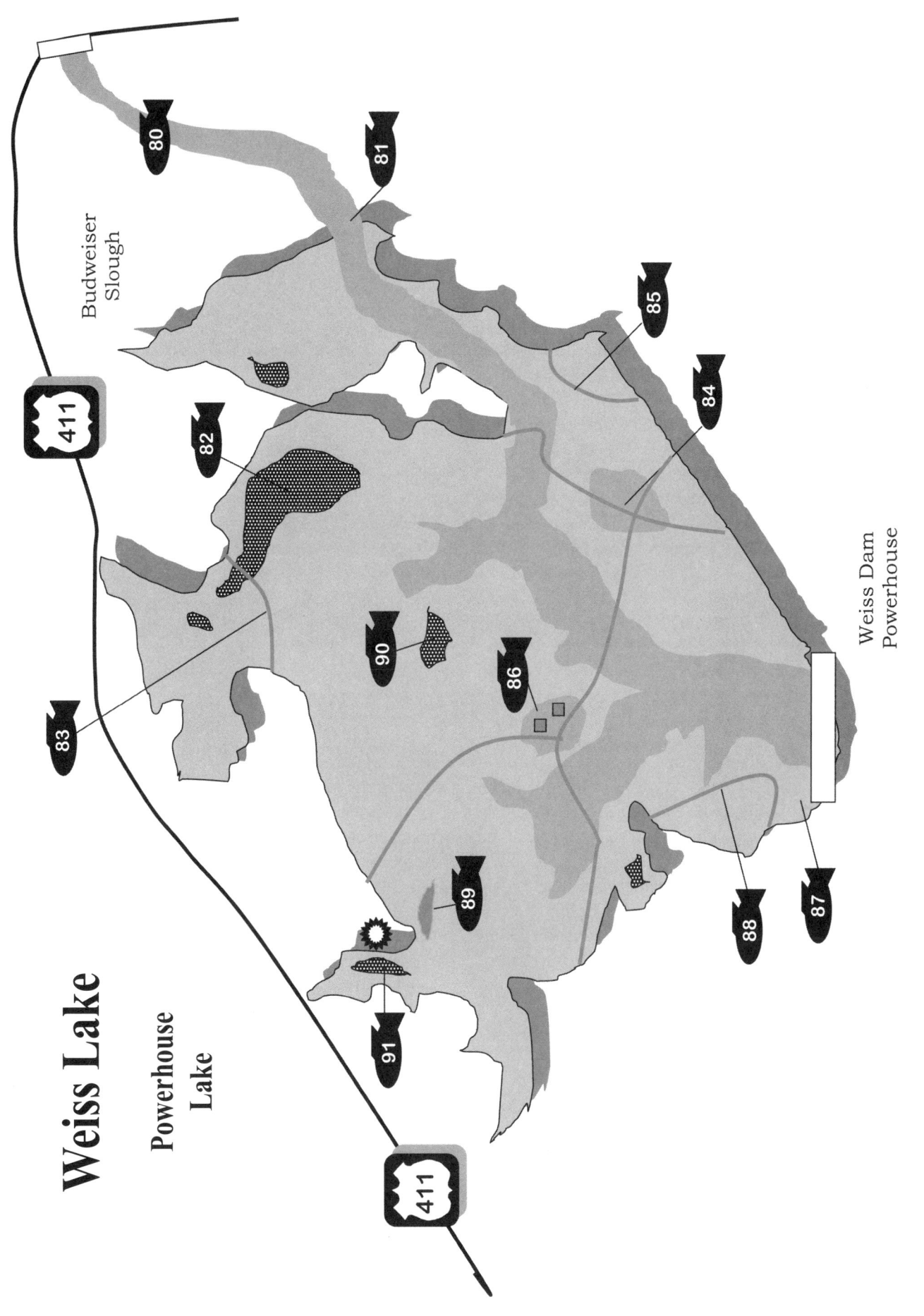

Powerhouse Lake

Budweiser Slough

Weiss Dam Powerhouse

Mike Bolton

411

411

80 81 82 83 84 85 86 87 88 89 90 91

77

Lake Neely Henry

There are few places in Alabama where a businessman in his suit and tie walk out of the office for lunch and in a matter of minutes be catching fish. Lake Neely Henry, which winds through downtown Gadsden, offers that rare opportunity.

Lake Neely Henry offers excellent bass fishing, but its downstream neighbor, Lake Logan Martin, receives the majority of the bass fishing publicity and the resulting pressure. Its crappie fishing is top-notch, but crappie anglers from around the U.S. flock to upstream neighbor Weiss Lake.

Lake Neely Henry offers the best of both worlds without the fanfare.

Spring bass fishing

Weiss Dam to Gadsden bridges

The upper end of Lake Neely Henry, which offers such poor fishing during the summer months, offers just the opposite during the spring. During the spring, this end of the lake consistently gives up larger fish than any portion of the lake any time of the year.

Numerous pockets and small feeder creeks enter the main river channel all along the upper section. Most of these pockets and creeks have stump rows running the length of their original channels. Cove Creek, Henley Creek, Ballplay Creek and Turkey Town Creek offer excellent spring bass fishing, but no small pocket should be passed up because almost all have pretty good amounts of water and can be productive.

Early in the morning and late in the evening, bass move into these creeks to feed along small ledges supporting stump rows. But as on all creeks on the Coosa chain, remember that action in creeks is best when current is moving in the river. The best possible situation is when water is moving out of the creeks.

The creeks fill when Weiss Dam generates power and empty whenever Weiss Dam ceases generating power, or when Neely Henry Dam is moving water and Weiss Dam is not.

Traditional spring baits such as spinnerbaits, buzzbaits and jig-and-pigs work well in these areas.

Gadsden bridges to Southside bridge

If an angler likes topwater action, or if he enjoys spinnerbait fishing, Lake Neely Henry's middle section offers the spring grass beds that make this type of fishing so productive.

Minnesota Bend, the lake's most pronounced bend, is full of grass beds during spring. A lot of small pockets also can be found on this river section and they, too, are full of grass beds during the spring. These small pockets are the spring bedding area for bass.

For those who don't like top water or spinnerbait action, this area offers a few alternatives. The eastern side of Minnesota Bend offers a good rocky bank which provides excellent fishing for spotted bass. A brown jig-and-pig, tipped with a black pork frog, is often murder when water is being pulled through the area.

Night fishermen have success in this area in late spring by throwing the same bait.

There's action for the crankbait fisherman, too. Spring bass often school in the flats when water is being pulled and gold or shad- colored, blue-back Rat-L-Traps often catch large stringers of fish.

Southside bridge to Neely Henry Dam

The lower third of Lake Neely Henry boasts the large feeder creeks which supply much of the backwater of the lake. Canoe Creek is by far the largest and most productive because it offers almost everything any angler could want, including spawning areas, boat docks, rocky banks, stump rows and grass beds.

One of the most effective methods of fishing Canoe Creek during the early spring is using a crawfish-colored crankbait, a worm or brown jig-and-pig to work the deep banks where fish have begun schooling prior to spawn. The back of Canoe Creek past Canoe Creek Marina offers grass beds and stumps and either spinnerbaits or flippin' a jig-and-pig can be productive.

Beaver Creek can be fished the same way as Canoe Creek and although it lacks numbers of fish, it often produces big fish in the spring.

Shoal Creek offers a road bed across its mouth and this road bed traditionally holds spotted bass in spring. A Carolina-rig fished atop the road bed will produce fish.

Rocky banks extend back into Shoal Creek and these banks hold spotted bass. Crawfish-colored crankbaits and brown jig-and-pigs should be worked along the banks.

A large number of downed trees and brush extend to the back of the creek and offer good cover for pre-spawn largemouths. Worms and jig-and-pigs are effective.

Bridge Creek also has a road bed in its mouth. Rocky banks, timber and deep water can be found in this creek.

Hot weather bass fishing

Weiss Dam to Gadsden bridges

The upper portions of Lake Neely Henry, cursed by its river-like makeup and its lack of major cool-running creeks, generally offers poor bass fishing during summer months.

The Coosa River doesn't overflow its original channel in the upper portion of the river and the result is a lack of ledges that provide such good summer fishing elsewhere on the lake.

The mouths of Henley Creek, Cove Creek, Ballplay Creek and Turkey Town Creek offer the best opportunity to catch summer bass, but these few spots are well known and receive extreme fishing pressure during the summer. Bass anglers find limited success by tossing deep-running crankbaits, worms or jig-and-pigs in the areas.

Other patterns offer limited success, too. On the occasions when water is being generated, shad tend to congregate in the small river cuts and river bends where logjams have formed. Bass will occasionally slip into these areas for an easy meal. Worms or swimbaits will sometimes evoke a strike.

Another semi-productive area during summer months when there is current is the rocky and debris-strewn banks directly below the Weiss Dam powerhouse. Largemouths and spots gather there to chase shad. Imitation shad and shallow-running crankbaits are the best baits there.

Although the upper end of Lake Neely Henry offers poor summer fishing, never doubt the fish are there. Fish finders will show schools of bass suspended in the main river channel on hot summer days. Many anglers fish for them with jigging spoons, but that fishing is too tough and too erratic when much better fishing is so close by.

Gadsden bridges to Southside bridge

As much as summer fishing is poor in the upper reaches of Lake Neely Henry, it is good in the middle portion. The area is full of ledges, ditches, cuts, riprap and log jams that offer excellent cover for summer bass.

If sheer numbers of bass excite you, this portion of Lake Neely Henry offers probably the best summertime fishing in Alabama. It is possible to catch numerous keeper-size bass per day during 90-degree temperatures by fishing the ledges.

In fishing the summer ledge pattern, anglers should use their depthfinders to locate the original river channel's western bank. It is at this point that 18 to 20 feet of water will be adjacent to a ledge with 7 to 9 feet of water. This ledge is full of stump rows the length of the western bank from the Gadsden bridges to approximately four miles south of Southside Bridge.

Marker buoys should be dropped along this ledge to allow the river channel and stump rows to be located. During the summer, bass school and suspend in the deep water offered by the original river channel. Whenever a hunger pang hits, they dash onto the ledge to pursue shad that find refuge in the stump rows.

Plastic worms fished on a light spinning outfit offers the most deadly weapon, but shad-colored crankbaits are a close second. Almost all of the fish caught by this method will be tournament-measurable fish 12 inches or longer. This pattern rarely gives up fish three pounds or larger, however.

The second most productive method—and probably the most exciting—is fishing the same area with topwater lures shortly after daybreak. Buzzbaits, chuggers and stickbaits will be effective until the sun clears the tree line.

Big Will's Creek is one of the few creeks on the lake to have enough influx of springs to make a difference in water temperatures. Its cooler waters during the hotter part of summer naturally draw fish. Worms and crankbaits work well throughout its length. This area continually gives up better fish during summer.

Southside bridge to Neely Henry Dam

Approximately four miles of river south of Southside bridge offer the ledges like mentioned above. They, too, offer excellent fishing. They should be fished exactly as you would fish the ledges in the middle portion of the lake.

The most widely used and most successful summer pattern for bass fishing in the lower section is fishing the creek channel in Canoe Creek, however. Large, thick, standing timber runs approximately four miles into the creek along the original creek channel and this creek channel is well defined and easy to find. The lake's better summer fishermen use the trolling motor to ride the deepest portion of the creek channel while casting deep-diving crankbaits and worms to the timber that lines both sides of the original creek channel. The summer months offer lots of schooling activity in these areas.

The mouth of Big Canoe Creek is full of stump rows and worms or deep-diving crankbaits will take an occasional fish there. Canoe Creek also hosts a large number of boat houses and piers. Anglers may find success by flipping a jig-and-pig into the Christmas trees and other brush placed around many of the structures. A worm skipped under the structures will take bass, too.

These same patterns will work in the other major creeks on the lower end such as Beaver Creek, Shoal Creek and Ottery Creek, but Big Canoe Creek offers the best fishing by far.

Cool and cold weather bass fishing

Weiss Dam to Gadsden bridges

The small feeder creeks in Lake Neely Henry's upper end are excellent spots for cool- and cold-weather fishing as bass move into the stump rows in 10 to 15 feet of water.

The timber along the original creek channel was cut off to stump level (2 to 3 feet) with chain saws before the lake was impounded. What remains is stump rows running the length of most of the original creek channels. Plastic worms, slider worms and jig-and-pigs seem to work in these areas equally well.

The warm-water discharge at the Alabama Power Co. steam plant attracts all species of fish during cool and cold months, but the location is nobody's secret and fish there receive extreme pressure. Bass fishermen find success by fishing crankbaits such as medium-depth, lipped crankbaits or Rat-L-Traps in open water, or worms or jigs and pigs along the banks.

Some anglers choose to fish the main river channel in winter by using their depthfinders to find schools of fish and then using jig-and-pigs and spoons to catch them. But the main river channel normally stays dingy during winter months, which forces bass into the feeder creeks. That's one of the main reasons the feeder creeks in the upper end offer such good fishing during winter months.

These creeks will muddy during strong rains, but clear quickly and bass fishing quickly turns good again.

Gadsden bridges to Southside bridge

The middle portion of Lake Neely Henry offers poor winter fishing because of the lack of creeks that run into it. Big Will's Creek offers only fair winter fishing in the middle portion.

Southside bridge to Neely Henry Dam

The number of large feeder creeks that dump into Lake Neely Henry's lower third during cool and cold months offer excellent bass fishing. One of the key areas is Beaver Creek which offers not only deep water next to stump rows, but three road beds and flooded home sites as well.

The road beds and home sites traditionally hold numerous large bass during the cool and cold months. Jig-and-pigs and slow-fished worms work best on and around the old road beds.

The next best winter fishing area in the lower third is Big Canoe Creek. This creek offers stump row-lined creek channels and deep brush throughout its length.

A road bed that runs through the mouth of the slough above Shoal Creek, extends through the mouth of Shoal Creek, crosses the river and into the slough on the east bank. It offers excellent winter fishing. Bridge Creek also has a road bed across its mouth and it too will hold winter bass.

Other types of fishing

Crappie fishing

Only a dam separates Lake Neely Henry from the Crappie Capital of the World, Weiss Lake, and it doesn't take a rocket scientist to know the major difference between the two lakes is publicity. In fact, one Weiss Lake guide told me that on occasion he'll bring out-of-state fishermen to Lake Neely Henry and they never know the difference.

Studies by the state have shown that Weiss Lake has an unexplainably large number of crappie for an Alabama reservoir and admittedly, Lake Neely can't compare to that. But what the powers-that-be on Weiss Lake don't want you to know is that when it comes to big crappie, studies show Lake Neely Henry has the edge.

While crappie fishermen from around the U.S. flock just upstream to Weiss Lake, Lake Neely Henry regulars sit back and enjoy their own little secret.

Lake Neely Henry lacks numbers of large, wide creeks, so a few creeks are left with the burden of handling the spring crappie spawn. Shoal Creek, Big Will's Creek, Canoe Creek and Beaver Creek bear the brunt of the spawn which means fish are concentrated and easy to catch.

Crappie will begin leaving the main river channel in late February and early March and will use the creek channels as their route to shallow water. These spring crappie will first hold in the stumps along the ledges awaiting the 68-degree water temperature in which they prefer to spawn.

The fish won't be tight on structure, so catching them can be relatively easy. Drift fishing is the best pattern. Either spincast or spinning tackle should be used.

First, tie either a pearl-white or chartreuse jig to 8-pound test line on as many poles as legal. Second, using the trolling motor and not using fishing rods, make a run parallel to the creek ledge as close as possible to where it drops in the river channel. Make a mental note of the shallowest water you located and remember exactly where it was.

Next, start in the main creek channel and drop your lines two feet deeper than the average water depth you found along the ledge. Using the trolling motor on medium speed, run the boat along the path you took before. The speed of the boat should pull the jigs above the ledge and prevent them from becoming entangled. You may have to experiment some.

When you come to a hump or other obstacle that you found on the preliminary trip, switch the trolling motor to high and this should lift the jigs high enough to avoid entanglement. Switch back to medium speed when you clear the obstacle.

This type of fishing is a two-person pattern with one running the trolling motor and the other reeling in fish. When only one person is available, he will need fewer rods and will have to steer the boat into the creek channel to remove the fish to keep the other lines from dragging bottom.

Another pattern that works well is to use a live minnow with a float and bobber stop. Several bobber stops are available on the market ranging from small rubber beads to nylon string that can be tied onto your line. A bobber stop comes with a float that has hole completely through it and this float runs up and down your line freely, stopping only at the bobber stop.

It is possible to set the stop at any depth and this stop can be wound up inside the reel. The float hangs down against the hook when casting, thus making casting easy. When the bait hits the water, it sinks, pulling the line through the float. The line stops when the bobber stop hits the top of the float, putting the bait at the correct depth.

When the water temperature hits the 68-degree range, usually in mid-April, the crappie move from the ledges and head to the banks. Any structure from blowdown timber to a single stick stuck in the bottom will hold fish.

Many anglers choose to use spincast or spinning

tackle with jigs or minnows, but a long, telescoping fiberglass pole is much more effective. The crappie will be holding tight at this time and getting the bait right next to them is important. A fiberglass "cane" pole allows you to do that.

Following the spawning period, the crappie will return to the ledges where they were during the pre-spawn period and they can be caught with the same pre-spawn patterns. They will stay there until the warm water temperatures force them deep.

Summer fishing for crappie is much more difficult than spring fishing because the crappie are deep. Many crappie will stay in the creek channels year round, but the better crappie move out to the river channel during the summer and take up residence in the logjams and brush piles found on the inside bends of the original river channel.

One of the more productive patterns at this time is pretty much a secret among the better fishermen.

They use a 1/4-ounce curly-tail jig and rig it weedless Texas-style, just as you would a plastic worm. This allows the jig to be jigged in the logjams and brush.

A depthfinder is a must when fishing for Lake Neely Henry crappie any time of year other than spring. During the summer months, the depthfinder is needed to locate the structure in the river bends. If that structure isn't showing fish along with it, don't spend much time trying to fish it. Look elsewhere.

During the fall, you're likely to find crappie anywhere on this lake. Some move into the creek channels and suspend in the middle of the channel. Those fish are difficult to catch. Other crappie move into the mouths of the many ditches that run from the ledge of the original river channel to shallow water. These ditches were dug before the lake was impounded as drainage and irrigation ditches.

The crappie you'll find here are easier to catch. Use the depthfinder to locate both the mouth of the ditches and the fish in them. They can be caught by drift fishing live minnows or jigs.

During the winter months, the crappie will move back to deep water, usually 20 feet or more. They pile up in the outside bends of river ledges and can be caught with jigs or minnows.

Lake Neely Henry's best fishing spots

1. Hokes Bluff Island. Stump rows on both ends. Produces lots of big fish with current moving. Worms or crankbait in spring, Rat-L-Trap or worm in summer.

2. Mouth of Cove Creek. Deep crankbaits or worms in summer.

3. Stumps rows run the length of right bank going in. Spinnerbaits in springtime.

4. Boat docks and piers on left bank going in. Lots of brush in water around them. Good most of year.

5. Small ledge in mouth of Henley Creek. Deep water and well-defined creek channel. Hot weather crankbait or worm.

6. Weed beds in Henley Creek. Spinnerbaits in spring.

7. Scattered stumps in mouth of Henley Creek. Shallow-running crankbait or spinnerbait in spring. Worm in summer.

8. Warm-water discharge of Alabama Power Company steam plant. Fantastic cold weather fishing. Swim baits, spoons, shad-colored crankbait in winter.

9. Old furnace pond or old Conservation Lake. Excellent spring spawning area. Shallow water and weed beds. Buzzbaits and spinnerbaits.

10. Riprap banks from Goodyear plant to Gadsden bridges, both banks. Spring and fall fishing area for spots and largemouth. Buzzbaits, worms, jig-and-pigs and crawfish-colored crankbaits.

11. Bridge pilings. Fish suspend in this area during summer months. Grubs and light line.

12. Pockets. Spinnerbaits or flippin' in spring or winter.

13. Mouth of junior college slough. Deep crankbaits and worms in summer.

14. Grass beds in junior college slough. Spring spawning area. Spinnerbaits and buzzbaits in spring.

15. Mouth of Big Will's Creek. Cold-water creek. Hot weather patterns such as deep crankbaits or worms.

16. Highway 411 bridge and riprap under it. Good spring and fall fishing.

17. Big Will's Creek original creek channel. Spring and hot summer months.

18. Boat docks and piers in Big Wills Creek. Lots of brush around and under them. Worms or jig-and-pigs.

19. Grass beds and multi-channels. Good spring and summer areas. Spinnerbaits in spring. Worms in June, July and August.

20. WAAX radio tower area. Excellent summer and fall schooling area. Rat-L-Traps, Little Georges and medium-diving crankbaits.

21. Grass beds and flats in cut. Fish both with spinnerbaits in spring.

22. Sand pits. Sand was once pumped from this area causing mini-ponds and lakes. Although fairly deep water, a super area for big spotted bass in spring.

23. Keeling's Island. Standing timber around island produces lot of schooling activity during summer. Crankbaits and worms.

24. Mouth of Honey Creek. Stump rows gives up summertime bass. Grass beds hold spring bass.

25. Back of Honey Creek. Boat docks in fall hold fish around brush.

26. Large grass beds in Minnesota Bend. Spring spinnerbaits and buzzbaits.

27. Rocky banks in Minnesota Bend. Lots of spotted bass schooling activity during summer and early fall.

28. Stump rows in Minnesota Bend. Good spinnerbait fishing on summer nights and in early fall.

29. Cuts and small feeder creeks. Good summertime fishing in mouths with crankbaits.

30. Grass beds. Good spring fishing with spinnerbaits and buzzbaits.

31. More sand pits. Surrounded by grass beds and stump rows. Good fall fishing.

32. Pockets. Spring fishing in back of pockets with spinnerbaits and buzzbaits; summer and fall fishing in stump rows in mouths with worm or jig-and-pig.

33. Bridge area. Fish suspend around pilings in summer and into fall. Crankbaits, grubs and jigging spoons.

34. Best ledge fishing begins. Use depthfinder to find original creek channel. Stump rows run length of original channel on both sides, but west bank is best. Great summer fishing with worms or crankbaits along ledge for four miles south.

35. Marked standing timber running length of original channel to Neely Henry Dam. Well-marked and fishy-looking, but surprisingly gives up few fish. Standing timber is best.

36. Ledge fishing continues this far.

37. Broughton Springs. Prime spring fishing area. Grass beds and boat docks. Spinnerbaits and buzzbaits.

38. Mouth of Broughton Springs. Good summertime area as fish school. Worms and crankbaits.

39. Excellent summer ledge fishing area with worms and crankbaits.

40. Cooper's Branch. Good spring spawning area. A few weed beds, but mostly pea gravel. Spinnerbaits and worms.

41. Standing timber in Canoe Creek stretches one mile from the mouth up the right bank. Crankbaits in spring through fall.

42. Weed beds and piers in Canoe Creek stretch to the bridge. Buzzbaits and spinnerbaits in the weeds in spring, worms and jig-and-pigs in pier brush in summer, fall and winter.

43. Above bridge. Good fall and winter fishing area. Jig-and-pig or worm.

44. Road bed. Runs parallel to east bank between Broughton Springs and Green Creek. Good winter schooling area with worm or jig-and-pig.

45. Standing timber. Good schooling area in summer or fall. Worms or crankbaits.

46. Road bed. Crosses Green Creek. Good winter jig-and-pig.

47. Green Creek. Impossible to cross under bridge, however. Must launch behind bridge with small boat. Good spring bedding area. Spinnerbaits or buzzbaits.

48. Mouth of Beaver Creek. Stump rows. Good in summer with crankbaits. Fish school there in spring, too.

49. Road bed. Crosses mouth of Beaver Creek, turns sharply east and crosses main channel. Good in summer with worms or crankbaits. Roadbed section in main channel occasionally good in winter with jig-and-pig.

50. Another road bed in Beaver Creek. Crosses middle of creek and parallels west bank before exiting out of cut. Carolina rigs or worms in summer, jig-and-pigs in winter.

51. Another road bed in Beaver Creek. Crosses small slough. Same as above.

52. Big fish area. Jigs and worms in spring and in winter.

53. Another road bed. Crosses Beaver Creek. Same as No. 50.

54. Road bed. Crosses cut. Same as No. 50.

55. Road bed. Crosses cut above Shoal Creek then crosses Shoal Creek, hangs a sharp left across main channel and enters feeder creek on east bank. Same as No. 50.

56. Road bed crossing Ottery Creek. Fish same as No. 50.

57. Road bed crossing Ottery Creek and passing directly under bridge. Fish same as No. 50.

58. Good spring spawning area for spotted bass. Spinnerbaits.

59. Mouth of feeder creek. Good standing timber in mouth.

60. Good summer and fall schooling area. Crankbaits and worms. Road bed crosses cut in feeder creek. Fish same as No. 50.

61. Deep-water pockets full of logjams. Good in late spring until water warms and fish move to ledges. Worms, jig-and-pigs and crankbaits.

62. Road bed in mouth of Bridge Creek. Excellent schooling area for white bass, hybrids and spotted bass.

63. Rip-rap around dam. Excellent topwater area in spring; swim weightless worm around dam in summer.

Lake Neely Henry

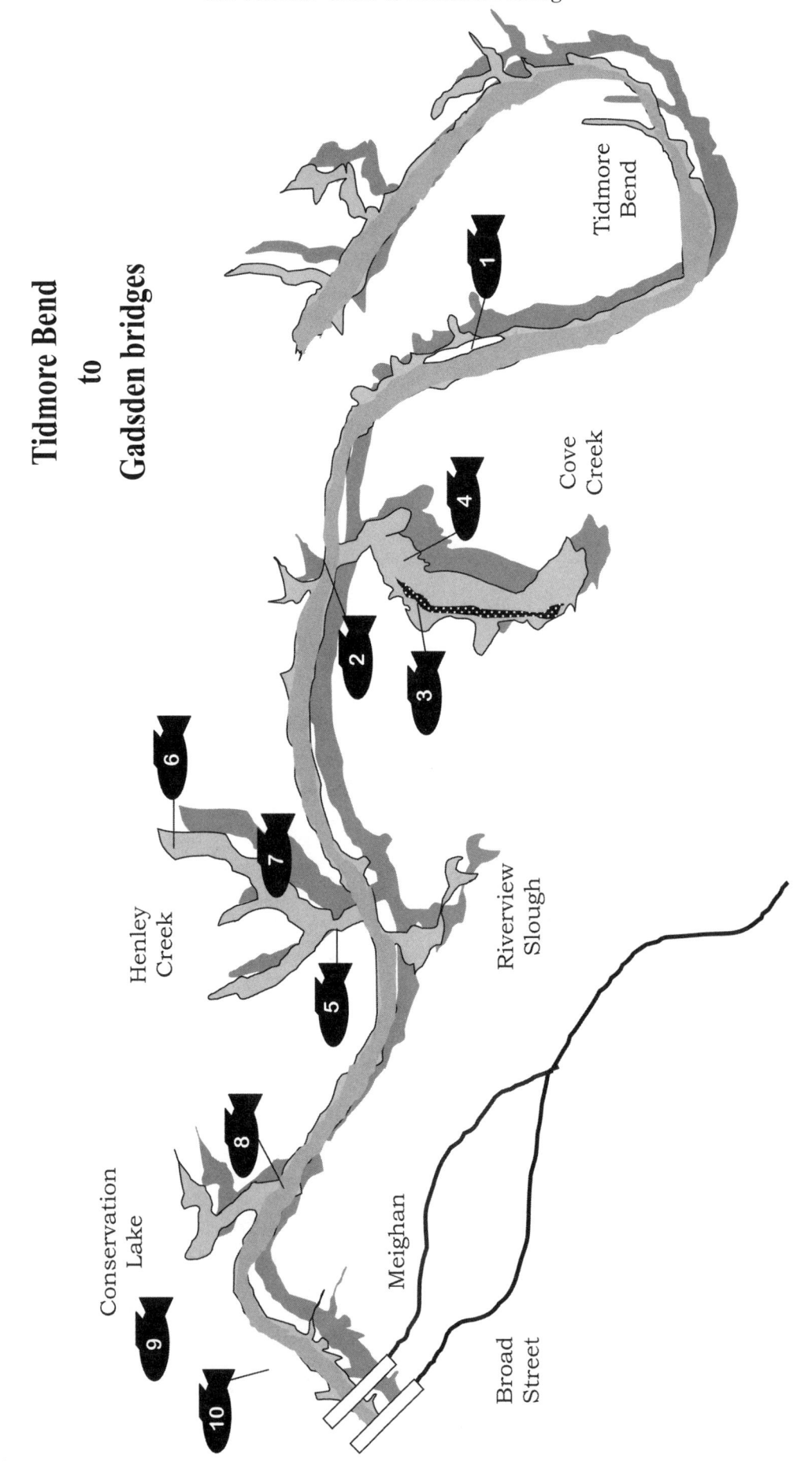

Tidmore Bend
to
Gadsden bridges

Tidmore Bend

Cove Creek

Henley Creek

Riverview Slough

Conservation Lake

Meighan

Broad Street

The Ultimate Guide to Alabama Fishing

Lake Neely Henry

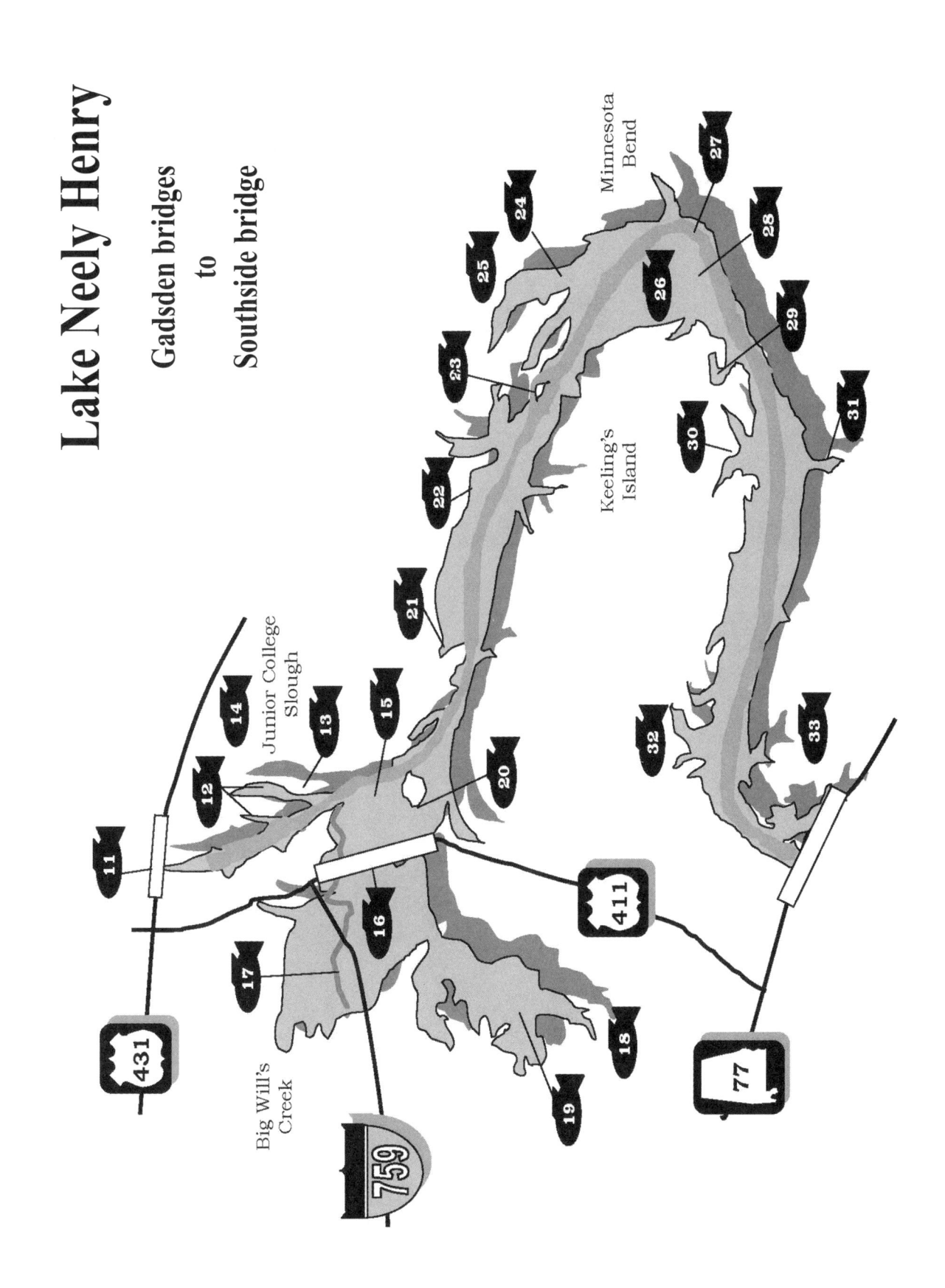

Gadsden bridges to Southside bridge

Lake Neely Henry

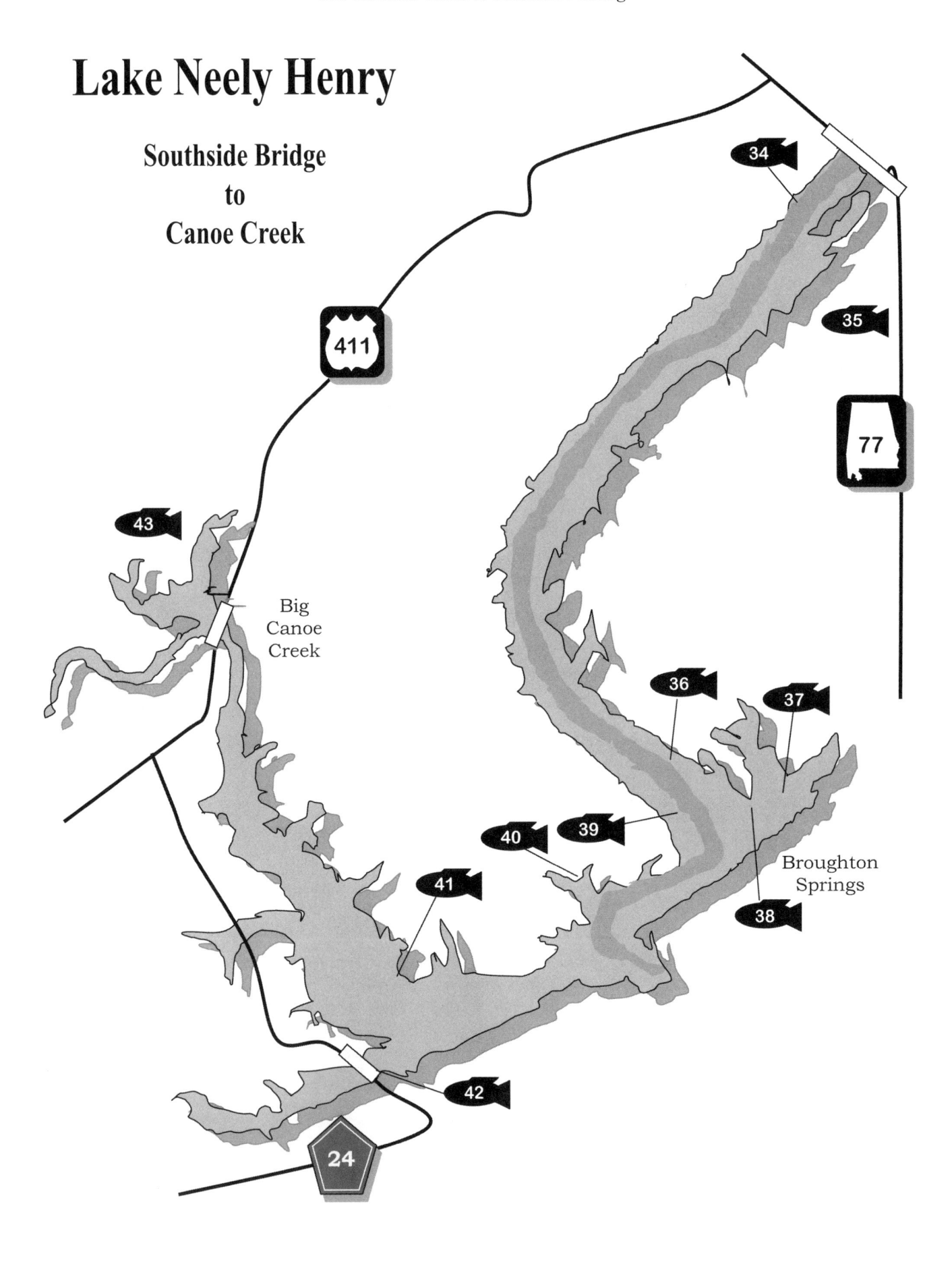

Southside Bridge to Canoe Creek

86

Lake Neely Henry

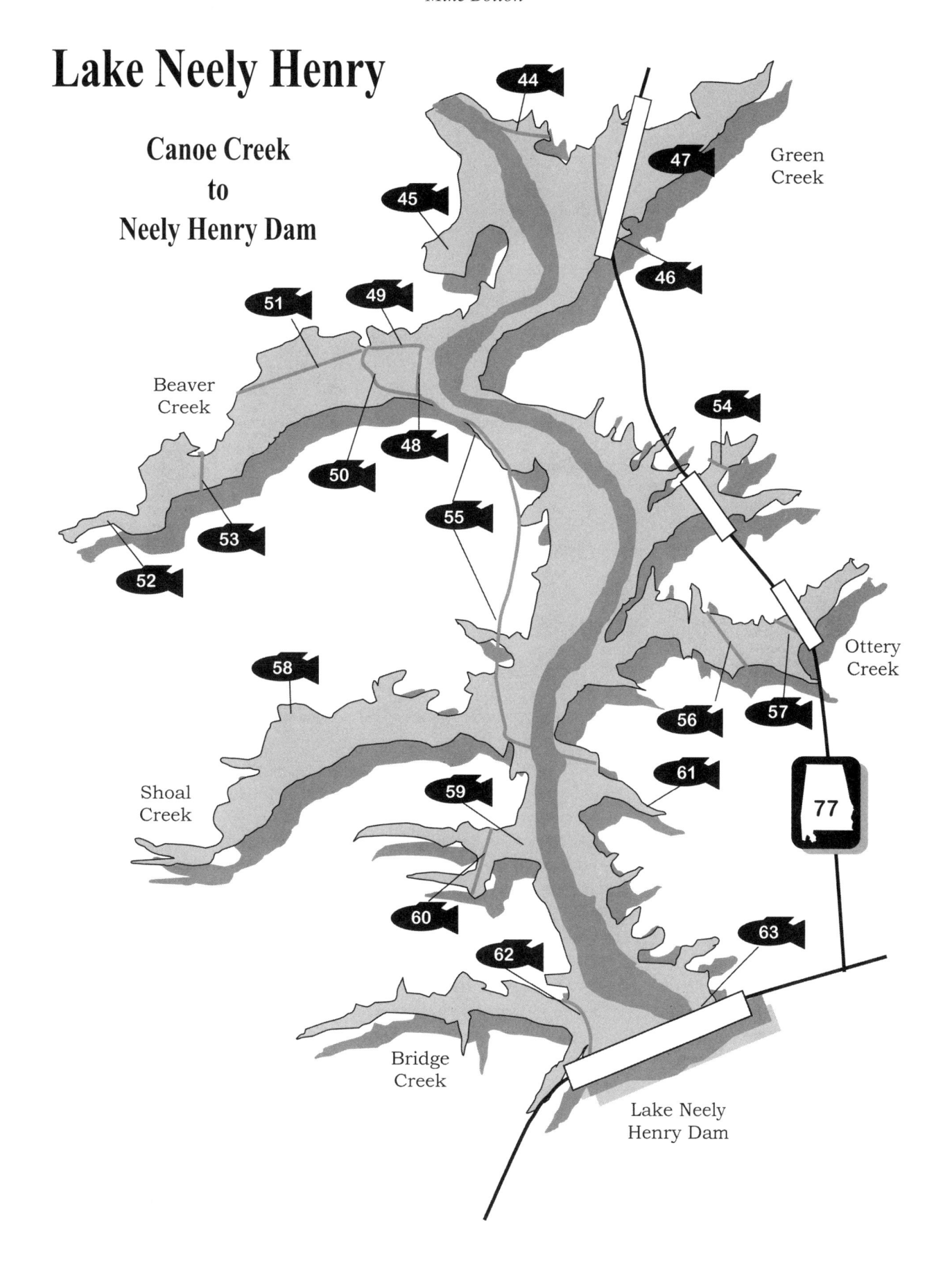

**Canoe Creek
to
Neely Henry Dam**

Lake Logan Martin

Lake Logan Martin is living proof that a major reservoir receiving extreme pressure from major urban areas can produce excellent fishing.

Lake Logan Martin is near the top of the list of the state's most fished lakes, yet many anglers will argue it offers bass fishing as good as any lake in the state. Its large population of better-than-average spotted bass, a good population of largemouths, incredible numbers of catfish and excellent fishing for hybrids and stripes makes Logan Martin a near-perfect lake for the urban-based fisherman.

Spring bass fishing

Neely Henry Dam to I-20 bridge

Spring fishing in the upper portion of Logan Martin is feast or famine depending on whether rain supplies the water to fill the creeks and accompanying flats. If the rain comes, spring fishing in these areas can be excellent.

Keep in mind the cool-water creeks such as Ohatchee Creek and Acker Creek will be too cool for spawning during the spring and anglers should seek warm areas such as Alligator Creek, Broken Arrow Creek and Blue Springs. These areas are full of stumps, weed beds and flats—the type areas bass bed in.

Other good areas that many anglers overlook are the stump rows in the main river channel. Fish that aren't quite on bed or coming off the bed will hold in these areas. Slow-crawling a spinnerbait through these stump rows will produce fish.

I-20 bridge to Stemley Bridge

Unlike the upper portion of Logan Martin, which is

really nothing more than a river, the mid-section from the 1-20 bridge south to Stemley Bridge is a full-fledged reservoir full of creeks and flats that make excellent spring spawning areas.

Some of the most notable areas are Seddon Creek, Fish Creek, much of Choccolocco Creek, Poorhouse Creek and Stemley Cove.

When warm weather approaches after the long winter, the small male bass are the first to seek these areas while the females will hold in the main river channel or creek channels. Two or three days of warm weather will normally warm up the shallow water enough to spur the male bass into their first aggressive feeding of the year. Worms, jig-and-pigs and occasionally a shallow-running crankbait or Rat-L-Trap will draw strikes.

The warming eventually stretches toward the creek mouths. The main river is always the final area to warm. After water warms to approximately 68 degrees, the big females laden with eggs move in to spawn. Poorhouse Creek and Choccolocco Creek are the best spring fishing areas in the mid-section because the larger female bass are relatively easy to find. The top pattern for this April through May fishing is to use jig-and-pigs to fish the piers and docks where these large females like to find the brush tops that homeowners have placed there.

Another key spring bass area is the junction of the original river channel and Choccolocco Creek. A large underwater island surrounded by brush is located there and bass ready to spawn or in the spawn gather there.

Stemley bridge to Logan Martin Dam

Like the middle section of the lake, the lower area also boasts flats and incoming creeks that offer fine spring fishing. Top spring spawning areas include the Redstone area south of Powell's campground, the sunken roadbeds and flats in the red barn area, much of Clear Creek, the upper end of Rabbit Branch, Bum's Slough, the first large pocket on the east bank above Logan Martin Dam and the riprap along the dam itself.

The bass during this period like to move up to feed on flat sandbars and warming pea gravel banks in 5 feet of water or less. These areas almost always have a little stained water during these times so shallow-running chartreuse crankbaits or chartreuse Rat-L-Traps work well. These are the areas that spring bass first move into.

The water on the main river is normally a little cool during this period so fishing isn't good on main river points, but secondary points will often hold larger fish because they are warmer. The secondary points with the most exposure to the spring sun will almost always hold the most fish. This is the northwest side of the lake. Following the spring spawn, the bass will situate in 8 to 12 feet of water on humps or stump rows and that's one of the main reasons this area is considered so poor for those who like to fish the banks.

Literally hundreds, maybe thousands, of brushpiles have been placed in this area by tournament fishermen and almost every point or underwater hump has at least one or two on it.

There are two April and May patterns which consistently take fish off of these locations. The best is a spinnerbait with a No. 5 or No. 6 willow-leaf blade slow-crawled through these brushtops and tree tops. Most people consider these good crappie locations and they are, but after the water warms past 69 or 70 degrees, the crappie move off to the channels and the bass move in.

For those who don't like spinnerbait fishing, a jig-and-pig will work just as well in these areas but it doesn't cover ground as quickly.

Another excellent spring fishing area is the old U.S. 231 (Old Dixie Highway) roadbed through the Cropwell Creek area. The Hannon Lake and Avondale Lake beds also are good. All of these areas should be fished with willow-leaf spinnerbaits or jig-and-pigs.

If you're looking for the perfect spring bass fishing situation, two or three days of bright sunshine the week before a full moon will make fishing as good as it gets on Logan Martin.

Hot and warm weather fishing

Neely Henry Dam to I-20 bridge

Although less popular as a fishing spot, the upper reaches of Lake Logan Martin offer excellent hot weather fishing.

Several hot weather patterns will work, but it is important to remember that for these patterns to be successful, current is a must for day fishing and a lack of current is a must for night fishing. One of the most effective methods is known as "fishing the cuts," the cuts being the tiny sloughs and pockets that sprout from the river bends.

Every single cut along this length of the river has stump rows in its mouth and offers the perfect place for shad to escape the currents when power is being generated.

Naturally, with shad being in these cuts, bass are sure to follow. Shallow-running crankbaits or plastic worms are effective and when conditions are right, as many as 15 or 20 bass can come from a single cut.

Another popular summer pattern in the upper end is fishing the cool-water creeks. Cane and Acker Creeks, as well as the back of Ohatchee Creek, are spring fed and will often be several degrees cooler when the mercury hits 80 degrees elsewhere. Naturally, bass bunch up there.

These creeks all have stumps in mid-channel and bass congregate at the base of the stumps and down in their roots. Slow-moving lures such as a swim bait or jig and pig should be pulled on bottom and bounced over these stumps and roots.

Another lure that works well early and late in the day is a buzzbait. The bass are much more active in these creeks because of the cooler temperatures and will attack a noisy topwater bait.

It is important to understand two important things about fishing cool-water creeks in this area:

A. Bass are most active when water is being pulled through Logan Martin Dam—not when it is being pushed through Neely Henry Dam—and the water level in the creeks is falling.

B. It is possible to be caught sitting in too shallow water if you keep your boat in these creeks too long while the creek level is diminishing.

The third pattern—a late summer pattern that works well at night—is fishing the rock shoals in the main river channel with spinnerbaits. Every year during August, September and October, shad migrate to Neely Henry Dam for the fall and the bass follow. The area between the dam and Cane Creek offers the best fishing.

A spinnerbait, tipped with a contrasting-color pork frog to make it more visible and more buoyant, and slow-crawled in 5 feet of water, can produce tremendous bass during this time of the year.

This area is full of boulders including a cofferdam, which is a dam that diverted the water flow while Neely Henry Dam was being built, and a fish trap, a rock formation in a V-shape that Indians once used.

An interesting note is that the river, from the CSX Transportation railroad trestle to Neely Henry Dam, is almost solid with rocks and on summer nights, bass could be anywhere in the area.

I-20 bridge to Stemley bridge

Make special note of the map of the middle portion of Lake Logan Martin. This map shows the original river channel (dark area) and where the water overflowed the channel to create the lake. It is important to know that when impoundment preparations were being made, trees along the river channel that would be covered with water fell to the chainsaw.

Many stumps remain in rows down both sides of the main channel for the length of the river.

Hot weather patterns in this area are much like the area above it in that you need current and stumps to be successful. Just as in the upriver areas, fishing the cuts can be productive, but the cuts in this area have two stark differences. Not only are they much larger, they also boast ledges at their mouths. It is common to find 8 feet of water in the cut itself and then 25 feet at the mouth.

This is where an understanding of ledge fishing is important.

Baitfish will hole up in these cuts to get out of the current and the bass will follow. They will suspend in anywhere from 5 to 15 feet of water.

Spinnerbaits, worms, or jig-and-pigs thrown into the mouths and pulled toward deeper water will most often provoke a strike.

It is also important to remember that bass in these areas are not always active and failure to draw an immediate strike is not necessarily a sign that no fish are there. If fish are showing on the fish locator, it may take patience to cast until the lure is dragged right across a fish's face or body before a strike comes.

Another hot weather pattern that works in the mid-lake area is fishing the back end of Choccolocco Creek from the Alabama 77 bridge to the Jackson Shoals area. This area is often referred to as the Rushing Springs area. A cold-water spring near Jackson Shoals causes cooler water as far west as the Alabama 77 bridge and makes for good hot-weather fishing.

This area boasts plenty of structure and shallow-running crankbaits or spinnerbaits work best.

Just as in the creeks in the upriver area, Choccolocco Creek should be falling to be most productive. It is also important to note that numerous large stumps are just below the surface between the Alabama 77 bridge and Jackson Shoals. Extreme caution is needed in navigating that area.

Another hot weather method that works well is using electric red plastic worms to fish the shady piers and boathouses. Baitfish are attracted to Christmas trees and brush placed at the end of piers and docks, and the smaller bass feed in these areas at dawn and again at sunset.

In bright sunlight, bass will position themselves underneath these structures to avoid the rays. Bass usually avoid floating structures and choose permanent docks, boathouses and piers because they not only offer shade, but the pilings offer structure as well. Expect to find the bigger bass hovering near the most centrally located, most hard to get to poles.

Smaller bass often can be caught on the fringes and

around outside poles.

Stemley bridge to Logan Martin Dam

The lower third of Logan Martin is probably the most productive portion simply because it has more water. Prime fishing areas are Cropwell Creek (Town and Country, Big Bull, Aqualand, Pine Harbor area), Rabbit Branch and Clear Creek. Many top anglers also believe this portion has a larger concentration of spotted bass.

The plastic worm is the primary summertime weapon here. Motor oil, pumpkinseed and electric red are consistently the most productive colors.

Any irregularity on the bottom such as a hump, hole, sunken roadbed, etc., as long as it has some structure such as a brush top or stump row, will likely hold summer fish.

Deep stump rows can be found in the mouth of almost all incoming creeks and the lake's most serious fishermen often shine during June, July and August as they run deep-diving, big -lipped crankbaits through these areas in 12 to 15 feet of water. They crank these big baits down so they bump brush tops, rocks and stumps and cause a commotion.

These baits also churn up sediment as they bump structure and many anglers think the bass believe they are attacking a crawfish, one of their favorite foods.

A pattern that works in this area for the anglers not necessarily hunting the biggest fish is flipping or casting worms or jigs around and under pier pilings.

Cool and cold weather fishing

Neely Henry Dam to I-20 bridge

The most productive winter pattern in this area is one that many anglers have refused to believe until they see for themselves.

This pattern works best from October to February when the lake is full and Alabama Power Company is forced to open floodgates and turbines. What happens is this: The swift water forces the shad, which cannot fight the current, to head for the banks to seek refuge in the nooks and crannies, logjams and other structure to escape the current. Naturally, the bass will flock to these areas for an easy meal.

An angler can position his boat in this swift water and use his trolling motor to keep his boat facing the onrushing water. While drifting backwards, he can cast a 1/2-ounce or l-ounce spinnerbait or jig-and-pig against the bank and, while keeping a tight line, let the lure wash down in the current.

The key to this pattern is that there must be an abundance of water coming through Neely Henry Dam. The shad will school behind the large boulders in the middle of the creek and will do so in moderate current. It takes extreme current to force them to the banks.

Anglers should note that many times each little nook and cranny can have several bass in it, so if you catch several fish, crank the big motor and go upstream and drift back over the exact area. Do not continue drifting downstream because you have already located the fish.

A pattern that works when there is no current is flipping. The stump rows in the mouths of the creeks and cuts, stump rows in the mid-channel of warm creeks and the logjams in the river bends are excellent areas.

I-20 bridge to Stemley bridge

Although several winter patterns will work in this section of the lake, the most notable is sunken roadbed fishing because it is there that bass gather and can be found the easiest.

Several sunken roadbeds can be found in the mouth of Choccolocco Creek and Poorhouse Creek. Bass will school in these areas during December and January. Jig-and-pigs dragged across the roadbeds and dropped down the sides will catch fish for the patient fisherman.

Several of the lake's better fishermen use hair jigs on a spinning outfit because this not only produces numerous spotted bass, it is possible to catch crappie at the same time.

During the fall, bass fishing often becomes erratic as the fish turn on and off for no apparent reason.

During the early fall, the fish are gorging themselves, but as the first cold fronts come in, they'll turn on and off at a moment's notice. In the early fall, when the fish are schooling and chasing shad, shallow-running crankbaits, Rat-L-Traps are best for checking out where the fish are. The bass will almost never be found in the back of creeks, sloughs or pockets during this time. The bass will always be close to the river off points or flats.

One of the best-kept secrets among top anglers during this time is the use of spinnerbaits. A small, double-bladed, 1/4-ounce, white spinnerbait thrown into the schooling fish and retrieved slowly is absolute murder.

Stemley bridge to Logan Martin Dam

Fall fishing in the lower end is good when bass can be found, but many anglers have difficulty finding them because of wide expanses of water and greatly varying depths.

The quickest pattern is locating schooling bass because they like to bunch up on the main river points. Grubs or an imitation shad work wonderfully. Unfortunately, this pattern works only when current is moving and anglers may find current only on certain days.

Winter bass fishing in the lower end is poor and many anglers crappie fish instead. But with the water down during winter pool, it gives the angler an excellent opportunity to see structure he normally doesn't get to see.

Other types of fishing

Stripe and hybrid fishing

The introduction of hybrid and saltwater striped bass into Logan Martin has opened up a new world for many anglers. For many fishermen, the hybrid or striped bass—not the black bass—has become the largest freshwater fish they've ever caught. The sheer number of hybrids and stripes stocked in this lake through the years has livened up many an angler's otherwise dull fishing life.

Summertime stripe and hybrid fishing has caught on like wildfire in recent years on Logan Martin and for good reason. The fish are plentiful when other species are not.

The secrets to good summertime fishing are healthy live shad and locating cool-water springs. Threadfin shad, which make excellent stripe bait, live for only a few minutes in a common minnow bucket.

They must have a round bait bucket or they'll pile up in the corner of a square bucket and die from lack of oxygen. The bait bucket must have aerated and flowing water.

The second gadget needed to do the job correctly is a water temperature gauge, and not a surface temperature gauge like many boats are equipped with. This gauge must be attached to a cable so it may be lowered into the lake in varying depths to find thermoclines.

Threadfin shad are easily caught with a cast net as they school by the thousands in the mouths of sloughs and creeks. The often appear as a large dark cloud.

Finding a thermal refuge in which to fish these live shad is the key to successful fishing. Cool-water springs such as those located in Choccolocco, Ohatchee and Cane creeks attract hybrid and stripe and offer the best fishing.

Hybrids are less susceptible to the heat and light than stripes and are more easily caught in these areas. Stripe tend to move in the later afternoon period as the temperature cools and the sun drops.

Open-face reels equipped with 14- to 17-pound test line are best and the live bait can be hung beneath a float. Water in the 72- to 76-degree range is where the big stripe are normally found. An angler can expect to catch his share of small hybrids before catching a big hybrid or stripe, however.

Hybrid and stripe also can be found chasing shad in shallow water flats during the summer and winter.

Hybrids and stripes will be found in the same creeks in the spring where they are found in the summer, but water temperatures are not nearly as critical. These fish seem to have a dislike for lively bait during this period and cut baits (threadfin shad with their heads cut off) work better than live shad.

Almost any bend in the creek that has rocky banks will have stripes and hybrids schooling on them during this time of year.

Although many stripe and hybrid stay in the same creeks to enjoy the warm water that springs emit in December through February, a great number move into Cropwell Creek near the Pine Harbor, Aqualand and Harmon Island areas.

Many anglers watch for gulls chasing the shad to give away the location of hybrids and stripes. Topwater lures such as jerkbaits or shallow-running crankbaits such as Rat-L-Traps and Rattlin' Spots will catch these fish.

The extreme upper end of the lake in the Neely Henry Dam tailrace offers the easiest fishing for hybrids. Unlike the Logan Martin Dam tailrace, the area below Neely Henry Dam is smooth and void of dangerous rocks and boulders. When the dam is moving water, an angler needs only to run his boat up to the dam, kill the motor and drift backwards while bumping a 3-inch grub or small white spoon along the bottom.

Another popular method is anchoring outside the swift water and casting a pearl-colored, 3-inch Sassy Shad with a 1/4-ounce or 1/2-ounce lead head into the swirls. The best time for either of these patterns starts when the dogwoods bloom and extends into the fall.

The hybrids along the banks in the tailrace will often feed on topwater, especially when the water has just been turned on or off. Jerkbaits will catch these fish.

Crappie fishing

Lake Logan Martin has become one of Alabama's top crappie fishing lakes over the last decade.

Like all lakes, spring crappie fishing is best because the fish move shallow and are accessible to more fishermen. When the water warms for the first time of the year, both species of crappie begin looking for shallow water in sloughs and creeks that have blowdowns, stickups and logjams in which to spawn.

The water color in these areas can vary drastically from year to year, so the jig fisherman should vary his baits accordingly. Chartreuse, lime green and orange bodies with red heads work well when the water is dingy, while pearl, white and smoke colors work well when the water is clear.

Fiberglass crappie poles is the choice of the better crappie fishermen because they allow the angler to stay a distance away while jigging a jig or minnow tight against the crappie-holding structure.

Many anglers don't like losing the tackle that is lost fishing this method and choose rather to fish submerged structure or boat houses with a float and jig or a float and minnow. Piers with brush around them are prime areas for this type fishing. A float and jig can be an effective method. The angler should cast the rig well past the structure, then "pop" the float back to the boat over the structure. This popping draws the attention of fish and the float holds the jig out of the structure where it won't become entangled. Rabbit Branch, the Town and Country area and Choccolocco Creek are the lake's best areas for spring crappie fishing.

Once the spawn is over, the whites and blacks split into different areas. It should be noted that the white and black crappie on this lake segregate with the whites being oriented to deep-water points and blacks oriented to deep-water rocks and stumps. The whites tend to weigh more, but the lake has produced several black crappie in the 3-pound range.

These crappie will move to water 7 to 12 feet deep until late June or early July. The better crappie fishermen bump bottom for the fish at this time, using a minnow on a bottom-bumping rig outfitted with a l/2-ounce to l-ounce weight.

Summer fishing for crappie is almost non-existent on the lake, but these fish can be caught during this time. These summer crappie move to river ledges in water up to 25 feet deep. The river ledges between Town and Country and Stemley bridge are excellent for summer crappie fishing as are the ledges between Town and Country and Clear Creek. The mouths of big creeks all along the lake also hold summer crappie.

Lake Logan Martin's best fishing spots

1. Large boulders remaining from cofferdam use to divert water flow when Neely Henry Dam was being built. Good bass fishing with spinnerbaits on summer nights when no current is flowing.

2. Railroad trestle. Good summer bass fishing day or night with spinnerbaits slow-crawled around pilings. Plenty of rocks and rock shoals in the area. Good with either current flowing or off. Train going over will scare you to death, however.

3. Rock rows on the western shore in the river bend. Lots of large boulders. Good spring and fall bass fishing when bass chase shad that school there. Best fished with jigs, spinnerbaits or worms when no current is flowing.

4. Rock banks. Good drift-fishing area during the winter for spinnerbait fishing for bass.

5. Ohatchee Creek. Excellent cool water creek. Good summertime fishing for all species.

6. Cane Creek. Also an excellent cool water creek for all species.

7. Acker Creek. Excellent spring spawning area for bass. A swim bait works well in warming months. Also an excellent flipping area for bass during the winter.

8. Old Indian fish trap. Boulders placed in main channel in V-shape. Good spinnerbait fishing early morning or late evening during hot weather months. Best when current is off. Located approximately I/2-mile above Alligator Creek.

9. Alligator Creek. Excellent spring spawning area.

10. River cuts. Good hot weather fish as shad gather in cuts to escape currents. Crankbaits or worms.

11. Rocky banks adjacent to deep water. Good crankbait area spring, summer and fall.

12. Broken Arrow Creek. Good spring shallow fishing area back in the creek when water is falling. Lots of stumps and roots. Spinnerbait fishing.

13. Excellent spring spawning areas. Worms or Lunker Lures.

14. Location of old dam, now partially collapsed. Sits about three to five feet underwater at full pool. Worms or jigs year round.

15. Thick stump rows adjacent main channel. Good year-round fishing with worms and jigs with no current flowing.

16. Blue Spring Creek. Excellent spring spawning area. A long string of thick stump rows run northeast from the mouth of the creek halfway to the back. Spinnerbaits, jigs and worms.

17. Railroad trestle. Summer fishing is good around pilings and rocky banks with crankbaits and spinnerbaits with current on or off.

18. U.S. 78 bridge. Good warm-weather crappie fishing area at night with lights. A rocky point from eastern bank juts under the bridge and offers occasional good bass fishing.

19. Pipeline crossing. Large rocks dumped into seven feet of water. Good summertime bass fishing with worms. Also excellent trotline or jug fishing area.

20. Stump rows, adjacent to old river channel. Good bass fishing area year round with worms or jig-and-pigs.

21. One of very few weed beds in upper end. Small islands with weed beds on top and around them. Spring topwater area. Caution: shallow area that doesn't always appear shallow. Can be dangerous.

22. Good ledge-fishing area. Water dips from eight feet to 25 feet. Lots of stumps in the area.

23. Seddon Creek. Good spring spawning area.

24. Powerline tower island. Rocky. Good spring area for several species as rocks catching sunlight warm surrounding area.

25. Underwater hump. Good summer bass fishing with crankbaits and worms when current is moving.

26. Fallen trees and man-made fish attractors in mouth of the slough.

27. Old Lock No. 5, now partially collapsed. Concrete wall and rubble running from north bank clear into the channel. Good worm or jig-and-pig fishing year round.

28. Logjams and fallen trees on southern bank of river bend. Spinnerbaits, jig-and-pigs and worms. Summer, fall and winter.

29. Mouth of Choccolocco Creek where old river channel and old creek channel meet. Several humps and rises. Spring and summer worm, crankbait or spinnerbait area.

30. Good point with fallen trees and brush tops. Good worm or spinnerbait area.

31. Large, warm underwater spring. Good fishing for bass, stripe and crappie in winter.

32. Highway 77 bridge to Jackson Shoals. Late summer or early fall area. Shallow water. Lunker Lures or other top water when creek is falling.

33. Old roadbed in Poorhouse Creek. Excellent winter fishing area for bass and crappie using spin outfits with jigs.

34. Back of Poorhouse Creek. Excellent spring spawning area for bass and crappie.

35. Old Stemley Bridge Road, a sunken roadbed. Excellent jig fishing on roadbed sides for winter crappie and bass. During spring and fall, fish the top of roadbed with crankbait or worm when there is current.

36. Hump known as "stripe heaven." Stripe and hybrid often school here by the dozens during the summer. Topwater baits or swim baits, Rat-L-Traps, etc. Lower edge of hump is also surrounded by fallen tree tops and is good bass area year round.

37. Shallow spring fishing area. Bass school here often chasing shad.

38. Excellent summer stripe fishing area. Shallow sandbar. Swimbaits or topwater baits when the fish are chasing shad.

39. Red Barn area. Sunken roadbed not shown on any map crosses area. Good spawning area, too.

40. Rocky bank leading to deep water. Good trolling area in spring. Worms and jig-and-pigs in winter.

41. More rocky banks. Worms and jig and pigs in summer and winter.

42. Old U.S. 231 (Dixie Highway) roadbed running north-south in the Cropwell Creek area. Well-known hotspot. Fish roadbed dropoffs in summer.

43. Old Avondale Lake, now underwater. Fish lake with spinnerbaits during summer.

44. Old Harmon Lake Dam. Fish area with spinnerbaits during summer.

45. Flagpole hump. Good area for various species. Good night fishing for crappie and stripe. Lots of brush placed in area. Good early morning jig-and-pig fishing for bass.

46. Brush tops and fallen trees. Crappie and bass.

47. Hump loaded with brushtops. Fish for summer bass at night with spinnerbaits and jig-and-pigs.

48. Deep, rocky bank in river's bend. Night fishing for bass with spinnerbaits.

49. Old homes flooded when Logan Martin impounded. Rubble and foundations still stand. Good spring and early summer spinnerbait, worm or jig area.

50. Rocky point. Good spring point with crankbaits or worms. Good summer night point on spinnerbaits.

51. Rocky point. Same as above.

52. Road bed crossing Clear Creek. Excellent winter crappie fishing.

53. Clear Creek causeway. Summer night fishing with spinnerbaits and worms.

54. Rocky point. Worms and crankbaits in spring, worms, jigs and slow-crawled spinnerbaits in summer.

55. Deep pit area. Good summertime hole for bass, crappie.

56. Excellent spring spawning flats.

57. Underwater rubble from dam. Buzz baits and other topwater in spring.

58. Deepwater ledge. Jigs, worms in winter, crankbaits in summer.

59. Upper end of Rabbit Branch. Excellent spring spawning area.

60. Rabbit Branch causeway. Fish summertime bass with spinnerbaits at night.

61. Above water islands. Fish summertime bass with

spinnerbaits at night.

62. Causeway. Fish summertime bass with spinnerbaits at night.

63. River points. Good night spinnerbait fishing.

64. Rocky banks falling into deep water. Good trolling area with deep-diving crankbaits.

65. Burn's Slough. Excellent spring spawning area.

66. Rubble left by house that was flooded by impoundment.

67. Good spring spawn area.

68. Underwater pit. Good summer and winter bass spot, good summer stripe location.

69. Good points for summer night fishing with worms, jigs and spinnerbaits.

70. Riprap above the dam. March through April. Crankbaits or worms.

71. Logan Martin tailrace area. Excellent stripe and hybrid fishing.

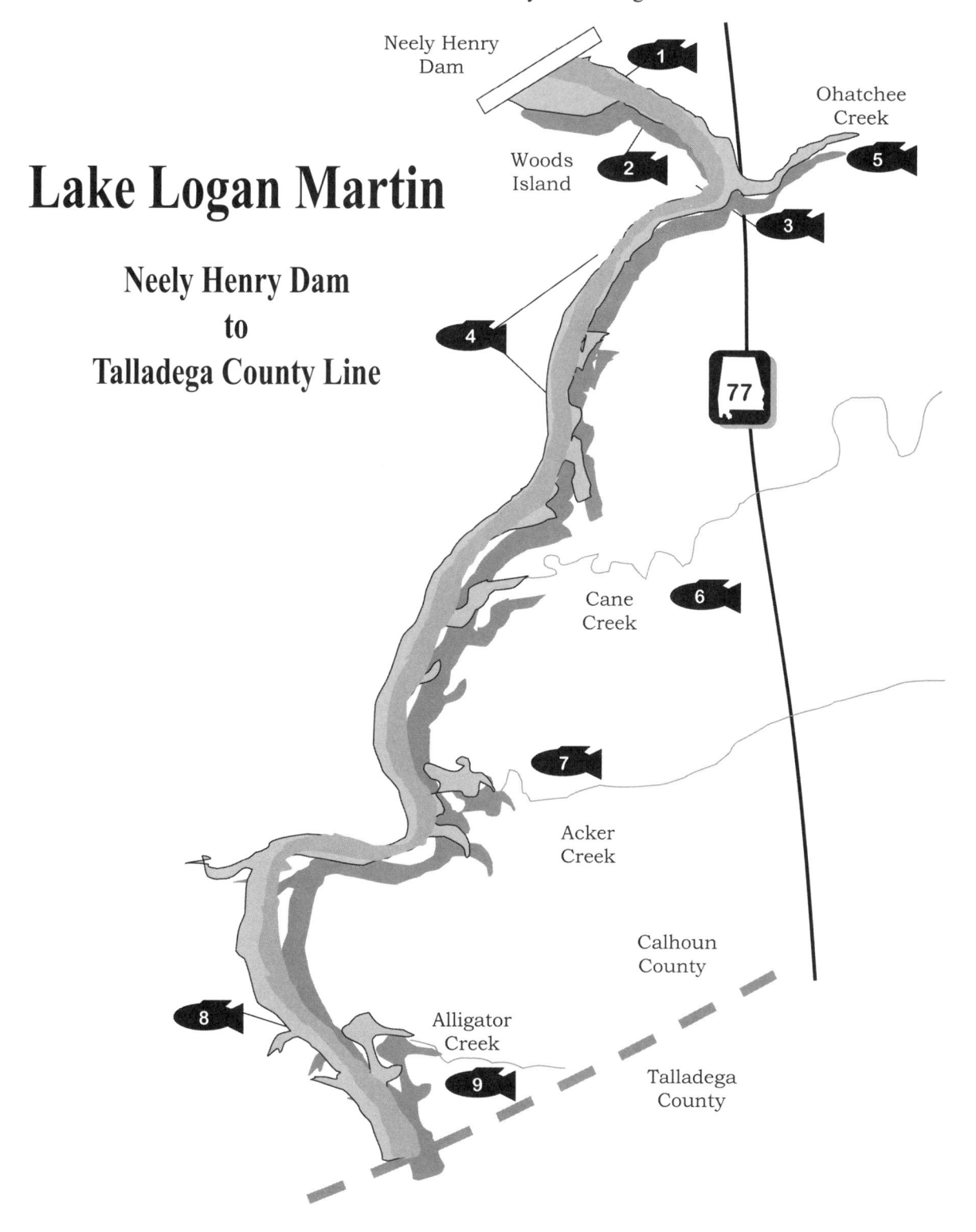

Lake Logan Martin

Neely Henry Dam to Talladega County Line

The Ultimate Guide to Alabama Fishing

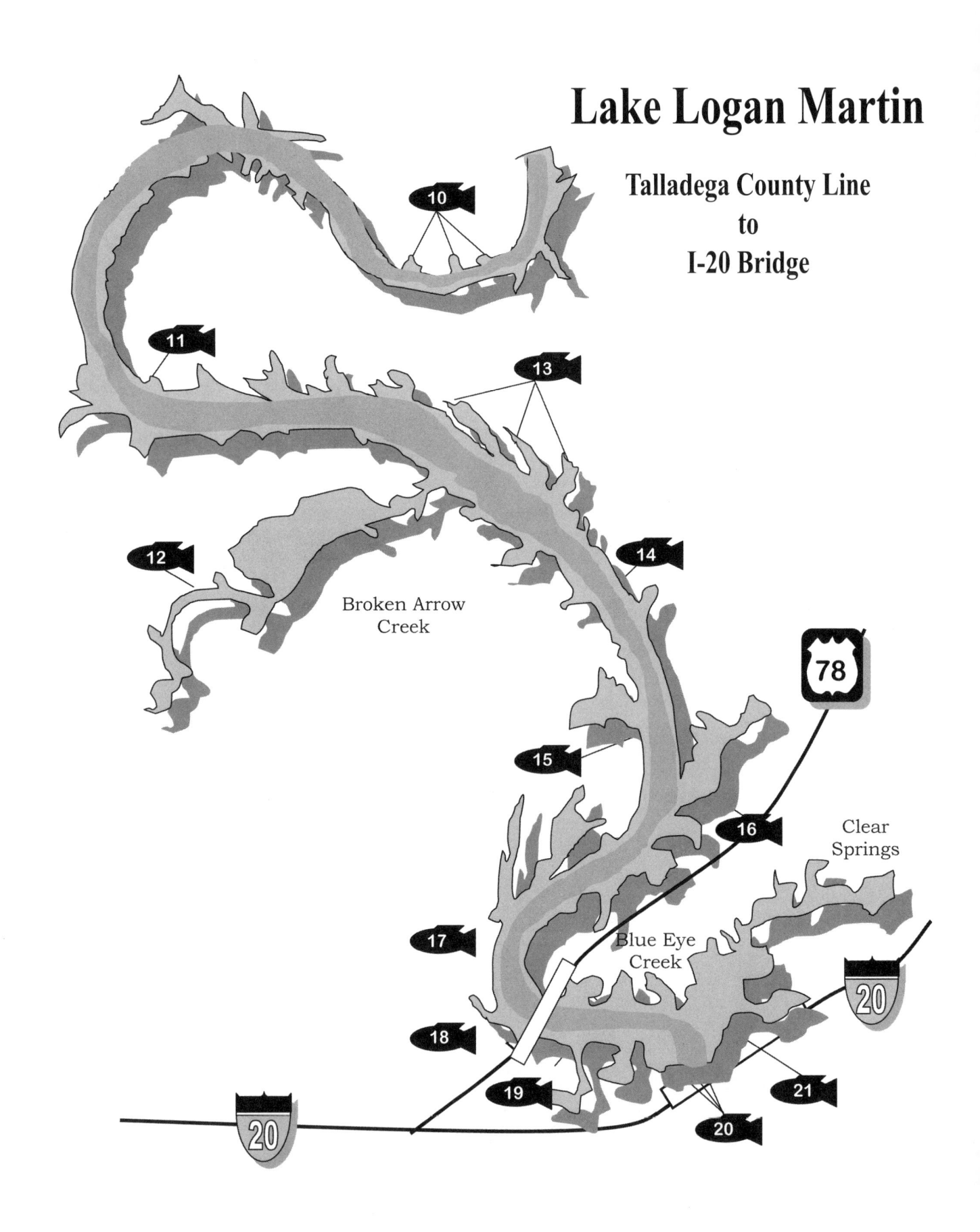

Lake Logan Martin

Talladega County Line to I-20 Bridge

Broken Arrow Creek

Clear Springs

Blue Eye Creek

96

Mike Bolton

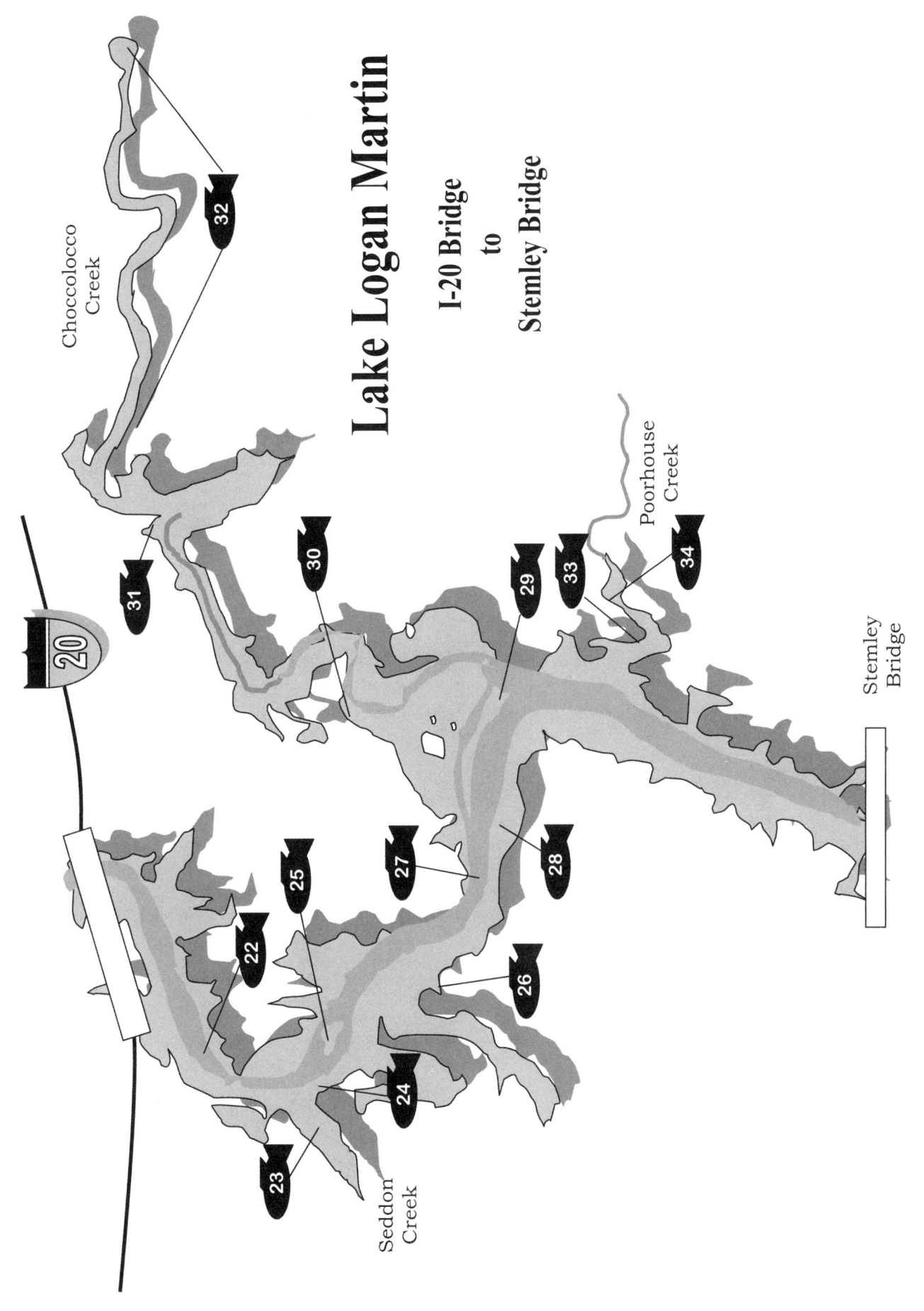

Lake Logan Martin

I-20 Bridge to Stemley Bridge

Choccolocco Creek

Poorhouse Creek

Seddon Creek

Stemley Bridge

97

The Ultimate Guide to Alabama Fishing

Lake Logan Martin

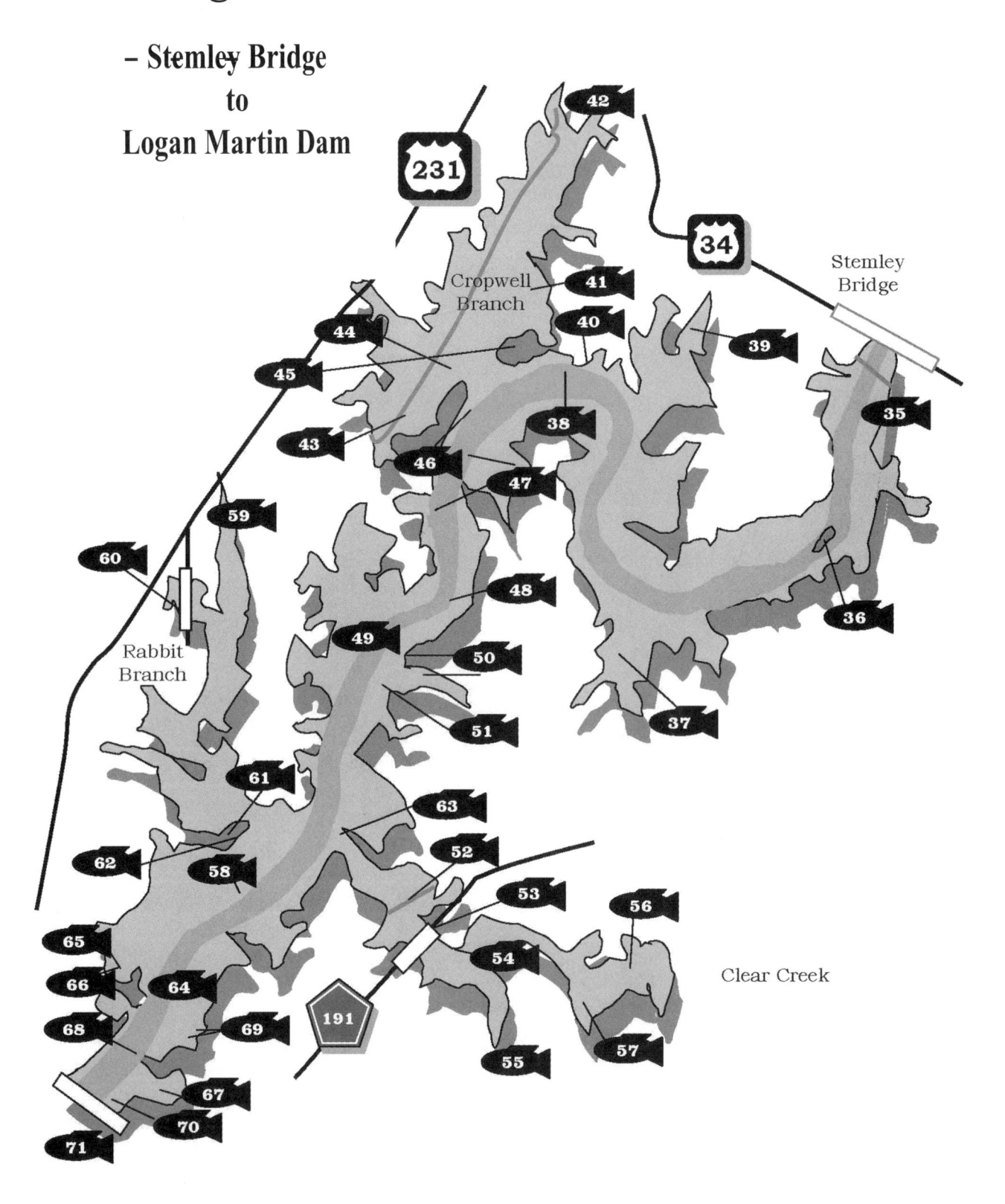

– Stemley Bridge to Logan Martin Dam

Lay Lake

As one of the oldest lakes in Alabama (first impounded in 1914), Lay Lake had become somewhat stagnated by the mid-60s, but a new dam was built in 1968. That flooded an additional 6,000 acres of farmland and timber. The water rose 14 feet and virtually created a new lake.

The lake immediately became a haven for bass and crappie and despite the fact it receives tremendous fishing pressure, that excellent fishing is carrying on until now. The lake is as fine a lake as you'll find so near a major metropolitan area.

The Logan Martin Dam tailrace at the head of Lay Lake is not only the state's top spot for hybrid bass fishing, it's an incredible bass fishery as well.

It was here in 2002 that pro angler Jay Yelas taught local fishermen a thing or two about tailrace fishing for bass. Yelas used a 3/8-ounce Berkley Power jig and fished a 200-yard stretch in the tailrace to win the Bassmaster Classic.

Catches of limits of hybrids in the 6-, 7-, 8-pound class almost year round are foregone conclusions for most of the better tailrace anglers.

The miles of grass beds that can be found in this lake is a plus, too, for they tend to hold bass shallow where they are accessible to more fishermen.

Spring bass fishing

Logan Martin Dam to U.S. 280

The upper end of Lay Lake offers few flat, sandy areas which bass can use to spawn, but that doesn't mean spring bass fishing can't be excellent there.

When there is current in the river, pre-spawn bass tend to gather along the river banks underneath the hundreds of willow trees along the banks, or just inside cuts or behind logjams, A lack of current is rarely a problem in the spring because the turbines and/or floodgates are often working overtime to accommodate spring flooding. An angler should position his boat in the swift current near either bank and use the trolling motor to keep his boat headed upriver.

With his boat drifting backward, a spinnerbait pitched underneath the overlapping willow trees will sometimes produce tremendous stringers of spotted bass.

Fishing this pattern can be tricky in that it takes a knack to properly place the spinnerbait upriver of a willow tree and then allow it to drift with the current under the tree all the while the boat is drifting backward at a different speed.

This pattern also can be tricky in that the angler may fish a quarter-mile stretch of bank and not get a bite and then get 10 strikes in the next 50 yards. The pre-spawn bass tend to school and a number of strikes may come in a short distance. When that happens, crank the big motor, run upstream and fish the productive area again and again.

Also remember that these bass are against the bank in the small eddies to escape current. Rarely will you find one further than five feet from the bank. Once your lure is five feet from the bank, crank it in and cast again.

U.S. 280 to Yellowleaf Creek

During the early spring, largemouth bass and spotted bass segregate on this lake to a certain degree. The spots tend to gather on main river points and humps, while largemouth move to the secondary points inside sloughs and creeks. The spots aren't nearly as grass-oriented as the largemouth. It's not unusual to find dozens of spots schooled up on these points in the pre-spawn stage.

Several baits can be productive at this time. The Carolina-rigged lizard has become this lake's No. 1 bait in recent years and for good reason. The slow bait has been catching tremendous numbers of fish. Jig-and-pigs, spinnerbaits and Texas-rigged worms also will work.

These fish will eventually move up into the shallow water of creeks, and they'll look for bottoms with sand or pea gravel.

Beeswax Creek, where Kevin VanDam won the 2010 Bassmaster Classic, is a prime spring spawning area, as are the sloughs northwest of Beeswax on both sides of the river. Look for the largemouth in the thick matted grass during this time. They'll bury themselves in weeds that seems impossible for them to be in. Sometimes it becomes necessary to flip a jig in a l-inch hole to get to them.

The normal spring baits—spinnerbaits, buzzbaits, worms, jig-and-pigs and twitch baits—can be productive in these shallow water areas.

Yellowleaf Creek to Lay Dam

Fishing in this area during the spring is basically the same as fishing the lake's middle section, the only difference being the lake's lower end lacks shallow ledges. Most spring fishing takes place on the points, humps and old roadbeds.

Spring Creek, Blue Springs, the back part of Waxahatchee Creek and Paint Creek are prime spring fishing spots. Paint Creek has about every type of structure imaginable for spring fishing.

Warm and hot weather bass fishing

Logan Martin Dam to U.S. 280

The upper end of Lay Lake is much like the upper end of all of the lakes in the Coosa River chain in that it is nothing more than a river. And like the upper ends of Weiss Lake, Lake Neely Henry, Lake Logan Martin, Lake Mitchell and Lake Jordan, this river-like upper end of Lay Lake can offer good summertime bass fishing at night.

One of the best-kept secrets among Lay Lake's best bass fishermen is the art of night fishing the upper end during summer months. When Logan Martin Dam ceases generating power in the afternoons, the shad that have gathered below the dam lazily float down the river to gather in the creek mouths, around the islands and in the hundreds of logjams directly below the dam.

As a result, night fishing on Lay Lake's upper end is often tremendous. Spinnerbaits, worms, jig-and-pigs and even topwater lures offer superb fishing when the sun goes down.

A spinnerbait tipped with a pork frog is probably the best all-round night time bait because it can be fished in so many different ways and because it covers so much territory so quickly.

It can be thrown into the mouths of the large creeks, allowed to settle to the bottom and slowly retrieved through the stump rows and standing timber. It also can be fished in and around the stump rows that encircle the islands on the upper end, dragged across logjams or skipped underneath the willow trees that overhang the banks.

The worm or jig-and-pig can be fished in the same locations with much the same success, but neither are able to cover the territory as quickly as a spinnerbait and, as a result, are less productive.

Another exciting alternative is fishing top water at night. Fishing the same areas with medium to slow topwater baits such as Jitterbugs, Hula Poppers and Nip-I-Diddees will sometimes produce when spinnerbaits, worms or jig-and-pigs won't.

During the day, swim baits and Rat-L-Traps are unquestionably the kings of lures on Lay Lake's upper end. Anchored 30 yards from shore on either bank when the water is turned on, the angler can take a light spinning outfit and a pearl-colored, 3-inch swim bait and often catch bass by the dozen.

An angler should cast the artificial shad upriver and allow it to drift downriver in the current. A bass running the banks in search of shad find this bait too tempting to pass up. The only problem with fishing this bait—that is, if you consider it a problem—is the number of hybrid bass that will be caught in addition to largemouth bass.

Another lure that works well when the current is running during the daytime is a plunker-type topwater lure with a trailing horsehair or bucktail jig. This lure combination, thrown into the eddies around the islands and logjams, often produces large numbers of smaller bass.

Unfortunately for the average angler who works for a living, Alabama Power Company rarely generates power on weekends during summer months. The weekend angler must opt to fish the upper-end ledges with swim baits, grubs, worms or medium-running pearl-colored crankbaits when no power is being generated. This type of fishing is average at best.

Although the upper end of Lay Lake offers several excellent cool-water creeks such as Kelley Creek, Beaunit Creek and Tallasseehatchee Creek, these creeks surprisingly do not offer good summer bass fishing. Keeper-size bass are rarely found further than a quarter-mile from their mouths during hotter months.

U.S. 280 to Yellowleaf Creek

Once the bass angler finds himself south of the U.S. 280 bridge, he'll find three distinct changes in Lay Lake that will affect fishing. One of the most notable changes is that the river widens and spills the banks of the original river channel, thus creating shallow ledges next to deep water.

Another noticeable difference is the number of large, wide creeks which pump fresh water into the lake.

The third, and maybe most important difference, is the grass beds. No matter what type of structure a bass angler likes to fish, the middle section of Lay Lake offers it. There are ledges, humps, stump-row-laden creek mouths, roadbeds and railroad beds, islands, cool-water creeks, collapsed bridges, riprap, weedbeds, piers and docks and a host of other fish-holding structure.

Just as the case on the five other reservoirs in the Coosa River chain, a summertime fisherman who wants to be successful on Lay Lake during the dog days of summer better have an understanding of ledge fishing.

Beginning where the river bends just south of Wilsonville and directly across from Heaslett's View, the good bass fishing ledges begin. Bass gather here by the dozens during hot summer months and these schooling bass are a fool for worms and crankbaits. They also will attack buzzbaits and topwater plugs during the early-morning and evening hours.

Another productive ledge area is directly across from Sally Branch. This ledge is a strange one in that it sits at least 85 yards from the bank. A bass fisherman should use his depthfinder to find the ledge. He should then place his boat in 17 or 18 feet of water and cast into five feet of water with a worm. Such an incredible amount of structure rests on the bottom here that other lures will constantly stay hung up.

Another good ledge area is directly below that on the same side approximately a quarter mile to the south.

Ledge fishermen will find good summertime fishing in the mouth of almost every creek or branch entering the middle section. Bully Creek, where George Cochran won the Bassmaster Classic in 1996, Dry Branch, Pope Branch, Beeswax Creek, Mud Branch, Flat Branch, Kelly Branch and Cedar Creek all boast humps and ledges where summer bass school.

The hot-weather bass fisherman in Lay Lake's middle section wouldn't be complete, however, without a good understanding of fishing weedbeds.

You can generally catch bass in these weeds year around, although the better fish move out by mid-summer. Spinnerbaits and buzzbaits early in the morning will take fish, as will flipping worms and jigs.

Another hot-weather pattern that works is fishing around the large number of boat houses and piers in Bully Creek. Almost all of these piers and boathouses have Christmas trees and/or brush tops around them. Bass tend to hold in these brush piles on the shady side of the structures.

Yellowleaf Creek to Lay Dam

Current, if you can find it, is a key to river fishing on this end of the lake during the summer. Bass that are sluggish during the hot part of the summer spring to life when the water starts moving. Fish suspended on points, or even bass buried in the weed beds, can be affected by current. This current starts shad to moving and this is apparently what triggers the bass' feeding instincts.

Points, ledges and dropoffs hold the majority of these fish and crankbaits, worms or jigging spoons are needed to catch them. A good way to locate the better areas is to look at the bank and assume the points will come off into the water at the same angle. This gives a mental picture of what's under the water. If the bank has a steep angle, the point probably goes deep into the water. If the bank has a sharp curve, the original river channel probably does too. Look for these points and ledges that have some type of structure on them and you'll find summer bass.

Cool and cold weather bass fishing

Logan Martin Dam to U.S. 280

Cool weather fishing in the upper end is much like spring fishing there in that bass tend to school along the banks underneath the overhanging willow trees when current is being generated.

Spinnerbaits can be used during the cooler months, but a crankbait such as a Fat Rap or Shad Rap will work equally well when the water has cooled.

Flipping also works well at times in the upper end during winter months. Some bass will move into Kelley Creek, Beaunit Creek or Tallaseehatchee Creek for their warmer spring-fed waters and a jig-and-pig flipped along stump rows, trash or logjams will produce an occasional fish.

Flipping also produces a few winter fish from the stump rows around Ratcliffe's Island, Buzzard Island and Prince's Island.

U.S. 280 to Yellowleaf Creek

In terms of numbers of fish and big fish, fall fishing on this section of the lake is as good as it gets. Granted, a lot of big fish are caught from this lake in spring, but spring bass have loving on their minds. Fall bass have feeding on their minds and are much easier to catch.

By the end of October, the feeding frenzy is usually well underway. These fall bass school heavily in the creeks and it's not unusual to find 50 or more bass schooled together. To make things even better, these bass can usually be caught in 10 feet of water or less. Peckerwood Creek, Cedar Creek, Spring Creek, Beeswax Creek, Bully Creek, Dry Branch and Kahatchie Creek are good fall fishing creeks, with the first three being the best.

The fish school up on the ledges and points and the key to finding them is locating shad. This is normally easy as huge schools of shad can be seen roaming in the flats and creeks, their unmistakable small ripples giving them away. These shad don't always indicate bass are under them, but you can usually bet they are there. Largemouths rarely bust the top of the water when chasing shad like spotted bass do.

Once the schools of shad are located, the first bait to try is a fast-running lipless crankbait like a Rat-L-Trap. Shad-colored lures are good, but if the water has a little stain, a chartreuse bait might be needed.

Cast well past the school of shad and rip the lure through them. Once the temperature dips below 68 degrees, the metabolism of the fish also drops. The fish that were so active a month before will be back out on the main river channel on points, humps and roadbeds. That's when the jig-and-pig becomes the top bait again.

Yellowleaf Creek to Lay Dam

The only difference in fall fishing in this area and the area above it is that it often becomes necessary to change lure colors due to water clarity. The clearer water in this area makes shad-colored and mirror-colored baits the best bet.

When using a worm, go with a pumpkinseed or June bug color. Almost every creek on the lower end offers good fall fishing, with Blue Springs, Spring Creek and Waxahatchee Creek being best.

Other types of fishing

Crappie fishing

Although Lay Lake isn't that well known as a crappie lake, studies show it boasts a crappie population as healthy as any Coosa River chain lake other than Weiss.

Spring is the best time for crappie fishing. The downed trees in the back of Kelley Creek offer spawning cover, as does the old bridge in Cedar Creek. Cedar Creek and Blue Springs Creek get some of the earliest warm water of the year, thus crappie spawn in these creeks earlier.

Look for crappie around the standing timber along the banks. Curlytail 2-inch chartreuse grubs are best, with a 1/32-ounce or 1/8-ounce lead head. Use the 1/8-ounce to fish deep or when it's windy, and the 1/32 to fish shallow or on calm days. Plan on using 6-pound-test line with spinning tackle for best results. Cast to the structure, allow the jig to sink to the desired depth, then begin the retrieve. If the jig hangs up, shorten the count before the retrieve. If it doesn't, go deeper. Strive to pull the jig just above the structure.

Crappie go deep in Lay Lake during the summer months. Night fishing is best. Plan to use lanterns and minnows around old bridges and railroad trestles in 15 to 20 feet of water. The old trestle in Sulphur branch, the old trestle in Peckerwood Creek and the second bridge in Cedar Creek are among the best areas holding summer crappie.

Fish can be found during the day in deep brushpiles. Always remember in summer months that crappie tend to seek shade and stay close to areas that hold shad. Also remember that crappie are always in or near deep water and always near structure.

January to March is the best crappie time on Lay Lake if you're looking for big crappie. Fish the same way as in summer, only deeper and slower. Cedar Creek and Beeswax Creek are always the best winter crappie areas. Fishing the main river channel from Beeswax Creek to Sulphur Branch can be productive, too.

Stripe and hybrid fishing

If catching big fish is your dream, you probably have no better opportunity in Alabama than fishing for stripe and hybrid in Lay Lake in the Logan Martin Dam tailrace. It is not uncommon for anglers to catch 25 fish a day with each fish topping five pounds.

Unfortunately, fishing the Logan Martin Dam tailrace is among the most dangerous fishing in Alabama. Large boulders dot the area directly below Logan Martin Dam and, along with the incredibly swift current, an angler can find himself facing trouble in a split second. Anglers considering fishing the area should first contact an angler experienced in tailrace fishing this area.

Plastic artificial shad imitations are the best baits for this area. A popular pattern is to anchor in the quiet water to the right of the powerhouse (facing upriver) and let a Sassy Shad equipped with a lead head wash downstream on a tight line.

Another method that works well is to anchor directly in the swift water and cast the same lures behind the large visible boulders. Stripe and hybrid hide there out of the current while waiting on a shad crippled by the dam's turbines to wash by.

Fishing the tailrace is normally good year round, but spring is best. A good rule of thumb is that when the dogwoods bloom, bring a big cooler.

Although it's possible to catch an occasional stripe in the tailrace, (it's almost impossible to land a 20-pound-plus fish in the swift current, however) stripe fishing further down Lay Lake is best. In late spring, these big stripes move from below the dam into the mouths of creeks. Kelley Creek, Talladega Creek and Beaunit Creeks are among the best. Live shad fished 15 to 20 feet deep offers the angler the best chance to hang into one of the big ones.

Once hot weather arrives, the stripe move into the creeks seeking cooler water. Look for deep holes in the creeks. Stripe will gather there. Live shad floated above them with a cork produces the best fishing.

Once fall arrives, the stripes begin their migration out of the creeks into the main river channel and back up to the dam to feed for winter.

Live shad fished in the creek mouths will once again catch them. Fall is also the time hybrids begin surfacing to attack shad on topwater. Almost any white or shad-colored topwater bait that resembles an injured minnow, or a crankbait such as a Rat-L-Trap, will produce hybrids at this time.

The key to this type of fishing is watching for seagulls diving to feed on the shad schools. Hybrids will almost always be in the area. The mouths of Peckerwood and Spring creeks are especially good for schooling hybrids.

Hybrids can also be caught on topwater during the summer.

Approximately 1/2 mile below Logan Martin Dam, the hybrids will attack topwater lures on the west bank of the river when power is first generated. The hybrid race upstream at the first inkling of current and think topwater lures are dead shad floating downstream. This type of fishing is popular with those afraid to get into the immediate tailrace area.

Lay Lake's best fishing spots

1. Boulder-strewn tailrace. The state's best fishing for big hybrids. When turbines are running, Sassy Shads thrown into swift current produce consistent stringers of hybrids 5 pounds or more. Summer and fall good; spring excellent.

2. Alabama Power Company fishing platform. Enables anglers to fish tailrace without a boat. Good for hybrids, striped bass, catfish, drum and carp year round.

3. Steep muddy bank on west bank of river, west of first tiny island below the dam. You can often see shad trying to jump out of water here. Big stripe move in here during colder months. At least one 22 pounder has come from this spot.

4. For those who don't like to fish the swift tailrace, the water is more stable here. Hybrids swim along these banks to and from the dam when water is cut on or cut off. Topwater baits produce explosive action. Late spring, summer best. West bank always better than east bank.

5. Mouth of Kelly Creek. Good drop going into the mouth. Fish of all species congregate here. Excellent spotted bass, hybrid and crappie fishing.

6. Deep hole in Kelly Creek. Appoximately 200 yards below big rock. Watch depthfinder. Live shad fished in deep hole produce big stripe, hybrids and spotted bass which gather in cool water. Creek temperature is at least

six degrees cooler than main lake in July and August.

7. Big rock in Kelly Creek. Deep water between rock and right bank going in. Deep, cool pool holds big stripe, hybrids and spots in summer. Live shad fished below a cork is best.

8. Upper end of Ratcliffe's Island. Good stump rows and logjams. Upper end is best for flipping when current is off; spinnerbaits when current is on.

9. Overhanging willow trees along Ratcliffe's Island's east bank. Good spinnerbait and buzzbait fishing in spring and fall.

10. Underwater rock pile 15 yards below railroad trestle. Thirty feet off of west bank. Spotted bass holder. Spinnerbaits.

11. Jetties and cuts on both sides.

12. Buzzard Island. Upper end good; lower end fair.

13. Mouth of Beaunit Creek. Temperature is 10 degrees cooler in July and August. Worms or Sassy Shad in mouth during summer months.

14. Sharp bend in Beaunit Creek. Twelve-foot hole 20 feet to right directly off point. Stripe and hybrid lay here awaiting shad traveling the creek.

15. Narrows in Beaunit Creek. Only five feet wide. When creek's flow reverses (current flows into creek) as a result of Logan Martin Dam generating power, hybrid go on a feeding binge here. Live shad is best.

16. Fork of Beaunit Creek (Railroad Lake) and Little Blue Creek. Excellent summer schooling area for spotted bass and hybrid. Worms for spots, live shad for hybrids.

17. Excellent stump rows, rock piles and ledge 22 yards from east bank adjacent to old powerhouse. Good spot schooling area. Worms in warmer months; jig-and-pig in cooler months. Live shad bumped on bottom will also produce hybrids when current is moving.

18. Mouth of Tallasseehatchee Creek. Real good with Sassy Shad or worm during late spring and summer.

19. Tallasseehatchee Creek. Good logjams and stump rows the whole length. Buzzbaits and spinnerbaits.

20. Prince's Island. Surrounded with stumps. Excellent flipping area when current is moving. Good year round.

21. Small, unnamed island. Good flipping area year round when current is moving.

22. River cuts. Mouths are good in summer with Sassy Shad or worm. Back of cuts early and late with buzzbaits or spinnerbaits.

23. Kahatchie Creek. Spring spawning area. Shallow. Weedbeds in late spring and early summer. Spinnerbaits and buzzbaits.

24. Yellowleaf Creek. Mouth of creek good year round with worm, jig-and-pig. Weedbeds in back of creek good early and late with spinnerbaits and buzzbaits in late spring.

25. Wilsonville Steam Plant discharge.

26. Standing timber around Bullock's Islands adjacent to Three Island Shoals. Good largemouth fishing with spinnerbaits or worms.

27. Mouth of Hay Spring Branch. Occasionally holds hybrids. Some spring bass spawning in back area.

28. Cuts in river bend. Shad use this as an escape area when current is on. Use worm or crankbait in the mouths for bass that follow shad.

29. Beginning of excellent ledge fishing area. Bass school here in summer to chase shad up on ledge. Water is 8, 10 or 12 feet on ledge and falls off to 65 feet or more. Lots of riprap on west bank, too. Holds some spotted bass. Worms, crankbaits and jig-and-pig all work here.

30. End of ledge fishing area.

31. Mouth of Bulley Creek. Worms, crankbaits.

32. Back of Bulley Creek. Weedbeds back in creek in spring. Good early morning fishing with spinnerbaits, buzzbaits or floating worm.

33. Mouth of Pope Branch. Good hump in mouth with brush. Worms, jig-and-pigs.

34. Mouth of Beeswax Creek. Sassy Shad, smoked grubs.

35. Bridge in Beeswax Creek. Good riprap. Crankbaits, worms almost year round.

36. Creek channel in back of Beeswax Creek. Stump rows. Good flipping.

37. River ledge 100 yards out from mouth of Sally Branch. Drops from 6 feet to 18 feet. Covered in stumps and rocks. Deep-diving crankbaits good, but you'll get hung up easily. Worms next best.

38. Hump in mouth of Mud Branch. Kimberly-Clarke Penthouse slough. Recognize by large house with golf course. Good hump 6 feet on top that falls to 30 feet on one side, 20 on the other. Covered in brush piles. Excellent jig-and-pig fishing.

39. Good ledge area. Worms and Sassy Shad year round. Big bass producer.

40. Road bed. Old Perkins Ferry Road. Worm, crankbait or smoked grub.

41. Hump in mouth of Flat Branch. Worm, year round. Sassy Shad or Smoked Grub spring and summer.

42. Dual points leading from Flat Branch out into river channel. Hump in mouth with a few stumps. Good summer schooling area for bass. Rat-L-Traps work well when fish are hitting shad on topwater. Sassy Shad; worms other times.

43. Back of Kelley Branch. Left bank has downed

trees that hold good largemouth in spring.

44. Deep hole. Spots, hybrids, stripe and crappie can be found here from time to time. Site of old Fort Williams Ferry.

45. Mouth of Cedar Creek. Where creek channel meets river channel is good summer schooling area for both bass and hybrids. Worm, Sassy Shad or smoked grub for bass, top water plunkers for hybrid.

46. Island in Cedar Creek. Left of main creek channel. Underwater point southwest of visible point good with Sassy Shad, smoked grub and occasionally a crankbait.

47. Collapsed underwater bridge. Must find with depthfinder. Pilings, fallen timbers still remain. Big bass area in hot summer months with jig-and-pig.

48. Riprap in Cedar Creek under bridge near Cedar Creek Marina. WFF shocked up more big bass here in tests than in any area of Lay Lake. Large crawfish population likely responsible.

49. Island in back of Cedar Creek surrounded by weedbeds. Floating worms work well.

50. More weedbeds, but a dangerous area if you stray out of creek channel. Good flipping or with buzzbaits and spinnerbaits.

51. Standing timber begins here. Good flipping area.

52. Tiny rocky points inside tree line off of main channel. Good with Sassy Shad, worm.

53. $600 point. Named that because that's where Birmingham's Nolan Shivers sat to win his first tournament ever. Downriver side of point extends into river. Excellent schooling area for bass year round.

54. Mouth of Sulfur Branch. Downriver point is where old railroad track hugged bank and turned into Sulfur Creek. Down-river point best.

55. Old bed of railroad only six feet under water. Holds bass year round. Worm, shallow-running crankbait, jig-and-pig, Rat-L-Trap.

56. Remains of old railroad trestle. Fifty yards in creek from point. Good weed beds. Some of Lay Lake's best bass have come from these beds. Flippin', worms, spinnerbaits, buzzbaits. Year round.

57. Same railroad bed mentioned above crosses mouths of small cuts.

58. Shelby Shores flats. Big, flat grassy area behind standing timberline. Early and late with buzzbaits often produces good fish. Low water over some humps. Danger area.

59. Mouth of Peckerwood Creek. Same railroad bed mentioned above crosses here. Remains of dynamited trestle are next to standing timber. Can be found with depthfinder. Good flipping, crankbaits and worms.

60. Peckerwood Creek. Treacherous running area. Standing timber near bend. Good with spinnerbaits and buzzbaits. Worms work well around piers and fallen trees.

61. Sunken railroad bed exits river adjacent Okomo Marina. Sign proclaiming "dangerous area" marks exit point. Good schooling point for bass year round. Sassy Shad, worms, Shad Raps and Rat-L-Traps work well.

62. Mouth of Spring Creek. Points on island offer good spotted bass fishing. Worms by far the best bait. Standing timber surrounds island. Buzzbaits and spinnerbaits work well.

63. Bridge area in back of Spring Creek. Riprap under bridge always good for a few small bass. Worm, jig-and-pig or Sassy Shad.

64. Small humps in flats in Spring Creek. Must be located with depthfinder. Humps adjacent to deepwater is good schooling area.

65. Bridge crossing Kates Branch. Good riprap. Crankbait, jig-and-pig or worm.

66. Back of Kate's Branch. Loaded with stumps 3 to 8 feet under water. With Polarized glasses, you can usually see them. These stumps often hold bass. Worm or spinnerbait in day. Buzzbait early.

67. Good points at entrance to Blue Springs Creek. Good for spotted bass.

68. Blue Springs Creek. Good spring spawning area. Lots of standing timber which bass school around. Worm or Rat-L-Trap.

69. Ledge next to entrance to LaCoosa Marina. Standing timber has broken up and collapsed and is hard to see. Good schooling area for bass. Worms best.

70. The narrows. Some of the lake's best spotted bass fishing. Steep rocky banks next to 125 feet of water. Flipping, worms and jig-and-pigs on ledge.

71. Mouth to Waxahatchee Creek. Good fishing on points of river ledge with worms, crankbaits, smoked grubs.

72. Points of pockets. Good all the way up creek with spinnerbaits, buzzbaits or flipping.

73. Paint Creek. Good cool creek in summer. Small island in mouth good with jig-and-pig. Clear water. When everything in state is muddy, this creek will be clear and holding fish. Deep water throughout its length. Worm is best along banks.

The Ultimate Guide to Alabama Fishing

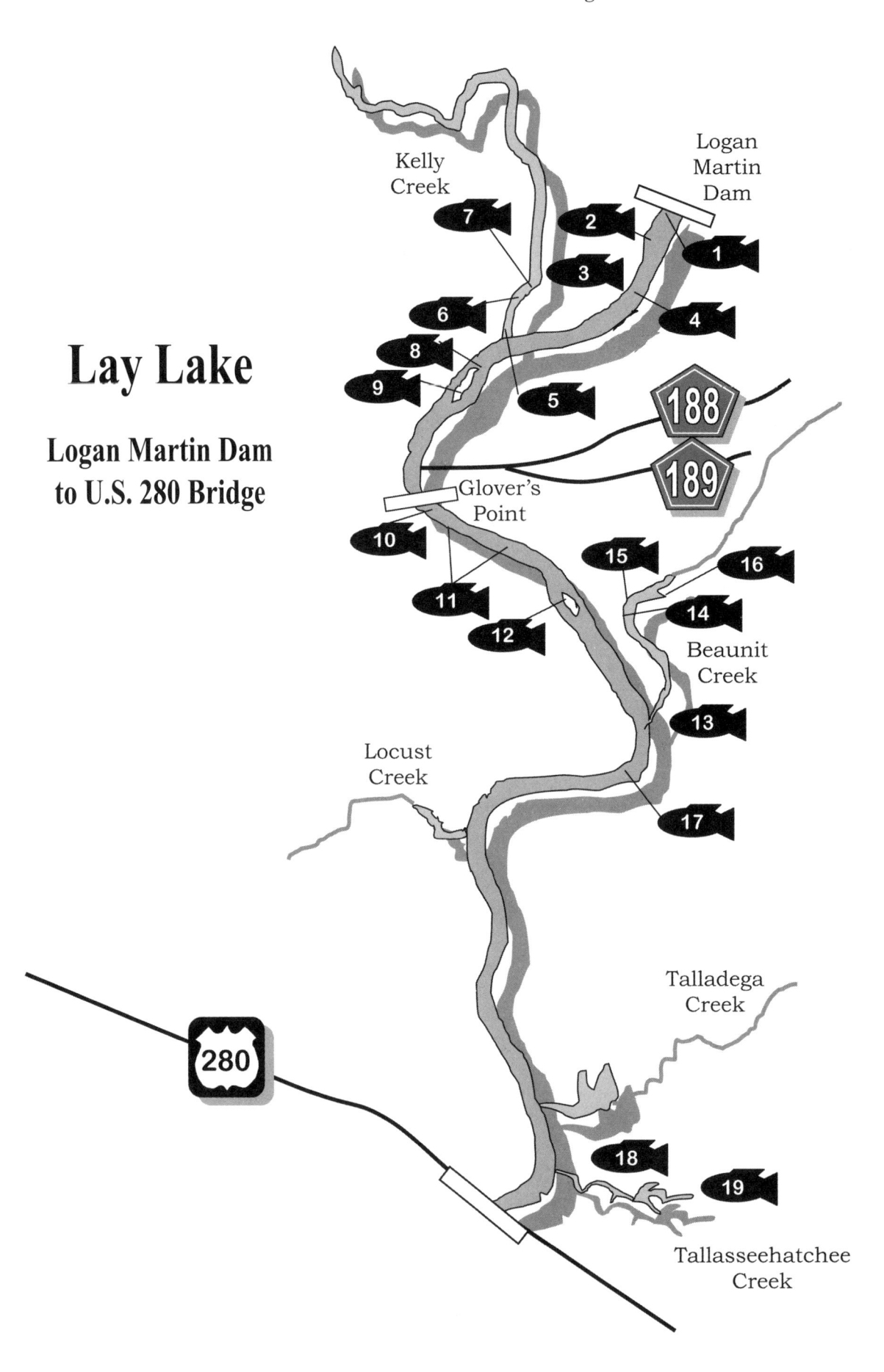

Lay Lake

Logan Martin Dam to U.S. 280 Bridge

Kelly Creek

Logan Martin Dam

Glover's Point

Locust Creek

Beaunit Creek

Talladega Creek

Tallasseehatchee Creek

Mike Bolton

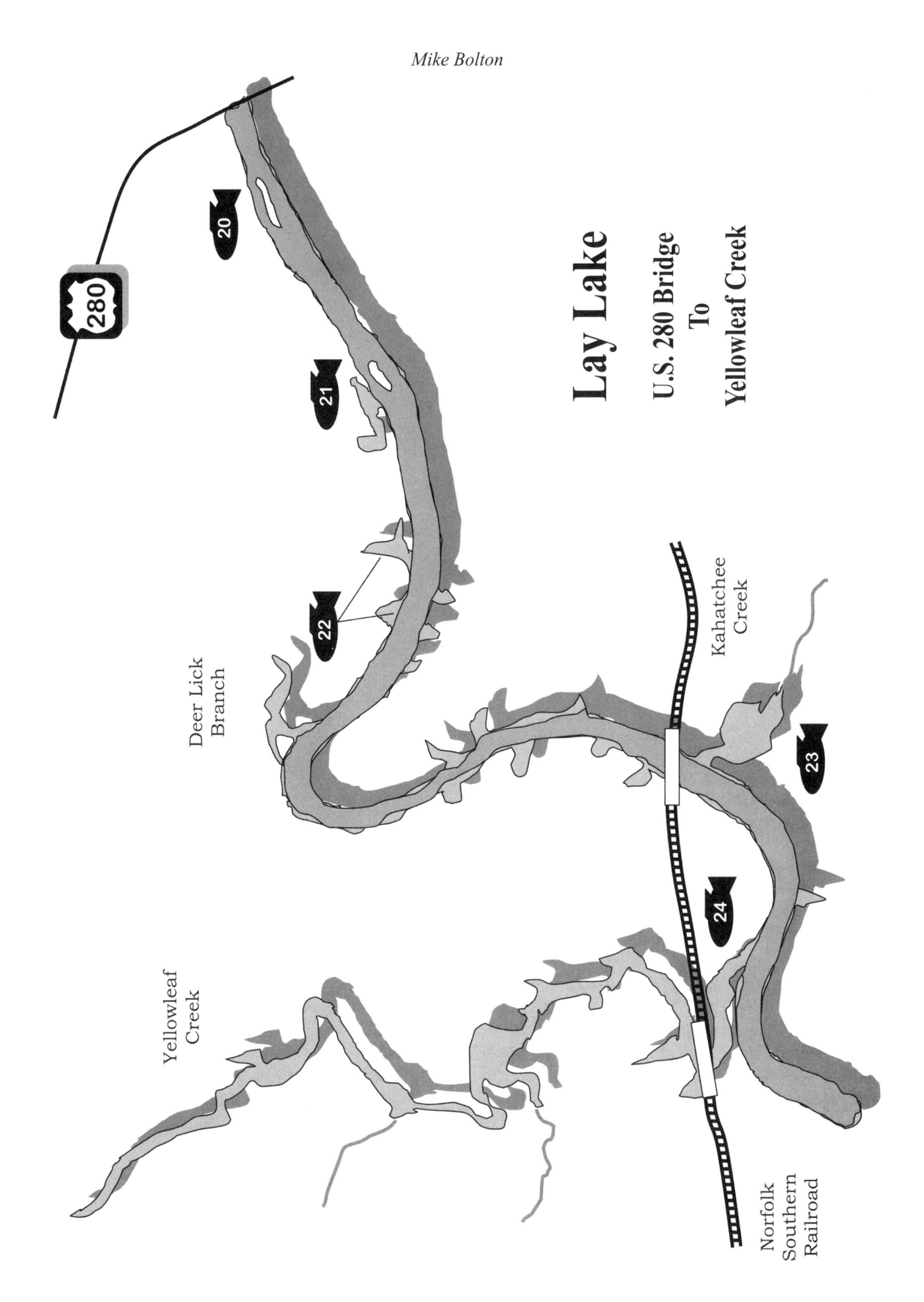

Lay Lake
U.S. 280 Bridge
To
Yellowleaf Creek

107

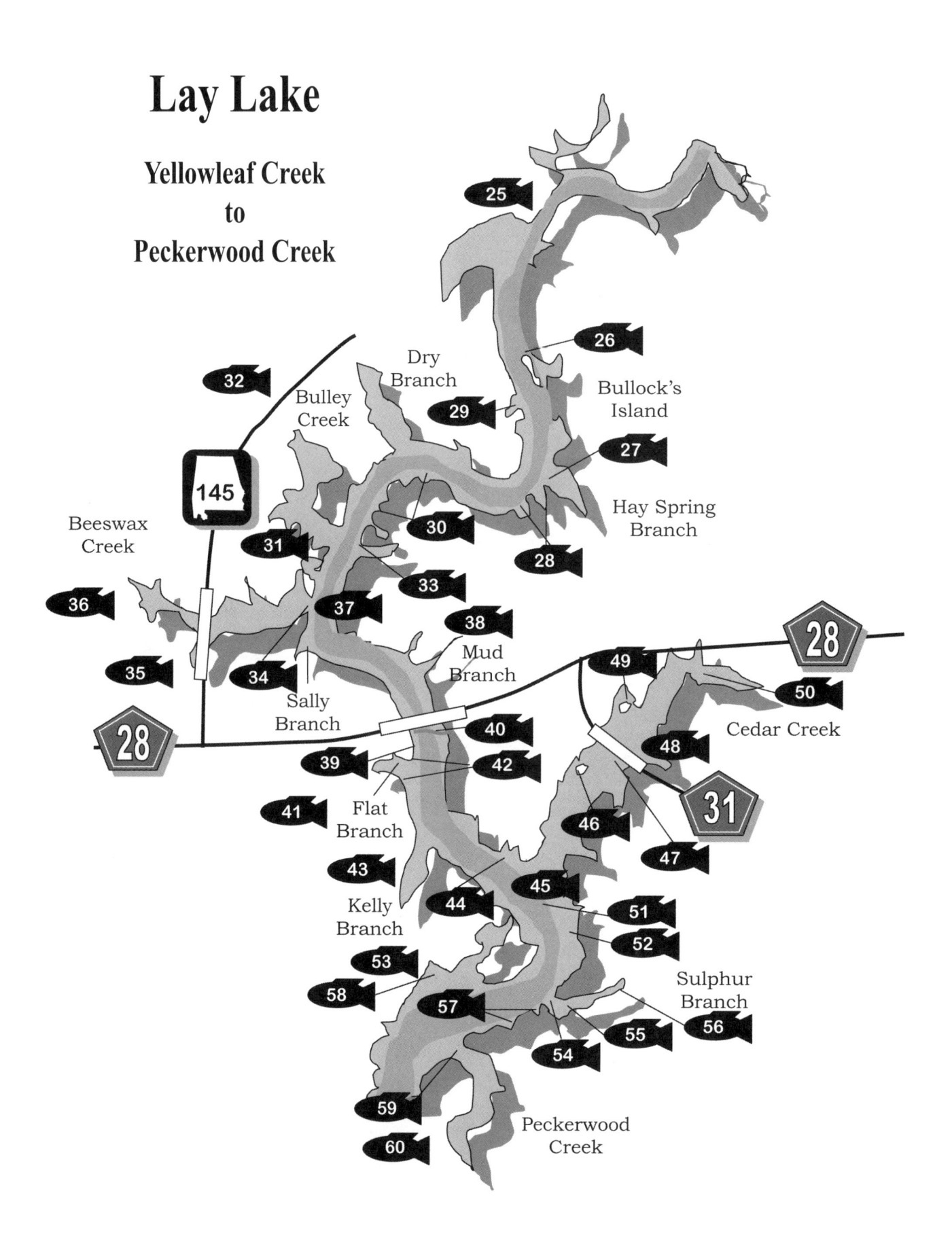

Lay Lake

Yellowleaf Creek to Peckerwood Creek

Mike Bolton

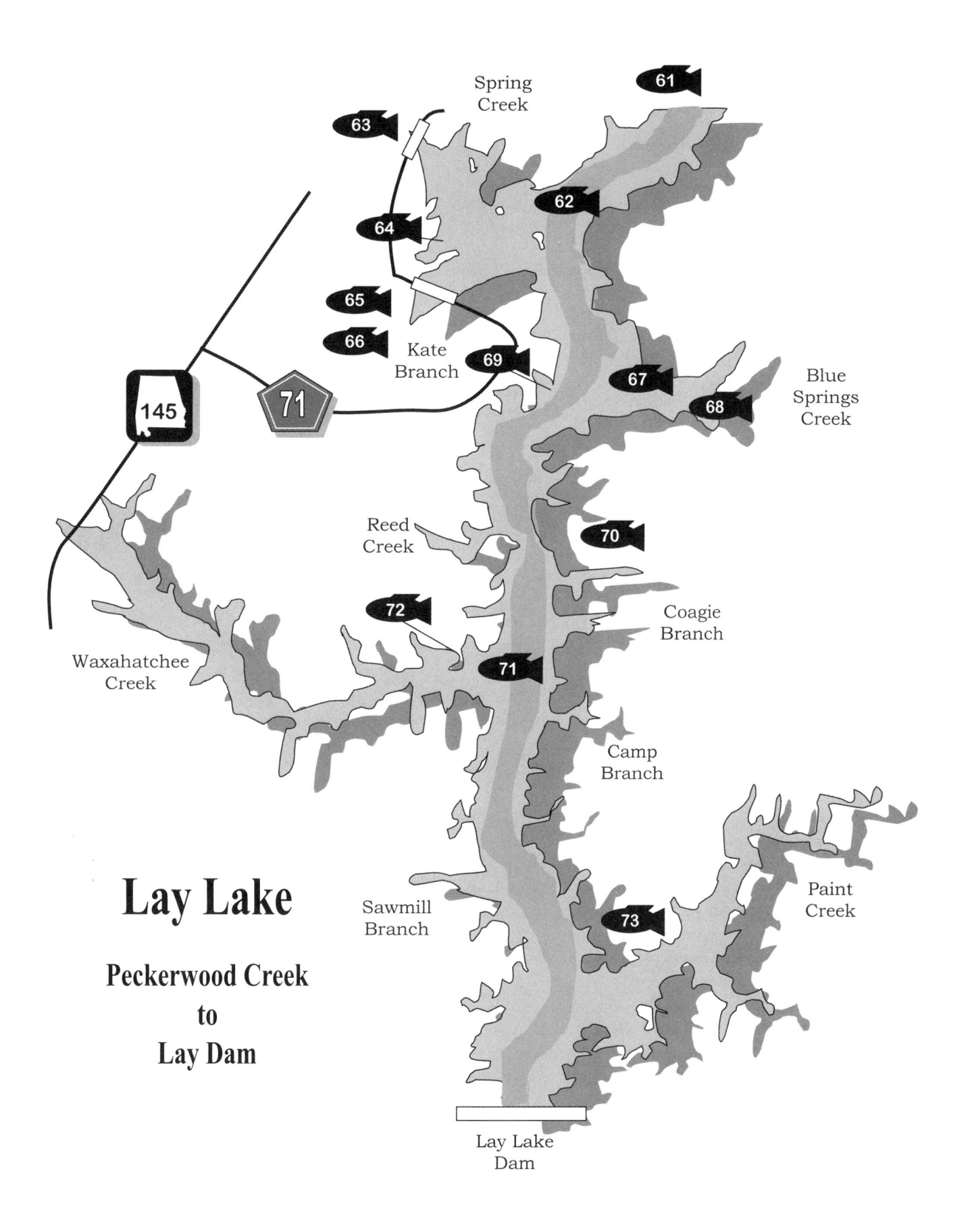

Lay Lake

Peckerwood Creek to Lay Dam

Lake Mitchell

Lake Mitchell has become a common stop for bass clubs all across central Alabama for good reasons. The launch facilities at Higgins Ferry are among the best in Alabama. The lake also provides diverse structure enabling a bass fisherman to fish the patterns he's best at duplicating. Simply put, bass fishermen love Lake Mitchell.

Because of its somewhat remote location it doesn't get the amount of fishing pressure seen by its upstream neighbors Lay Lake or Lake Logan Martin. It's too far away from Birmingham or Montgomery to have ever hosted a Bassmaster Classic or a FLW championship. It has never received the nationwide publicity like its sister lakes. Those who fish Lake Mitchell are perfectly fine with that

Impounded in 1921, Lake Mitchell is one of the oldest man-made reservoirs in Alabama but the older the fiddle the sweeter the music. It's a lake that is not very cyclic like some lakes and has provided quality bass fishing non-stop for almost a century now.

Lake Mitchell also offers a variety of other types of fishing including catfish, crappie and other species.

Spring bass fishing

Lay Dam to Cove Branch

The effect moving water has on the quality of bass fishing has no better example than the upper end of Lake Mitchell. Spring, summer, fall or winter, the fishing is always better on this end of the lake when water is flowing through Lay Dam.

The reasons for this are simple. Besides the obvious increase of oxygen in the water, swift water forces baitfish from the security of vast expanses of open water. Unable to withstand the current, these baitfish rush to the banks where pockets, sloughs, log jams and other structure offer a haven from the current.

The bass, as well as other species such as stripe, hybrid and crappie, find the hemmed up baitfish easy pickings. This sends these fish to the banks in search of food, and there, they become easily accessible to fishermen.

During spring months, something slightly different happens, however. Heavy rains that normally come during February, March and April force dams across the state to open their floodgates. When that happens, largemouths and spots are forced to seek refuge behind structure just the same as the baitfish.

For years, the opening of floodgates sent 90 percent of the bass anglers to the house for the week, but times have changed. The better bass fishermen now realize these conditions mean fishing suddenly becomes excellent.

The pattern for these fish is simple. The outboard motor is used to put the boat as close to the dam as deemed safe (sometimes no closer than 1/4 mile) and the outboard is turned off. The trolling motor is then used to keep the boat pointed into the current as the boat drifts backward down the river.

The trolling motor, besides keeping the boat straight, also slows the boat's backward movement enough that whenever a pocket, creek mouth, slough or eddy is found, the fishermen can cast into that area.

It should also be noted that because of the trolling motor operator's struggle to keep the boat straight and because of the back-seat fisherman's position, this is one of the rare times the fisherman in the back of the boat has the best opportunity to catch fish.

Several baits work well for this pattern. A white spinnerbait with a single spin No. 5 nickel blade is the best bait in dingy water, but a chartreuse spinnerbait with a gold blade is needed in muddy water. A 1/2-ounce or 5/8-ounce jig-and-pig tipped with a pork frog is close behind. A black-on-black jig-and-pig works best in dingy water while a brown-on-brown or brown-on-orange jig and pig works best when the water clears some.

Some anglers have had success with a crankbait but the number of crankbaits lost doesn't justify using them. If you use a crankbait, use a chartreuse bait when it's muddy and a shad-colored bait when the water is dingy.

The best time for this pattern is usually three days after the floodgates have been open. By that time, the water flow has sucked all of the muddy water out of the creeks and the water directly against the bank of the main river channel has also cleared some.

A straight-running bank with a few indentations or points tends to hold the most bass on this pattern. It should be noted that most strikes come within two feet of the bank because it is directly against the bank that the bass find the best refuge. Once the bait is six to eight feet off the bank it is time to reel in the bait and re-cast.

The number of spotted bass that can be caught using this pattern can be surprising.

A number of anglers often have a difficult time trying to keep up with their baits in the swift water and more than one has thrown up his hands in disgust. It takes patience to learn the pattern and practice is the best teacher.

Anglers who master the pattern and suddenly catch 10 or more quality fish in a 10-yard stretch find their efforts worthwhile, however. This pattern works all the way from below Lay Dam to Cove Creek. The back sides of Gilchrist and Ware Islands also hold fish on this pattern.

One other pattern will work when the floodgates are open. This is for those who don't like the idea of tackling the incredible swift water.

When the floodgates open, the first refuge many fish can find is Yellowleaf Creek and Cove Branch. These two incoming water sources act much like the log jams, points and cuts on the main river. Fish find calm water there.

These two locations—especially Yellow leaf Creek —are extremely muddy and virtually worthless as fishing spots the first three days after the floodgates are opened, but on the fourth day the mud is sucked out and bass by the dozens can be found stacked up in the mouth.

The pattern will also work in Cove Branch, but doesn't work in Clay Creek for some reason. For those who don't like getting on the lake when the flood gates are open, drifting the main river channel pattern will work when the floodgates are closed. Water must be moving through the dam, however. The diminished water flow that comes with the floodgates being closed also means diminished catches.

Cove Branch to Mitchell Dam

Unlike the river-like upper end of the lake, the lower end of Lake Mitchell offers not only wide expanses of water, but beautiful grass beds as well. Blue Creek, Walnut Creek and the public boat landing area at Higgins Ferry offer some of the prettiest and fishiest-looking grass you'll encounter on the Coosa River chain.

The grass beds hold the majority of spring bass and the angler should attack this end of the lake with that in mind. Several patterns will work in the grass this time of the year, the most notable being flipping and buzzing the grass with a swimming jig.

The swimming jig has become popular only in the last decade. Besides being weedless, this bait also rides the grass better than most other lures.

For those who like to flip, this lower end of the lake is prime real estate. The area is full of grass beds in 3 to 5 feet of water. These grass beds will be matted and laid over in early spring and punching through them to get to fish is a tough chore. The lake's top anglers have solved that problem by switching to a full 1-ounce jig.

A jig of this weight will punch a hole through the thick grass and fall quickly. Black-on-black or black-on-brown jigs work best.

The angler who likes to flip should concentrate his efforts in the grass beds upriver from Higgins Ferry Public Launch, and from Higgins Ferry Public Launch downriver to the mouth of Blue Creek. The grass beds on the same side of the river as the launch tend to hold the better fish.

Hatchet Creek

For all practical purposes, the Hatchet Creek area on Lake Mitchell is a lake in itself. The large creek carries nearly as much water as the remainder of the lake. Pennamotely, Weogufka and Hatchet creeks are a trio of major feeder creeks that supply water to the big body of water and a trio of creeks that bass fishermen need to be aware of.

When the lake's better bass fishermen want to win a tournament in the spring, this is where you'll find them. There's plenty of spotted bass in this area, but the larger, tournament-winning largemouth bass are abundant here, too.

The sheer numbers of big largemouth, coupled with the fact the weedy water is dingy and moving almost every day, makes this a bass fisherman's delight.

The sight of terribly muddy water in this area often chases anglers away in the spring, but the backs of the three feeder creeks usually are much clearer and are full of 3- to 4-pound largemouths.

The successful tournament angler will usually start the tournament by running to points in this area and catch a limit of smaller spotted bass that will measure. These smaller spots will take worms and swim baits readily. The tournament anglers then start looking for "kickers," or the larger largemouths in the backs of these creeks.

They do this by flipping a jig-and-pig or swimming a jig. Most go with a black/brown jig-and-pig A green frog pork chunk also makes a good trailer for some reason.

It should be noted that the above patterns are daytime patterns. The large influx of largemouth in the area don't seem to bite as well at night. The fishing is better out on the main river channel after dark.

Hot weather bass fishing

Lay Dam to Cove Branch

Unfortunately for the working man, the quality of summer fishing on the upper end of Lake Mitchell depends on water flow, and water flow is almost non-existent on weekends when most people are able to fish. Alabama Power Company normally moves water through its turbines on the peak user days of Monday through Friday, but the lack of need for power on Saturdays and Sundays leaves the turbines off on those days.

The best pattern to use during those weekdays in the hot part of the summer surprisingly is the same pattern you should use during winter months. With the current running, let your boat drift backward down the river while using the trolling motor to keep the front of the boat facing the current.

Instead of a spinnerbait or jig-and-pig that works there in the spring, a plastic worm is needed for summer fishing. A 6-inch blue worm is ideal for crystal clear water, while an electric red worm produces better in dingier water. Spinning tackle outfitted with 12-pound test line will complete the rig.

With the boat drifting backward at a slow pace, cast the worm directly against the bank and bring it slowly to the ledge and let it fall. Bass hold on these ledges and occasionally scoot up on the ledge to feed on shad, and they find the falling worm. Some anglers find similar success with swim baits or Rat-L-Traps white, silver or pearl in color.

Approximately 90 percent of the anglers who use the lake can only do so when the water is off, however. Locating fish during this time is a much more difficult chore.

With the water off, the main river channel varies from 9 to 15 feet in depth, but an occasional hole 20 to 25 feet deep can be found and that is where summer bass will be holding. Several drops and waterfalls caused these holes in the river beds many years ago and the current has kept these holes from silting in.

A depthfinder is a must in locating these holes, but the effort is well worth the results. It is not uncommon to drift along and see the bottom plunge from 10 to 25 feet and see bass stacked on top of each other on the depthfinder. Worms or Rat-L-Traps will take these fish.

When the water is off, the smart angler attacks this portion of Lake Mitchell at night. The bass will leave these hard-to-find deep holes and chase shad onto the ledges near the banks and these fish are more accessible to the average fisherman. Worms and single blade spinnerbaits cast against the banks and slowly retrieved off the ledge will produce fish.

Cove Branch to Mitchell Dam

Just like the upper end of the lake, current is a must to have any real chance of catching bass on the lower end of Lake Mitchell. Summer bass suspend themselves between 20 and 35 feet in 50 to 80 feet of water and become downright tough to catch. Even if these fish are located, getting the bait down to them and enticing them to strike is a shot in the dark. If the current is off in the hot summer months, plan to fish the lake above Cove Branch.

If you find one of those rare instances where current is moving during the summer, don't hesitate to take advantage of it, however. Light spinning tackle, light line and pearl swim baits or plastic worms fished off points and ledges adjacent to grass can offer good bass fishing.

Hatchet Creek

The lack of deep water needed by bass in summer makes the area that was so good in spring just an average place to fish for bass during the hot summer months. The quality spotted bass get scarce in this area in the summer and the 12-inch or better largemouth aren't much more common. The larger fish move out of the area onto the ledges of the main river channel where they can find cooler water and baitfish.

This isn't to say an angler can't pull into this area and catch a limit of bass during the hottest part of August, but it will be rare. Don't waste your time here if you're looking for a good day of fishing. The numbers of bass and quality bass come from the main river channel this time of the year.

Cool and cold weather bass fishing

Lay Dam to Cove Branch

Starting in November, the bass in the upper end of Lake Mitchell begin a feeding frenzy to store fat for the long winter. Fortunately, fishing for these bass doesn't require that the water be moving, although moving water does provide better fishing.

Since the bass are feeding almost entirely on shad, a shad imitation bait or a 2 1/2-inch grub with a 1/4-ounce jig head is ideal. An angler needs only to rig a spinning reel with 8-pound line and the bait of his choice. Eight-pound line is the perfect choice because it's thin enough not to scare fish, but strong enough to straighten a hook when the lure is hung up.

The angler needs only to go point-to-point or on the backside of islands to find bass feeding. The bait should be fished just as you would fish a worm. Cast it to shallow water, allow it to sink and hop it back to the boat.

A keen eye helps in locating where to fish. Any sign of shad activity or bass hitting shad on top of the water is a signal that fish are in the area.

The mouth of Mountain Branch is the top spot for this pattern this time of year. This pattern will hold most of the late fall and, depending on the temperature, on into December or early January.

Cove Branch to Mitchell Dam

Once the stifling heat of August turns into crisper mornings and eventually into evenings that require a sweater, Lake Mitchell bass begin leaving their deep, rapidly cooling summer home to move into shallow, more tolerable water.

Every angler knows cover is a must for bass and the lower end of Lake Mitchell offers the type shallow-water cover bass like most—grass. Besides cover, the grass offers protection from the sun, it produces oxygen (during its growing cycle) and is a natural magnet for baitfish that also are looking for cover. The bass can bury themselves deep in the grass and have every need met without moving more than a few yards.

Because of the amount of grass and the number of fish that can be found there in the fall, the lower end of Lake Mitchell is buzzbait heaven.

Before choosing your buzzbait, first check the water clarity. If the water is dingy as it normally is this time of year, use a chartreuse on white bait. If the water is clear, stick with a white bait. This patterns works early and late in normal conditions, and all day if there are overcast skies.

As the weather turns colder, the lake's better anglers realize that the shallow grass can't hold bass forever. Not only does the water begin to get uncomfortably cool for the bass, the dying grass is also robbing the water of oxygen. The bass move out to the points and down the rock bluffs in 12 to 20 feet of water and a jig-and-pig becomes the weapon of choice at that point. Go with a black jig and brown pork frog trailer.

The fish at this time are beginning to fatten up for winter. Not only are they more active at this point, they also tend to be larger fish.

Hatchet Creek

The fish that make Hatchet Creek their home in the summer months stay there during the fall and are joined by numbers of fish that come out of the main channel. Most of these fish seek out the weed beds and flats because it is there that the baitfish are located this time of year.

Buzz baits and spinnerbaits work well when the fish are feeding heavily in anticipation of winter. The good news is that some October and November days are dreary and overcast, meaning the fish stay shallow all day long.

The angler fishing this area in the fall should concentrate in Hatchet Creek because that not only has the deepest creek channel, but easy access to the main river channel as well. The upper reaches of other creeks in this area rarely produce numbers of fish or quality fish during the fall because there's no deep water nearby where the fish can escape during the sudden changes in pressure that always come this time of the year.

Other types of fishing

Hybrid fishing

Lake Mitchell was without hybrid bass until 1982 when the state stocked 65,616 there, but the stocking of more than 300,000 hybrids since 1982 has made Lake Mitchell a first-class hybrid fishery.

The hybrids on this lake behave much like they do on the other lakes on the Coosa Chain. Spring sees these fish make artificial spawning runs (they can't reproduce) upriver to the tailrace and into various creeks and long sloughs.

Tailrace fishing is excellent usually about the time the dogwoods bloom. These hybrids get into the swift water below Lay Dam to feed on the shad that are sucked through the turbines and shredded into bite-size pieces. These hybrid are easily caught by using either artificial or live bait. Live-bait fishermen anchor adjacent to boils and use bass rods outfitted with 20-pound test line to cast live willow grubs into the boils.

Artificial bait fishermen use the same fishing outfit with bucktail jigs or Sassy Shads cast into the boils.

This type of fishing can be exciting because the angler is just as likely to hang into a saltwater stripe, catfish, drum, carp or crappie.

For the angler who doesn't want to wrestle with the swift water in the tailrace, he can find the spring hybrids in various creeks on the lake. Cargile Creek, Blue Creek, the mouth of Hatchet Creek, Bee Branch in Hatchet Creek and Bird Creek are known for their influx of spring hybrid.

Anglers should locate baitfish in these areas to find hybrids. This can be done visually on days without wind, or with a depthfinder on windy days. These baitfish will gather on long points and humps. A Rat-L-Trap or a spoon thrown well past the school of baitfish and jigged back to the boat will produce hybrids, small saltwater stripes and white bass.

Hybrid fishing gets tough during the summer months, but a few anglers have unlocked the pattern. These fish head back to the main river channel to suspend themselves on the ledges during the summer, but they make periodic feeding runs to the top of the ledges. A jigging spoon works well on these fish.

The end of September usually signals the hybrids' return to the creeks. They return to the same areas they used in the spring and can be caught using the same lures. These fall hybrids tend to school more this time of the year and feed a little more heavily, however.

Lake Mitchell's best fishing spots

1. Exposed rocks below dam. Dangerous navigating. Dumpster-size rocks. Spinnerbaits produce super spots at night in summer when water is off.

2. Bridge at LaVada's Camp. Sassy Shads or grubs around pilings in fall produce numbers of spotted bass.

3. Yellowleaf Creek. Good escape area for fish when current is swift. Fishing is excellent for 1/4-mile in creek.

4. Northern tip of Airplane Island, or Ware Island. Long string of rocks run out upper end and lower end. Dangerous navigating. Good spinnerbait fishing.

5. Lower end of Ware Island. Good area when current is moving.

6. Down river point of Mountain Branch. Stump piles in 7 to 15 feet of water. Fish schooling area.

7. Cove Branch. Bass school up in back of creek in spring and fall chasing shad.

8. Clay Creek. Far back as possible. Waterfall area holds spotted bass in spring and fall.

9. Big Rock Island. Fast current makes back of island a fish haven. Spots gather in numbers.

10. Mouth of Clay Creek. Upriver side has steep, sharp rock bank. Fish are there in numbers or not at all.

11. Long grass bed 1/2-mile. Big spots on Lunker Lures in fall.

12. Excellent drifting bank when current is moving. From LaVada's Camp half mile downstream.

13. Excellent drifting bank when current is moving. From mouth of McSwain Bank to 1/4 mile downstream.

14. Great drifting bank when current is moving. Sheer rock bluffs.

15. Excellent drifting bank when current is moving in spring.

16. Excellent drifting bank when current is near Gilchrist Island.

17. Good sheer rock bluffs. Good spinnerbait fishing at night in spring, summer and fall.

18. Good grass beds above Walnut Creek.

19. Good pockets. Rat-L-Traps in spring and fall.

20. Good schooling area for hybrids and stripes.

21. to 28. Grass beds.

29. Good fishing point. Upriver point of Cargile Creek.

30. Good secondary point in Cargile Creek.

31. Good point in back of Cargile Creek.

32. Rocky Island in Cargile Creek.

33. Grass beds. Good spinnerbait area in summer.

34. Back of Cargile Creek. Excellent bass schooling in fall.

35. and 36. Grass beds.

37. Cuts. Excellent in spring, early summer and fall.

38. Grass beds on upriver point.

39. Back of Blue Creek. Good bluff fishing.

40. Grass beds.

41. Grass beds in "Y" Slough.

42. Grass beds on point between "Y" Slough and downriver pocket.

43. Grass beds on point.

44. Grass beds in Little "Y" Slough.

45. Grass beds in mouth of Hatchet Creek.

46. Grass beds in shallow water.

47. Good pockets and points.

48. Long cut. Good spring fishing.

49. Good pocket opposite mouth of Pennamotley Creek.

50. Good cut in Penamotley Creek.

51. Right bank of Pennamotley. Full of cover.

52. Left Bank of Pennamotley. Same as 51.

53. Back of Penna motley. Excellent schooling area in fall.

54. Last 300 yards of Bee Branch. Excellent cold weather and early spring fishing with worms and jig-and-pigs.

55. and 56. Weed beds.

57. Good point.

58. Good creek point in Weogufka Creek. Back of creek tremendous spring and early summer area.

59. Early spring area for flipping or small spinnerbait.

60. Afixico Creek. Same as 59. March to May. Rough navigating.

The Ultimate Guide to Alabama Fishing

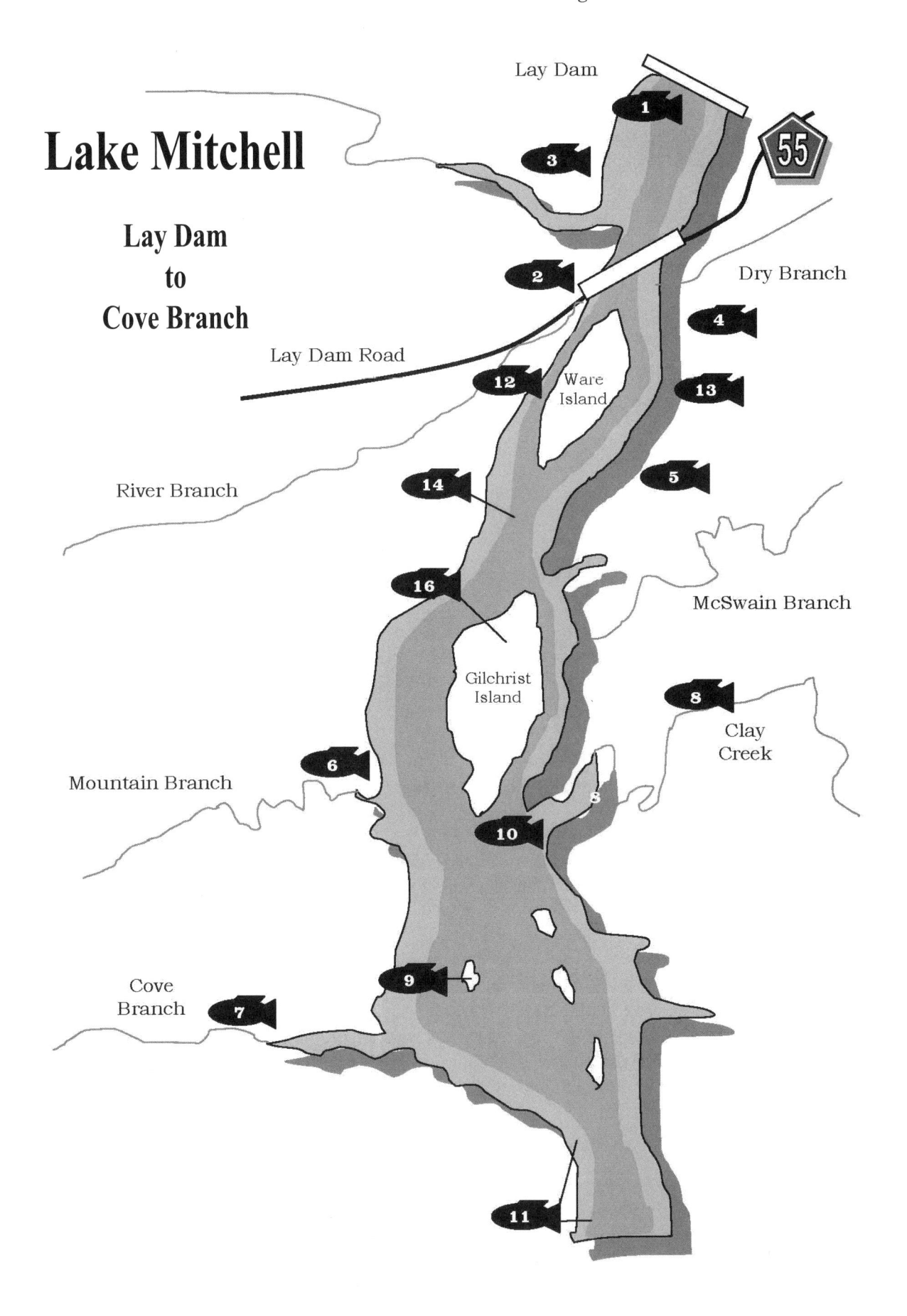

Lake Mitchell

Lay Dam to Cove Branch

116

Mike Bolton

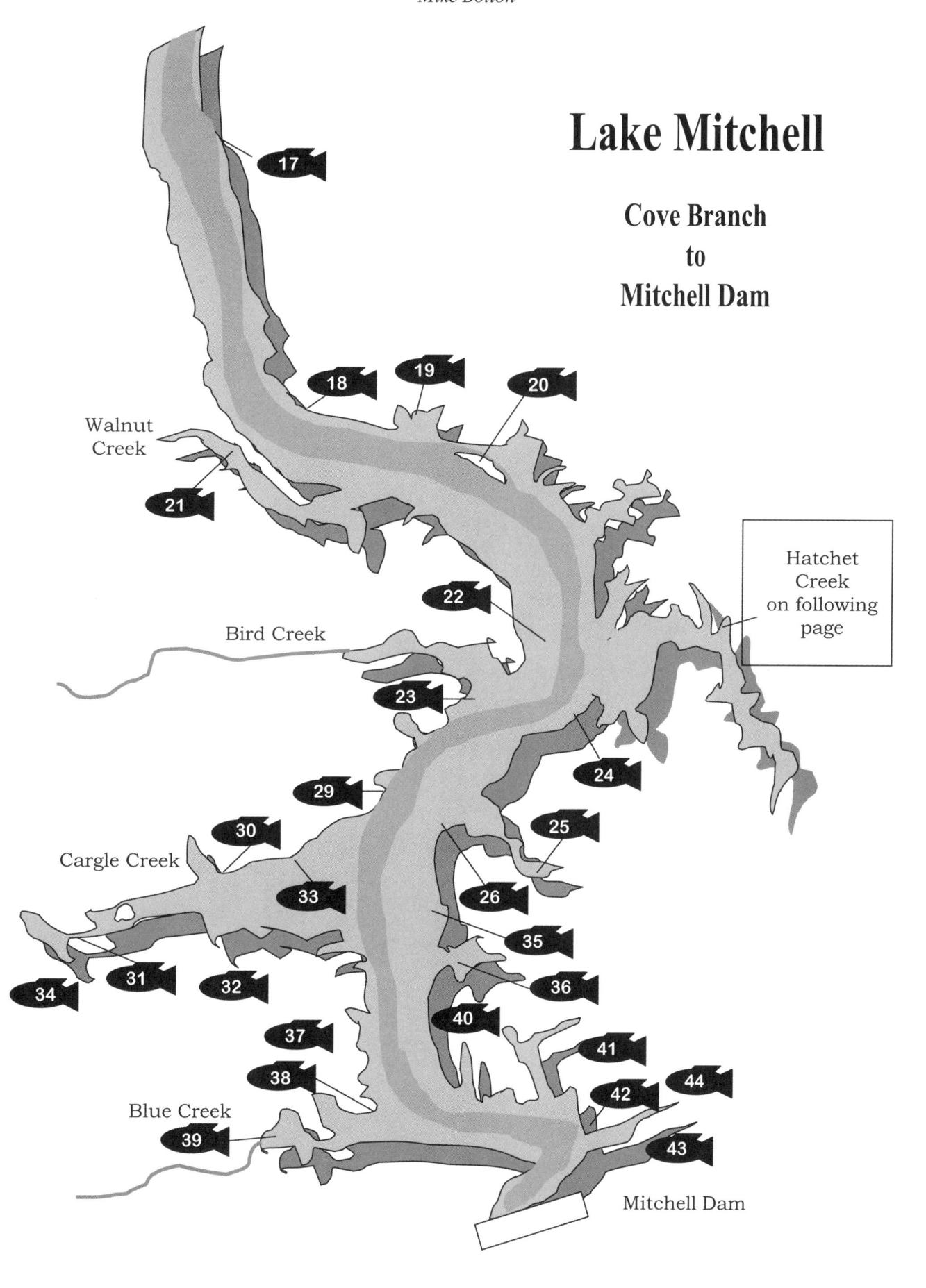

Lake Mitchell

Cove Branch to Mitchell Dam

Walnut Creek

Bird Creek

Hatchet Creek on following page

Cargle Creek

Blue Creek

Mitchell Dam

117

Lake Mitchell
Hatchet Creek

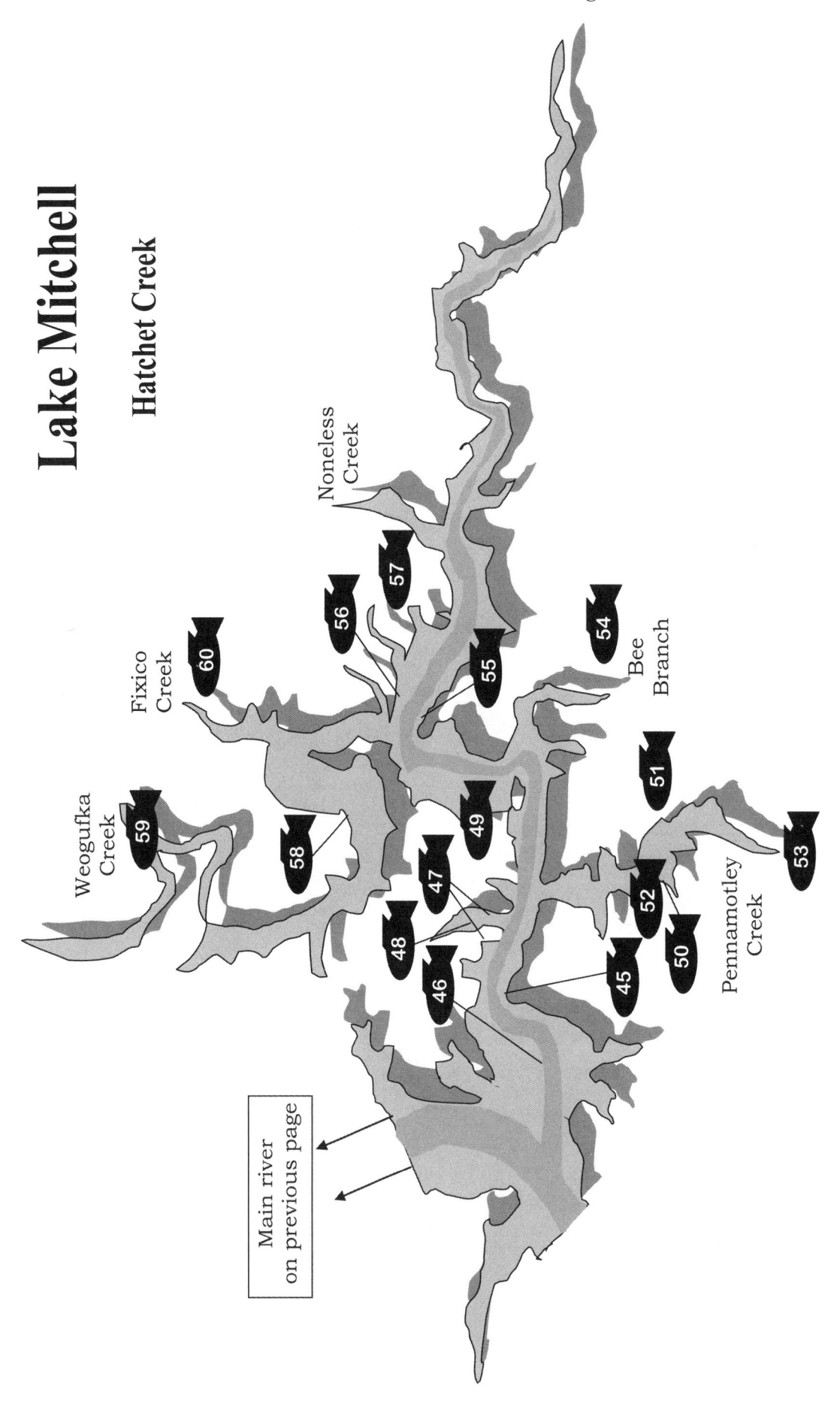

Lake Jordan

Lake Jordan is the smallest reservoir of the Coosa chain but its combination of deep and shallow water and its variety of structure make for an excellent fishery for largemouth bass, spotted bass, crappie, hybrid and catfish.

This is a lake all types of anglers can enjoy. Its large amount of shallow structure makes for excellent fishing for bank fishermen and anglers with a small aluminum boats without depthfinders, while its deeper water offers excellent fishing for the serious angler with the $60,000 rig.

Like all the Coosa River impoundments, Lake Jordan has excellent spring bass fishing, but its fall bass fishing is equally good. It also offers excellent summer bass fishing at night.

The tailrace below Mitchell Dam offers excellent year-round fishing for bass, hybrid, crappie and catfish, and thankfully lacks the big boulders and rocks that make tailrace fishing so dangerous beneath other dams on the Coosa River.

The lower end offers much wider water which makes for excellent spawning grounds.

Spring bass fishing

Mitchell Dam to Blackwell Slough

The lack of large creeks and big sloughs for approximately three miles below Mitchell Dam makes for an unusual fishing situation on Lake Jordan, but a situation most anglers find favorable. The lack of escape areas on the upper end means fish are forced to gather on bank structure and on the stump-lined original channel. These are areas even the angler without a depthfinder has little trouble locating.

Just as on the upper end of sister lakes Lay and Mitchell, fishing the upper end of Lake Jordan offers some tremendous spring bass fishing opportunities. The rocks, stump rows, humps, rock piles, grass, standing timber and log jams along this stretch of the lake plays home to both spotted bass and largemouths. This visible structure holds numbers of fish.

It should first be noted that the area between Highway 22 bridge and Mitchell Dam offers an unusual current situation. This area lacks the dangerous boulders found below many dams, but it does have a large amount of reverse current, or backflow.

The water exiting the turbines of Mitchell Dam immediately swings to the east bank. The path of the original river channel in conjunction with some underwater rocks make a swirling motion and actually pushes current up the west bank. An angler can go down to the Highway 22 bridge, position his boat on the west bank and be pushed upstream to the dam while the current is going downstream.

This isn't a dangerous situation, but is noted because this west bank offers tremendous fishing when this happens.

During the spring months, an angler can take a brown and orange 1/2-ounce jig with a No. 11 brown pork frog trailer and fish the jig-and-pig in this area. First, ride the current down the east bank with the boat pointed upstream approximately 25 yards off the bank while allowing the jig-and-pig to bounce down the underwater rocks on the east side of the boat.

If this fails to produce fish, move to approximately 30 yards above the bridge, and use a white or white/charteuse 1/4-ounce spinnerbait with double gold willow leaf blades to fish the west bank as you float upstream. Again, position your boat 25 yards off the bank and fish between you and the west bank.

The tendency is to cast these baits in the eddy areas and that is fine, but expect to get strikes in water that doesn't look productive, too.

This area is relatively shallow 10 yards off the bank and there are underwater rocks that you can't see (too deep to be a boating hazard) that hold fish. Stay alert and fish the lure back to the boat. This is a pattern that works best when power is being generated at Mitchell Dam and water is being passed through the turbines. It is not for when the floodgates are open in heavy spring rains.

Drifting down the banks will work to a degree when the current is off, but not nearly as well.

There are some occasions, such as when all turbines are running or a floodgate is open, that the current will be too swift for some boats. During this time, it's possible to pull your boat into a deep pocket, deep slough or in Chestnut Creek, Proctor Creek or Pollis Creek just far enough that a long cast will clear the mouth of the creek.

Current too swift for a boat to fight with its trolling motor is too swift for fish to fight for long and the fish in this area will be forced to seek refuge out of the current. Sometimes hundreds of bass will gather behind the primary point of a slough or narrow creek. Throw a Sassy Shad or a grub upstream as far as possible and allow it to bounce downstream on a tight line. This pattern will sometimes produce tremendous numbers of big spotted bass. If the floodgates are open, that means the water is muddy and the fishing will be poor for several days. The swift muddy current will eventually suck the muddy water out of the three creeks in this area after a few days. Never miss the chance to fish the stain line where muddy water meets clear water in the creeks. Schools of bass following this line out of the creeks are easily catchable on swim baits or Rat-L-Traps.

Except for the underwater rock pile under the overhead gas line and the underwater gas line (see map) downstream of the bridge, the rest of this area to Blackwell Slough offers only average spring fishing. Spinnerbaits, buzzbaits, stickbaits and worms fished in the grass pockets along the stretch of river from the last gas line to Blackwell's Slough will produce some fish.

Blackwell Slough area

If you were a stranger fishing Lake Jordan's upper end, then were to fish its middle section another day, it might not dawn on you that you were on the same lake. The river-like upper end turns into wide water in the middle section and the patterns needed to fish the two areas are considerably different.

The middle section offers the wide, shallow water more conducive to the spring spawn, thus is an excellent spring fishing area.

Blackwell's Slough, as well as the Weoka Creek/Holiday Shores area across the lake, offers excellent spring fishing.

The bass will stage out of the deep water by moving up the many points. It's at this time that what may be the lake's best bait is needed.

A chartreuse or lime-colored crankbait fished on these points in February will often produce large numbers of big fish. Blackwell's Slough, fed by Hogbed Creek and Shoal Creek, has a tiny opening but expands into a big area. It is basically a lake in itself.

The bass move into this area in March and April. Points and sea walls will hold the fish at first, but they'll eventually move into the grass where spinnerbaits, buzzbaits and flipping come into play.

Another key area is the three pockets below the mouth of Blackwell Slough. Due to the bend in the lake in this area, the current drives directly into all three of these sloughs. This forces thousands of shad into the back of the pockets and bass have a field day there. Plan to fish Rat-L-Traps, swim baits and medium-running crankbaits.

The Holiday Shores/Weoka Creek area across the lake is another prime location for spring bass fishing. The islands, piers and sea walls draw shad, and schooling bass feed on them. In Weoka Creek, buzzbaits work exceptionally well. Almost all of the points and sloughs will hold bass sometime during the spring.

Below Blackwell Slough

Fishing the lake's lower end is almost identical to fishing the middle section during the spring. Spinnerbaits, lipless crankbaits, worms and swimbaits will take fish.

The only real difference is the presence of the canal. This man-made canal holds spring fish in its mouth and throughout its length. Spring bass love the canal because it almost always has some type of water movement. The canal has several cuts, nooks and crannies and these areas hold fish, but so do the slick banks.

This is an excellent area for spinnerbaits, buzzbaits, Rat-L-Traps and worms.

Many sloughs and pockets on the main river channel also offer good spring fishing. Many of these pockets have grass, as well as piers surrounded by brush.

Sofkahatchee Creek is one of the lake's top spring spawning areas. A plastic crawfish or lizard floated around grass is deadly here. The reservoir above Walter Bouldin Dam offers spotty spring fishing.

The road bed (see map) across the reservoir is a good location to try with jig-and-pigs or crankbaits when current is moving, as are the three underwater humps in the area.

Hot weather bass fishing

Mitchell Dam to Blackwell Slough

Summer fishing during the day is poor except for early morning and late afternoon hours when worms and an occasional spinnerbait will take bass off ledges or points. Let the sun go down and it's a different story.

From May through November, night fishing can be excellent. A dark (black, brown or red) 1/2-ounce or 1/4-ounce spinnerbait with a dark pork frog trailer and a No. 5 nickel blade is the bait of choice for the better anglers.

The angler should position his boat 25 yards off the bank and cast the spinnerbait to the bank and retrieve it slowly back to the boat, always being alert for a sharp strike from one of the river's feisty spotted bass.

A black light and fluorescent line makes for improved fishing, allowing the angler to keep tabs on his line.

Blackwell Slough area

Because Blackwell Slough is so shallow, the quality fish vacate in hot weather. The same pretty much holds true across the river at Weoka Creek, although it does have some deep water. Some bass will stay close to the grass in these areas all summer, but they are usually lesser bass. The larger fish move into the river channel area and gather on the deep points and ledges.

A worm, grub or a deep-running crankbait can be used to take these deep fish. Water flow is a big advantage, although you won't see water flow as often as you like.

Night fishing is a good way to fish this area in July and August. The bass will move somewhat shallower at night, often gathering around lighted piers and seawalls. A good rule of thumb is to look for piers with some signs of fishing, such as rod holders, bass boats, etc. These piers normally will have some brush around them. If a pier has a diving board, avoid it.

Below Blackwell Slough

Summer bass fishing in this area isn't that good. Worms or spinnerbaits fished on the river points early in the morning or late in the evening will produce both largemouth and spots, but rarely larger than two pounds. Fishing here is tough like it is on most other Alabama lakes.

Cool weather bass fishing

Mitchell Dam to Blackwell Slough

The coming of fall means bass will be on a feeding binge in anticipation of winter. It's not unusual to see huge schools of shad scurrying in fear of their lives. It's a good rule of thumb that if you see shad jumping out of the water, they are being chased by largemouth. If you actually see the shad being hit on the surface, they're being chased by spotted bass.

If you can throw a swim bait, jig, grub or lipless crankbait, you will catch fish on this lake during the fall.

Points, stickups, concrete sea walls, piers—just about anything irregular—will hold shad this time of year and bass won't be far behind. Sea walls and the stump rows along the original river channels are areas to concentrate on.

The easiest way to fish is "jump fishing," or watching for schooling shad and rushing to them. A pair of binoculars in the boat during this time is a big plus.

When the cold of winter arrives, this area also offers good fishing, although baits must be slowed down considerably. Hair jigs, jig-and-pigs and worms on ledges, deep points and rocks are prime areas to look for.

Blackwell Slough area

This area that offered such good bass fishing in the spring does likewise when the bass return to shallow water again in the fall. This is an excellent area for willow leaf spinnerbait fishing.

The shad school heavily in the pockets and cuts on both sides of the river and the bass follow for an easy meal. Rat-L-Traps will work when shad are spotted.

The three pockets on the west bank below Blackwell Slough, the same pockets that produce such good catches in the spring, also concentrate shad and bass in the fall. Don't pass up these sloughs.

During the cold part of the winter, bass in this area will sometimes be found in surprisingly shallow water. Anytime there are several days of bright sunlight, the concrete sea walls will hold fish. These walls attract and hold heat and radiate it into the nearby water. Bass somehow know that and concentrate there.

If you were to tell friends you caught bass on spinnerbaits in five feet of water during January, they will probably snicker behind your back. Let 'em snicker.

Below Blackwell Slough

Once again, fall bass use the shallow water sloughs and pockets in these areas as a dead end street for trapping shad. Bass will travel in large schools this time of year in Sofkahatchee Creek and in the mouth of the canal and will be roaming in search of shad. The best method is to watch for signs of shad schooling on topwater and head to the area armed with Rat-L-Traps, Sassy Shads or Little Georges.

This area is also a wonderful place for night fishing in October and November. Spinnerbaits fished on the banks on grass fringes, sea walls, stump rows and other structure will sometimes result in true trophy bass.

Many of this lake's better fish come from this end of the lake at night in the fall.

Lake Jordan's best fishing spots

1. Rock outcropping on west bank below dam. Holds fish year round. Bass, hybrid, crappie.

2. Rock outcropping on east bank below dam. Same as No. 1.

3. West bank below dam and above bridge. Excellent year-round fishing, but exceptionally good in spring. Rocks, blowdown timber, log jams.

4. East bank below dam. Same as No. 3.

5. Bridge pilings under Highway 22 bridge. Good year-round fishing. Good crappie area at night in late spring and summer.

6. Long stump row on east bank. Old creek channel. Stumps run length of channel. Approximately 20 yards off bank. Good year-round fishing.

7. Long stump row on west bank. Same as No. 6.

8. Large rock pile. Sits in middle of channel directly under overhead pipeline. Approximately 35 feet of water around rock pile. Rock pile comes up to 12 feet. Tremendous fall and winter gathering location for spotted bass.

9. Long stump row 40 yards off bank. Runs along ledge of old river channel. Good year-round fishing. Especially good in summer.

10. Underwater gas line. Signs mark spot. Rocks cover gas line. Gas flowing through pipe cools surrounding water in summer. Excellent bass fishing spot in July and August with jigging spoons, crankbaits or worms.

11. East bank. Excellent spinnerbait and worm fishing area year round.

12. Mouth of Chestnut Creek. Stumps on upper point, rocks on lower point. Good year-round fishing.

13. Hole in first bend of creek. Five feet of water falls off into 13 feet of water. Good summer hole for hybrid and crappie. Some bass.

14. Mouth of Pollis Creek. Good year-round fishing.

15. Mouth of Pinchoulee Creek. Good year-round fishing.

16. Excellent stretch of bank on east side. Lots of stickups and stumps. Good in spring and fall with spinnerbaits, crankbaits and worms.

17. Double pocket. Holds lot of fish. Spots and largemouth. Good spring and fall.

18. Mouth of Welona Creek. Good summer and fall schooling area for bass and hybrid.

19. Excellent bank for night fishing with spinnerbait.

20. Same as No. 19.

21. Excellent spring spawning pocket. Good buzzbait area.

22. Good fall crankbait bank.

23. Excellent worm point year round.

24. Never pass up this point for bass fishing. Crankbaits, worms, jig-and-pigs and Sassy Shad.

25. Good spring and fall fishing bank. Exceptional in March and April. Lime-on-lime crankbaits, spinnerbaits, worms.

26. Same as No. 25.

27. Bridge pilings. Crappie fishing hotspot year round.

28. Spring fishing bank in Blackwell Slough. Grass. Good flipping area. Buzzbaits and spinnerbaits.

29. Good worm bank in spring.

30. Super pocket. Water flow pushes right into pocket. Shad get trapped there and are easy prey for bass and hybrid. Best spring fishing cut on the lake.

31. Same as No. 30.

32. Same as No. 30.

33. Underwater hump in deep water. Excellent summer fishing. Stripe and bass schooling area.

34. Island. Bass and crappie fishing. Bass good around island; crappie under causeway.

35. Good fall schooling area for bass. Some spring fishing.

36. Same as No. 35.

37. Excellent point. Good year round.

38. Pockets and points. Late afternoon fishing with worms on points. Piers have brush. Overall good area.

39. Good pocket for spring fishing.

40. Spring fishing area. Shallow water. Rapalas and buzzbaits.

41. Sofkahatchee Creek. Points on left bank going into creek has stumps. Good spring area for bass and crappie.

42. Back of Sofkahatchee Creek. Excellent spring spawning area.

43. Good fall fishing area on points and in pockets for bass. Some stripe school in pockets, too.

44. Mouth of canal. Excellent area for hybrids and bass. Live shad for hybrids; Rat-L-Traps for bass.

45. Canal. Good worm, spinnerbait or jig-and-pig area in spring.

46. Walter Bouldin Dam powerhouse lake. Good March, April and May Sassy Shad fishing.

47. Underwater road bed. Excellent area for Shad Rap when water is being pulled.

48. Underwater island. Good fishing around island when water is being pulled.

49. Underwater island. Same as No. 48.

50. Underwater island. Same as No. 48.

The Ultimate Guide to Alabama Fishing

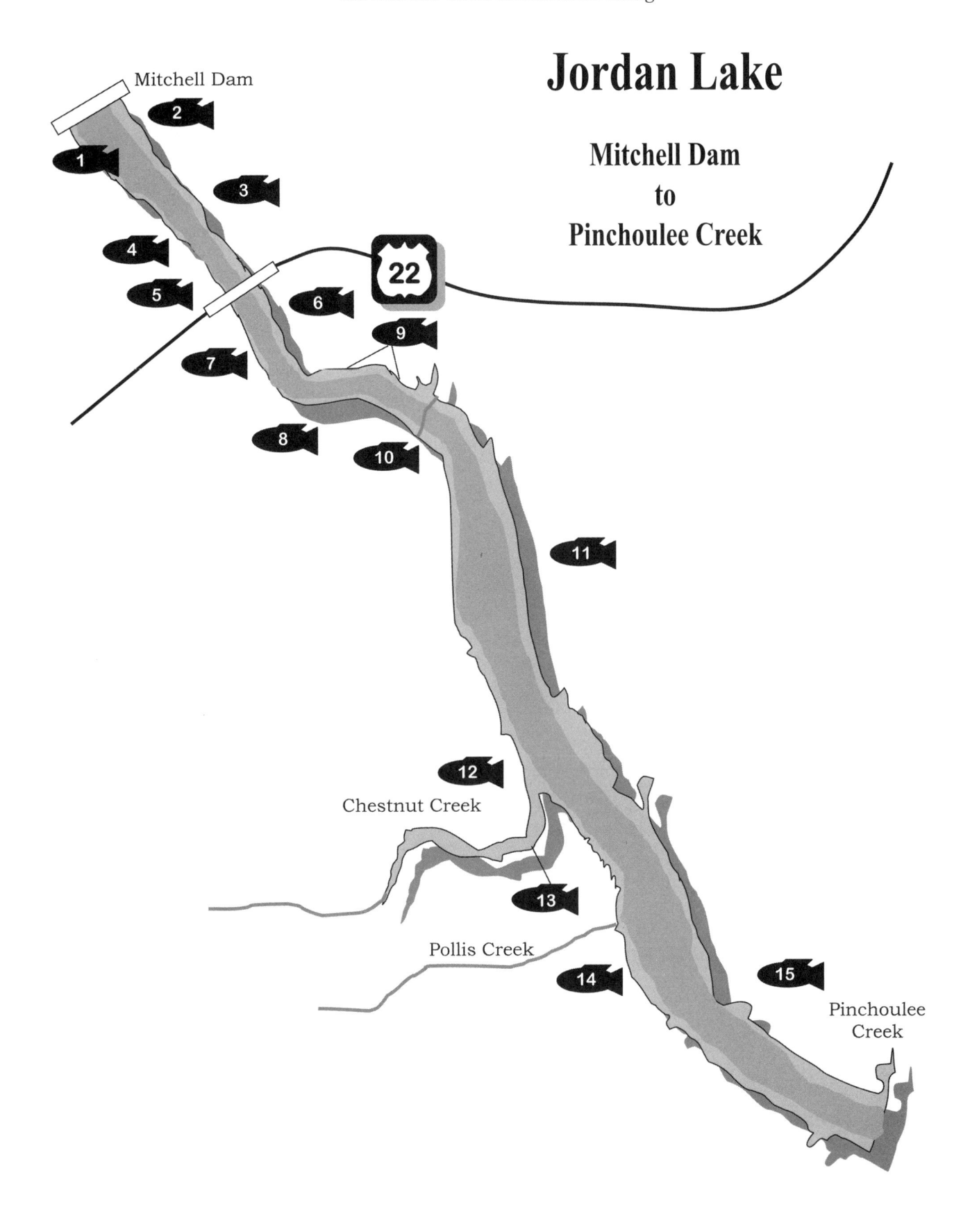

Jordan Lake

**Mitchell Dam
to
Pinchoulee Creek**

Mitchell Dam

22

Chestnut Creek

Pollis Creek

Pinchoulee Creek

124

Jordan Lake

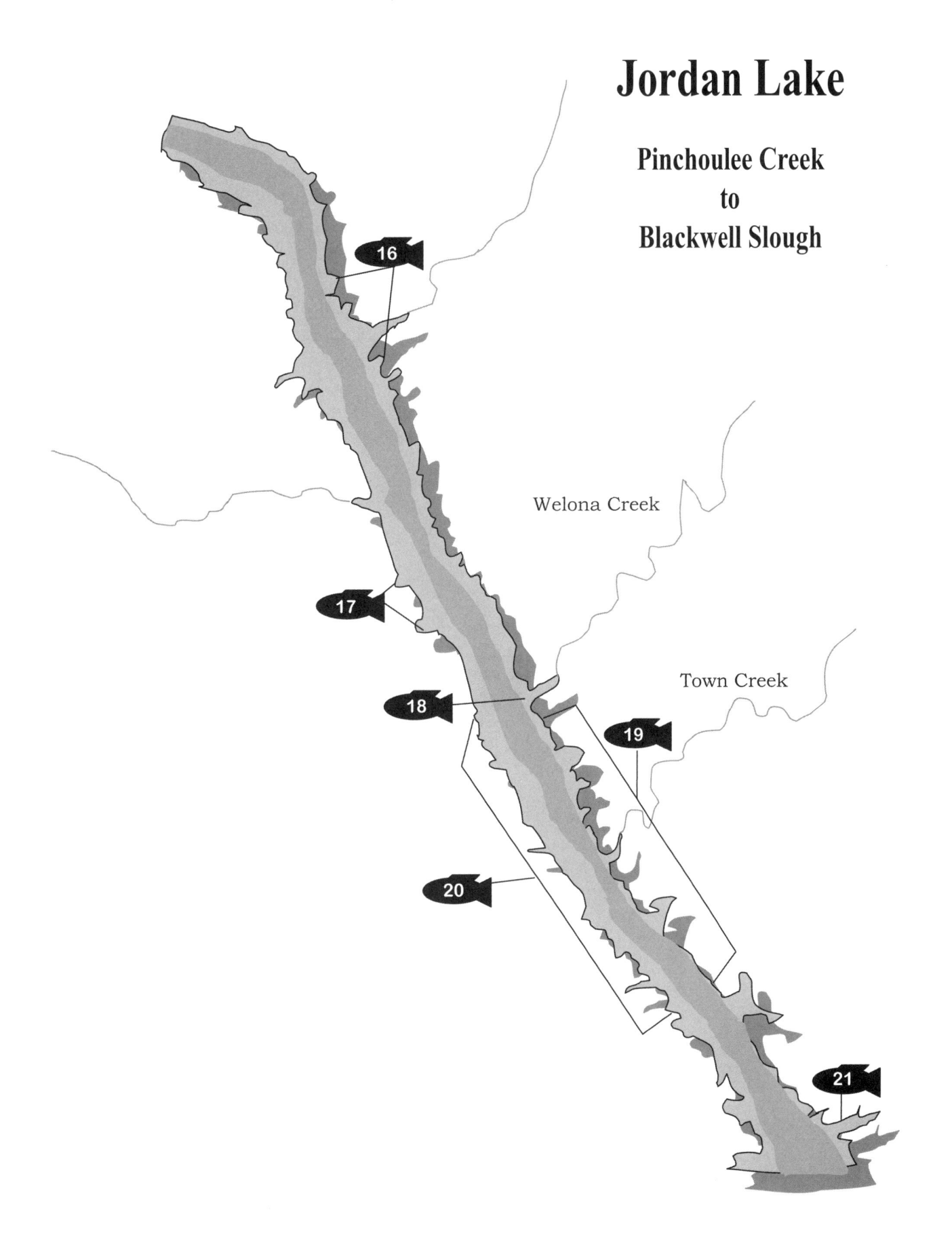

Pinchoulee Creek to Blackwell Slough

Welona Creek

Town Creek

Jordan Lake

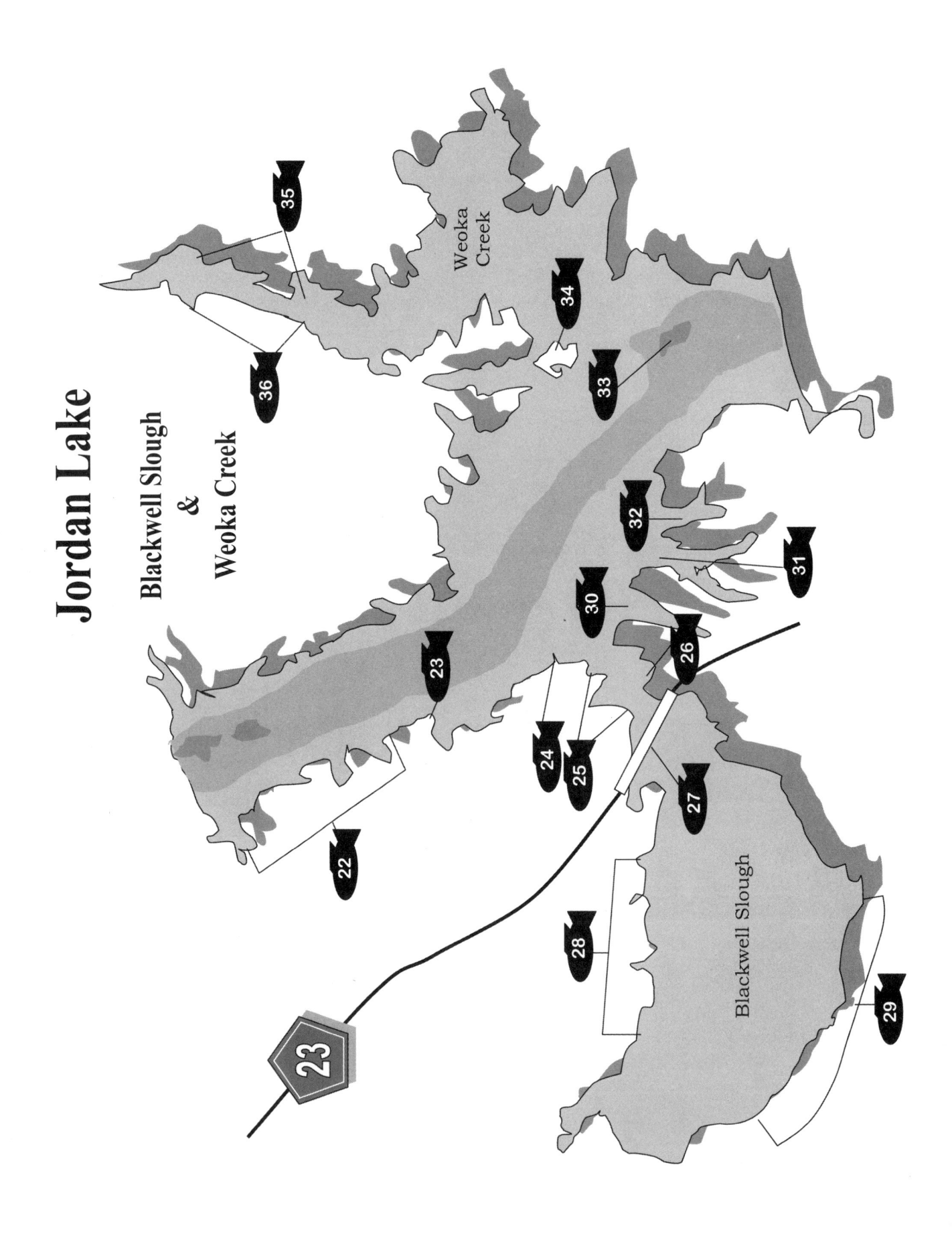

Blackwell Slough
&
Weoka Creek

Jordan Lake
Below Blackwell Slough

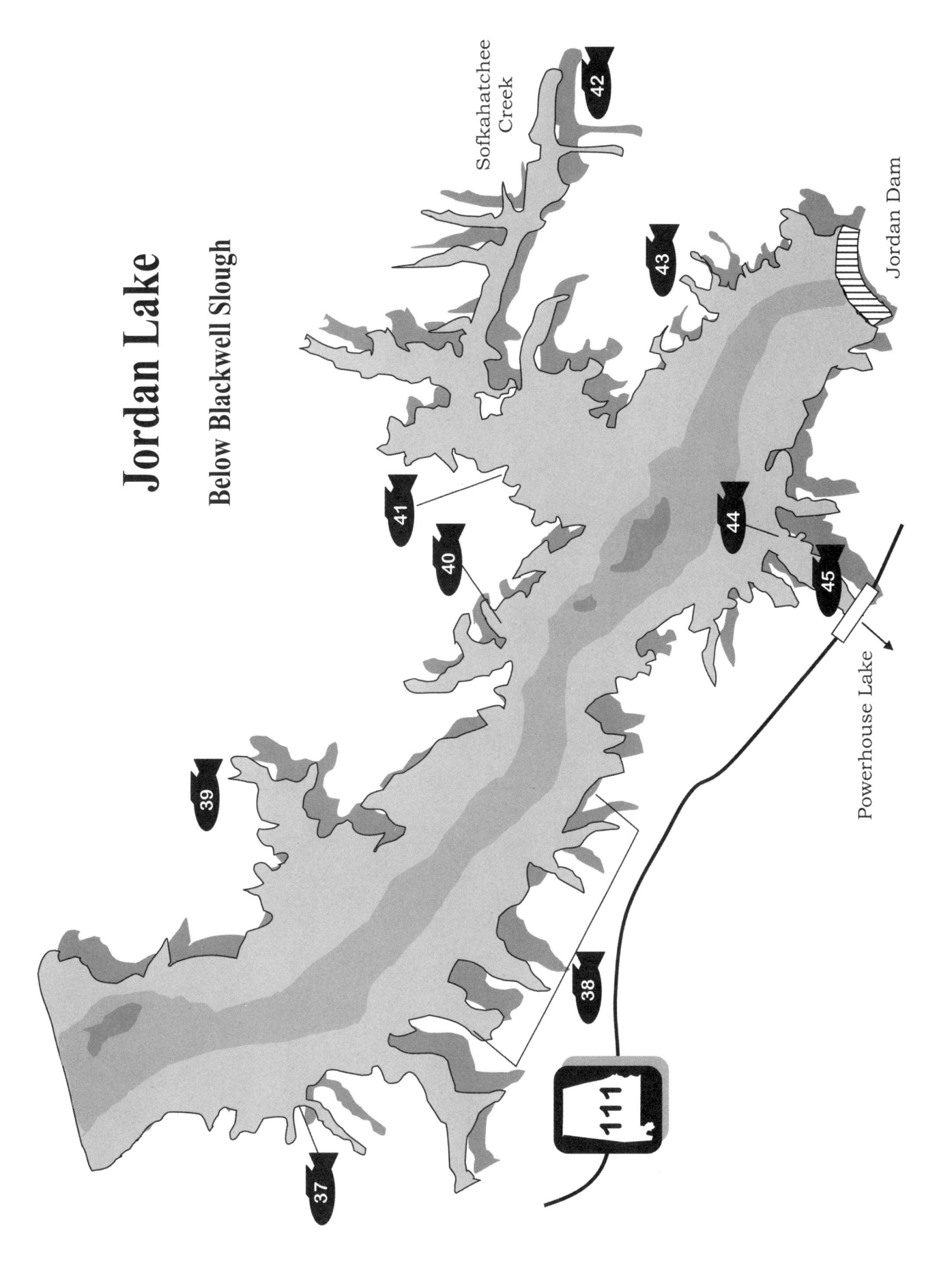

Jordan Lake

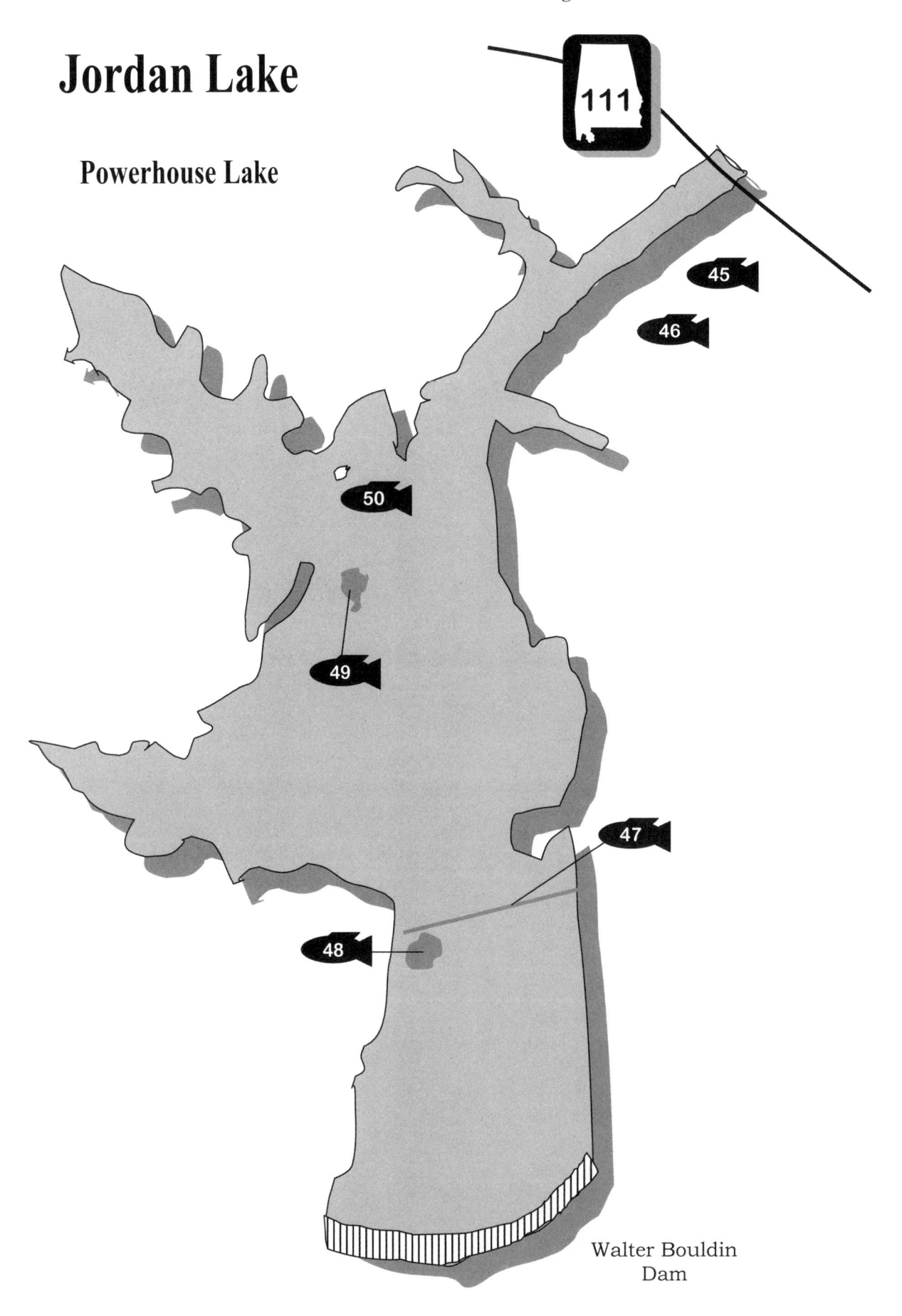

Powerhouse Lake

Walter Bouldin Dam

128

Mike Bolton

The Warrior River

Smith Lake

Bankhead Reservoir

Smith Lake

There's a story that a scuba diver once went to the bottom of Smith Lake above the dam in search of an outboard motor that had fallen off a boat. Upon returning to the boat, the diver was asked if he had seen anything.

"No, I didn't," the diver said. "But the strangest thing happened. I could have sworn I smelled rice cooking."

In other words, this lake is pretty danged deep.

If a poll were to be taken among Alabama's fishermen as to what Lake in Alabama is the most difficult to fish, Smith Lake would unquestionably get the majority of the votes. Even the pro fishermen struggle here. Several Bassmaster Southern Opens have been held on the lake in recent years with over half the field typically failing to get a five-fish limit each day.

Deep water and clear water are two situations which Alabama anglers aren't normally accustomed to and it shows when it's fish-catching time on this lake. It's not uncommon to find bass 40 feet deep in August, and there are times when striped bass can be located in 50 feet of water. For bass fishing in summer, night time fishing is a must. When the stripes go that deep, you simply have to wait.

Bass fishermen who fish Smith Lake regularly, however, have made great strides in techniques to catch fish in these deep, clear waters in the past 20 years. The catches at wildcat night tournaments seem to be producing better and better catches.

If you accept Smith Lake's bass fishing challenge, forget what you know about shallow water fishing and plan to fish deep.

Spring bass fishing

Above Big Bridge

Simpson Creek offers the most shallow of all water in Smith Lake and as a result, it is the first water to warm as the lake shucks the bonds of winter. This is always the first location on the lake you can catch a fish on a spinnerbait when the new year rolls around.

If you are an angler who feels he must fish the banks, this is the only location and time of the year you can get by with that tactic on this lake. The shallow cuts, pockets and sloughs running off of Simpson Creek tend to hold large numbers of fish on the banks in spring.

The large concentrations of smaller bass will be way back in the pockets in the bushes, while the larger bass will be in the deeper cuts where they have quick access to deep water.

These patterns will work for daytime fishing until about the second week of April. At that time, the better bass fishermen switch to night fishing.

Between Big Bridge and Duncan Bridge

Rock Creek, Crooked Creek and White Oak Creek offer good spring bass fishing in the lake's middle area. Crooked Creek offers both the numbers of fish and the larger fish, but that's nobody's secret and it receives the most spring fishing pressure.

White or white/chartreuse spinnerbaits with tandem blades (No. 7 willow leaf), a brown jig-and-pig with a black pork chunk and crankbaits are the best baits for this area.

Jigs should be thrown in 10 feet of water and worked down into 25 feet of water. Another good pattern is casting a crankbait parallel to the broken rock banks on Rock Creek's upper end.

Above Duncan Bridge

After the spring rains raise the water in Smith Lake, Brushy Creek, Rockhouse Creek and Hoghouse Creek have flipping fish that are ready to be caught. The dingy water covers bushes and other structure on the banks and the bass gather there for their annual spawning ritual. A black/brown jig-and-pig with a pork frog trailer of a contrasting color flipped with 20-pound test into this structure will catch these fish.

Another pattern that works well is a big-bladed spinnerbait "buzzed" along the bluff walls in the area following the spawn. A chartreuse or chartreuse/lime spinnerbait with a long twister tail worm trailer should be cast parallel to the bluff walls and retrieved quickly enough to "V" the water. This pattern will give up big spots and largemouths.

Hot weather bass fishing

Above Big Bridge

If night fishing or fishing deep isn't your forte, many will find fishing on Smith Lake pretty tough. It's clear, deep water makes daytime fishing a day's work.

As early as June, temperatures force bass into 20-foot water. By the dog days of August, bass anglers regularly catch fish 35 feet deep.

A spinnerbait at night is the method most of the lake's best fishermen use to catch bass during the hot summer months. A chartreuse spinnerbait with a No. 11 green spotted pork rind or a black spinnerbait tipped with a No. 11 green pork frog are tops.

When the lake was impounded, a great deal of timber in Ryan Creek and Simpson Creek was bulldozed and piled into wide rows adjacent to the original creek channel. These trash piles still remain and they make excellent fish-holding habitat.

Most anglers have an urge to fish the banks during the hot summer months, but doing so on this end of Smith Lake is a waste of time. Bass anglers should ignore the banks and pay close attention to their depthfinders in these creeks. Once 45 to 50 foot depths in the main body of water are located, the angler will know the boat is in proper position. This will allow him to fish the spinnerbait away from the boat in water 10 feet deep or deeper.

In June, throw the spinnerbait into 10 feet of water, allow it to settle to the bottom, then slow roll it through the debris on bottom to about the 30-foot level. A lot of big bass call these piles home during this time of the year, so don't be shy in continually throwing to the same area and fishing it hard.

In July and August, forget the 10-foot water and start throwing into 15 feet of water. Work the spinnerbait down to 40 feet of water before ending the retrieve.

Although these piles of debris offer some of the best fishing in this area this time of the year, don't overlook ledges and long points in the same areas. These deep points and ledges also hold quality fish on occasion. They should be fished in the same manner and with the same baits as the brush piles.

Between Big Bridge and Duncan Bridge

This area of the lake is known by several names, some of which are unprintable in a family-type book such as this one. This area, which offers unbelievably clear, deep water (315-plus in some locations) is sometimes known as the "well water," the "swimming pool" or the "gin bottle."

No matter the name, bass fishing in the main body of water running from bridge to bridge is extremely tough during summer months. During the hot days of June, July and August, you will find bass 10 feet deeper here than any other location on the lake. If you're catching bass in

25 feet of water elsewhere on the lake, the bass in this area will be 35 feet deep.

So much for the bad news. The good news is that during the full moon of June, you'll find no better area on the lake that will produce 3-pound-plus spots. Surprisingly, some of these early-June spotted bass are late spawners and you'll find them full of eggs. Many anglers believe these fish bury themselves so deep (45 feet-plus) during winter months that they mess up their spawning times. These spots normally gather around any of the long clay points located near steep bluffs. Jigging spoons, deep-diving crankbaits (20-plus), Rat-L-Traps or even spinnerbaits can be used to extract them.

After these spotted bass disappear from the area some time in June, the best bass anglers usually consider the area pretty much useless until late fall.

The exception to this rule of thumb is the Rock Creek area. Unlike the main body of water between the bridges, this creek offers good summer fishing. Of course, night fishing is the key.

Ledges and long-points are spread throughout the creek and offer the best bass-holding structure. Many anglers make the mistake of fishing too close to the bank and as a result, are usually sitting directly atop the fish.

Just like in the area above Big Bridge, use the depthfinder to put yourself in 45 to 50 feet of water and let your casts land 15 to 20 feet off the bank. Allow the spinnerbait to settle and bump the structure on a slow retrieve back to the boat.

Many of the banks will have extremely deep water immediately, but many times a rock pile or other structure juts 20 yards off the bank and this is the type of structure that produces good stringers of bass.

Above Duncan Bridge

Once again, finding the points and ledges away from the bank in deep water is the key to catching summer bass on the area of Smith Lake above Duncan Bridge.

The stretch of the Sipsey River from Duncan Bridge to the mouth of Brushy Creek offers the area that holds the most summertime bass. Just as in fishing other areas of the lake during summer, the angler should fish at night. He should back off of these points into deep water and cast into 10 to 15 feet of water and begin his retrieve with slow-rolled spinnerbaits. Casting any closer to the bank may result in an occasional small bass, but rarely produces a keeper fish.

Cool and cold weather bass fishing

Above Big Bridge

The cooler temperatures of October and the months that follow give Smith Lake anglers some relief from the heat, but more importantly, relief from the necessity of fishing strictly at night.

Both Ryan and Simpson Creek offer good daytime fishing this time of the year. Jig-and-pigs (flipping), spinnerbaits, buzz baits and crankbaits all will put bass in the livewell, but the angler serious with his flipping or fishing a Carolina rig seems to fare the best.

Ryan Creek offers the best bass fishing of the two creeks, and anglers will find something rare for Smith Lake there this time of the year—water color.

Ryan Creek plays home to a large amount of stumps, brush and logs and other type structure that flippers love. Ryan Creek won't produce large numbers of fish this time of the year, but hookups will result in a high percentage of quality fish. During fall night tournaments on the lake, it is not uncommon to see the creek filled with fishermen flipping jig-and-pigs with great haste as they try to pull a big bass out before darkness falls.

The back end of Ryan Creek (above Speegle's Slough) is a good fall area for those who like to continue with their night fishing. Spinnerbaits fished in shallow water will produce fish.

The back end of Simpson Creek is a good bass area starting in late October. Jig-and-pigs flipped in debris in medium-depth water also produces good fish.

Between Big Bridge and Duncan Bridge

October at night is the angler's best opportunity to catch big fish in this area of Smith Lake. Several anglers have dominated fall tournament fishing on the lake by fishing their spinnerbaits in Rock Creek. Start by fishing spinnerbaits in 10 feet of water and fish the baits to 25 feet of water. Long points with stumps are the best areas to fish, but the better tournament fishermen narrow down these areas by visiting the lake during the day and watching for schooling shad.

The fish are schooling heavily during this time of the year and finding fish chasing shad is not a problem. The long points on which you find shad during the day is the points that will hold shad at night.

Above Duncan Bridge

For those who have found that fishing season doesn't necessarily end on Labor Day, fall and winter bass fishing on this area of the lake can be a pleasure. Better yet, night fishing isn't a must.

The blown-down timber above the dam offers good buzz bait fishing and other types of topwater action early in the fall.

When November arrives, a surprisingly good pattern is ripping a spinnerbait parallel to the rock walls. For some unexplainable reason, the spots will leave a slow spinnerbait alone. Instead, a spinnerbait cranked fast enough to make the water "V" drives them crazy. Just as surprising, clear water seems to produce better than dingy water with this pattern.

Other types of fishing

Stripe fishing

Never in Alabama has the introduction of a new species of fish into a lake had the impact of the introduction of the saltwater striped bass into Smith Lake.

The 3-inch fingerlings first placed in the lake in 1983 had transformed into 30-pound-plus monsters by the summer of 1989 and Smith Lake, considered by many to be the lake in Alabama to stay away from, was suddenly offering the most exciting new freshwater fishery in the state.

The deep, clear, cool, infertile water found in Smith Lake is a natural for the implanted saltwater striped bass. The deep standing timber also provides the type structure the fish needs to thrive.

Interestingly enough, although Smith Lake was probably the most natural location for the saltwater striped bass, it was the last Alabama lake to receive the fish. And for good reason.

The state kept the lake free of stripes and hybrids until 1983 as the basis for an experiment. While other reservoirs in the state were receiving the Atlantic Coast strain of striped bass, Smith Lake was stocked with the Gulf Coast strain of striped bass starting in 1983. The rapid-growing Gulf Coast strain was believed to be more tolerant of warm water and how it fared in Smith Lake would be a factor in the fish being introduced into other state lakes.

Lake Martin was the foremost striped bass fishery in 1987, but it didn't take long for Smith Lake to steal that thunder. By the summer of 1989, a gaggle of guides had unlocked the secrets of fishing for the stripes and 15-pound to 20-pound fish were almost a daily occurrence.

Other fishermen began to bank on the practices these guides utilized and soon their techniques were common knowledge. Striped bass fishing on Smith Lake was off and running.

The introduction of the striped bass into Smith Lake meant many pluses for the lake. No. 1, it meant a new fishery for those who lived in the vicinity. Smith Lake is full of bass, but the deep, clear water means tough bass fishing for most people not accustomed to that type of fishing.

No. 2, it meant an economic plus for the Cullman area and the stores and marinas on the lake. Stripe fishing began to bring in a wave of new fishermen who had traditionally avoided the lake.

No. 3, and probably most importantly, it meant a boon to bass fishing. Not only did stripe fishing take some pressure off the bass, the striped bass themselves were a help. The tremendously large (too large for bass to eat) shad suddenly disappeared and fishermen were suddenly discovering the largest schools of small shad they had ever seen on the lake.

Biologists attributed this to the demise of the large, sickly shad that were in no condition to reproduce. The smaller, healthy shad that remained were suddenly back in the reproduction game.

Many bass fishermen cried foul over the introduction of the striped bass into the lake, saying they were eating the bass, but the facts are that it takes more pounds to win a Smith Lake bass tournament than ever before.

Catching Smith Lake's striped bass is either easy or difficult, depending on how you look at it. If you understand stripe fishing and have the right equipment, it can be a snap. If you're taking up stripe fishing from scratch, you may find it difficult.

Ideally, the stripe fisherman will have (1) a bait tank designed to keep shad alive, (2) numerous (up to six) stiff backboned rods with limber tips, (3) a wide-spooled reel with a clicker, (4) a boat with rod holders and (5) a graph depth recorder to locate fish.

Those items make for a perfect setup, but stripe can be caught with other setups.

Stripe bass fishing on the lake is almost entirely done with live bait, so a live bait tank is important. Some have had luck in keeping shad alive in boat live wells and in foam coolers, but the most successful fishermen are the ones who have bait tanks. These tanks can vary from a custom-made $400 rig, to a plastic drum with the top cut off.

Tanks should always be round (shad will pile up in

a corner of a square tank and smother) and they should be white to reflect heat. An aerator is important, and a filter to remove scales and slime is nice if you keep the shad for a long day of fishing.

Another key consideration is the rod and reel. A custom-made striper rod is nice, but a bass rod with strong backbone (a flippin' stick, for example) will suffice. On the rod itself, measure 24 inches from the front of the reel to the front of the rod and mark this spot with a piece of white tape wrapped around the rod. You'll read later why this is important.

A wide-spool reel with a clicker is the reel of choice of the better fishermen, but again, other reels will work. A baitcaster with the drag set lightly will do.

Because of the clear water, relatively light line (15 lb. to 20 lb. test) is needed. That may sound light for a 20-pound-plus fish, but almost every fish is caught in open water and landing one on a well-set drag isn't that difficult. A four-foot leader of the same line is needed, as is a strong swivel, a 1-ounce or larger weight and an orange bead. Hooks vary between fishermen, but a 4/0 is probably the average.

Rig your outfit by placing a 1-ounce worm weight on the main line. Put the bead on next and then tie on the swivel. The bead keeps the worm weight from cutting the knot.

Tie on the leader and the hook next and you're ready to fish. Shad can be hooked in several ways, but in one nostril and out the next works best. Another popular method is placing the hook through the throat and out the mouth, but a shad hooked in this manner doesn't live as long as one hooked through the nostrils.

Stripe need to be located with a depthfinder. When they are located, lower your bait right on top of them. This can be done by counting out line using the 24-inch mark you made on your rod. If the fish are 30 feet deep, for instance, subtract the 4-foot leader and pull your line out of the reel to the mark 13 times (26 feet) that will put your bait exactly at the 30-foot level.

If the stripe are shallow (20 feet or less)—and they will be in the spring and fall—the presence of the boat will often spook the fish. This can be avoided by balloon fishing. In balloon fishing, you rig the same way except you tie a balloon (blown up to the size of a baseball) on to your line. Tie it the distance of the fish above the hook. (If the fish are 20 feet deep, tie it 20 feet above the hook). Allow the balloon to drift away from the boat and this will greatly increase your chance of hookups.

Another method the better stripe fishermen use in spring and fall is using at least one free line. A free line is nothing more than a hook tied to a main line with no other terminal tackle involved. The free line is let out 30 feet or more with the shad allowed to swim free; This is a good method if the stripe are staying deep and occasionally coming up to feed. This free-swimming shad may range from 30 feet of water to five feet and offers an enticing meal.

Another pattern that offers tremendously exciting fishing in the spring and fall is fishing a Redfin. This 1-foot-long artificial bait is deadly on stripes that feed on topwater and hell hath no fury like a 28-pound stripe exploding on a topwater bait.

Many of the lake's top guides keep a Redfin tied on a rod at all times and are ready to strike when the fish suddenly start hitting topwater.

The striped bass in Smith Lake, like all striped bass, travel great distances and knowing where to look for them is important. The fish spend most of their lives in the deep water areas directly above the dam, but spring and fall will find them on the move.

During spring months, the stripe attempt to spawn even though it is impossible for them to do so. They will make long spawning runs as far as the shallow water in the back of Ryan Creek. The spring months will find these fish scattered throughout the shallow-water creeks all over the lake.

Smith Lake's deep water makes it very difficult to catch shad from the lake. Several locations on the lake sell shad commercially.

The hump directly out of the mouth of Speegle's Slough, the shallow flats around Goat Island, Burr's Island and Miller's Flats and other relatively shallow water in the area have traditionally held the most spring fish. The fall months will see some fish make a less pronounced run to the same shallow-water areas.

During the hot summer months and the cold winter months, the deep water within five miles of the dam is where you're most likely to find the most fish. The rocky points in the lower end of Ryan Creek, the intersections of smaller creeks and Ryan Creek (such as Coon Creek and Lick Creek), the mouth of Rock Creek, the deepwater piers in Rock Creek and a host of other areas hold summer fish.

Mike Bolton

Smith Lake's best fishing spots

1. Upper end of Ryan Creek. Good colored water, stump rows, rocks. Good flipping area spring and fall bass.

2. Southeast bank of Ryan Creek. Excellent night fishing for bass. May to July.

3. Goat Island. Good spinnerbait fishing in spring. Trophy spotted bass area.

4. Speegle's Slough, also known as Brushy Creek. Back end good for spinnerbait fishing for bass in spring. Excellent spring fishing area for crappie. Lots of brush.

5. Bailey Holler. Excellent crankbait fishing in February.

6. Upper end, Simpson Creek. Bass move here earlier in spring than any other location. Start with crankbaits early February, move to spinnerbaits early March.

7. Rocky bank between Calvert's Marina and Misty Harbor Marina. Good night fishing for bass in fall.

8. Cuts on north bank of Simpson Creek. Late spring spinnerbait area.

9. Miller's Flats. Good stripe fishing area. Fish school in this area much of year.

10. North slough, Miller's Flats. Spring fishing. Good spawning area for bass, crappie.

11. Northeast slough. Excellent schooling area for bass and stripe.

12. Mouth of Miller's Flats to castle. Good stripe fishing, summer months. Big stripe school at either end of wider water.

13. Adjacent to Calvert Island. Night fishing in hot summer months. Spinnerbaits.

14. Calvert's Slough. Spring, summer bass fishing. All bank fishing at night. Good schooling for bass and stripe.

15. Pigeon Roost Creek. Lot of rocks. Good summer bank fishing for bass. Night fishing.

16. Castle Rock area. Good stripe fishing area at night during summer months. Live shad.

17. Big Bridge. All columns will hold spotted bass or stripe at times.

18. East bank of Ryan Creek. Holds bass only in late spring. Fish rocky bluffs with buzzbaits and spinnerbaits. You'll find 315 feet of water here. Bass and stripe schooling area.

19. Back end of Willoughby Creek. Early spring fishing. Big rains dump lots of water into feeder creeks. Spots and largemouth school in mouths following big rains.

20. Coon Creek. Same as No. 19.

21. Good, rocky points. Standing bluffs. Winter fishing with light line and grubs, jigs.

22. Upper end Spring branch. Good run-in fishing where spring rains enter from feed creeks. Stripe school here in fall.

23. Lick Creek. Good run-in fishing in early spring with buzzbaits and spinnerbaits. Stripe school in the mouth of creek in the fall.

24. L.M. Smith Dam. Good stripe area in summer.

25. Feeder creeks on northeast bank. Late fall and winter bass schooling area. Lots of open water. Stripe school here in winter.

26. Mouth of Rock Creek. Big spot area in early summer with spinnerbait and jig. Bass and stripe school on up in creek.

27. Mouth of Little Crooked Creek. Good spot schooling area.

28. East Sipsey Fork. Good bass fishing year round from Little Crooked Creek to Bailey's. Points and a few ledges. Jig-and-pig February to March; good night fishing until fall with spinnerbait.

29. Mouths of White Oak and Crooked Creek, Bailey Bridge. Good year round fishing for crappie.

30. Crappie holes. Good spawning flats and brush.

31. Clay points in White Oak Creek. Good bass fishing with crankbaits in March. Also good spring spawning area for crappie.

32. Rock Creek. Good year round creek. Deep water for winter fishing with jig-and-pig. Good spring with spinnerbaits and crankbaits.

33. Underwater timber.

34. Underwater island.

35. Underwater island.

36. Dismal Creek. Trophy spot area mid-February to early spring. One of most under-fished creeks in lake. Bass and stripe move from mouth of creek to upper end during spawn.

37. Deep water (225 feet) area from mouth of Dismal Creek to Duncan Bridge. A few points, a few ledges. Good night summer fishing for bass.

38. Bridge pilings, Duncan Bridge. try around pilings year round.

39. Schooling area for bass and stripe summer and fall.

40. Underwater island.

41. Clear Creek. Logjams and blowdowns in mouths of pockets.

42. Underwater timber in Butler Branch. Spring crappie.

43. Underwater island. Good schooling area for bass, stripes in schooling months.

44. Bluffs. Good fall fishing with buzzbaits and spinnerbaits.

45. Mouth of Big Bear and Little Bear Creek. Good night fishing in summer. Spinnerbaits, jig-and-pig.

46. Mouth of Brushy Creek. Good for a small bass or two at any time with spinnerbait or jig-and-pig.

47. Brushy Creek. Good year round. Spinnerbaits or flipping early in year, buzzbaits early spring. Lots of willow bushes in area. Good crappie fishing.

48. Upper end Brushy Creek. Good fall of the year with spinnerbaits.

49. Yellow Creek. Flat creek with very little structure. Good early spring with spinnerbaits.

50. River points. Good at night in summer with worms or jig-and-pig.

51. Hoghouse Creek. Good fishing early spring with crankbaits and jig-and-pig.

52. Rockhouse Creek. Few fish, but an occasional big bass in late spring.

53. Sipsey River above Houston Recreation Area. Best crappie fishing on lake in spring. Blowdowns and treetops coming off the banks.

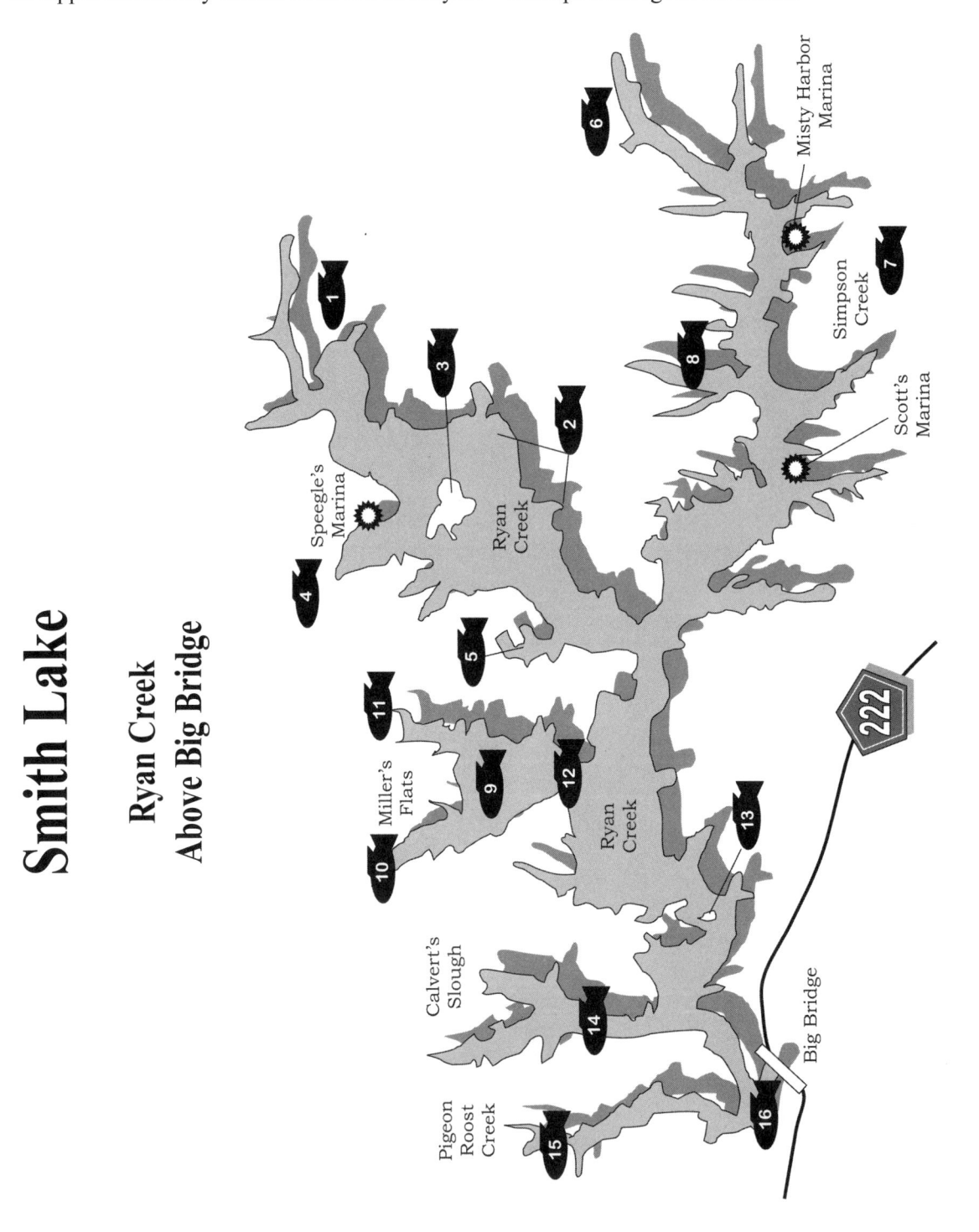

Smith Lake
Ryan Creek Above Big Bridge

Mike Bolton

Smith Lake

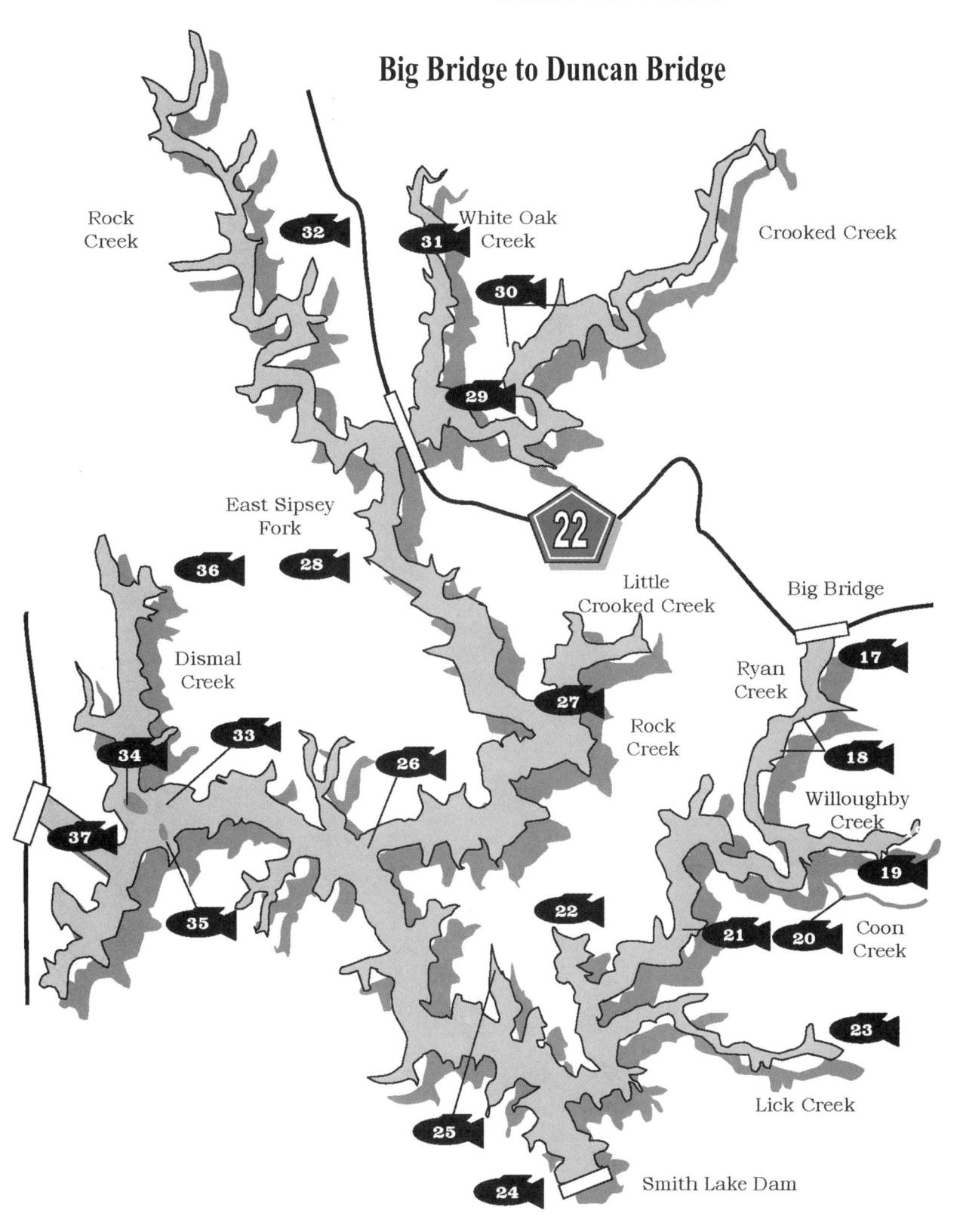

Big Bridge to Duncan Bridge

The Ultimate Guide to Alabama Fishing

Smith Lake

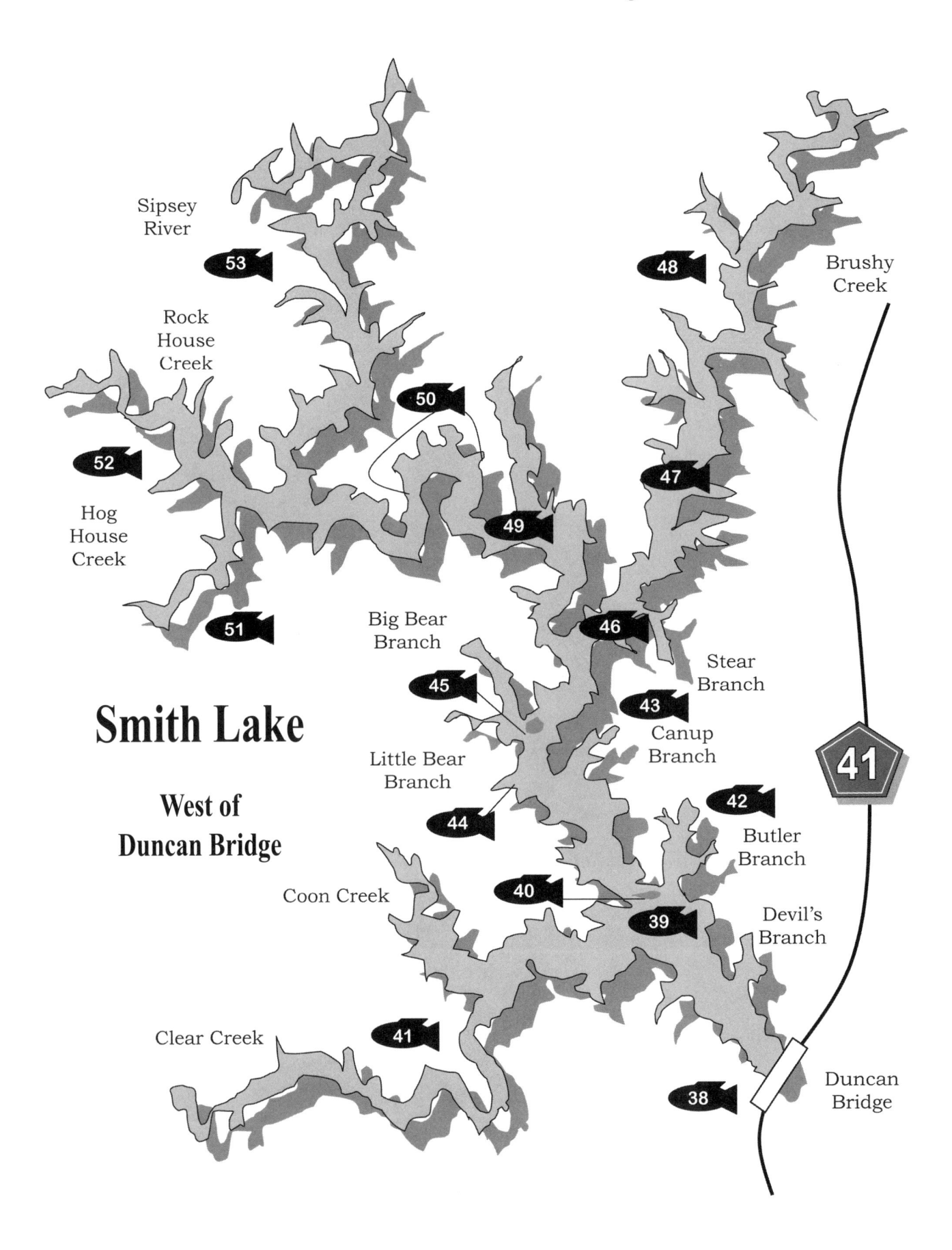

West of Duncan Bridge

Sipsey River
Rock House Creek
Hog House Creek
Big Bear Branch
Little Bear Branch
Coon Creek
Clear Creek
Brushy Creek
Stear Branch
Canup Branch
Butler Branch
Devil's Branch
Duncan Bridge

41

138

Mike Bolton

Bankhead Reservoir (Warrior River)

Just glancing at a map, you may think Bankhead Reservoir, or the Warrior River as most who fish it call it, is nothing more than a stretch of river. On second glance, however, the lake has more wide water than first appears.

The 9,200-acre reservoir begins in the tailrace of Smith Lake in Walker and Cullman County and stretches 77 miles to Holt Reservoir near Northport in Tuscaloosa County.

In that length there is a good bit of wide water off the original river channel. This wide, shallow water is full of structure that includes grass, brush and blow-down timber. The angler who has a particular pattern he likes to fish can find the type water in which to fish it in this lake.

The lake hides its age well. It was one of the first rivers in Alabama to be impounded. It came to be following the completion of the John Hollis Bankhead Lock and Dam in 1916.

One of Alabama's top bass fisheries it is not. It is a challenging lake for bass fishermen, however, and a common stop for bass clubs across Alabama.

Spring bass fishing

Below Howton's Camp

Much of the water on this reservoir's lower end is deep and clear, thus spring bass fishing is limited. The bass, both spots and largemouth, begin their year deep in the main river channel, but begin stair-stepping themselves up on structure on the way to shallow water.

Points are the main route for the migration. Light line and small baits are the key to success on this end of the lake during the move. Tube-type crappie jigs, 4-inch plastic grubs or 4-inch plastic worms are best bait for this finesse fishing. Six- or 8-pound test line is needed.

The baits should be cast shallow, then gingerly hopped down into deeper water.

This pattern catches the majority of fish, but not the bigger ones. The bigger fish are usually taken by fishing a jig-and-pig in the brush tops just inside the main pockets. Keep in mind that northeast banks receive the most sunlight and heat the quickest. A water surface temperature gauge is critical.

For those wanting to fish for spotted bass, deep river bluffs hold the majority of them in spring. The spots school at this time and a limit of keeper fish is possible.

When spawning time arrives, bass move into shallow water in creeks and pockets. Spinnerbaits are impossible to beat at this time. A 1/4-ounce, lime-colored spinnerbait with a single No. 4 or No. 5 gold willowleaf blade is best. These spawning bass gather in flat pockets on the blow-down timber to build their beds and spawn, and the spinnerbait works well there.

Topwater baits also will work where the brush and trash aren't too thick. Either a black/gold Rapala or a similarly colored jerk bait should be used.

These spring fish are spooky because of the clear, shallow water. An angler needs to back off and make long casts with a spinning outfit.

Yellow Creek is full of shallow cover that is excellent for spawning. It has a good water flow and lots of flow off the creek. Fryley's Creek is much the same way. Both of these creeks are areas to concentrate on. Walker County Shoal Creek has lily pads and grass in the back which holds some good fish. White Oak Creek is much like Walker County Shoal Creek. Shoal Creek and Camp Creek offer only fair spring fishing. Short Creek has plenty of shallow water structure that holds spawning bass.

Many of the post-spawn bass will stay shallow in grass in a few feet of water throughout the summer. A Lizard in plum or June bug colors flipped into the grass works wonderfully as do swimming jigs.

Above Howton's Camp

The water above Howton's Camp is shallower and dingier than the water on the lower end of the lake. This allows the angler to fish patterns that take much less finesse. Most creeks off the main lake on this end are shallow and full of grass or lily pads. This makes for excellent spring fishing which most anglers find easy.

The bass start off deep in the channel, then in mid- or late-February begin their move toward the pockets and creeks. Points are their primary migration routes. A jig-and-pig or a single-bladed spinnerbait slow-rolled here is needed to catch these cold fish.

Once the bass are in shallow water, they begin seeking cover. Many anglers make the mistake of thinking bass want thick cover to spawn in, but that isn't the case. The bass need sunlight to penetrate the weeds to warm the water and hatch their eggs. That's why small isolated patches of grass or lily pads on this end hold more spring bass than thick patches. Thick, matted grass won't usually hold bass until after the spawn.

Try Rats, frogs, spinnerbaits, buzzbaits and jig-and-pigs in grass.

When the bass move onto the points, Bluff Creek is a key spring area to focus. The log jams in the mouth hold fish coming and going from the creek. The brush and wood up in the creek hold spawning bass. Lost Creek is excellent and good anglers can usually catch a limit of fish in there in an hour or less in the spring.

Another area to concentrate on in spring is the section of river that stretches for at least a mile below the steam plant at Gorgas. There is always a radical change in water temperature below the steam plant and the bass gather there in numbers. It, too, is a key area. The blowdowns along both banks hold fish.

Kisner Slough, Miller's Slough and Richardson Slough all have fair spring fishing.

The little river area lacks the shallow water and pockets needed for good spring fishing and most good anglers stay out of it during the early part of the year.

Hot weather bass fishing

Below Howton's Camp

Because of the deeper water, the lower end of the Bankhead Reservoir offers relatively good summer bass fishing. The mouths of the creeks, the rock-pile points and sheer rock bluffs hold the majority of these hot-weather bass.

Grubs, worms or crankbaits can catch bass in this area during summer. If you choose to grub fish, a blue or black works well. Worm fishermen need a blue or black 6-inch worm, and crankbait fishermen need a deep-diving crankbait. Black on silver foil works well in the crankbaits.

It is possible to catch some topwater bass early in the morning or late in the evening. Most gather on the rocky points and will fall to jerk baits and buzzbaits.

The mouths of the three pockets on the left side of the bank above the dam offer good summer fishing, as does the mouth of Yellow Creek and all the rocky points with cedar trees growing on them. Fryley's Creek is excellent.

Above Howton's Camp

This section of the lake can offer some good fishing during the spring months. The grass and lily pads in the area often grow in five or six feet of water and they put off oxygen, hold baitfish and offer shade. It's no secret these offerings attract bass.

Summer fishing around this greenery is exciting, too. A 1/2-ounce white on white buzz bait with a white grub trailer works well in the grass, while spinnerbaits and jerkbaits work on the fringes of the grass. It's also possible to flip this grass with a black jig-and-pig. Prescott Creek, which flows into the little river, and Lost Creek on the big river, are excellent areas for this type of fishing. Both creeks have good water depth, as well as a large amount of grass and lily pads.

Cool and cold weather fishing

Below Howton's Camp

By September, shad have started running heavily in the creeks and the bass move in hot pursuit. Spinnerbaits are difficult to beat at this time. A double-bladed chartreuse/blue spinnerbait with a No. 3 gold blade in the front and a No. 5 gold blade in the back works best. Any grass-covered secondary point is likely to be holding bass at this time. The spinnerbait should be cast up to the bank and rolled slowly down the point.

Topwater baits work well in the fall, too. Jerkbaits and noisy baits work good early and late and all day when there is a cloudy sky. A 1/2-ounce buzzbait with a gold blade also works well. Yellow Creek, Walker County Shoal Creek and White Oak Creek are all good fall fishing areas. The mouth of Camp Creek with its grass-covered flats and blowdowns is also good.

Above Howton's Camp

Fall fishing in this area is basically the same as in the area below it. Spinnerbaits fished in pockets is the choice of the better fishermen, but Rat-L-Traps fished in the middle of schooling fish is good in that situation, too

The blow-downs on the main river also will hold bass in the fall. Worms are excellent in this fallen wood.

Once winter arrives, the ledges off the mouth of creeks will hold large numbers of fish. A black/blue jig-and-pig tipped with a No. 11 pork frog is the best bait for winter bass because it's slow and falls rapidly to the deep water where the fish are.

Bankhead Reservoir's best fishing spots

1. to 3. Excellent spring pocket. Mouth is excellent in early spring with grub. Up in pockets for shallow-water during spawn.

4. Mouth of Yellow Creek. Good spring and fall staging area.

5. Deep water bend in Yellow Creek. Rocks hold lots of crawfish. Good jig fishing area. Brushtops out in water.

6. Tan Troff Branch. Good boathouses and piers with brush. Jigs and grubs in spring.

7. Shoal Creek. Excellent point. Good year round.

8. Camp Creek. Good blowdowns. Shallow water in back. Some spring bass.

9. White Oak Creek. Left fork has grass and blowdowns. Good in spring.

10. Walker County Shoals Creek. Good bluffs in mouth. Grubs in mouth in winter.

11. Back of creek. Lots of grass, lily pads. Good spawning area.

12. Hurricane Creek. Tremendous number of piers and boat docks. Jig-and-pigs in brush year round. High banks block out wind on windy days.

13. Short Creek. Full of pine and hardwood blowdowns. Grass for 200 yards on right bank going in. Excellent spinnerbait and flipping area. Both deep and shallow water.

14. Fryley's Creek. Lots of flats. Right bank has lily pads, grass or weeds all the way to back. Good in spring, early summer and fall.

15. Mouth of Friley's Creek. Good year round.

16. Coal Bed Creek. Narrow creek with deep water; grass and blow downs both sides. Good topwater fishing under overhanging bushes. Good fishing year round.

17. Good pocket. Deep water with blowdowns. Limited grass. Log jams in mouth. Good year round.

18. Islands. Often passed up. Good grass. Good fishing on all sides of islands.

19. Prescott Creek. Good spawning area. Left bank good pockets with blow downs and grass.

20. Good bay. Warms good in early spring. Weeds, lily pads, blow downs and boat docks. Sleeper hole.

21. Big bend in little river. Good island. Heavy cover and rocks in deep water. Good year round.

22. Pocket in bend of river. Good water depth. Good year-round fishing.

23. White Branch. Mouth has grass on both sides. Good spring fishing with jigs and spinnerbaits.

24. Back of White Branch. Good crankbait area in spring. Flat of lily pads in back. Back pond warms up early in spring.

25. Mouth of Bluff Creek. Good brush pile in mouth with weeds on right side. Excellent in dead of winter. Good worm or jig-and-pig.

26. Piers and boat docks in Bluff Creek. Good brush around them.

27 to 30. Excellent pocket. Spring spawning area. Shallow water with grass.

31. Mouth of Richardson Slough. Right bank has two good pockets just inside mouth. Good spring fishing in grass.

32. High bank in Richardson Slough. Deep water. Overhanging bushes. Good topwater fishing.

33. Large grass and lily pad patch on left bank of Lost Creek. Big fish area.

34. Pockets with grass below steam plant. Good area for big fish in early spring and in dead of winter.

35. Steam plant. Area just below discharge good year round with crankbait.

The Ultimate Guide to Alabama Fishing

Bankhead Reservoir

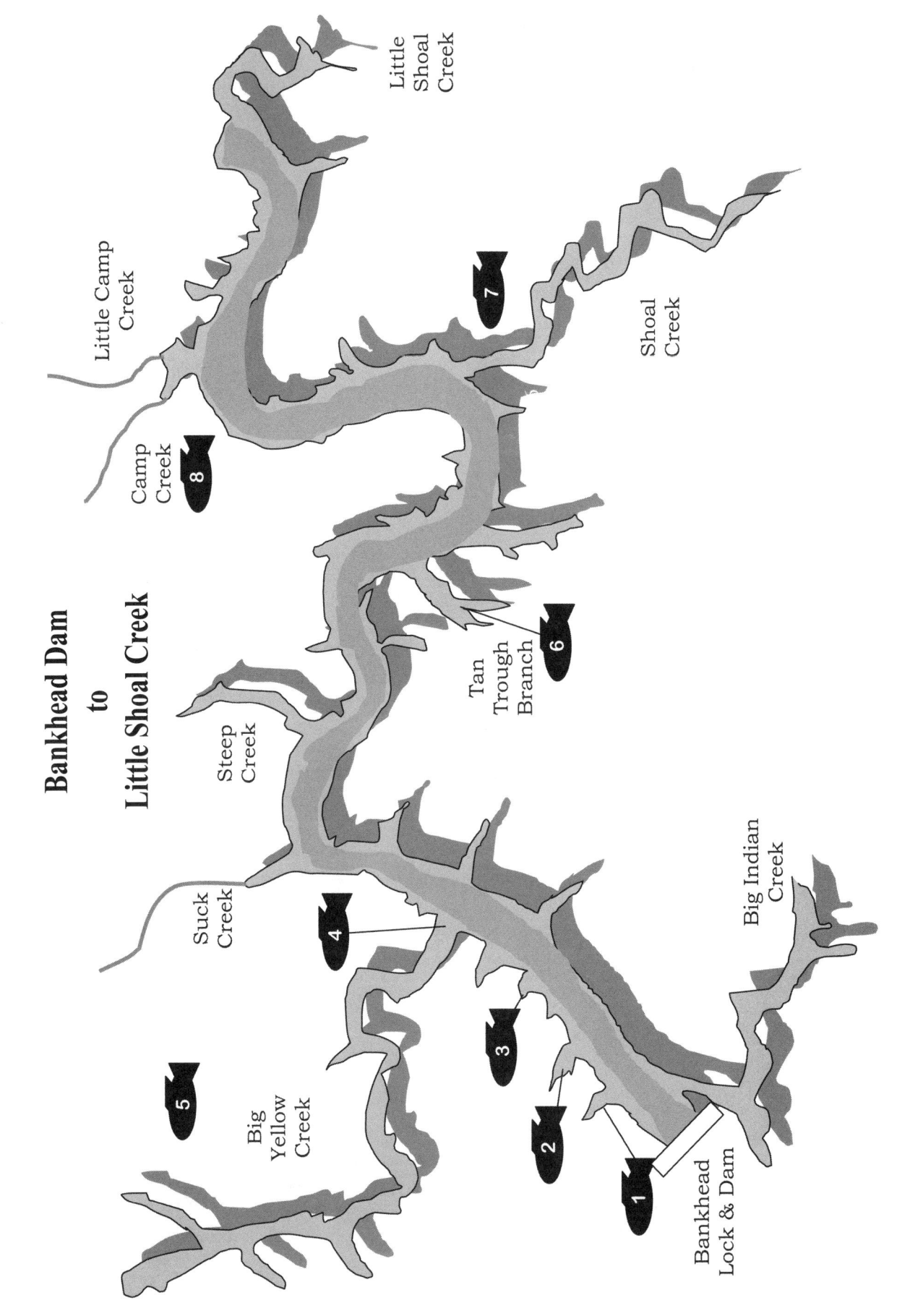

Bankhead Dam to Little Shoal Creek

- Little Shoal Creek
- Little Camp Creek
- Shoal Creek
- Camp Creek
- Steep Creek
- Tan Trough Branch
- Suck Creek
- Big Yellow Creek
- Big Indian Creek
- Bankhead Lock & Dam

142

Mike Bolton

Bankhead Reservoir

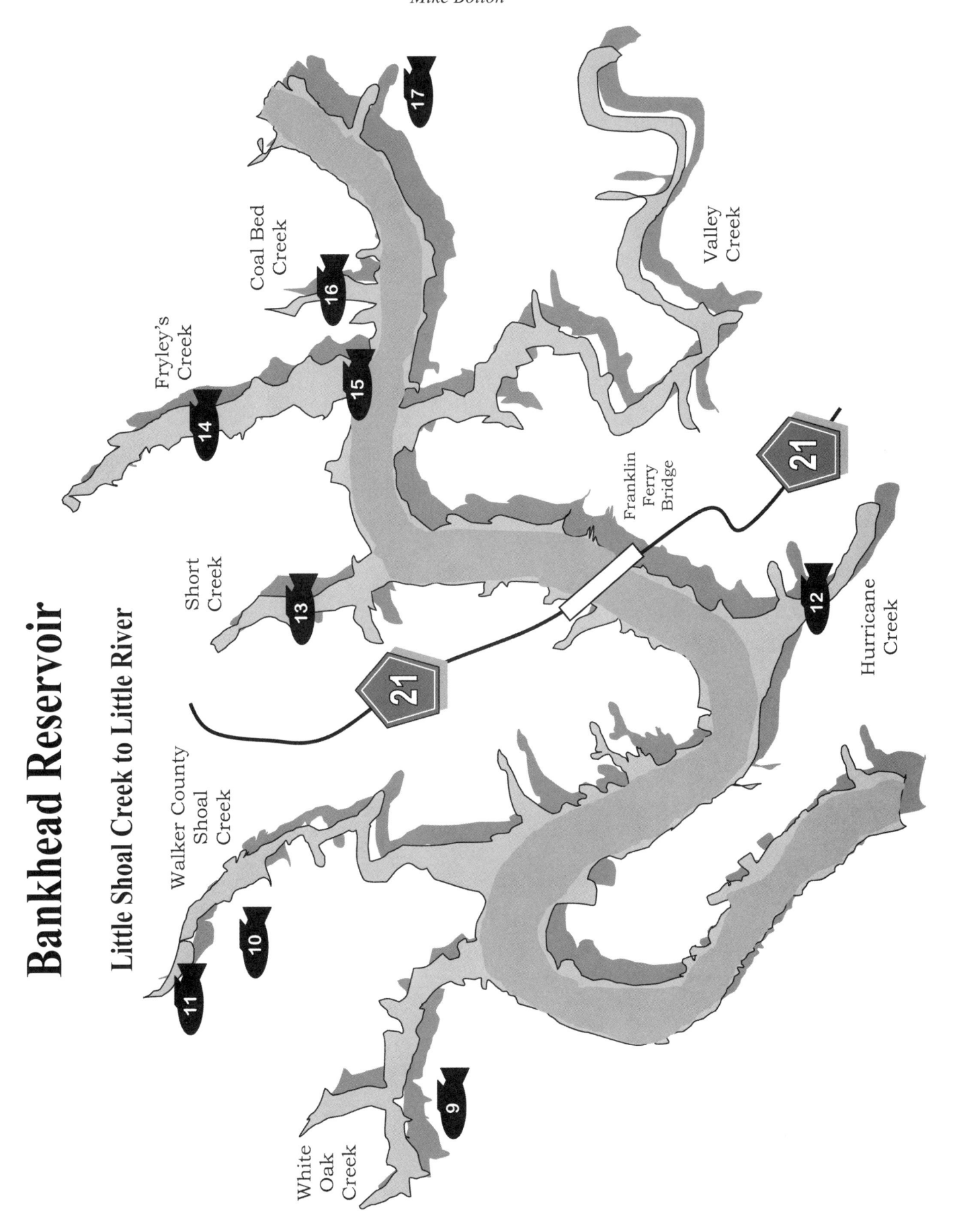

Little Shoal Creek to Little River

Valley Creek

Coal Bed Creek

Fryley's Creek

Short Creek

Franklin Ferry Bridge

Hurricane Creek

Walker County Shoal Creek

White Oak Creek

143

Bankhead Reservoir
Little River area

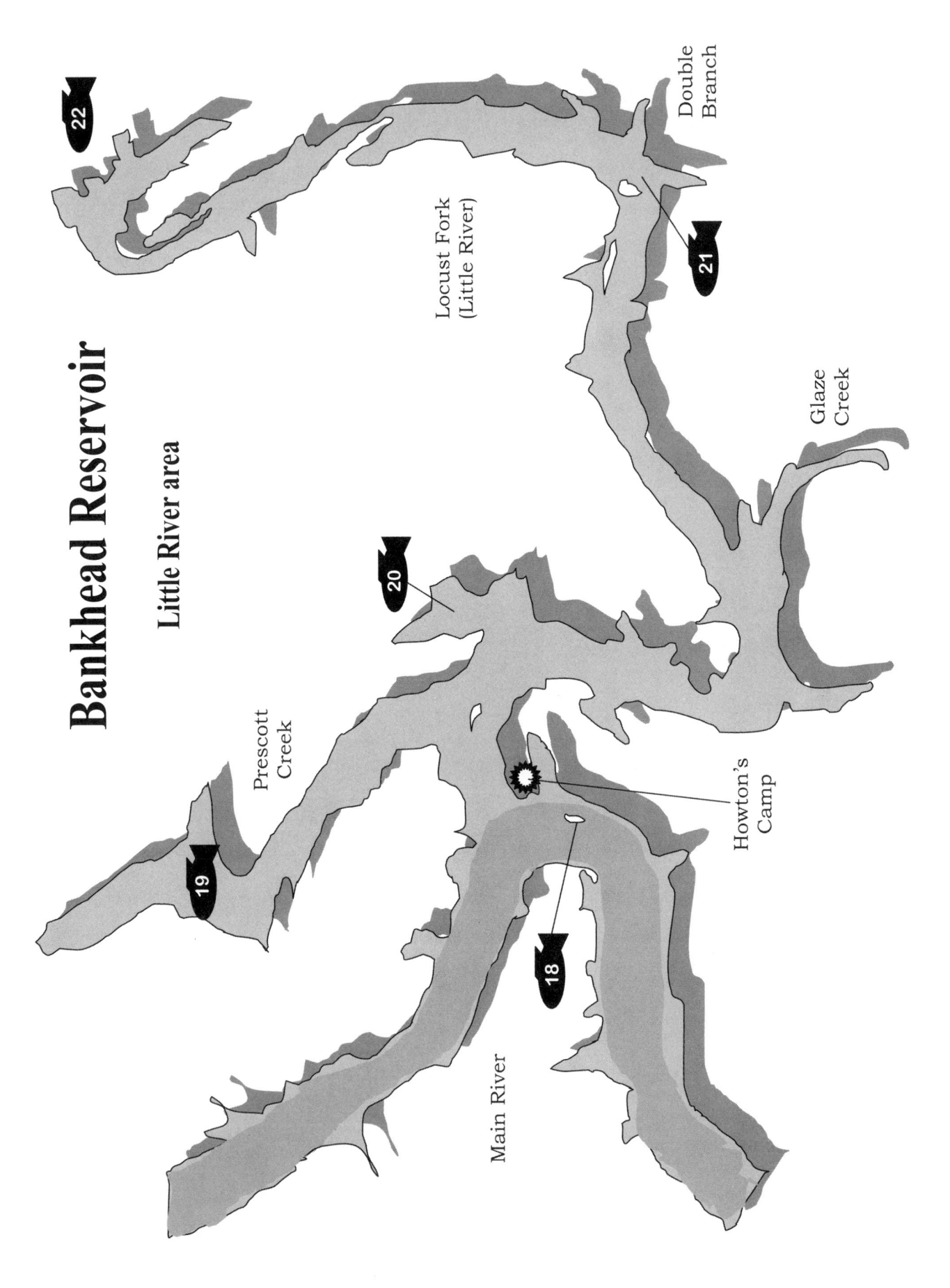

Mike Bolton

Bankhead Reservoir

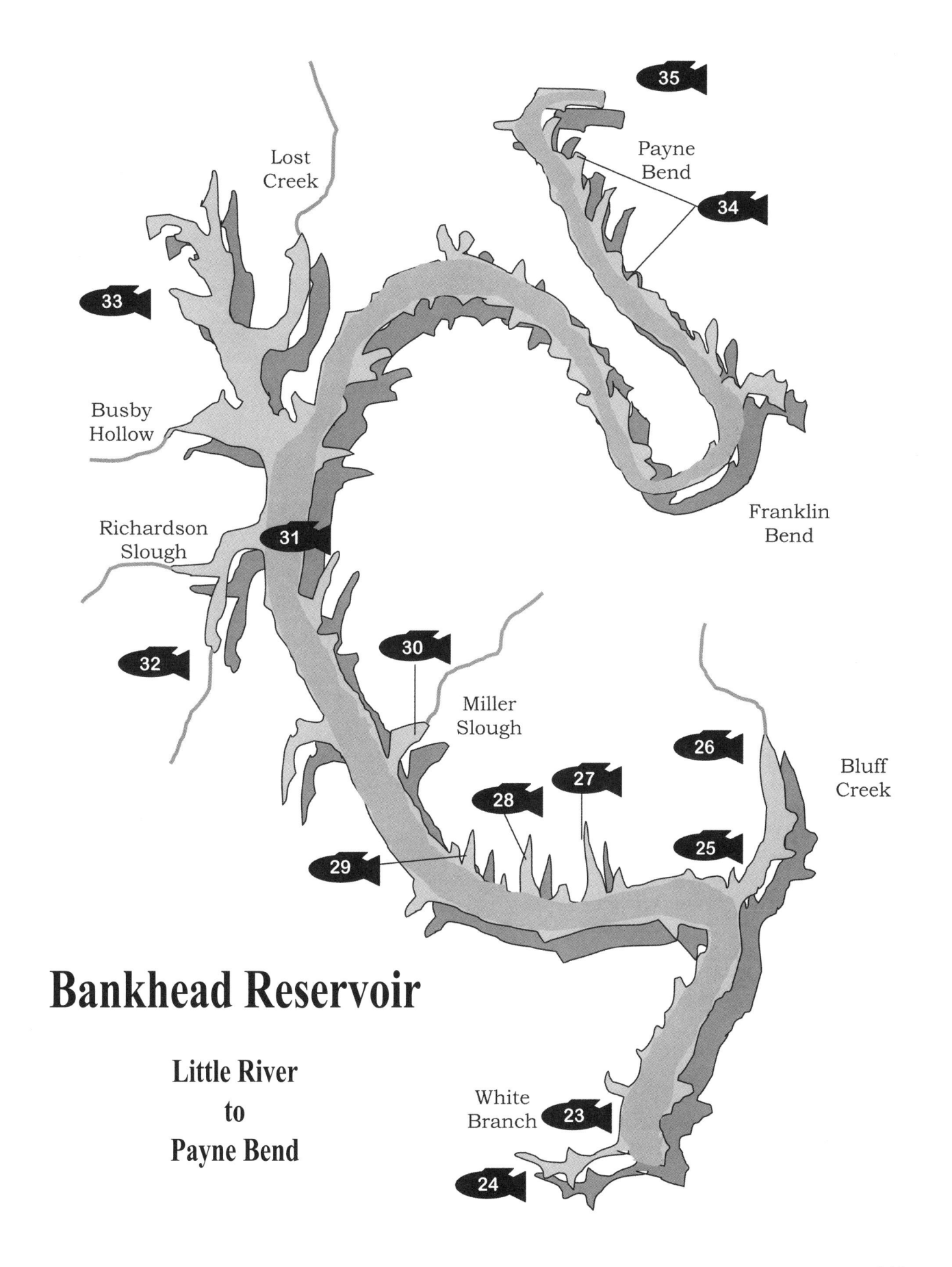

Little River to Payne Bend

Lost Creek
Busby Hollow
Richardson Slough
Miller Slough
Payne Bend
Franklin Bend
Bluff Creek
White Branch

The Ultimate Guide to Alabama Fishing

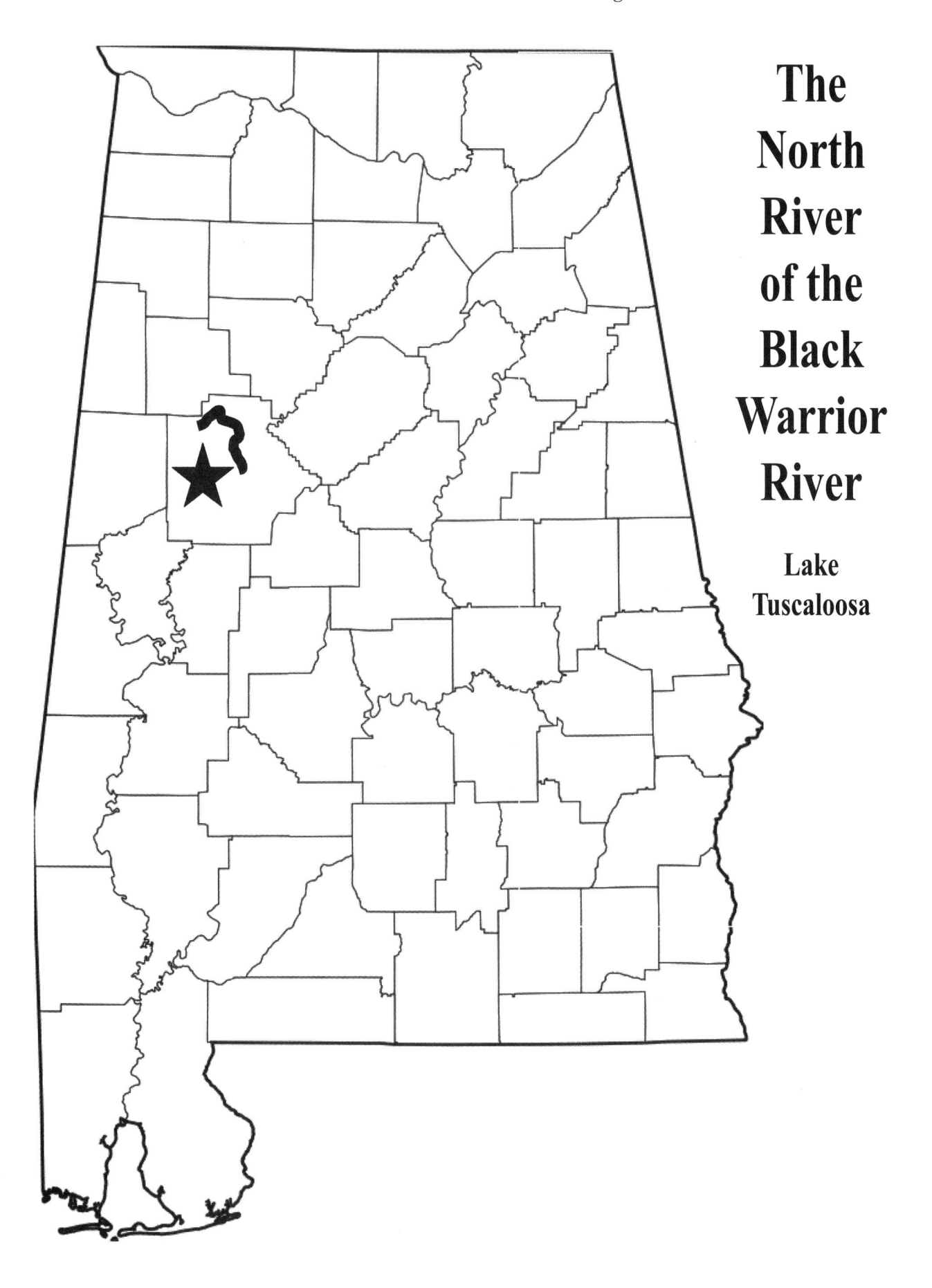

The North River of the Black Warrior River

Lake Tuscaloosa

Lake Tuscaloosa

Lake Tuscaloosa wasn't created with fishing in mind but it has been a decent bass lake through the years. The clean, clear waters of North River were captured and impounded to become a water source for the city of Tuscaloosa. All timber was cut, piled up and burned before impoundment. The result is a lake almost void of natural structure.

There are many good bass fishermen that have found that scenario very much to their liking, however. That situation enabled them to build their own private structure that isn't there for every fisherman to see. Those fishermen excel on the 5,885-acre lake.

The problem with Lake Tuscaloosa is the fishermen rather than the fish. Fishing deep, clear, structureless water is something most anglers are not accustomed to and after a few unsuccessful tries, most give up and go elsewhere.

The result is a lake that receives little fishing pressure. That, coupled with deep, clear water computes into numbers of big fish.

Those who have learned to properly fish the lake are more than glad that most anglers think of it as a lake for waterskiing.

Spring bass fishing

Hagler Bridge Road to Binion Creek

Flat, wide, dingy water greets the angler fishing this end of Lake Tuscaloosa during the spring. Weed-filled creeks, stump-filled flats and islands dot the landscape. This area is as close as Lake Tuscaloosa anglers get to fishing a reservoir that resembles other state reservoirs.

Three creeks offer good spring fishing. Binion Creek offers relatively shallow, dingy water and is an excellent spring spawning area. It boasts a few weed-filled sloughs. Chartreuse/black spinnerbaits with a single gold willowleaf blade work well in the weed-filled sloughs, but medium-running, pearl-colored crankbaits and Carolina-rigged plastic lizards are needed for much of the rest of the creek.

Turkey Creek is a wide, shallow creek that offers weed beds on the back side of Treasure Island, as well as weed beds in the back of the creek itself. Spinnerbaits and buzzbaits work well in the grass areas, as do plastic worms. Shallow-running crankbaits such as Rapalas work well on the fringes of the grass in early spring when the fish aren't yet buried in the grass.

Dry Creek, which can be found further north, is a shallow creek filled with grass beds throughout its length. The creek has excellent color in spring and offers excellent buzzbait fishing when the bass move shallow. This creek is home to a large largemouth bass population during the spring.

Once you go north of Dry Creek, you find nothing but the original run of North River, but the area can offer good spring fishing. This water is safe to navigate until you reach Skelton Bend. North of there, your boat motor's lower unit will discover rocks. The area above Skelton bend is littered with large boulders.

It's in this area you'll find is what is usually the lake's only moving, aerated water. Spinnerbaits, buzzbaits and floating worms are excellent spring baits here.

Binion Creek to Carroll's Creek

The portion of the lake between Carroll's Creek and Alabama 43 bridge is much more spread out than portion of the lake below it. There's still plenty of 90-foot water in the main river channel, but this area offers plenty of flats and ledges around the banks, especially above Alabama 69.

Spring fishing is better than it is in the section of the lake below because of the shallow water and the fact this area will normally have some stain during the spring.

Deep-diving crankbaits are good in February and early March as the bass begin staging out of the deep water and up the points and ledges. Plastic worms and jig-and-pigs can be effective. During this time of the year, the bass will suspend in deep water along the ledges and points and occasionally race up on the flat to feed. Once the urge to spawn becomes strong, the fish will spend more and more time on the ledges.

Once the bass reach water 10 feet or less and stay there for long periods of time, spinnerbaits become the preferred bait. Many flats will be covered in dormant grass beds and this underwater cover offers excellent protection to baitfish and crawfish. A chartreuse spinnerbait with a willow leaf blade matched to the size of the bait and fished over the grass beds will produce the lake's best fishing of the year.

This section of the lake also boasts the lake's largest population of largemouth bass. This area has traditionally produced the lake's largest fish during spring.

Carroll's Creek to Lake Tuscaloosa Dam

The lower end of Lake Tuscaloosa is the last place in the lake that takes on any color in the spring, that is if it does take on any color. It is not uncommon for this end of the lake to remain crystal clear year-round. When the water surface temperature reaches 55 degrees, which is usually in late February or early March, the bass begin leaving their deep-water haunts and moving into shallow water in anticipation of the spawn. The spawn on this lake normally takes place from late March or early April through mid-May.

The bass begin their heavy feeding at this time, dining heavily on the distressed shad population, minnows, crawfish (their No. 1 food in this lake) and small sunfish.

Due to the lack of structure, the spring bass begin staging on the many points in the area. Keep in mind that bass in this lake are point-oriented year-round.

These bass are prone to backing off this time of the year at the slightest weather or sunlight change. They may be found on a shallow point one day, and suspended in schools 30 to 40 feet deep the next.

Spinnerbaits and buzzbaits aren't as effective on this end of the lake during spring months as you might expect. The bass want a slower topwater bait and a torpedo bait fits perfectly. This lure thrown around shallow points and in shallow slough mouths this time of the year will usually produce fish.

On the days when the fish are backed off and suspended over deep water, trolling with deep-diving crankbaits and jigging with spoons is sometimes effective.

If your depth recorder shows schooling bass are staging 15 to 30 feet deep along the many stair-stepped ledges in early fall, a slow-moved worm or jig-and-pig can be effective.

When these bass actually go into the spawn, they do so in the extreme back of pockets and in creeks. Pockets with red clay or sand are best. Carroll Creek is one of the best creeks for spawning activity.

Hot weather bass fishing

Hagler Bridge Road to Binion Creek

When the spawn is over and the water begins to heat rapidly, the days of finding fish with topwater baits end quickly. The bass leave the creeks and head for the cooler temperatures that can be found along the original North River channel.

Although the surface water may seem dead this time

of year, a slight water flow can often be detected in the original channel.

Large spots and an occasional large striped bass can be caught by jigging spoons during the summer.

Night fishing is best during the summer as fish do sometimes move away from the channel toward shallow water. Deep-running crankbaits, spinnerbaits run along the bottom and jigging spoons are good baits for fish that you first mark with a recorder.

Binion Creek to Carroll's Creek

Summer fishing here is point-oriented, but anglers also have the advantage of being able to fish around boathouses and piers at night. These piers and boathouses boast excellent fishing because they offer a bit of structure in a structure-starved lake. Many boathouses and piers have Christmas trees and brush piles around their perimeters.

If an angler is fishing during the day, he must plan to fish deep-diving crankbaits, worms or jigging spoons for fish suspended along the points and ledges. These fish can be anywhere from 20 to 50 feet deep.

The angler choosing to fish at night will find much better fishing. The bass will move into the areas around the boathouses and piers to feed on the baitfish and crawfish that find protection in the brush piles. Worms, jigs-and-pigs and grubs are the top baits for night fishing.

From May to September most of the fishing takes place at night.

Carroll's Creek to Lake Tuscaloosa Dam

Once the spawn is over, bass head back to the deep-water points and suspend themselves there until fall.

Deep-diving crankbaits are effective in May and June, but in July and August jigging spoons are needed because bass are commonly found 35 to 60 feet deep.

Night fishing is best during summer months. One, it is more bearable to the angler because he doesn't have to deal with the sun. Two, the fish move to shallower water because they feed at night in summer months. Three, ultra-light line isn't needed because line is harder to see at night.

Spinnerbaits are excellent for night fishing. A black, double-bladed spinnerbait with a black pork frog works best. Position your boat in deep water and cast to the bank, and begin a retrieve just fast enough to make the blades spin. Using this method, the spinnerbait drops down along the bottom as it falls off.

A plastic worm can be fished in much the same way, but but a line that glows and a black light are needed so the angler can keep tabs on his line.

Lake Tuscaloosa is an excellent lake for night boating since there is nothing in the open water for the boater to come in contact with.

Cool and cold weather bass fishing

Hagler Bridge Road to Binion Creek

The cooling weather of September and October trigger bass to move toward shallow water to find baitfish and crawfish on which to gorge to store fat for the winter.

Crawfish or shad-colored deep-running crankbaits, jig-and-pigs and worms can catch these fish after they are located with a depth finder. The mouths of creeks and sloughs next to deep water tend to hold the fish better than other areas.

Binion Creek to Carroll's Creek

Fall fishing is similar to fall fishing in the lower part of the lake, except this area occasionally gets some color due to heavy rains. When this water gets stained, fall fishing can be excellent by Lake Tuscaloosa standards.

When August passes and cooler temperatures begin to creep in, the bass will begin a gradual move to shallow water. Worms, deep-diving crankbaits and jig-and-pigs work well.

The wider expanses of water between the Alabama 69 and Alabama 43 bridges offer the best fall fishing. Boathouses, the mouths of sloughs and weed beds next to deep water are typically the most productive areas.

When cold weather arrives, the fish move out on the ledges next to the natural river run and become hard to find. It's possible to find large schools of bass in 50 or 60 feet of water in December or January. A depth finder is needed to locate the fish.

Once they are located, jigging spoons or jig-and-pigs will sometimes take fish.

The strikes at that depth are gentle, so anglers must pay close attention.

Carroll's Creek to Lake Tuscaloosa Dam

When fall arrives, bass abandon the super-deep water and move into shallower water, usually 15 feet or less. This shallow water holds a good supply of oxygen, which also attracts baitfish and crawfish.

Primary areas for fall bass include the areas around rock walls and in the rocky portions of creek mouths.

The spotted bass is the major bass species in the lake and spotted bass like rock bluffs.

The 4-inch plastic worm in watermelon, apple, pumpkin seed or salt and pepper, fished with 6-pound test line, makes a perfect fall combination.

Lake Tuscaloosa's best fishing spots

1. Weed-filled, shallow pocket next to Hagler Road Bridge. Excellent spawning area for largemouth and spots. Good spring fishing with topwater. Last slough deep enough to fish in as you head north.

2. Mouth of weed- filled slough. Excellent fishing in early spring and October and November. Worms or spinnerbaits.

3. Weed-filled slough. Treetops and brush on left bank. Good spring spawning for bass. Only slough large enough to fish in until you reach Dry Creek.

4. Mouth of slough. Excellent fishing in early spring and early fall.

5. Dry Creek. A solid weed bed.

6. Weed beds next to deep water.

7. Trash and brush piles in entrance to Dry Creek.

8. Weed beds is shallow water.

9. Island. Trash and weeds surround island.

10. Shallow pocket above Bull Slough Bridge. Man-made trash piles. Good fishing in spring and fall. Excellent spawning area.

11. Bull Slough Bridge. Man-made brush piles adjacent to main river channel. Main river channel cuts directly against west bank under bridge.

12. Sharp bend in original river channel. Excellent fishing in early fall.

13. Weed beds around front of Treasure Island.

14. Weed beds in back of Turkey Creek. Good spring spinnerbait fishing. Excellent spawning area.

15. Good sharp points and deep sloughs. Fishing good on points with worm in summer. Back of pockets filled with weeds. Good spring fishing.

16. Area between island and bank. Fish like a creek mouth. Area comes off banks shallow and fall directly into river channel.

17. Stump-filled shallow area on back side of island. Adjacent to deep water. Good spring fishing.

18. Underwater road bed. End of original road is used for boat ramp. Old bridge that crossed the river there was exploded and collapsed into water. Concrete and steel rubble remains. Good night summer fishing with worm.

19. Alabama 43 bridge. Riprap under bridge along both banks. Good fishing year round.

20. Good slough. Riprap from bridge area curves up into slough for good distance. Good weed beds and brush piles. Good in early spring and early fall.

21. Bridge. Good spring and fall fishing around pilings. Creek above bridge is shallow.

22. Underwater islands in upper end of Binion Creek. Dangerous area to navigate, but good spring fishing can be found between islands and cuts. Weed beds and brush piles around islands.

23. Back side of island east of main river channel. Adjacent to deep water. Good fall fishing.

24. Boathouses. Treetops and trash in shallow water next to original river channel. Excellent summer fishing at night.

25. Sharp River bend. Underwater island well-marked with buoys. Good in summer.

26. The Narrows. Good stretch with boathouses and piers. Trash and brush around those structures. Good year round.

27. Weed-filled slough. Excellent spawning area.

28. Tierce Patton Farm Road bridge. Riprap along east banks under the bridge. Pilings also hold fish.

29. Deep bank with trash along west bank. Good at night in summer and late fall.

30. Boathouses. Good brush piles off end of structures in deep water. Good summer fishing.

31. Deep slough. Excellent fishing year round.

32. Lary Lake Road bridge. Bridge is narrow, but slough widens past bridge. Good weedbeds and trash in back of slough.

33. Points along south bank. Good fishing year round.

34. Boathouses next to main channel. Lots of trash around structures.

35. Good weed bed-filled slough.

36. Excellent stretch of lake. Deep water, trash, boathouses. Good year round.

37. Underwater road bed.

38. Horseshoe-shaped underwater dam. Located 100 yards off of north bank. Five to 10 yards wide. Located in 15 feet of water, but deep water nearby. Excellent summer fishing. Locate dam with depth finder and mark with buoys. Back off and fish with deep-running crankbaits or worms.

39. Beacon point. Weed-filled sloughs on both sides of point. Boathouses and islands, also. Excellent spring spawning areas. One of best fishing areas in the lake.

40. Excellent sloughs filled with boathouses and weeds. Adjacent to deep water.

41. Steep rock walls with trash and brush at base. Excellent summer night fishing. Adjacent to deep water.

42. Alabama 69 bridge. Riprap on both banks. Pilings hold fish year round.

43. Tierce's Slough. Long, deep, weed bed-filled slough. No boathouses. Good spot-holding area.

44. Boathouses and piers with brush.

45. Weedbed-filled sloughs. Also good points outside.

46. Good point. Early spring fishing. No trash or weeds, but good ledge.

47. Good, long-sloping point. Excellent summer fishing.

48. Good points for summer fishing.

49. Hamner Creek. Weed beds and trash. Good points in mouth of creek. Good year round fishing.

50. Island. Good fishing around island early spring.

51. Mayor's Point. Boathouses with brush. Excellent fishing in hot weather.

52. Carroll's Creek Island. Weed beds on back side. Good spring fishing.

53. Back of Carroll's Creek. Weed beds and boathouses. Good spring spawning area.

54. Excellent point for year-round fishing.

55. Underwater island. Located in 8 feet of water, but adjacent to 110 feet of water. Top of island filled with brush tops. Good year round.

56. Brush Creek. Lots of weed-filled pockets. Deep points.

57. Weed-filled pocket.

58. Weed-filled pocket.

59. Good point.

60. Wishbone Slough. Good pocket with grass.

61. Slough above North River Yacht Club. Good brush in back.

62. Crown Pointe Slough. Boathouses, some grass.

63. Good grass-filled pockets on east side of lake near dam.

64. Small footbridge to building over water. Good crappie fishing area.

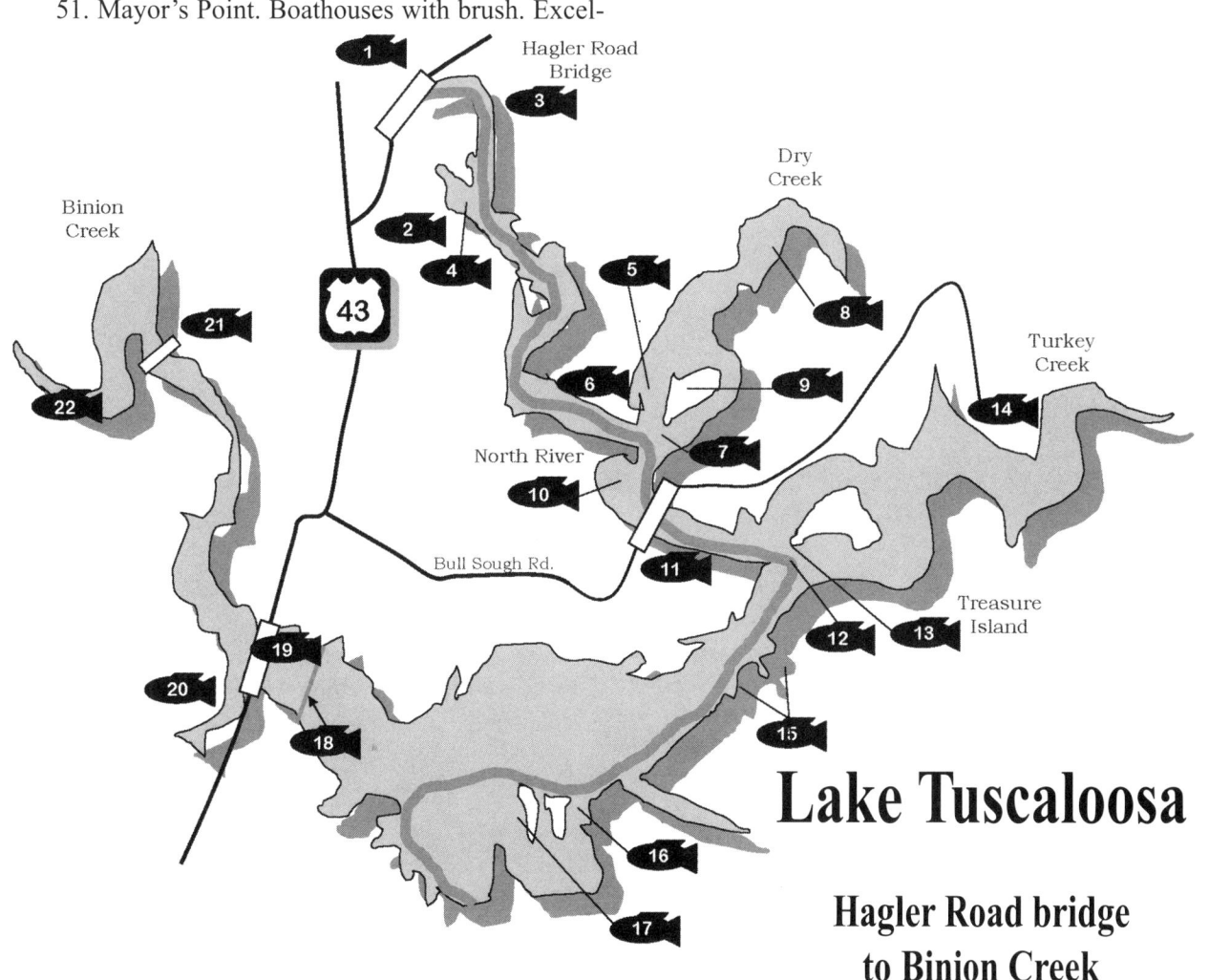

Lake Tuscaloosa

Hagler Road bridge to Binion Creek

The Ultimate Guide to Alabama Fishing

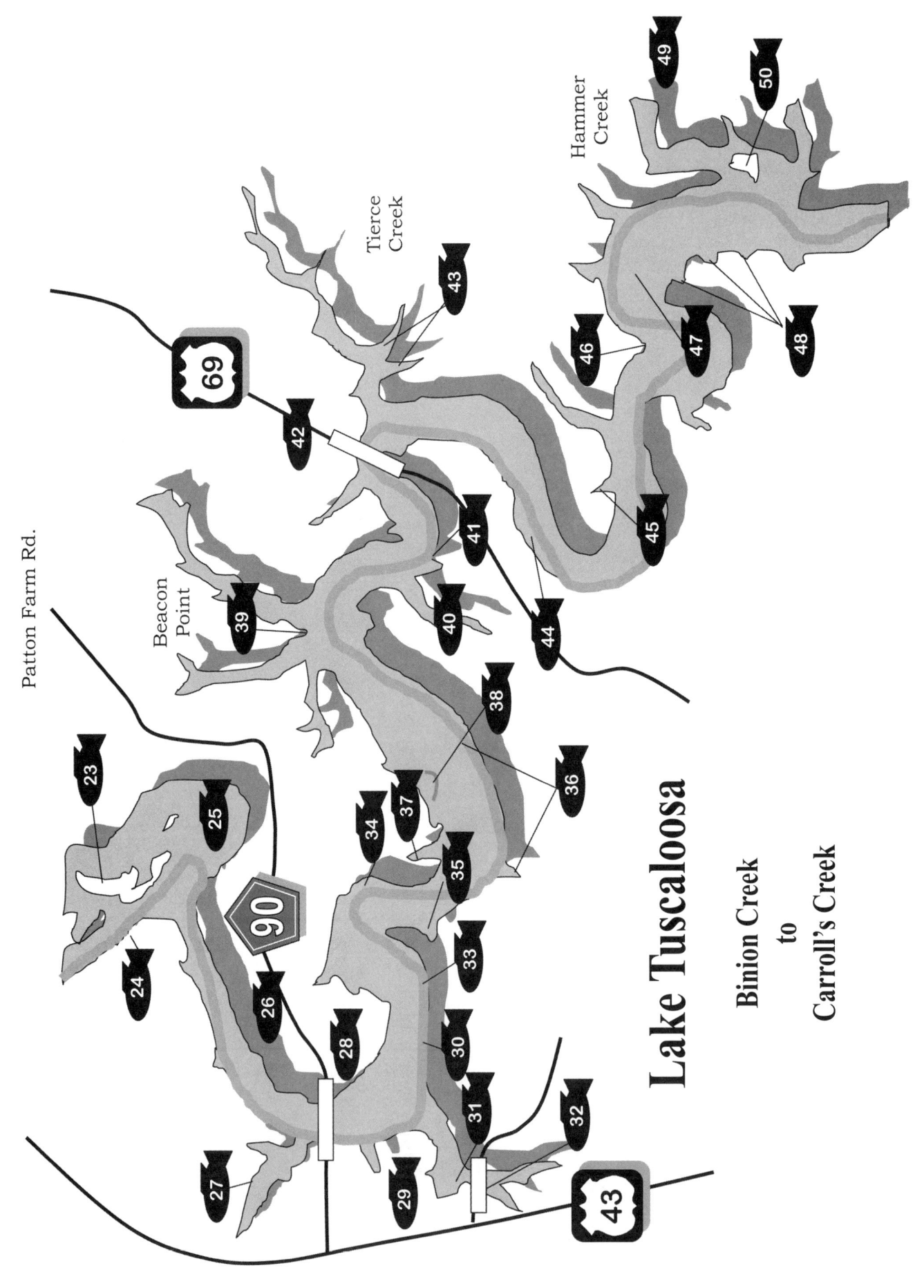

Lake Tuscaloosa

Binion Creek to Carroll's Creek

Lake Tuscaloosa

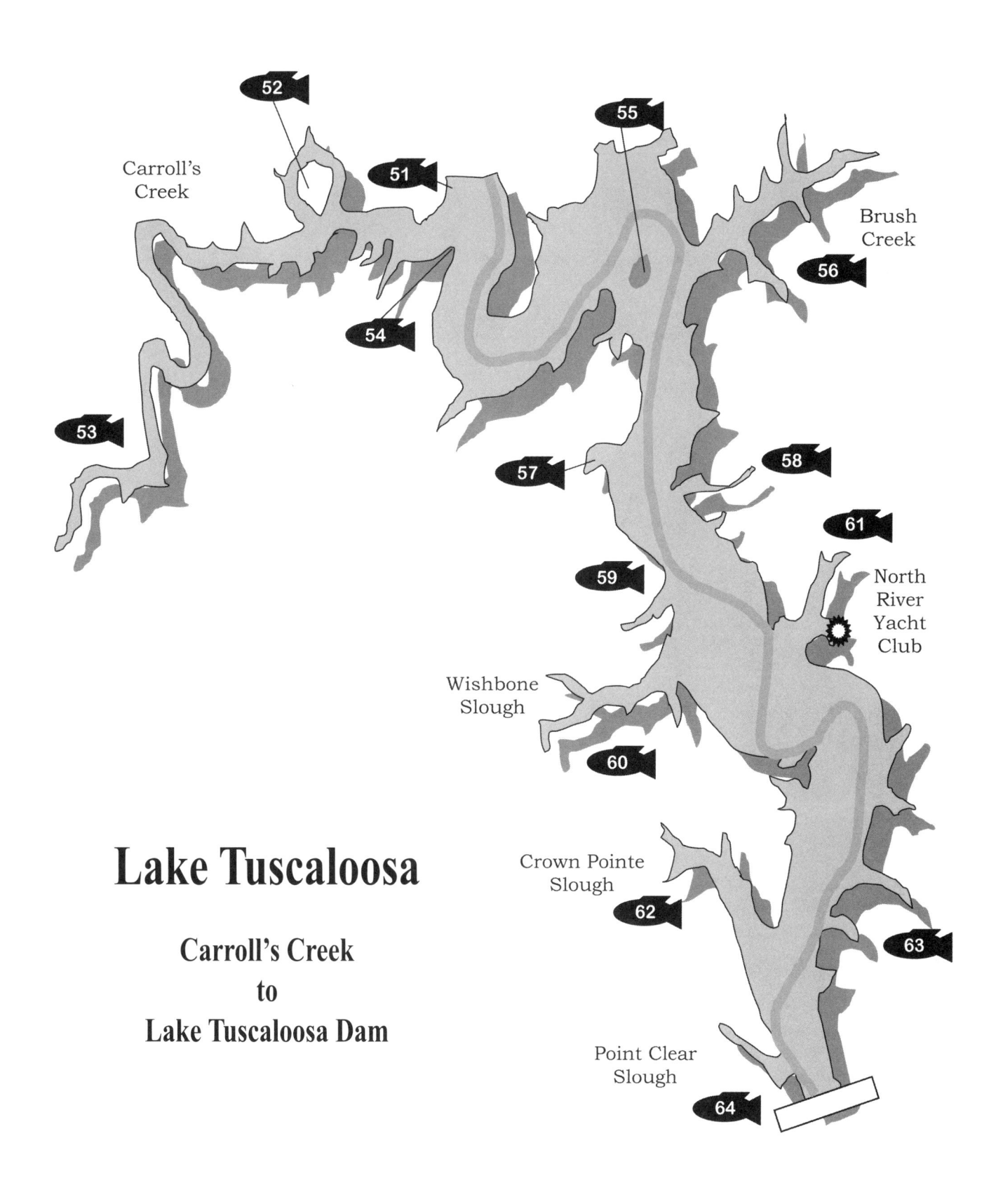

**Carroll's Creek
to
Lake Tuscaloosa Dam**

The Ultimate Guide to Alabama Fishing

The Chattahoochee River

West Point Lake

Lake Eufaula

West Point Lake

Spring bass fishing

West Point Lake may lack the reputation of its downstream neighbor, Lake Eufaula, but the anglers who frequent the Chattahoochee River reservoir find the easy-to-locate roadbeds, rock piles and underwater bridges and trestles more to their liking.

The 16-inch size limit imposed on the lake years ago has resulted in an excellent bass fishery. According to the state's Bass Anglers Information Team annual report, Lake West Point annually ranks near the top of state lakes in the production of bass 5 pounds or larger.

It should be noted that only a small portion of the lake is in Alabama, and a non-resident Georgia fishing license is needed by Alabama anglers to fish most of the lake. An Alabama fishing license is honored only on the small portion of the lake west of the Rock Mill Road (Highway 109) bridge. So when you've got the licenses straight, this becomes a lake to enjoy.

Chattahoochee River above Yellowjacket Creek

The presence of excellent springtime fishing in nearby Yellowjacket Creek and further downstream in Wehadkee Creek makes bass fishing almost a wasted effort in this portion of West Point Lake.

The only exception to that rule is in early spring when the bass haven't gotten into the creeks and sloughs to spawn and heavy rains causes the release of large amounts of water from upstream. The swift current resulting from the released water forces the bass to escape in the mouths of the numerous creeks, sloughs and pockets throughout the upper end, with this being more pronounced the further you go upstream.

These creeks, pockets and sloughs will get muddy immediately after the heavy rains and fish will be impossible to catch, but several days of swift water will suck the muddy water out.

Shad and other baitish will be forced into the pockets to escape the current, and the bass will follow. Brightly colored crankbaits (chartreuse or orange) work nicely in the mouths of these pockets when the water is dingy, but the color should be toned down all the way to white as the water clears.

Spinnerbaits with a pork chunk work nicely in the small pockets and logjams in the main river itself during this period.

Yellowjacket Creek

For the angler who loves fishing visible structure in spring, Yellowjacket Creek is just what the doctor ordered. The back end of the creek is full of grass and explosive topwater action is common there. There are numerous bridge pilings and acres of standing timber. Less visible, but easy to locate rock piles and deep points are also abundant.

Bass determined to spawn will start working their way up this creek sometimes as early as February. They don't have to travel far to find shallow pockets and sloughs in which to spawn, but many make a run as far back in the creek as they can go. The area between the U.S. 27 bridge and Young Mill Road bridge is full of standing timber and is a spinnerbait fisherman's heaven. A 3/8-ounce spinnerbait with a chartreuse and black skirt and a nickel willowleaf blade work in this shallow standing timber has won many tournaments. The same pattern will also work in the standing timber in the north bank of Beech Creek, in the back of Beech Creek and in any of the grass beds throughout the creek. Buzzbaits and floating worms also work in the grass beds.

Flipping in the grass beds and standing timber is also a good pattern in the early spring.

Yellowjacket Creek to West Point Lake Dam

Unlike many Alabama reservoirs, the lower end of Lake West Point boasts enough deep water outside of the original river channel that bass don't find it necessary to suspend themselves in the channel itself to find water warm enough to tolerate. It's not unusual to locate water 40- to 45-feet deep as much as 200 yards from the channel on the lake's lower half. This means some adjustment to the normal patterns that many anglers use in early spring.

Anglers fishing the late-winter/early-spring pattern on most Alabama reservoirs traditionally fish the ledges next to the original river channel. The fish suspended in the channel in winter begin to make frequent trips to these ledges as the water warms and their appetites improve.

Should a fish leave the original channel on this end of Lake West Point, however, it is likely to still find itself in as much as 65 feet of water. It's best to consider anything between the banks in this area the main river channel and pretend you don't know the original channel exists.

A multitude of long-running ridges, humps, road beds, old home foundations and even an old peach orchard can be found in water deep enough to satisfy bass during the cold winter months. It is these areas—not the main river channel—that the bass will begin leaving in the early spring.

A tremendous amount of easy-to-locate roadbeds (see map) lead directly from the main channel all the way up to the banks.

The route that automobiles once took to get to the river before the lake was impounded is now a natural roadway for bass (and other species) to migrate from deep water to shallow water during the spring.

In the late-winter/early-spring period, the better bass anglers first check the timber- and brush-lined humps, ridges and long-running islands with deep-running crankbaits and jig-and-pigs. The water is often dingy during this time, so a chartreuse or bright orange deep running crankbait makes a good weapon. If you go the jig-and-pig route, try a black with brown pork frog or a chartreuse pork frog trailer.

Should these areas prove to be unsuccessful, and they probably will once the weather gets tolerable enough to fish, it is then you start fishing the bass's natural routes to more shallow water.

Rainbow Creek, Wehadkee Creek, Indian Creek, Maple Creek and many of the large shallow pockets and shallow sloughs have roadbeds leading directly from deep water right into their mouths. You don't have to be a rocket scientist to understand the significance of that. An angler can use maps in this book and his depthfinder to locate these east-and-west running roadbeds and drop marker buoys along their length. He can then fish from deep to shallow water and back on these road beds to find the depth the bass are on any given day.

Deep-running crankbaits, spinnerbaits, jig-and-pigs and worms can be successful. A 3/8-ounce chartreuse or yellow-skirted spinnerbait with nickel blades will work better in dingy water, but don't hesitate to switch to a white bait with gold blades if the water is clear.

It should be noted that roadbeds aren't the only location to find moving spring bass, however. The lake's lower half is full of wide, long, sloping points that serve the same function as roadbeds. They can be fished with the same baits and the same patterns as the roadbeds.

Buck bushes are something of a novelty particular to Lake West Point, but don't ignore their presence when spring is in full force and the bass have worked themselves up to shallow water. These bushes are normally out of the water, but good spring rains will put one and two feet of water around them.

Bass love the bushes as a means of structure. Buck bushes can be found on the river and up in the creeks and they can offer tremendous topwater action with buzzbaits. On the main river, the bushes are often the last stopping point before the bass move into the creeks with spawning on their minds.

Wehadkee, Stroud and Veasey Creeks

Easy-to-locate structure is a feature that make many bass anglers fish this area and this area only. The back of Wehadkee Creek is full of standing timber and much of it is visible even in high-water conditions. There is standing timber in some pockets and sloughs off of Wehadkee Creek, including Caney Creek. There also are a few islands scattered throughout the length of Wehadkee.

Other fish-holding structure includes fallen timber, piers, roadbeds and underwater bridges. There is a limited amount of weed beds.

Bass begin moving into this area looking for spawning spots in early spring. The fish will move great distances, with many moving to the standing timber in the rear of Wehadkee Creek.

The standard springtime baits can be used to take these fish. A 3/8-ounce white spinnerbait with nickel blades is the best bet since the water is usually clear. A spinnerbait fished around the timber and stumps can produce quality fish.

Floating worms, buzzbaits, Rapalas, plastic worms and a host of other baits can also be successful. Before the fish get into the spawn good and after they come off, crankbaits fished on the points and around bridge pilings can produce fish. Key areas to look for include the Dewberry Access area in Wehadkee Creek (bridge pilings, an island, an underwater bridge, roadbeds and sloughs and pockets are in close proximity), the back of Wehadkee Creek (visible standing timber) and the Caney Creek area (an island, standing timber and a roadbed are close by).

An angler can find success in just about any area, however, if he applies what he knows about springtime bass fishing.

Hot weather bass fishing

Chattahoochee River above Yellowjacket Creek

This end of Lake West Point is virtually run of the river with only a few flats and no real widening of the river. Almost all of the creeks and sloughs in this end are too shallow to fish during the months when water levels are low.

There is relatively good fishing in this section of the river during the summer months, with the odds greatly increasing whenever there is some water flow. The mouths of the creeks and the ledges adjacent to flats is the best bet for summer fishing, with a swimbait, a plastic worm or lizard rigged Carolina style or a medium-running crankbait the best bet to take the fish.

The mouths of sloughs and creeks found in the bend of the river hold the most bass. Bass will school on these bends during summer months.

Fishing at daybreak or at night is the best opportunity to catch these fish when the water is off.

Yellowjacket Creek

Yellowjacket Creek is similar to most large creeks on Alabama reservoirs in that it will hold some bass year round. The relatively deep water found in the main channel of Yellowjacket Creek (45 feet deep a half mile into the creek), as well as relatively deep flats, tend to hold bass better than other such creeks however.

Boat houses, piers and docks can be found along the creek and some are in enough deep water to interest summer bass. The bass not only find shade there, but many docks have brushtops and Christmas trees around them which also attract baitfish. An electric red worm skipped under the piers will catch some of these fish.

Around bridge pilings in deep water, a spoon will also take some fish.

It is rare that you'll find the lake's better fishermen fishing this creek or any creek on the lake in summer, however. The lake's larger bass are enjoying the cool comfort of the deep, oxygenated water in the main river area.

Yellowjacket Creek to West Point Lake Dam

Even as late as the end of June and early July, you'll find a few post-spawn bass hanging around the mouths of all of the major creeks in Lake West Point's lower end. These natural funnels for baitfish are in unusually deep water (55 feet in Maple, Bird and Wehadkee creeks, 45 feet in Wilson, Alligator and Rainbow creeks). These extreme water depths in the creek mouths allow the fish to tolerate a creek mouth this time of the year.

Shallow-running crankbaits fished on the secondary points in and around the mouths of these creeks can sometimes produce tremendous numbers of bass. Spinnerbaits can also be highly successful. Don't doubt stories that some anglers catch good fish in these creeks year round. Deep-water piers and bridge pilings can produce bass during the hottest days of the summer and the coldest day of the winter, but these will usually be smaller fish than you'll find outside of the creeks.

Once hot weather arrives, most of the bass have begun their trek back out into the river. The road beds that the bass followed into these creeks are their highways back into deeper water.

Worms fished on top of and beside these roadbeds in water between 6 and 15 feet will catch good bass early in the summer. The roadbeds also offer excellent night fishing in late May and early June. Spinnerbaits slow rolled down the length of the roadbeds can be deadly.

Once the searing heat of late July and August arrives, the bass disappear from the roadbeds. They scatter to the deep, structure-covered mounds that parallel the main river channel and to the long, deep sloping points. Very few of these hot-weather bass will be found suspended in the original river channel.

Wehadkee, Stroud and Veasey Creeks

Whether you fish this area during the summer depends on whether you're playing or seriously trying to catch fish. Relatively deep water (up to 35 feet) can be found in Wehadkee Creek, Veasey Creek and Stroud Creek and this water will hold some hot-weather fish. Smaller bass will find refuge under the deep-water piers and docks during the hottest part of the day, then slip into deeper water early and late. Other smaller bass will find refuge around the bridge pilings under numerous bridges.

A good rule of thumb is that you will not find the quality bass in these creeks during the hot summer months, however. The better fish will have exited the creeks and taken up residence on the deep humps, islands and ledges.

Cool and cold weather bass fishing

Chattahoochee River above Yellowjacket Creek

Although this upper end isn't the best spot to find fall bass, whenever there's enough rainfall to allow a boat into the sloughs that have timber or stumps, flipping can produce some fish.

Wolf Creek and New River Creek both are filled with timber relatively close to their mouths and these areas hold bass in the fall. A black jig-and-pig with a brown frog pork chunk, 20-pound line and a flipping stick is all that is needed.

Yellowjacket Creek to Lake West Point Dam

Once the state gets a break from the unmercifully hot weather of August, the gradual cooling of the water in the lower end of Lake West Point sends bass to more shallow water. The anticipation of a long winter will send these fish into a feeding frenzy to fatten up. The fishing can be terrific.

These bass will slowly move from their deep-water haunts via the same routes they used to find shallow water in the spring: the many roadbeds and long, sloping points.

These bass won't seek out the same shallow-water spots, however, and actually the water they'll eventually settle into isn't shallow by spring standards.

These fall bass may make and occasional foray into 2 to 5 feet of water, (they've been known to even move into grass on the rare occasions the water is up that high), but finding fish at that depth isn't something an angler can count on.

These fall bass will settle onto the ridges in water 10 to 15 feet deep and hang close to rocks, stumps and fallen timber that offer both structure and an ample supply of food.

The ledges can be located with the use of a depthfinder and a good topographic map.

An angler should locate the ledges on the map, then pinpoint them with his depthfinder. The ledges should be traveled slowly until fish are marked. Marker buoys help pinpoint the ledges holding fish.

Several lures can take the fish. Crankbaits, worms, jig-and-pigs or Rat-L-Traps can be successful.

The bass' venture into more shallow water is short-lived, however. Once the water temperature begins to drop into the mid-50s and lower, the bass move to the deep-water humps, islands and off the end of some of the points that have drop-offs. Slow-moving lipped or lipless crankbaits can take these fish, but a jig-and-pig or spoon is your best bet. Locate the areas that have standing timber and you'll find cold-weather fish.

Yellowjacket Creek

If you ask a tournament fisherman if bass can be caught in this area during the cool- and cold-weather months, his answer will likely be a strong "no." The 16-inch fish he needs to weigh in during a tournament is a tough creature to locate here this time of the year. Fall fish will make a move in this direction, however, and can often be found in the mouth of the creek as far up as Cameron Mill Road bridge. The deep water found here along with access to shallow water ledges makes this a natural spot for actively-feeding fish to corner some shad.

Wehadkee, Stroud and Veasey Creeks

Fall and winter fishing here is much the same situation as summer fishing. It's possible you'll catch some bass, but these bass are generally going to be yearlings under 2 pounds. The larger fish are hanging in the deep water out on the main river.

The only exception to this is in the fall when the bass move into the mouth of Wehadkee Creek. These fall bass—and these are often quality bass—gather along the secondary points inside the creek mouth and are prime candidates for crankbaits, worms and jig-and-pigs.

The road bed (see map) running across the mouth of the creek starts in 15 feet of water on the north bank and runs southward as deep as 50 feet and back up to 5 feet of water on the south bank. You're likely to find fish at some depth somewhere on this roadbed during this time. This roadbed holds quality fish in fall and early spring, and some fish are taken off of it year round. Plan to use a jig-and-pig or spoon here during the fall and early winter.

Other types of fishing

Hybrid and stripe fishing

In a lake where a largemouth bass must be 16 inches long before taking it home to the frying pan, the annual stocking of hybrid and striped bass into West Point Lake is good news for meat fishermen.

But never doubt that those who fish for sport haven't been well pleased. Today, many anglers employing the services of one of the lake's guides want to fish for stripes and hybrids. The evolution of West Point Lake into one of the top hybrid and stripe lakes in the Southeast has been years in the making.

The lake has evolved into more of a stripe lake than hybrid lake in recent years. Those fish have a dual pattern in the spring months. A number of hybrids feel the urge to spawn, even though it's impossible for them to do so. These fish make a false spawning run upriver end end up at the head of the lake near Franklin. Anglers generally leave the fish alone.

Beginning approximately April 1st, the remainder of those stripes and hybrids begin suspending themselves over the standing timber above West Point Dam. These fish will be in 30 to 35 feet of water, but once a school of bait fish enters the area, the hybrids will rush to 12 to 15 feet of water to feed. These feeding fish must be marked with a depthfinder and by pinpointing them by lining up spots on the bank.

This is the perfect spot to try the Alabama rig. This umbrella-type rig trails five swimbaits or other lures behind it and is deadly in such situations. Bucktail jigs, Rat-L-Traps, swimbaits and spoons will take the feeding fish. The angler should throw well past the school and retrieve the lure back to the boat with a hopping motion.

Once summer arrives, the hybrids move from standing timber and position themselves over underwater islands, sandbars and roadbeds in 15 to 25 feet of water. The fish will lie dormant and refuse to take a bait until current is moving. The moving of water through West Point Dam triggers the fish into a feeding spell and bucktail jigs, jigging spoons, deep-diving crankbaits and live shad can catch them.

Fall and early winter is a transition period for the hybrids and fishing is generally poor, but a sudden drop in barometric pressure will send the schooling fish hitting topwater all across the lake's lower third.

Topwater jerk baits and chug baits will catch these fish, while a Rat-L-Trap or jigging spoon kept in the boat on another rod can catch fish when they submerge.

West Point Lake's best fishing spots

1. Excellent fall fishing ledge. Deep drop into standing timber. Worms or jig-and-pig.
2. Small underwater island in main river channel. Good spot for schooling bass and hybrids in summer. Lots of brush on top.
3. Excellent flat in bend of river. Lots of standing timber.
4. Mouth of Fish Creek. Points on both corners.
5. Excellent point for bass in late spring, summer or fall when water is moving.
6. Back of Fish Creek. Good spawning area.
7. Excellent point for bass.
8. Underwater road bed leading to rubble of collapsed underwater bridge. Bridge rubble directly in main channel.
9. Good Rat-L-Trap fishing in mouth of creek.
10. Highway 219 (Mooty Road) bridge pilings. Good riprap on both sides.
11. Underwater road bed.
12. Good shallow hole. Location of farm pond before lake was impounded.
13. Underwater road bed.
14. Two underwater islands in 25 feet of water located in sharp bend of river. Good brush on top. Excellent in summer for bass with worm.
15. Standing timber-filled slough. Excellent for spinnerbaits much of the year, topwater in late spring. Crappie fishing in summer.
16. Good creek. Standing timber throughout. Good spring spawning area. Topwater, spinnerbaits and worms for bass. Also find crappie here.
17. Underwater road bed.
18. Underwater road bed.
19. Underwater islands 35 feet deep in main channel. Covered with brush.
20. Underwater road beds.
21. Underwater road beds.
22. Mouth of creek. Good holding areas for bass in spring when floodgates are open. Fish escape here.
23. Standing timber area.
24. Good crankbait area in creek along south bank.
25. Good pocket for spawning bass. Topwater or spinnerbaits.
26. Where creek forks in back, excellent spinnerbait area. Lots of standing timber.
27. Small underwater island in main river channel in 25 feet of water. Excellent for schooling bass when current is being pulled.
28. Excellent pocket with lots of good points.
29. Underwater road bed.
30. Underwater hump. Main channel runs around both sides. Excellent for schooling summer bass when water is on or off.
31. Underwater road bed.
32. Good pocket with flats. Crankbaits or spinnerbaits.
33. Long, underwater island. Island splits main channel. Lot's of brush on top. In 30 feet of water. Excellent is summer with deep-diving crankbaits or in dead of winter with jig-and-pigs.
34. Underwater island. Brush piles on top.
35. Underwater island with standing timber on top. Adjacent to main river channel. Excellent spot for spooning in summer.
36. Location of small lake before West Point was impounded.
37. Small depression. Location of small farm pond before lake was impounded. Good topwater area in spring.
38. Underwater road bed.
39. Bridge pilings in Yellowjacket Creek. Lots of tournament-releasd fish gather here. Good year round.
40. Good coves with grass.
41. Grass beds at fork of creek.
42. Mill Road bridge pilings. Good riprap. Good crappie spot at night.
43. Small bridge. Same as above.
44. Underwater road bed.
45. Underwater bridge.
46. Underwater bridge.
47. Mooty Bridge Road bridge pilings. Tournament fish also gather here.
48. Standing timber area.
49. Bridge pilings in Beech Creek. Good spring fishing for crappie.
50. Underwater island with standing timber surrounding it in 5 feet of water. Excellent topwater in late spring.
51. Stumps on west bank.
52. Mill Road Bridge pilings. Located over location of small pond before lake was impounded. Good riprap. Good spring fishing for crappie and bass.
53. Underwater bridge.
54. Underwater road bed.
55. Good fishing around bridge pilings year round.
56. Lots of standing timber. Excellent spinnerbait fishing in spring. Also good crappie fishing in spring on

west bank.

57. Good spawning area for bass and crappie in back of Yellowjacket Creek.

58. Good spawning area for bass and crappie in back of the creek.

59. Site of large lake before lake was impounded. Excellent spring spawning area for bass and crappie.

60. Good spawning area in Half Moon Creek. Standing timber and grass. Good depression left by old farm pond before lake was impounded.

61. Whitaker Road Bridge. Good spring fishing around pilings. Deep water adjacent to shallow water. Good crappie fishing at night in late spring, early summer under the bridge.

62. Underwater road bed.

63. Underwater trees. Site of an old orchard before lake was impounded.

64. through 68. Underwater road bed.

69. Old houses flooded by lake's impoundment; houses are collapsed, but rubble and foundations remain. Good worming area.

70. Small, underwater island. Top of island in 35 feet of water. Good summer worming area.

71. Large underwater island. Top of island in 35 feet of water. Some brush. Crappie on top in summer. Good worming for bass.

72. Underwater road bed.

73. Location of farm pond before lake was impounded. Good spring spawning area for bass and crappie.

74. Underwater road bed.

75. Underwater road bed. Rubble from collapsed bridge in main creek channel 25 feet deep. Good summer crappie fishing.

76. Underwater road bed.

77. Location of farm pond before lake was impounded. Good spring spawning area for bass and crappie.

78. Standing timber in 15 feet of water. Good spring spawning area. Spinnerbaits for bass; minnows for crappie.

79. Antioch Road Bridge pilings. Holds pre-spawn and post-spawn crappie and bass.

80. Small underwater hump. Good location for stripers in summer.

81. Excellent pocket. Adjacent to main river channel. Fish run up on ledge in summer to feed.

82. Foundations of homes covered by impoundment.

83. Underwater railroad trestle. Across shallow water over main channel and back into shallow water. Old railroad bridge across main channel. Sometimes holds stripe in summer. Some bass year round.

84. Underwater road beds.

85. Long underwater island. Check for schooling stripers and hybrids in summer and winter.

86. Underwater road bed.

87. Standing timber area. Good spring spawning area for crappie and bass. Spinnerbaits and buzzbaits for bass, minnows and jigs for crappie.

88. Standing timber area. Good spring spawning area for crappie and bass. Spinnerbaits and buzzbaits for bass, minnows and jigs for crappie.

89. Excellent spring spawning area. Location of large lake before West Point was impounded. Holds both spawning bass and crappie.

90. Underwater road bed. Good spring pre-spawn and post-spawn holding area for crappie and bass.

91. Underwater road bed. Good spring pre-spawn and post-spawning holding area for crappie and bass.

92. Foundations of two old homes. Both in shallow water. Good spinnerbait and buzzbait area.

93. Underwater road bed. Runs from 10 feet deep to 45 feet deep. Can find several species up and down road year round.

94. Old home foundations. Some 10 feet deep, some 45 feet deep.

95. Highway 109 bridge. Good riprap. Year-round fishing for all species.

96. Underwater road bed.

97. Small underwater island in 45 feet of water. Top of island in 35 feet with some brush. Located close to main channel. Good hybrid and striper area.

98. Long, winding underwater road bed. Some brush on top.

99. Wilson Creek. Excellent year-round fishing. Good buck bushes.

100. Back of Wilson Creek. Good spring spawning area. Brush and standing timber.

101. Underwater road bed.

102. Underwater road bed.

103. Underwater islands in 45 feet of water. Top of islands 40 feet. Stripe and hybrids.

104. Shallow underwater road bed.

105. Pronounced small hump in 25 feet of water.

106. Rainbow Creek. Standing timber in mouth.

107. Stumpy area on east bank. Good buzzbait and spinnerbait area.

108. Underwater road bed.

109. Underwater road bed.

110. Good spring and fall area for bass and crappie. Small farm pond previous to impoundment.

111. Underwater road bed.

112. Underwater island that splits main channel. Good striper and hybrid area.

113. Underwater road bed.

114. Underwater road bed.

115. Underwater road bed.

116. Underwater islands in 65 feet of water. Shows good signs of hot summer stripers, but tough to catch.

117. Old home foundations.

118. Underwater road bed.

119. Underwater road bed in Veasey Creek

120. State Line Road bridge pilings in Veasey Creek.

121. Back of Veasey Creek. Good spring spawning area.

122. Good spring spawning area for bass and crappie. Location of old farm pond.

123. Underwater road bed.

124. Underwater road bed.

125. Good pockets in Veasey Creek

126. Old building foundations.

127. Underwater road bed.

128. Underwater road bed in Stroud Creek.

129. Underwater road bed with collapsed bridge in channel of creek.

130. Bridge pilings in Stroud creek. Good spring bass and crappie.

131. Back of Stroud Creek. Good spawning area.

132. Line Road bridge pilings. Good crappie area in spring.

133. Good bass point in Wehadkee Creek.

134. Underwater road bed.

135. Small depression that holds fish on sudden spring weather change. Site of old farm pond. Difficult to find. 15 yards due east of intersection of road bed and creek channel.

136. Small depression in 15 feet of water. Holds spring bass and crappie. Site of small farm pond before lake impoundment.

137. Excellent points in Wehadkee Creek.

138. Underwater road bed.

139. Underwater bridge.

140. Underwater road bed.

141. Underwater road bed.

142. Underwater road bed.

143. Spur 109 bridge pilings in Wehadkee Creek

144. Good bass and crappie spawning area.

145. Railroad trestle. Good spring fishing around pilings for bass and crappie.

146. Standing timber in back of pocket. Good spring spawning area for bass and crappie. Spinnerbaits and worms for bass.

147. Underwater railroad bed.

148. Underwater road bed.

149. Standing timber area. Good spring bedding area.

150. Underwater road bed.

151. Collapsed underwater bridge.

152. Large standing timber area. Good spring bedding area for bass. Topwater, worms and spinnerbaits.

153. McCosh Mill Road bridge pilings. Good in spring for crappie.

154. Good fishing around bridge pilings in spring.

155. Underwater road bed.

156. Underwater bridge.

157. Alligator Creek. Good spring area in back. Fishing good in mouth of creek almost all year.

158. Underwater road bed.

159. Numerous old home foundations in 25 feet of water.

160. Underwater road bed.

161. Underwater hump that splits main river channel flow. Good area for summer stripers.

162. Old home foundations in 15 feet of water. Holds small bass in fall.

163. Underwater road bed.

164. Underwater road bed.

165. Underwater road bed.

166. Long, underwater island in 55 feet of water. Good summer area for stripers using downriggers.

167. Underwater island that splits main river channel. Deep. Holds summer and winter stripers.

168. Underwater road bed.

169. Deep hole in 15 feet of water. Hole is 25 deep in deepest part.

170. Underwater road bed.

171. Numerous old home foundations in 25 feet of water. Excellent worming area for bass in summer.

172. Underwater road bed.

173. Deep hole in 35 feet of water. Site of old pond. Good worming area in August.

174. Underwater road bed.

175. Old home foundations.

176. Underwater road bed.

177. Site of old orchard in 65 feet of water. Adjacent to main river channel. Good summer and winter stripe area with downriggers.

178. Underwater road bed.

179. Old home foundations in 15 feet of water.

180. Good depression in shallow water. Good topwater area for bass and schooling hybrids.

181. Old home foundations in 5 feet of water. Topwater for bass.

182. Old home foundations.

183. Depression in 35 feet of water. Old farm pond.
184. Underwater road bed.
185. Old home foundations right of edge of road bed in 15 feet of water. Good worming area.
186. Large, shallow depression. Good area for hybrids and bass to chase shad. Watch for shad popping topwater.
187. Underwater road bed.
188. Glass Bridge Road Bridge. Good spring area for crappie and bass around pilings.
189. Good depression that holds fish on sudden spring weather change. Holds crappie and bass.

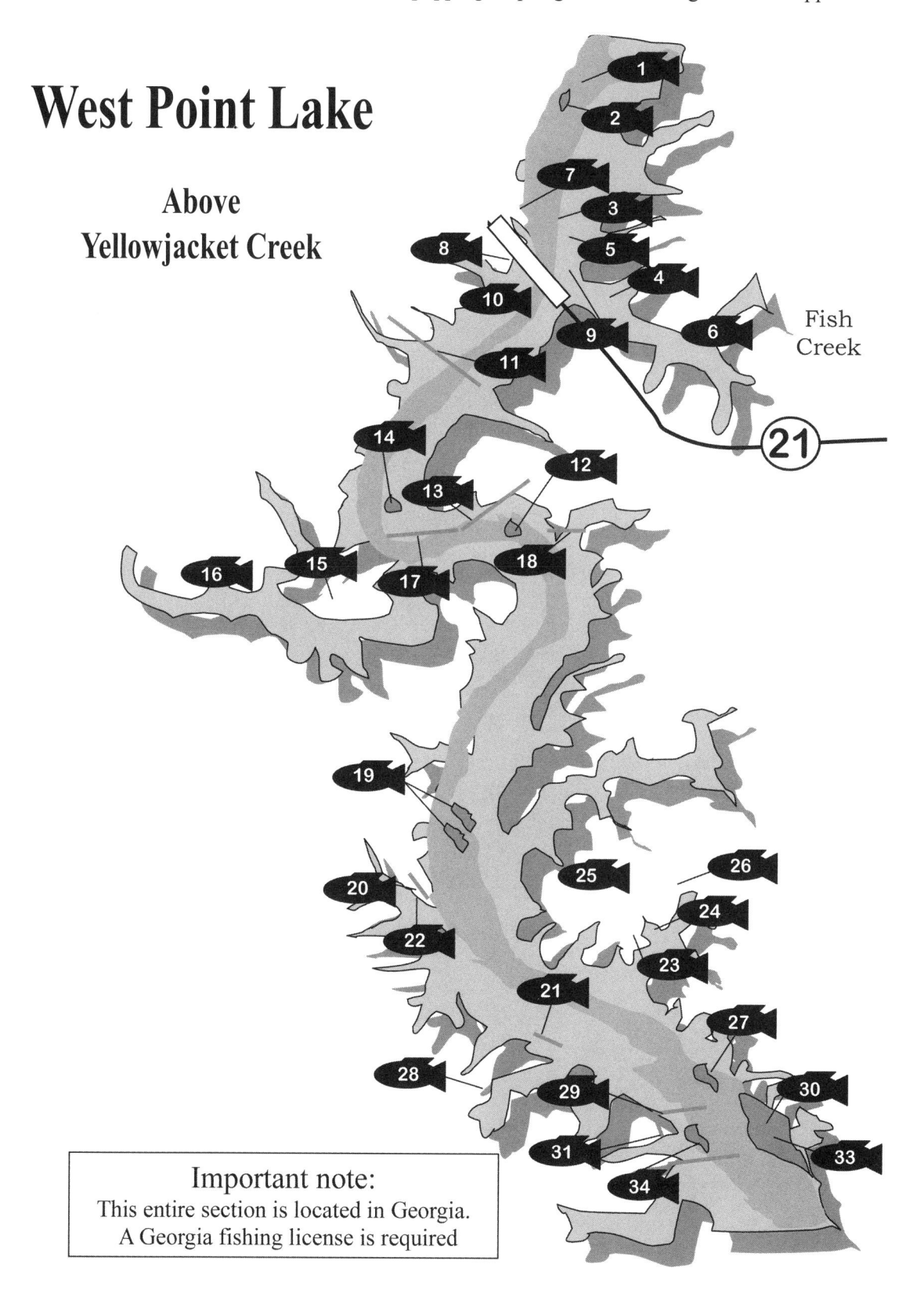

West Point Lake

Above Yellowjacket Creek

Important note:
This entire section is located in Georgia. A Georgia fishing license is required

The Ultimate Guide to Alabama Fishing

West Point Lake

Yellowjacket Creek

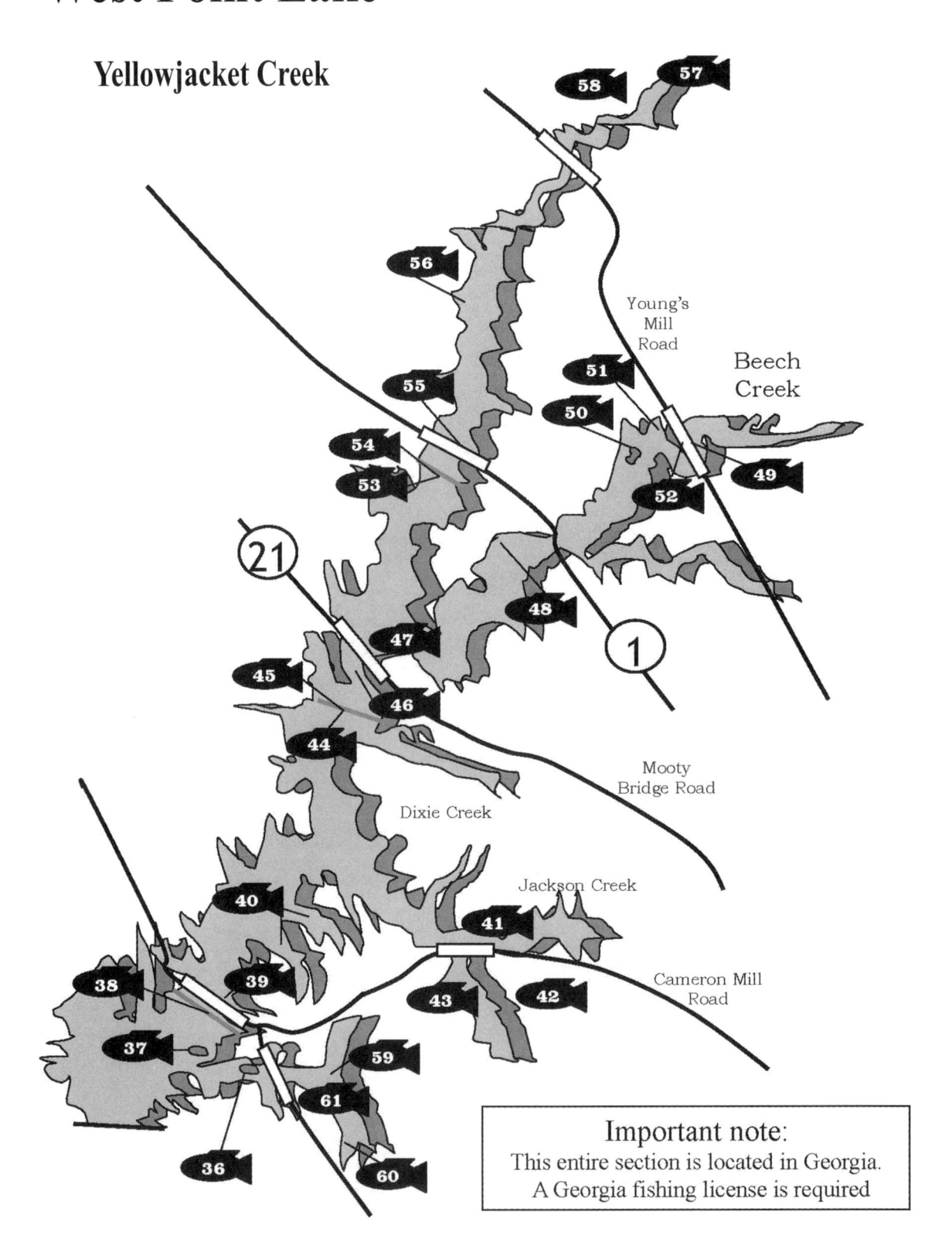

Young's Mill Road

Beech Creek

Mooty Bridge Road

Dixie Creek

Jackson Creek

Cameron Mill Road

Important note:
This entire section is located in Georgia.
A Georgia fishing license is required

164

Mike Bolton

West Point Lake

Highway 109
to
Yellowjacket Creek

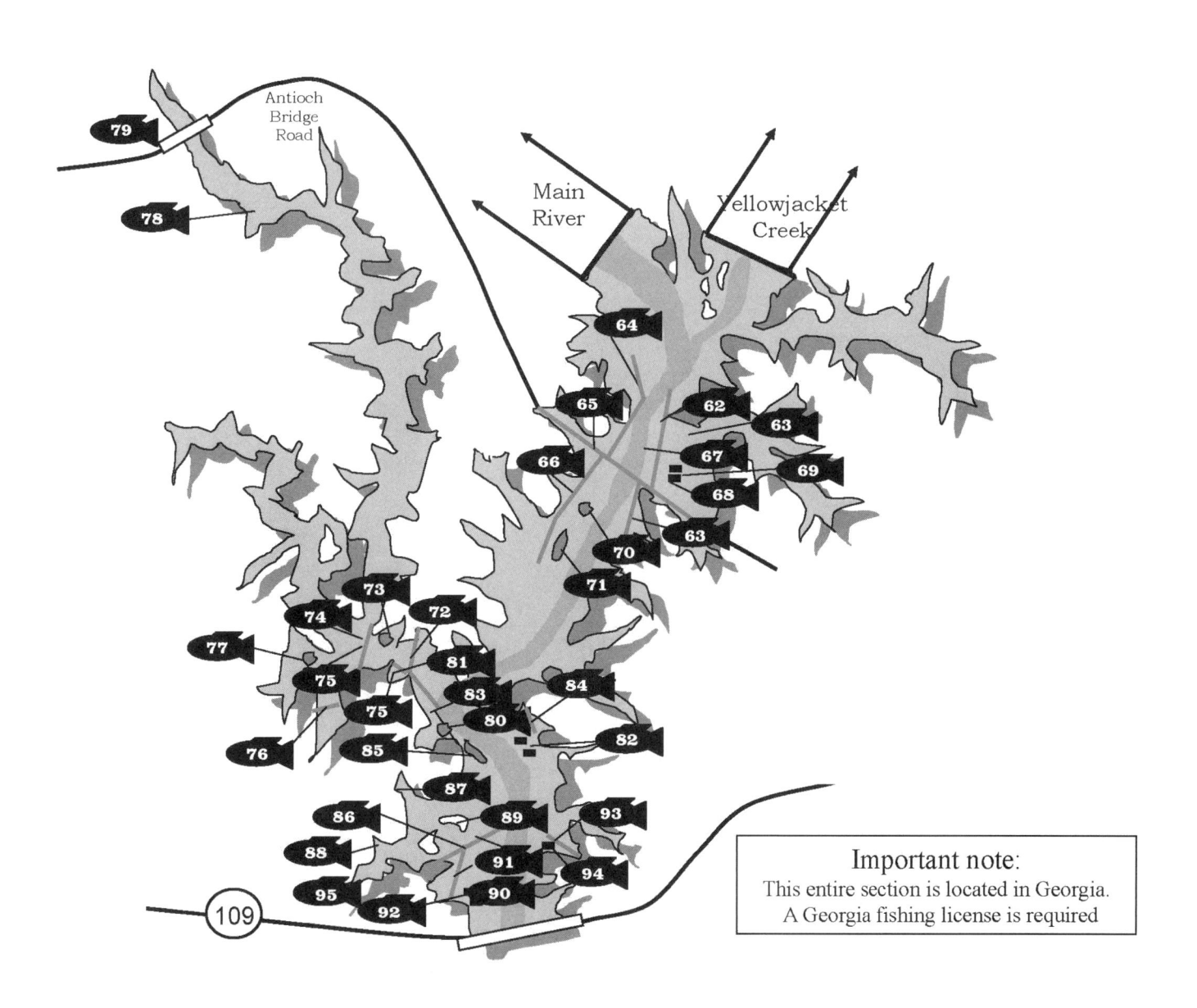

Important note:
This entire section is located in Georgia.
A Georgia fishing license is required

165

The Ultimate Guide to Alabama Fishing

West Point Lake

Highway 109 bridge to West Point Dam

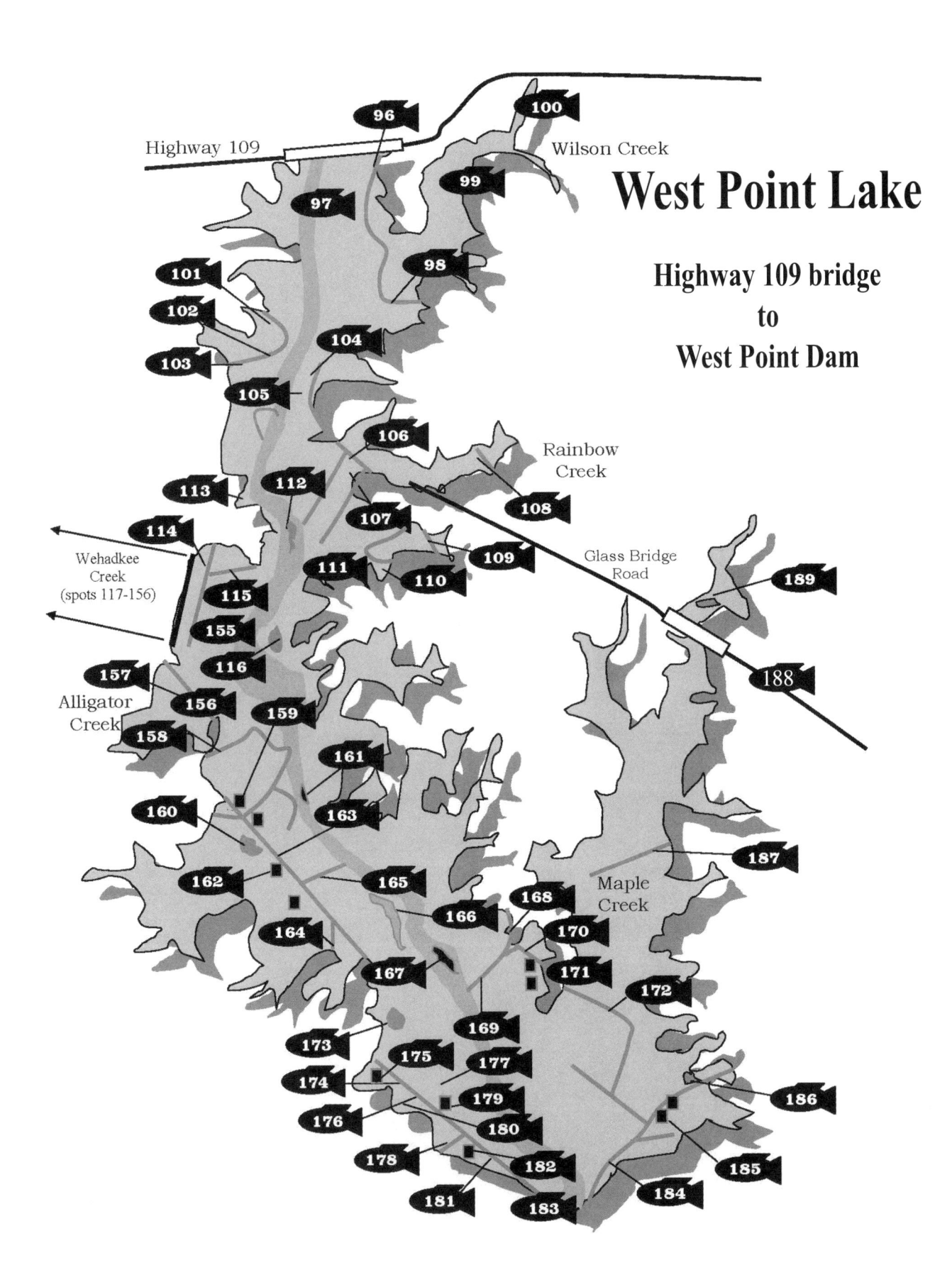

Mike Bolton

West Point Lake

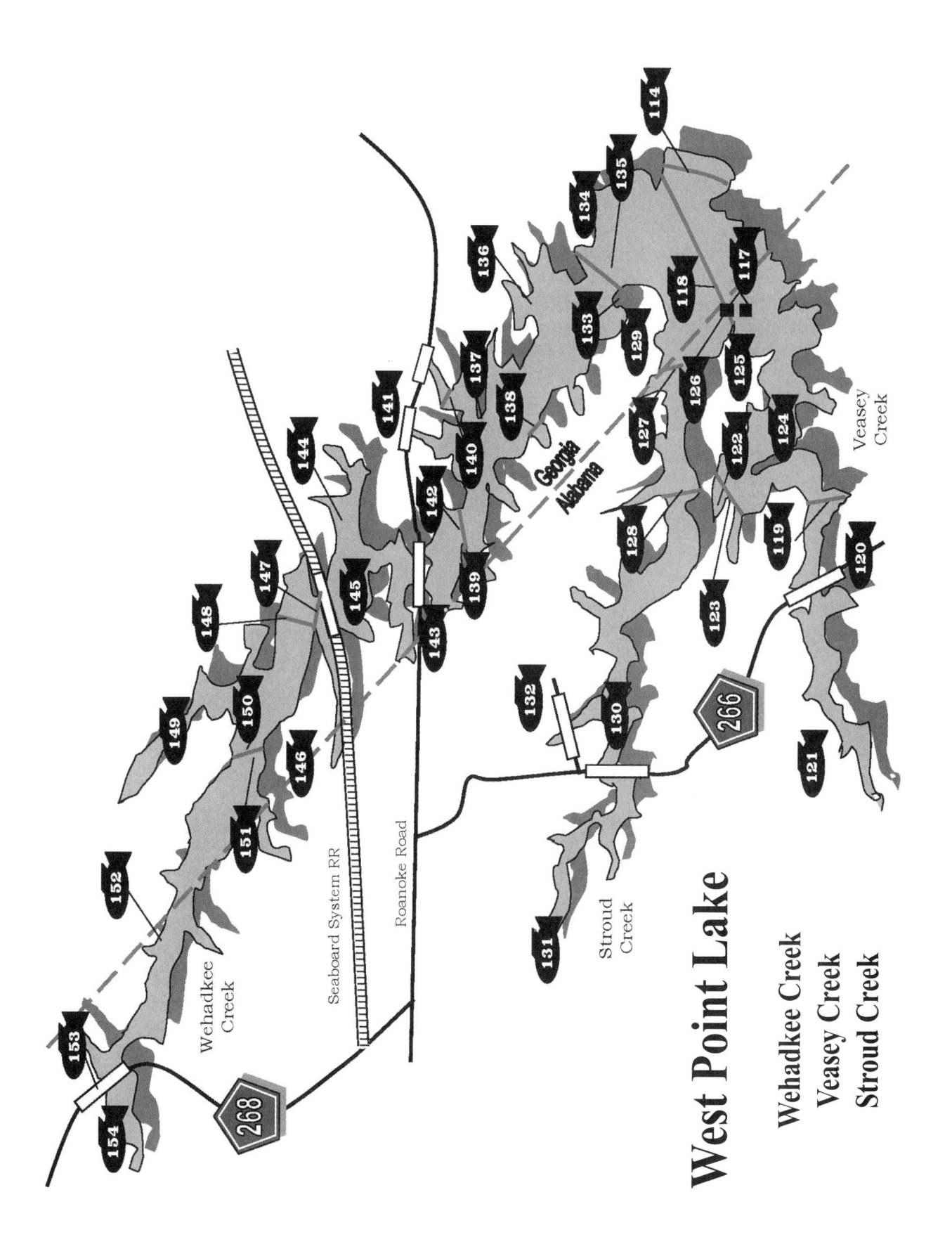

Wehadkee Creek
Veasey Creek
Stroud Creek

Lake Eufaula

Until the Bassmaster Classic began its run in Alabama on Lake Logan Martin and Lay Lake in the 1990s no lake in this state was better known to anglers across the U.S. than Lake Eufaula.

Technically this lake is Walter F. George Reservoir and the only Lake Eufaula is in Oklahoma. This lake has been called nothing but Lake Eufaula by fishermen from day one. Walter F. George's—whoever he was—relatives probably don't like that but everyone but fishery biologists call it Lake Eufaula. This book is calling it Lake Eufaula.

All major reservoirs offer their best bass fishing in the immediate years of first being impounded but in the case of Lake Eufaula that was especially so. The fish-rich, shad-rich Chattahoochee River was impounded in 1963, flooding thousands of acres of fertile farm land, much of it already loaded with fertilizer and animal wastes.

Sewage from Atlanta and other large cities from points north may have been pollution in those areas, but by the time it traveled down the Chattahoochee River to Lake Eufaula, it was fertilizer in its purest form. The large amount of "fertilizer" in the lake virtually made it the world's largest private pond. The nutrients resulted in a near-perfect food chain.

Couple all of these positive aspects with the fact that many, many farm ponds and private lakes already full of large bass were flooded and their fish released into waters already filled with big fish and you had a situation like Alabamians will never see again.

Lake Eufaula will never again see fishing like it had in its glory days but on occasions it still offers Alabama fishing at its best. The lake is very cyclic and rides a rollercoaster from great to average and back these days.

Spring bass fishing

Grass Creek to Cowikee Creek

When the first warm days of spring come to Lake Eufaula, a young man's fancy indeed turns to love—love of big bass.

It takes only a few warm days of spring weather to kick shallow water in this end of the lake up in the 55-degree range, and this sends the bass from their deep-water winter haunts into water much more accessible to fishermen.

The areas near the mouths of Cowikee Creek's North, Middle and South forks are full of shallow bars lined with stumps, as well as ditches filled with brush and other structure. The Chattahoochee River boasts similar areas, as well as a good number of points.

When these areas are adjacent to deep water, bass use the deep water as a spring staging area. The bass will hang in 12 to 15 feet of water, but will occasionally move up into the shallow water around structure to ambush shad. The cold fronts that roll through this time of the year always send the bass back deep, but only temporarily.

At this time of year, medium-running crankbaits fished slowly across the structure will produce bass. Fishing the crankbaits slowly is the key. These fish are still in water that makes them sluggish and a fast bait is impossible for them to catch.

When the water warms to 60-degrees, the bass will move closer to shallow water to feed and in anticipation of moving into their shallow spawning areas. Keep in mind that structure still is critical. Areas without stump rows, brush-filled ditches or grass will not have fish.

Spinnerbaits become the weapon of choice at this point. A chartreuse or chartreuse/white spinnerbait with a No. 4 or No. 5 Colorado blade does well. Fish a 1/2-ounce spinnerbait early because bass will still be holding on the outside of structure and you'll need a heavier lure that will drop quickly when the water suddenly deepens. When the fish move up into shallow water, throw a lighter 3/8-ounce or 1/4-ounce spinnerbait that can be worked in the grass.

Another key lure during this period is the buzzbait. Any noisy lure fished in or around weeds often will produce violent strikes. This offers some of the most exciting fishing that can be found on the lake.

When water temperature reaches 65 degrees, fish will disappear from these areas and head into the creeks and sloughs to spawn. The good part about fishing this area of the lake during the spring, besides the number of good fish it produces, is that whether or not water is being pulled is not important. This means fishing has the potential to be good most hours seven days a week.

Cowikee Creek to White Oak Creek

Early springtime will find bass in this section of the lake beginning to move from the main river channel and scattering to the ledges, points, creek mouths and the ditches that run from the main channel into the flats of the main lake.

These long ditches are one of the secrets of the lake's best springtime bass anglers. They are prime bass fishing areas because they act as interstates for fish moving from deep water to the grass.

The best way to locate these long ditches is to first buy one of the several detailed topographic maps of Lake Eufaula. The Lake Eufaula map published by Atlantic Mapping, Inc., is excellent. It is the map most commonly found in marinas and bait shops.

Locate these long ditches with your depthfinder by idling across the flats. After a ditch is detected, drop a marker buoy. Crisscross the area until you pinpoint the direction of the ditch. Mark its length with buoys.

Worms, spinnerbaits, buzzbaits, jig-and-pigs and medium-running crankbaits fished in these ditches will produce bass.

After the water temperature reaches 58 to 60 degrees and bass have traveled the length of the ditch and reached the grass, spinnerbaits, buzzbaits, worms and Rat-L-Traps work well.

The large creeks in the lake's middle section act much like bass highways just as the ditches do. White Oak Creek, Tabannee Creek, Cheneyhatchee Creek and Barbour Creek are excellent spawning areas. White Oak Creek offers the best grass beds.

White Oak Creek to Walter F. George Dam

If just catching bass, not trophy bass, is what takes the kinks out of your line, there's no better place to fish than on Lake Eufaula's lower end during the spring.

Unlike the area of the lake above White Oak Creek, the painstaking art of locating ditches and river channels isn't necessary. Bass gather on the visible, rocky points and this makes fishing relatively simple for even the most basic of bass fishermen.

This area has few creeks, so when spawning time arrives, most bass ease their way off the points and simply work their way into the nearest slough.

The Alabama side of the lake from White Oak Creek to the Hardridge Access Area boasts numerous points and sloughs.

Crankbaits are good in early spring when bass are staging on the rocky points, and worms and 1/2-ounce spinnerbaits work well when the bass move up the points into more shallow water.

Buzzbaits also will catch these fish, but usually only early and late. The bad part about fishing this part of the lake during the spring is that it's wide, and even the slightest wind deals havoc with efforts to fish. During March, this area can sometimes get downright choppy.

Hot weather bass fishing

Grass Creek to Cowikee Creek

It's common for anglers on this end of the lake to report catching smaller bass in the grass and shallow water until early- or mid-May. These are fish that are no longer spawning and have returned to shallows because of preferable water temperatures and access to shad. Other bass that have spawned by this time will have moved out to the points.

It's at this time that two patterns will work. Spinnerbaits will take fish in the shallows, while crankbaits are needed for bass on the points.

After the end of May, easy-to-get-to fish becomes a thing of the past. The largemouth bass retreat to the river and creek channels to escape increasingly warmer water temperatures. These bass bunch up in schools, frequently suspended along ledges in 12 to 18 feet of water. Bass like these areas because they not only provide cooler, oxygenated water, but because they also offer easy access to the stump-ridden ledges that hold shad.

Plastic worms and crankbaits thrown into shallow water and reeled off the ledge is the key to catching bass. Bass don't suspend themselves all along these ledges, however. The secret is finding something irregular along the channels, such as a sharp bend or a large concentration of brush or stumps. The sharp river bend (see map) directly across from Rood Creek, the sharp bend exactly one mile above Rood Creek and the sharp bend 1 1/2 miles up from Rood Creek are key areas.

Unlike in spring, pulling water is critical to good fishing during hot weather. Moving water sends the shad scurrying and sparks the bass into a feeding frenzy.

On the Cowikee Creek side, the entire crooked stretch of the creek channel across from Lakepoint Marina State Park is good.

For those unfamiliar with the lake, they should locate the ledges with a depthfinder and then drop marker buoys. This will ensure that you fish the areas you need to fish.

Cowikee Creek to White Oak Creek

By early June, the spawn is over and the fish have moved out of the creeks and flats and are searching for cooler water. The bass will gather in deep water near the mouths of White Oak Creek, Barbour Creek, Chenyhatchee Creek and Chewalla Creek, near the mouths of ditches that dump into the main channel and along the ledges of the main river channel wherever there's a bend with structure on it. Some of the better areas with such bends are directly out from Old Town Park, directly out from the mouth of Chewalla Creek, 1.3 miles above the mouth of Barbour Creek, directly out from the mouth of Barbour Creek and the area above White Oak Creek.

Crankbaits or worms thrown into shallow water, pulled through the structure and allowed to drop into the main channel, will draw strikes from these fish. It is imperative that water be pulled for this pattern to be successful.

A few summer bass can be found in the creeks because of their deep water, but they can rarely be found farther up than the first bridges an angler will come to. These areas should be fished just as you would the main river.

White Oak Creek to Walter F. George Dam

Point fishing is the name of the game on the Alabama side of the lake during the summer. The mouths of most of the creeks and sloughs on this side sit within a few feet of the main river channel and when post-spawn bass exit, they are forced to head directly to the deep water. These bass will align themselves in the deep water adjacent to points, and they'll briefly brave the warmer water to dash to the points and feed on shad.

These points should be fished as you would the ledges upriver—with worms or crankbaits. Most of these bass, which school in large numbers as they gather in the channel, lay dormant until power is generated. After the water is turned on, the bass rush to the backside of the points to chase shad that gather there to get out of the current.

On the Georgia side of the lake, it's different. The area from Pataula Creek to Sandy Creek is full of flats that you normally would avoid in summer, but a large number of fish-holding humps can be found in these flats. Once again, use a good topographic map to locate the humps, then use your depthfinder to pinpoint them. Worms and crankbaits on these humps can be deadly when water is being pulled.

Cool and cold weather bass fishing

Grass Creek to Cowikee Creek

The first signs of cooler weather in late September and early October once again release the bass from their deep-water haunts and send them in search of shallow water.

Much as in early spring, the bass in early fall will stage themselves in 12 to 20 feet of water and occasionally move up into structure-filled shallows to chase shad. Cowikee Creek is an excellent area for activity as the many stumps seem to hold winter shad. Medium-running crankbaits are excellent for catching these bass.

As the water temperature continues to drop, the bass will feed more to store fat for the winter. They'll also move into shallower water. Dingy water warms better than clear water, as does water over pea gravel. Look for these areas. The area between Gamage Bridge Road to the U.S. 431 bridge is excellent. On the Chattahoochee River side, check out Little Barbour Creek and Rood Creek. Both have gravel flats.

Cowikee Creek to White Oak Creek

The cooling temperatures of October will move the bass out of the main channel and head them into the creeks, much as in the pre-spawn period. These fish won't actually get in the grass, but they will seek shallow water and position themselves on long points.

The mouths of Barbour Creek and Cheneyhatchee Creek are the main areas for this pattern. The mouths of these two creeks feature ditches that criss-cross the area, as well as a lot of sand.

The flats between Barbour Creek and Cheneyhatchee Creeks also are excellent. Several ditches link these two areas.

White Oak Creek to Walter F. George Dam

The lower end of Lake Eufaula isn't a good place to bass fish during the colder months. Anglers can enjoy brief success during the fall as a large number of bass pile up on the humps between Pataula Creek and Sandy Creek on the Georgia side, but when cold, cold weather arrives, the bass seem to disappear.

This is the time of year when the bass fisherman needs to crank the big motor and run north.

Other Types of Fishing

Crappie

Lake Eufaula's long history of excellent bass fishing has overshadowed the fact the lake is an excellent crappie hatchery.

There is nothing quite as impressive as entering one of the creeks off the main lake during February and suddenly seeing tremendous schools of crappie on your depthfinder. These schools can be so large that many anglers have passed them off as schools of baitfish.

These crappie make their spring spawning run up the creeks, and there is no better method of catching them than trolling jigs through the schools. Several rods trolling chartreuse jigs at the 14-foot level almost guarantees success.

Chewalla Creek and Cowikee Creek are excellent. Most of your better spring crappie fishing is in the lake's upper end.

Another method that works well is using a heavy lead to bump bottom along the stump-laden ledges next to the deep creek channels.

Other spring crappie can be found in treetops near the bank. A small jig in these treetops works exceptionally well.

For some reason, after the weather turns hot, the area of the lake from Cowikee Creek to White Oak Creek heats up for crappie fishing.

The secret to catching crappie this time of year is locating the structure directly adjacent to the river channel. Any time you find structure, you'll find crappie. These fish will be anywhere from 12 to 25 feet this time of the year.

Winter fishing for crappie is identical to fishing for them during summer. They can be found at the same depths in the same structure on the ledges adjacent to deep water. Many will argue the point, but a spoon will put more crappie in the boat than will jigs or live minnows.

Hybrid bass

Lake Eufaula has been stocked with hundreds of thousands of hybrid bass by both Georgia and Alabama fishery officials through the years. Lake Eufaula offers tremendous fishing for this species.

The fish isn't capable of spawning, but it will make false spawning runs into the middle fork of Cowikee Creek in mid-March. These fish run up this creek as far as possible in their attempts to spawn.

The fish will situate themselves in incredibly shallow and narrow water, so the smaller the boat you use, the better. The shallow portions of these creeks will sometimes have a 20-foot depression in them and the hybrids will gather there. Live shad can be used as bait, but spoons and shallow-running crankbaits probably work better.

Spring hybrid also can be found in some smaller creeks such as Wylaunee Creek and Rood Creek, but neither area can compare with Cowikee Creek's middle fork.

During hot weather, hybrids will move to 18- to 25-foot depths and hang around bends in the main river channel. These fish will occasionally, usually at daybreak and late evening, scoot into the shallow flats to feed on shad. These shad often -give away their location by seagulls which dive into the schooling baitfish to pick up cripples. If the hybrids are hitting on topwater, throw a topwater bait such as a Pop-R . If they are feeding under the baitfish, throw a spoon or chrome Rat-L-Trap or similar bait. Allow the spoon or Rat-L-Trap to sink to the bottom and bounce it back to the boat.

The area from Cowikee Creek to the old railroad trestle below White Oak Creek is the best location to find these schooling hybrids. The flats across from Old Creek Town Park is the best area of all.

When fall arrives, many of the bigger hybrids move up to the shallower water in the narrow area between the U.S. 82 causeway and the railroad trestle. Cut bait fished from the bank produces a number of fish, but boaters find the best success with Rat-L-Traps and spoons.

Lake Eufaula's best fishing spots

1. Flats directly across from Florence Marina. Big bend in river. Lots of visible structure and incoming ditches into main river channel. Good year around for bass and crappie.

2. Sheer rock bank drops directly into river channel. Logjams. Flat water on both ends.

3. Rood Creek. Bass bed in back in April. Also spring crappie. Grass flats and stump rows throughout creek.

4. Location of ponds before river was impounded. Deep cut where dam was broken. Shallow flats. Underwater standing trees. Big bass in summer. Good spawning area in spring.

5. Wylaunee Creek. Dangerous area, but good crappie fishing in standing timber. Points produce good bass. Standing timber throughout middle of creek.

6. Underwater roadbed.

7. U.S. 431 bridge to intersection of Cowikee Creek and Chattahoochee River. Good year around area. Lots of flats with grass. Creek bends; 40-foot river channel. Bass and crappie year round.

8. Excellent fishing from South Fork to U.S. 431 bridge. Best producing area on lake. Next to Lakepoint Resort where most tournament bass are released.

9. South Fork Cowikee Creek. Good from Gamage Road bridge to mouth of Middle Fork. Deep water. Stump rows. Lots of visible structure.

10. Middle Fork Cowikee Creek. Good area for spring bass and crappie. Visible tree tops. Good all seasons except summer.

11. Underwater roadbed.
12. Underwater roadbed.
13. Underwater roadbed.
14. Underwater roadbed.

15. The clothesline. River hits rocks out from Old Creek Town Park and heads directly for Georgia bank. Excellent bass, crappie and hybrid fishing. Nearby flats are excellent spawning area; 65 feet of water jumps to 10 feet.

16. Chewalla Creek. Top early spring crappie area. Trolling for crappie in early February produces slabs. Also good area for big spring bass.

17. Highway 82 causeway. Excellent crappie fishing around pilings. Same for nearby railroad trestle. Good both day and night. Riprap along causeway good for bass and hybrids. Most big hybrids come from this area. Also produces some trophy bass.

18. Underwater brush pile. Thirty yards southeast of buoy 93.8. Line up buoy 93.8 with tallest pine across railroad on northwest horizon. 15 feet deep.

19. Underwater hump. Directly below trestle. Leave Alabama side and line up with second piling. Go straight south. 65 feet of water rises to 12 feet. Excellent for bass and crappie.

20. Underwater brush pile. 100 yards north of buoy 92.4 on old river channel. 15 feet deep.

21. Underwater road beds.
22. Underwater road beds.

23. Barbour Creek. Excellent spring crappie area to bridge. Good bass area. Sandy points. Crooked creek.

24. U.S. 431 bridge pilings. Excellent crappie fishing day and night.

25. Corps crappie plot. Marked with buoy.

26. Underwater road bed exiting Cheneyhatchee Creek.

27. Bridge pilings in Cheneyhatchee Creek. Excellent crappie fishing at night.

28. Corps crappie plot. Marked with buoy.

29. Underwater brush pile. Directly across from first slough south of Cool Branch Park. Line southeast corner post of boathouse in slough with tall white-colored tree. Also, line up black buoy with water tower on west side of lake. 20 feet deep.

30. Corps crappie plot. Marked with buoy.
31. Underwater road bed.
32. Underwater road bed.

33. Underwater brush pile 50 feet southeast of buoy 89.1 on old river bed. Line up three buoys to the north for correct alignment. 20 feet deep.

34. Big underwater brush pile. Go to first cove above White Oak Creek. Trees are east of the cove on old river channel (Alexander's Bend). Trees are on top of mound in 10 feet of water.

35. Underwater brush pile. Straight out of south point of Gator Hole slough. Line up south point with tallest pine in back side of slough. 15 feet deep.

36. Underwater brush pile. Northeast of the south point of Parmer Branch. 15 deep adjacent to dropoff.

37. Underwater brush pile. 150 feet northwest of buoy 87.8 on edge of old river channel.

38. Corps crappie plot. Marked with buoy.

39. White Oak Creek. Bridge pilings offer good crappie fishing. Riprap offers good bass fishing.

40. Underwater road bed.

41. Underwater road beds.

42. Underwater road bed across mouth of Drag Nasty Creek.

43. Underwater brush pile. Proceed west out of Drag Nasty Creek, along the south edge of the creek channel until the southwest corner post of the metal boathouse lines up with the large pine tree on the next point south. 15 feet deep.

44. Underwater road bed.

45. Corps crappie plot. Marked with buoy.

46. Underwater brush pile. South of first small island above the mouth of Pataula Creek. 13 feet deep.

47. Corps crappie plot. Marked with buoy.

48. Pataula Creek. Shallow. Excellent springtime spawning area for bass and crappie. Good fishing to bridge. Stumps and grass beds.

49. Old Clay County (Ga.) 39 highway roadbed crossing mouth of Pataula Creek. Excellent for big bass in spring.

50. Brushpile on top of old Clay County (Ga.) highway 39 roadbed, approximately 400 feet south of old Clay County (Ga.) 39 bridge.

51. Brushpile on top of old Clay County (Ga.) highway 39 bridge. Approximately 500 feet south of bridge.

52. Gopher Ridge island. Shallow water night fishing with buzzbaits.

53. Underwater road bed.

54. Underwater brush pile. Line up south point of Thomas Mill Creek with next point up creek. Proceed due east across old river channel. Brush located on Georgia side on top of huge mound in 23 feet of water.

55. Underwater cedar tree pile, 100 feet southeast of north point of Otho Branch. 16 feet deep.

56. Underwater brush pile 100 feet northeast of south point on Otho Branch. 16 feet deep.

57. Underwater road bed.

58. Underwater brush pile on Alabama side of main river channel, on second point above No. 60.

59. Underwater road bed.

60. Underwater brush pile. Immediately north of the rocky point at the north end of Holiday Shores subdivision. 15 feet deep.

61. Pockets off main river channel. Excellent bass fishing in back of sloughs in spring, excellent bass fishing in mouth where it falls into main river channel in summer.

62. Miles-long underwater road bed.

63. Sandy Creek. Flats, weedbeds. Good for bass and crappie in spring.

64. Corps crappie plot. Marked with buoy.

65. Underwater road bed.

66. Underwater road beds.

67. Underwater brush pile 50 feet northwest of big rocks at the point of the cabled off swimming area at Sandy Creek Park. 12 feet deep.

68. Hardridge Creek. Good bass fishing in mouth only.

69. Underwater road beds.

70. Underwater brush pile immediately south of the north point of Hardridge Creek. On edge of drop off. 16 feet deep.

71. Underwater brush pile. Go east of Pine Tree Point in Hardride Creek until you reach drop off. Trees are located on edge of dropoff. 16 feet of water.

72. Underwater cedar tree pile. East of the south point of Hardridge Creek. Follow 15 foot water until it drops off. Trees are at dropoff in 15 feet of water.

73. Underwater brush pile on point south of Hardridge Creek mouth. 15 feet deep.

74. Corps crappie attractor. Marked with buoy.

75. Underwater tree pile 40 feet north of the south point of the first cove north of the dam on the Alabama side. 12 feet deep.

76. Underwater brush pile 20 feet north of the dam on the Alabama side adjacent to the first "No Parking" sign from the west point of the dike road. 14 feet deep.

77. Underwater brush pile 40 feet northwest of riprap at south point going into east bank area boat ramp. 13 feet deep.

78. Riprap around dam. Good on both sides of the channel and along dam for hybrids and stripes. One of the first areas big fish move into during the spring.

79. Underwater brush pile northwest of second. point moving north from the dam on the Georgia side. 13 feet deep.

80. Underwater brush pile 50 feet west of first point north of the dam on the Georgia side. 13 feet deep.

81. Underwater brush pile 20 feet north of the dam at the first "No Parking" sign on the Georgia side. 10-14 feet deep.

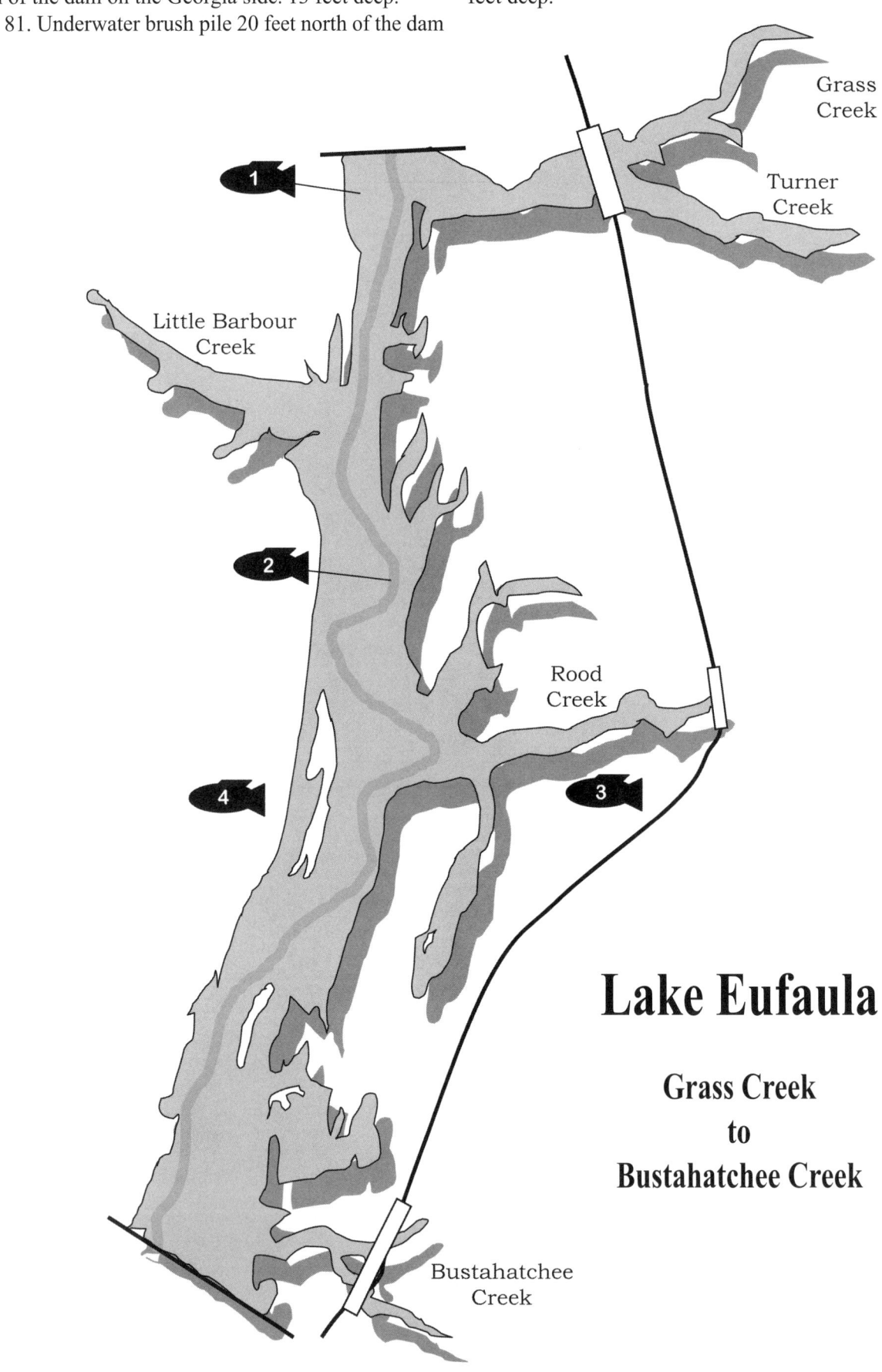

Lake Eufaula

Grass Creek to Bustahatchee Creek

Mike Bolton

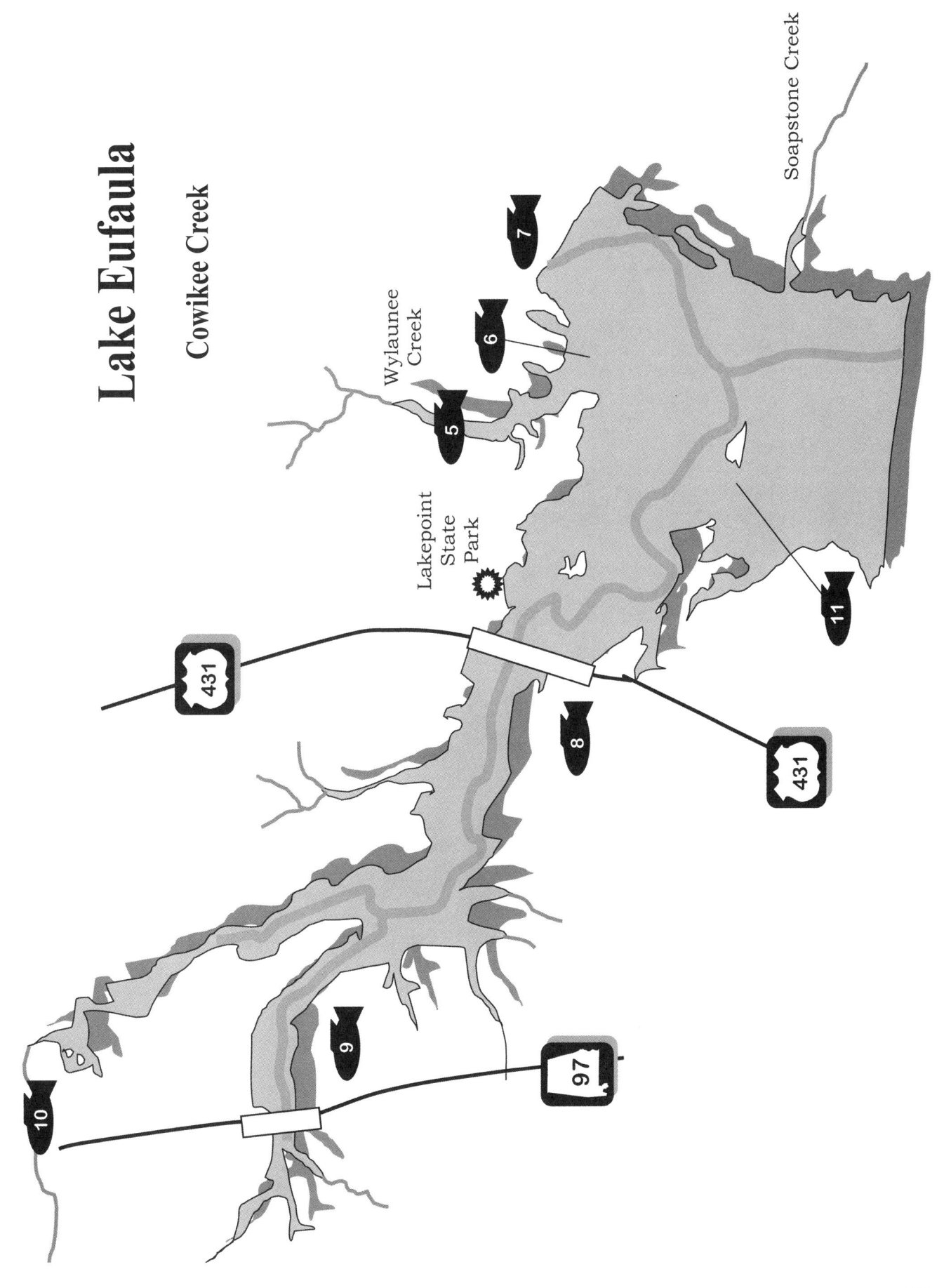

Lake Eufaula
Cowikee Creek

Soapstone Creek

Wylaunee Creek

Lakepoint State Park

The Ultimate Guide to Alabama Fishing

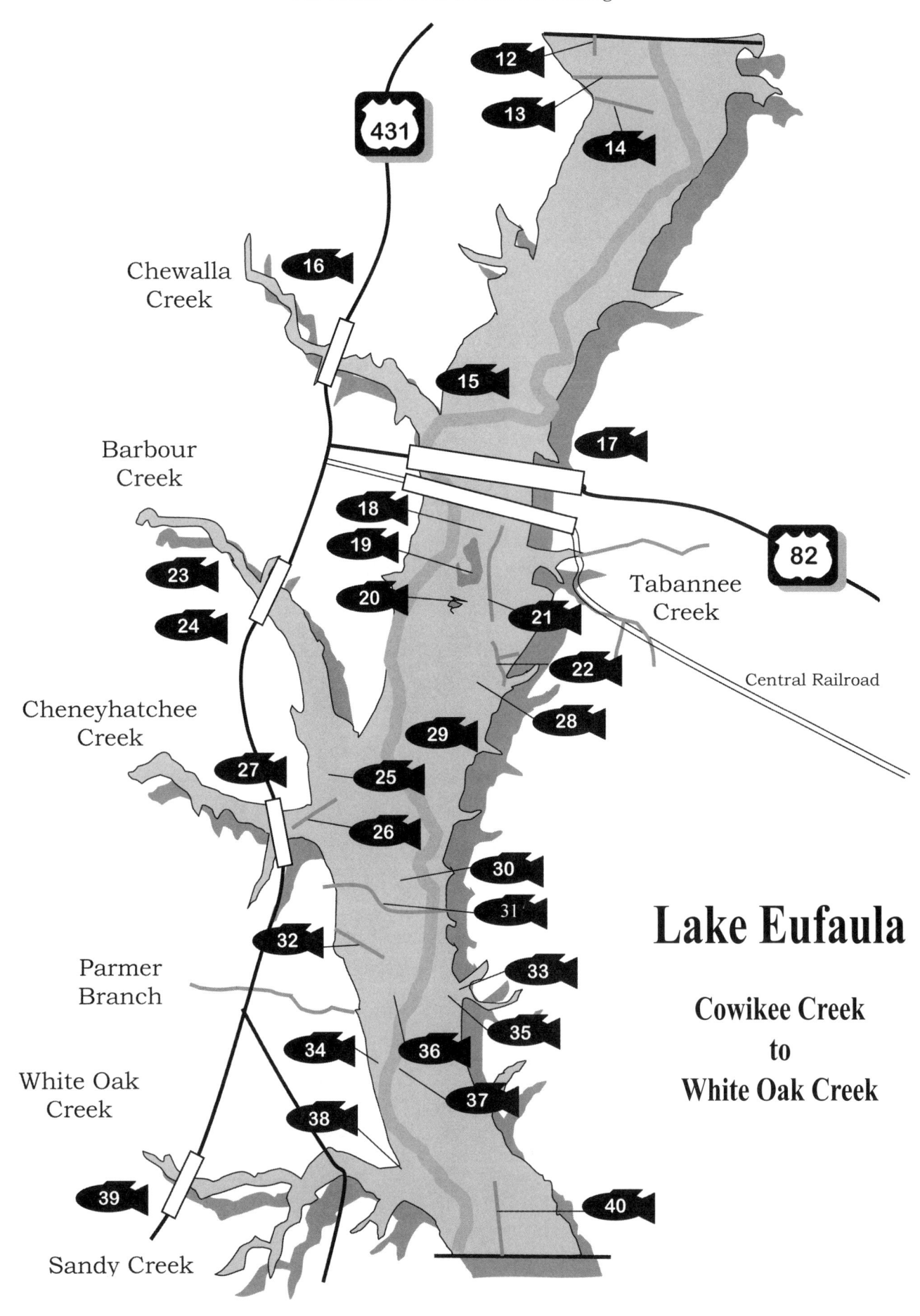

Lake Eufaula

Cowikee Creek to White Oak Creek

176

Mike Bolton

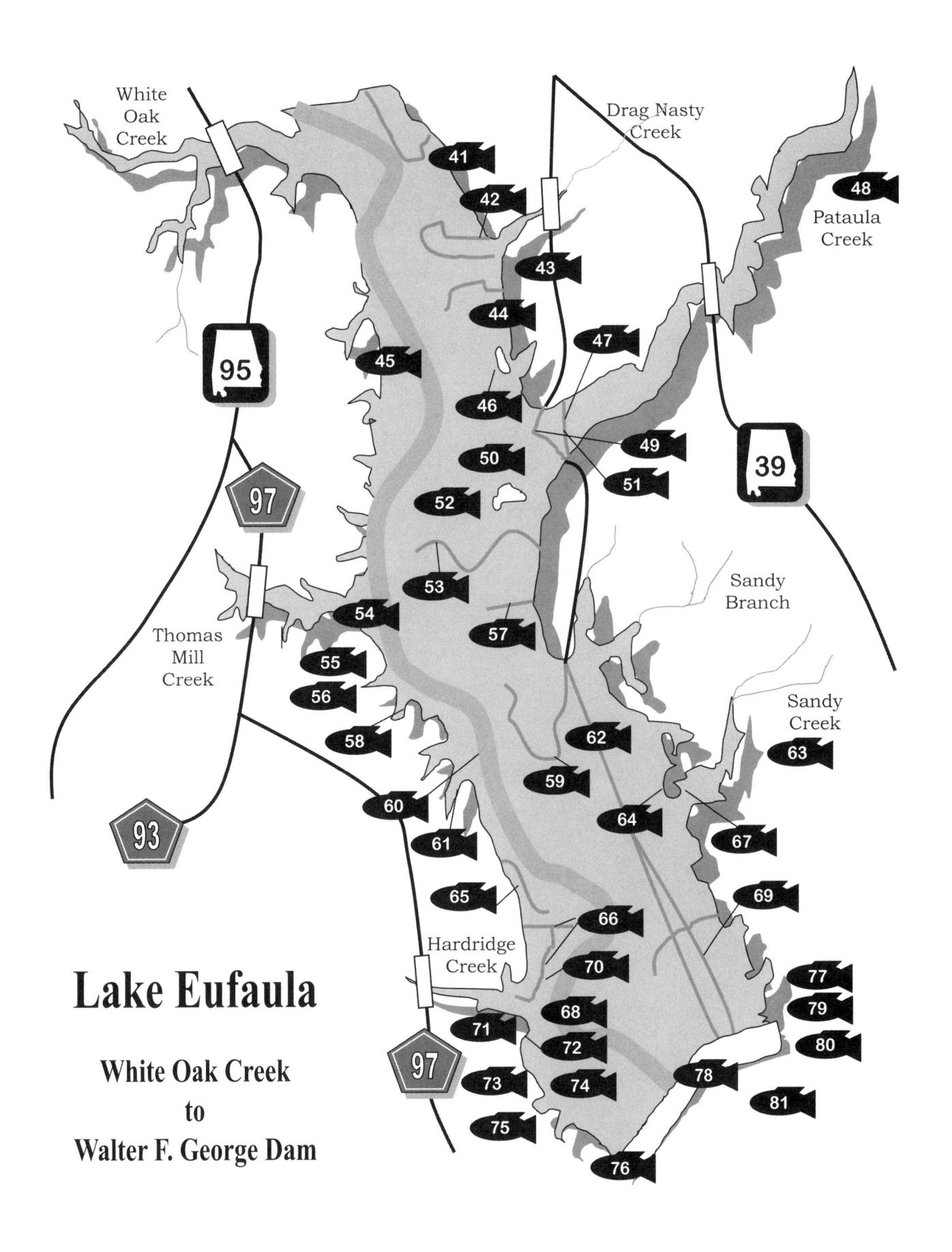

Lake Eufaula

White Oak Creek to Walter F. George Dam

177

The Ultimate Guide to Alabama Fishing

The Tallapoosa River

Lake Wedowee

Lake Martin

Lake Wedowee (R.L. Harris Lake)

Take a fish-rich river which receives agricultural runoff. Isolate it from metropolitan areas. Leave much of the surrounding structure intact. Impound the river. Stock it with more than 135,000 Florida-strain largemouth. Leave its already-thriving crappie population intact.

That's one danged good recipe for a tremendous bass and crappie fishing lake.

When it was impounded in 1983 Lake Wedowee had many of the same ingredients that made Lake Eufaula one of the nation's best bass lakes when it was impounded. Fortunately for Lake Wedowee, it doesn't get the publicity or the pressure Lake Eufaula has received.

The lake lacks a nearby chamber of commerce with a big bankroll to boast of the lake's exploits or local fishermen who like to talk. A small tackle shop in Lineville once had photos of 16-pound and 14-pound Lake Wedowee largemouths on its wall. The fishermen who caught the fish asked that the news and photos of the catches not be released to newspapers. The lake has given up several 12- and 13-pounders, too.

Spring bass fishing

Tallapoosa River

The steep rocky bluffs on the lake's lower end, the standing timber and the roadbeds and creek channels on the outside fringes of the standing timber, are the winter homes of many bass. It's from these areas that the bass will begin moving when spring arrives.

So few visiting anglers are accustomed to fishing deep, clear-water reservoirs that many anglers have trouble fishing pre-spawn bass on this lake. During construction of Harris Lake, much of the standing timber was left intact. Many anglers new to this lake tap their depthfinders in disbelief as they find themselves in 100 feet of water and can still see timber sticking out of the water.

Many fish suspend themselves 30 feet deep in these trees during the winter, while others may be 40 feet deep on the roadbeds on the lower end. It takes many anglers a tackle box full of jig-and-pigs or spoons and lots of practice before they master the feel of fishing that deep.

Fortunately, the warming water that comes with spring flushes the bass out of these deep haunts. They begin the staging process by moving to the points that have brush and timber. Slow baits such as a jig-and-pig or a big-bladed spinnerbait are needed in February when the fish are still sluggish and have yet to start their pre-spawn feeding frenzy.

A 1/2-ounce or 3/4-ounce brown or black jig with a weed guard should be used when fishing a jig-and-pig, and it should be tipped with a No. 5 brown or black pork frog. Many anglers opt for a full ounce jig—if they can find it—because of the water depth.

For those wanting to use a spinnerbait, a chartreuse/black 1/2-ounce or 3/4-ounce spinnerbait is needed. It should be outfitted with a large nickel blade. The larger blade forces the angler to fish the bait slowly. The angler has a tough pattern to fish here. He should position his boat in deep water and cast into 15 feet of water, then fish the bait slowly, allowing it to fall deeper and deeper. Many anglers fish the bait too quickly, not allowing it to fall back to the bottom on each retrieve.

These bass eventually will move into more shallow water where fishing smaller-bladed spinnerbaits and worms are more feasible. When the water temperature hits the 65-degree range, the bass will slide off the points and begin to look for spawning areas in the back of the warmer pockets filled with brush and timber.

Both topwater baits and spinnerbaits are strong at this time. A 5/16-ounce white or chartreuse/orange spinnerbait with a single-spin No. 5 Colorado nickel blade outfitted with a curly-tail grub trailer works well in these spawning areas. Almost any topwater bait, including buzzbaits and cigar-shaped baits, work well around the standing timber.

The lower end of the lake below the Alabama 48 bridge offers the best spring fishing, but the river as far as Ketchepedrakee Creek isn't bad.

Keep in mind that many of the pockets, because of the sunlight they receive and the shallow water in them, are much warmer than others in the spring. A water surface temperature gauge is a must.

There are several areas to concentrate during the spring. The three creeks entering Hunter's Bottoms (all are usually referred to by locals as Hunter Creek) is a key spawning area, as it Fox Creek. Triplett Creek is another good area. Mad Indian Creek is one of the best areas in the upper lake.

The Little Tallapoosa

The baits and patterns that work well on the Tallapoosa River also work well on the Little Tallapoosa. The main difference in fishing this part of the lake in spring is that it lacks the multitude of standing timber found on the big river. Anglers in this area must spring fish in areas much like those that hold spring bass in other Alabama lakes.

There are several key areas on the little river side. Wedowee Creek lacks standing timber, but is full of boat docks, piers, pockets and a brush-covered creek channel with many tight bends.

Allen Branch is a small creek with a good running branch in the back. It also has boat docks with brush around them. The series of pockets (No. 42 on the map) are also good for spring fishing.

Hot weather bass fishing

Tallapoosa River

Following the spawn, the bass will spend a brief amount of recovery time in the mouth of pockets and on the outside fringes of the standing timber in the pockets. They'll rest for a day or two, then feed heavily before going to deep water.

Spinnerbaits, top water baits and especially lipless crankbaits such as Rat-L-Traps are effective at this time.

Once the water temperature in the pockets and sloughs gets unbearable, the bass slide off into the deep water and take up residence in the standing timber, usually suspending themselves 30 feet deep or so. Fishing for these fish is relatively tough.

A plastic worm or plastic lizard flipped against the timber should be allowed to fall to the depth of the fish showing on the depthfinder. The bait should then be bounced around to attract the fish and entice them to strike.

A pattern that works well is doodling, the strange technique developed for use in the deep, clear lakes in California. This pattern uses light line—6-pound test or less—and a 3/16 bullet weight and a 4-inch plastic worm. The fish are first spotted on the depthfinder, then the worm is counted down to the correct depth. Once on target, the tip of the rod is shaken violently causing the worm to jump around erratically. It is a deadly pattern for these suspended fish.

The real drawback to the pattern is that the fish are on timber, and 6-pound test line is not conducive to cranking them out of it. The increase in the number of hookups with bass make the pattern worth using, however.

Some of the summer bass on this river will move up to the water surface at night, and during the early-morning or late-evening hours.

Topwater baits fished around the visible tree tops can be good at this time.

A few other fish hold on the few timber-covered points in 15 or 20 feet of water. Worms or spoons are good on these, but they are almost always smaller fish.

Key areas for this type of fishing on the lower end is the deep timber in Hunter Creek and Fox Creek. On the upper end, the standing timber on the right bank as you enter Mad Indian Creek is good.

The Little Tallapoosa

The summer patterns that work on the Tallapoosa River also work on the Little Tallapoosa River. Standing timber is evident on the left bank for approximately 2 3/4 miles as you enter the creek and the bass will suspend in that timber during summer months.

Wedowee Creek offers poor fishing during daylight hours because it receives so much recreational boat traffic. It does offer good fishing around the many points and boat houses and piers at night, however.

Cool and cold weather fishing

The Tallapoosa River
The Little Tallapoosa

If you're looking for exciting fall fishing, you can do no better than Lake Harris. The bass school heavily on the lake during this time and it's not uncommon for better anglers to catch and release up to 35 bass per day.

The bass begin chasing shad in the open water and are easy prey. A multitude of baits are good at this time, including jerk baits, Rat-L-Traps and spoons.

A pair of binoculars are beneficial. An angler can pull his boat into the middle of the lake and use the field glasses to scan the water for schooling activity. This is recognized by the ripple on the water surface punctuated by bass hitting shad.

When the angler spots this, he should run the boat to within 100 yards of the fish, then use the trolling motor to ease up on them.

The angler should keep at least two rods rigged. One should have a topwater bait and the other a Rat-L-Trap. When the fish are hitting topwater, the topwater bait should be cast well past the school and worked through it. When the topwater activity stops, a sinking bait should be cast well past the area and worked through it.

Surprising to most, much of this activity takes place right in the middle of the river. It is not unusual to catch these schooling fish 300 yards from the bank. This type fishing is best in the morning and evening, but on overcast days may go on all day long.

When the bass aren't schooling, they can be found in the same locations you found them following the spawn. The road beds in 15 to 20 feet on the lower end are good for worm fishing, while the outer fringes of the standing timber in the pockets is conducive to spinnerbait and topwater fishing.

This schooling activity and shallow water fishing decreases as winter arrives. The bass head back into the deep timber and suspend. Spoons, jig-and-pigs and worms should be used then.

Other types of fishing

Crappie fishing

Springtime crappie fishing on this lake brings back memories of the good old days. The multitude of standing timber provides the habitat crappie need to thrive.

Brushy pockets on the lower end hold the most easily accessible crappie in spring. These fish spawn in the timber in less than 18 inches of water. A live minnow fished on a No. 2 thin wire gold hook on 6-pound test line under a float is hard to beat. A 14-foot telescoping fiberglass pole allows an angler to gingerly swing the bait into place, but spinning tackle or spincast tackle will suffice.

The most deadly pattern during the pre-spawn and post spawn periods is spider fishing, however. A boat may be outfitted with as many rods as legal, each tipped with 6-pound test line and a jig. With the rods pointing out the side and the back of the boat, the trolling motor is used to troll the jigs through the water.

A 3/16-ounce jig with a white head and black body is excellent for this type of fishing.

Hunter Creek, Fox Creek and Mad Indian Creek are key spots for this type of fishing, as are the flats next to the standing timber on the left bank of the Little Tallapoosa River.

The angler working the trolling motor must watch his depthfinder carefully, speeding up to raise the jigs when humps and structure is sighted or slowing when the water deepens.

This type of fishing requires two anglers. One controls the bait while the other reels in fish. When a fish is on, the boat operator never slows. Allowing the boat to slow or stop will tangle the jigs in structure.

During the hot-weather months, the crappie move

deeper, suspending themselves in the timber or in the deeper creek channels.

Spider fishing works at this time, but most anglers fish a tight line.

Hunter Creek is one of the better creeks at this time. Fox Creek isn't that good. Triplett Creek and Mad Indian creek are also good.

The bridge pilings on any bridge on the lake will hold crappie at night during summer. Live minnows should be fished under a lantern in this pattern.

Lake Wedowee's best fishing spots

1. Good long, sloping point above the dam. Excellent spring spot for bass and hybrids.
2. Excellent timber pocket. Good in spring and summer.
3. Underwater road bed. Leads out from timber. Shallow part of road good in spring, deeper part good in summer.
4. Steep rock wall falls directly in river channel. Lot of spotted bass. Summer fishing.
5. Standing timber pockets. Good in spring and summer.
6. Ginhouse Branch of Hunter Creek. Good creek channel and brush; no standing timber. Good year round for bass and crappie.
7. Lane Branch of Hunter Creek. Right bank full of standing timber to the back. Good year round for bass and crappie.
8. Miles Branch of Hunter Creek. Good year round for bass and crappie.
9. Excellent point in Miles Branch.
10. Steep rocky wall in Hunter Creek. Excellent for bass in summer.
11. Good fishing in back of pocket. No standing timber, buts lots of brush.
12. Steep rock walls. Good in summer.
13. One Tree Island. River side of island is deep; back side shallow. Bass work their way around island different times of the year.
14. Back of pocket. No standing timber, but lots of brush. Good spring fishing.
15. Rock Crusher area. Big boulders. Summer night fishing for bass.
16. Underwater road bed. Original path that rocks were carried from crusher to dam.
17. Collapsed underwater bridge. Good bass fishing in August. Forty feet deep.
18. Road bed leading to bridge. Goes from shallow water to deep and back shallow. Good bass fishing.
19. Brushy pocket. Good spring fishing.
20. Fox Creek. Good year round for bass and crappie.
21. Standing timber in tight bends of Fox Creek. Good year round.
22. Underwater hump. Good year round.
23. River ledge. Good summer fishing.
24. Timber pockets. Good in spring.
25. Standing-timber-filled pockets. Good spring and fall.
26. Good summer schooling area in bends.
27. Standing timber in back of pocket. Good spring and summer.
28. Right bank of Triplett Creek. Good standing timber. Left bank steep, rocky bank.
29. Alabama 48 bridge. Good crappie fishing at night in summer.
30. Timber-filled pocket. Timber on both banks.
31. Standing timber on right bank of Mad Indian Creek. Good year round.
32. Sharp bend in creek channel of Mad Indian Creek. Good summer and winter area for crappie.
33. Back of Mad Indian Creek. Well-defined creek channel. Steep rocky bank on left that falls into channel. Good summer, fall and winter.
34. Standing timber in pockets on left bank. Good spring and fall.
35. Foster bridge pilings. Good crappie fishing at night in late spring, summer and fall.
36. Ketchepedrakee Creek. Good fishing in the mouth year round.
37. Stump rows. Good in summer for spotted bass.
38. Excellent river bend in Little Tallapoosa. Good shallow point. Good in spring.
39. Allen Branch. No standing timber, but good boat docks with brush. Good year round fishing. Spring bass and crappie in brush piles; summer bass on points and creek channel.
40. Excellent timber point sloping into river channel. Good year round.
41. Timber-covered island. Good spring fishing.
42. Standing timber pockets. Good spring fishing.
43. Underwater hump next to deep river channel. Also road bed on top. Excellent summer fishing for bass.
44. Wedowee Creek. Excellent points, creek bends and boat houses and piers with brush. Good year round

except during the day in summer. Much recreational boat traffic. Good at night in summer.

45. Good bass and crappie fishing in back of Wedowee Creek. Good brush around bridge pilings in creek. Good creek channel.

46. U.S. 431 bridge. Good bass fishing around riprap and pilings; good crappie fishing under the bridge at night.

47. Good stump rows in river channel bend. Crankbaits or worms year round.

48. Old U.S. 431 bridge. Same as 46.

Lake Wedowee

Below Mad Indian Creek

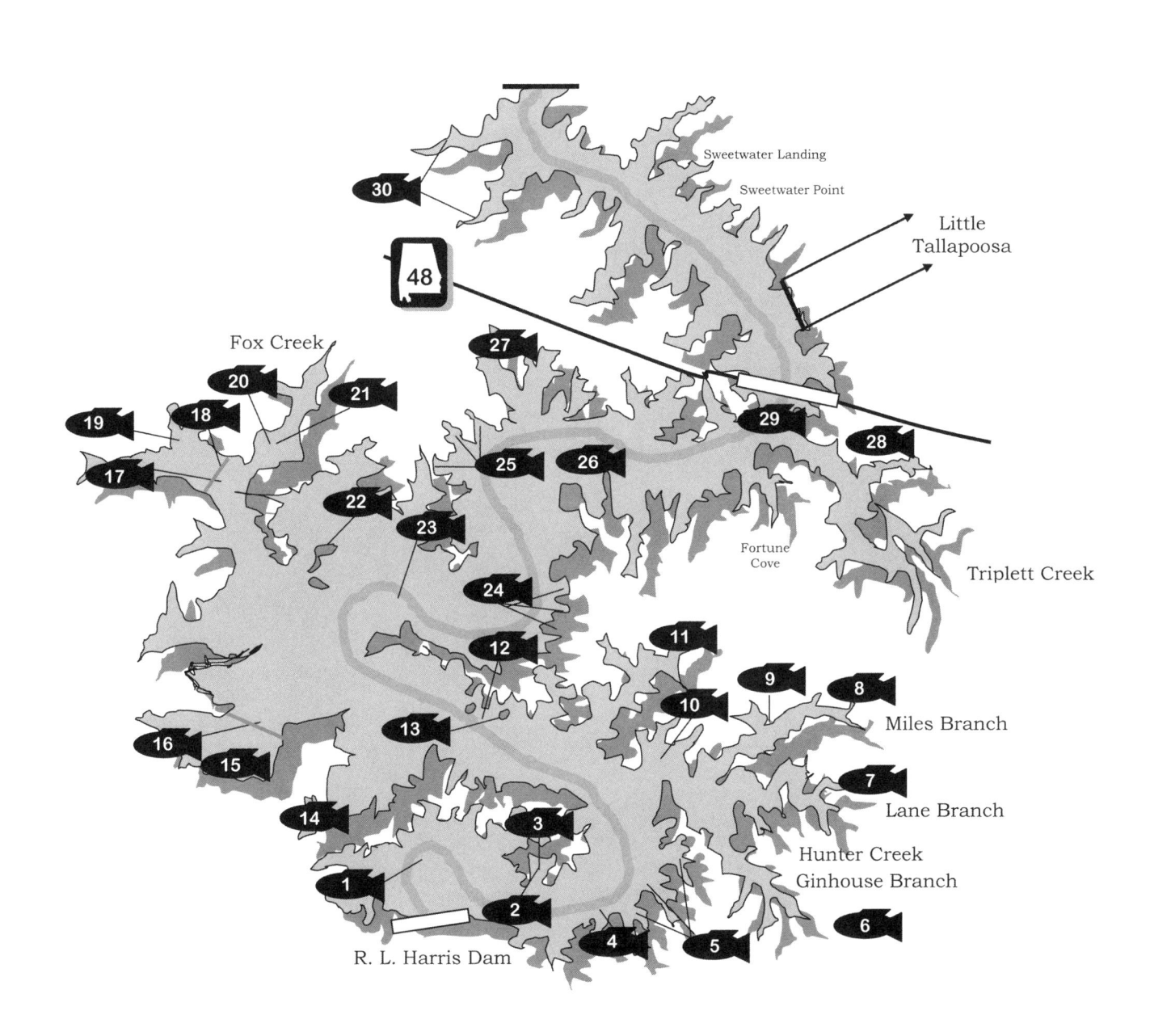

The Ultimate Guide to Alabama Fishing

Lake Wedowee

Above
Mad Indian Creek

Ketchepedrakee Creek
Lee Bridge
88
36
Little Ketchepedrakee Creek
Gobbler Creek
Foster Bridge
82
35
Fuller Creek
Outback Cove
Mad Indian Creek
34
31
33
32

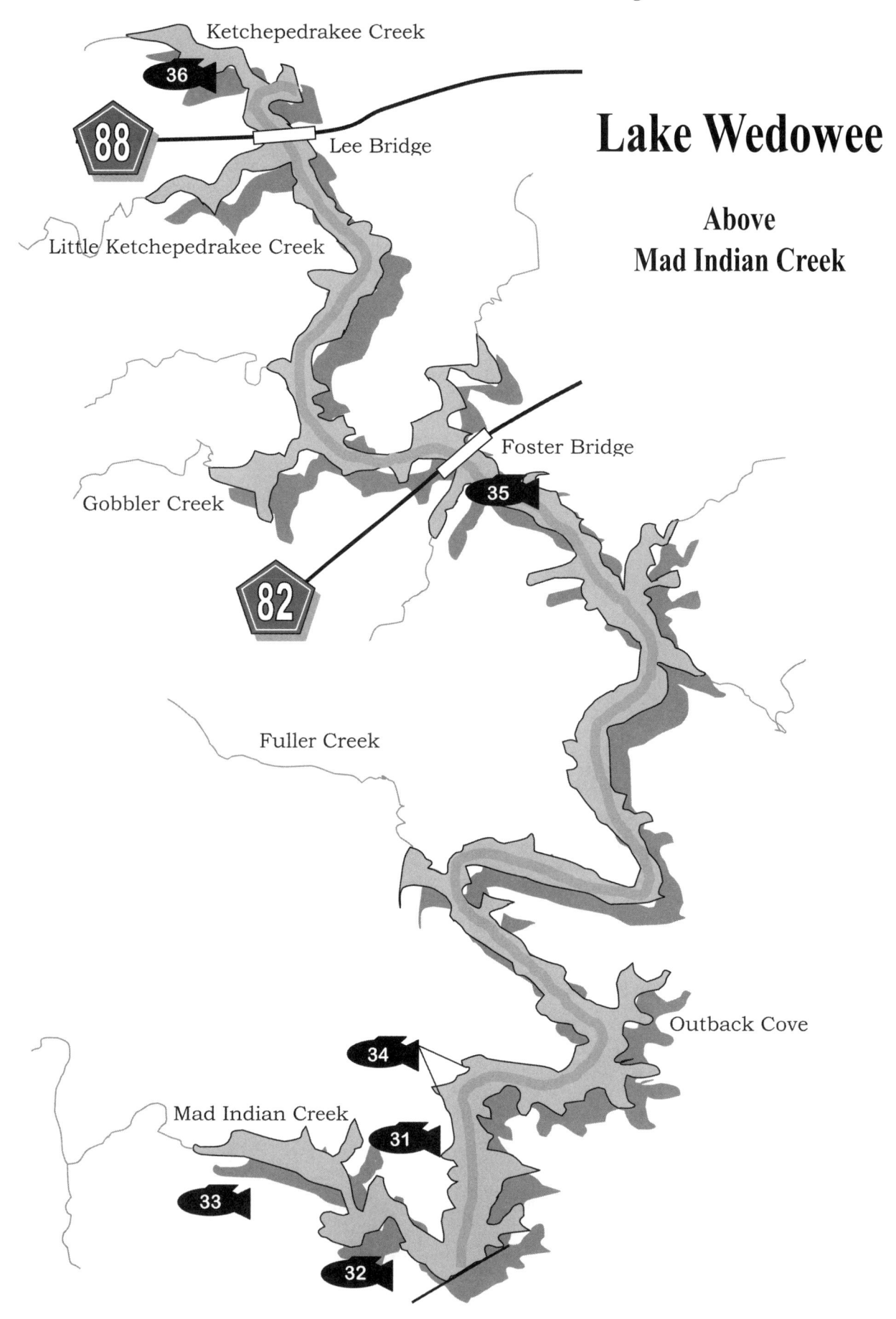

184

Lake Wedowee

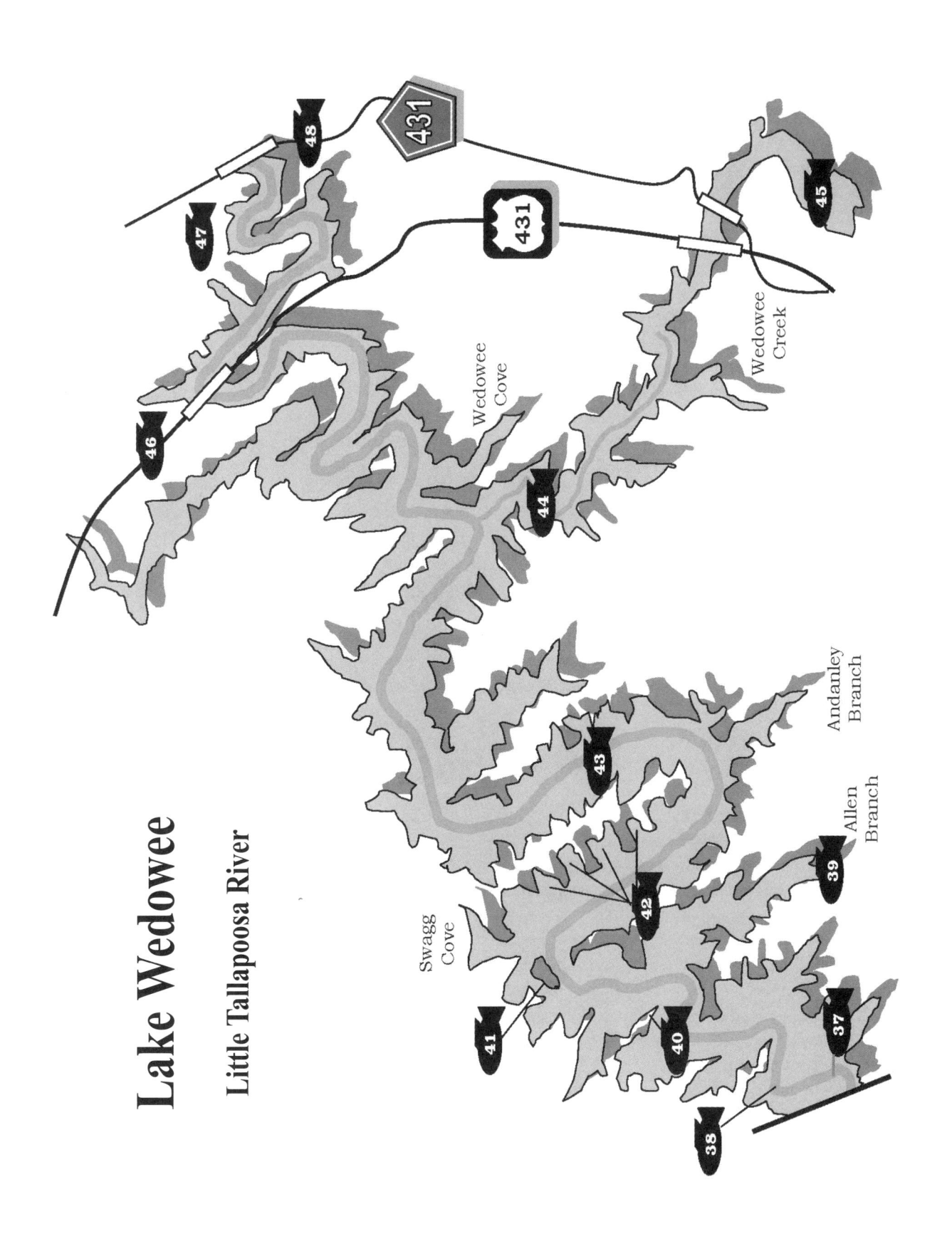

Little Tallapoosa River

Lake Martin

Lake Martin has been a lake Alabama fishermen have enjoyed for generations. Its bountiful supply of largemouth, spotted bass, crappie and catfish have since 1926 filled the creels of grandfathers, their children and their grandchildren.

The lake has seen some major changes in its days. It became the first Alabama lake to be stocked with striped bass in 1965 and in 1976 became one of the first non-county lakes to be stocked with hybrids. The stocking of both types of those fish have given the lake two exciting new fisheries.

Another change that affected the lake was the construction of Harris Dam in 1983. This dam blocked the lake's natural water flow and prohibited the flow of many nutrients that fertilized the water through the years.

As a result, Lake Martin has gotten clearer and clearer and fishing has become tougher. Anglers have adjusted well and fishing for a variety of species is still excellent.

Spring bass fishing

Above Pineywoods

The area of Lake Martin above the U.S. 280 bridge may or may not be a good location to fish in spring. It all depends on the water level. When the water level is at winter pool, spring fishing can be extremely difficult on the lake's upper end. This area contains a large amount of structure, but the pockets with structure may have only five or six inches of water during the early spring period some years.

During the spring when there are some good rains, these pockets can offer good fishing, especially when the water is rising. Chartreuse/white and chartreuse/black 1/2-ounce spinnerbaits outfitted with a No. 11 pork frog trailer (black or chartreuse) and a big willow leaf blade should be the first lure used in this area early in the spring.

Britt Creek, Sturdivant Creek and Coley Creek are the three creeks to concentrate on. The other creeks and sloughs above these areas often don't have enough water to hold fish. These three creeks don't have any weeds, but do have blow-down timber, man-made and natural brush piles and rocks. If you catch these areas at the right time, you'll find tremendous fishing. Catch them at the wrong time and you'll waste a day of fishing.

Pineywoods area to Lake Martin Dam

Lake Martin is a clear lake. You may find some stain early in the year, but rarely anytime else. They key to finding the big bass in this clear water during the spring is locating the right shallow rocky points. These points, almost always on the north bank, hold the heat and radiate it into the nearby water. The large largemouth bass will concentrate in these areas.

Find one of these rocky points with man-made brush piles and you will almost be guaranteed good fishing.

Being a good bass fisherman on this lake, or any lake for that matter, takes some effort. It's not unusual to come to this area of the lake in early January when it's too cold to be outside and find some of this lake's better bass fishermen out riding and looking, dressed like mummies. The lake's low water level during this time of the year exposes the man-made brush piles. The angler who comes to this lake in January with a GPS or even a notebook and pencil is the angler who will have success here in the spring.

Another key location for spring bass in this lake for late February and early March is the red clay points that have a few scattered fist-sized white rocks on them. No one has ever been able to successfully explain (or at least they don't want to explain) why the points with these rocks hold fish like they do. One guess, and it sounds far-fetched, it that at one time an angler somewhat unfamiliar with the lake filled his boat with these rocks and whenever he caught fish on one of the points, he threw these rocks on them to mark them. If that's the case, the man was fishing from a battleship because there are a lot of these points and an awful lot of these rocks.

Nevertheless, the bass use these points for staging areas this time of year.

Piers can offer some good fishing in the spring, but most of these piers are too far out of the water this time of the year to offer any help. The brush around the piers, if deep enough, can hold some fish, however.

There are several good area to concentrate on below the old Pineywoods Marina. Wind Creek, despite the fishing pressure and the amount of boat traffic, is one.

Madwin Creek and Manoy Creek are also good spring fishing areas. The area behind Woods' Island, if you can catch it right, is also good.

Sandy Creek is a large creek that, if you're willing to learn it, can hold your fishing attention all day in the spring. Its real drawback is the fact it's usually one of the first creeks to muddy in the spring and it will sometimes get too muddy to fish. It also holds the stain longer than most creeks on the lake. Its points, brush piles and piers hold numbers of fish, however.

All four of these creeks offer excellent bass fishing in the spring, but the key to being successful is fishing them enough to learn where the man-made brush piles are. This is where you'll find your quality fish.

A variety of baits will take fish in this part of the lake. A white spinnerbait is probably the most deadly, but only after the water temperature is over the 55-degree mark. Small, shad-colored crankbaits, Carolina-rigged lizards, jig-and-pigs, buzzbaits, topwater twitch baits and worms will all take fish here. Worms will take fish on this lake year round.

Kowaliga Creek

Kowaliga Creek is basically a lake in its own. The massive creek is nearly as large as the remainder of the lake. It contains both deep and shallow water, but it has one major drawback. It normally stays extremely clear year round. Fishing there can be dog-gone tough.

There is usually a 10-day span in the spring (usually right before Memorial Day) when this area offers tremendous fishing—a jubilee, if you will. The water rises significantly during this 10-day period and covers the brush under piers. The largemouth quickly move into this brush and a jig-and-pig flipped into the structure often produces numbers of big fish. Spinnerbaits are often effective at the same time for spotted bass on the fringes of the structure.

Several areas of this creek can offer good spring fishing. The best area is the back of Little Kowaliga Creek, past Real Island. A large number of the pockets have man-made brush piles and fallen timber.

Also, look for piers with rod holders and lights. These piers usually have brush around them. Buzzbaits, spinnerbaits and Rapalas all produce fish in these areas.

Always keep in mind that light line is an absolute must all times of the year.

Hot weather bass fishing

Above the Pineywood area

The lake's upper end offers little good summer fishing, even at night. It's possible to catch a few ledge fish in the mouth of Hillabee Creek, around the rocks near the back of the island downstream of Hillabee Creek, and around rocky points, but you're better off fishing below the Pineywoods area.

Pineywoods area to Lake Martin Dam

Combine deep clear water, high surface temperatures, crazed Jet Ski riders and water skiers and an unrelenting sun and you have a recipe for bass fishing at its worst in Alabama. Lake Martin unfortunately has all four ingredients. Bass are spooky and stressed in this lake during the summer and it's no mystery why. During summer days, an angler with a depthfinder will often locate small schools of bass off the deep rocky points on the main river channel, but finding these fish and putting them in the boat are two different things. These fish are not only deep, they are non-aggressive and spooky too.

Anglers have attempted to entice the fish to strike for decades with a limited amount of success. Worms and crankbaits have been the main weapons.

Lake Martin's bass anglers have improved their success ratio recently, however, by taking a cue from California bass fishermen, for whom fishing deep, clear, warm reservoirs is a way of life. The pattern they use is known as "doodling." Doodling" requires a light spinning outfit, 6- or 8-pound test line, a 4-inch worm, a 3/16-ounce weight and an orange glass bead. The bead is placed above the weight where it will bump it and make noise. The worm should be rigged with an exposed hook rather than being rigged Texas-style.

Once small schools of bass are located on deep points, the boat should be positioned directly on top of the fish. The plastic worm should be counted down to the exact depth of the fish and the rod tip shaken up and down quickly. This causes the worm to bounce erratically right in the face of the fish and it drives them crazy. Most believe the fish strike out of anger rather than wanting to feed, because a worm dropped the same way and not shaken most often won't evoke a strike. This pattern will produce good numbers of summer fish.

All the water skiers and bright sun in the world aren't going to keep bass from eating, however. That's why when the sun goes down on Lake Martin in summer, it's time to get out the fishing rods to do some serious fishing in the lake's deeper creeks.

A tremendous number of piers on this lake have brush around them and a large number have lights. These lighted piers with brush attract baitfish and the bass are rarely far behind.

Worms, small crankbaits and an occasional topwater bait are the lures to use. The boat should be positioned as far as possible away from the piers in the dark water. Long casts are the key.

Many of the better anglers also fish deep in the creeks during the summer at night and catch good fish doing so. Worms and crankbaits fished on deep points with brush will produce these fish.

Another pattern—a strictly-for-fun pattern—is fishing a flyrod around pier lights at night. A white fly with wide white wings is similar to the small moths that fly around these lights. Small bass wait until these moths hit the water and then hit them hard. Plan to use a long, 4-pound leader, however.

Kowaliga Creek

Skiers, big boats and the crystal clear water makes Kowaliga Creek a difficult place to fish between Memorial Day and Labor Day. The best advice is to fish elsewhere on the lake at this time, but an angler with a little patience can catch some fish there.

A doodling worm fished 25 feet deep on the deep rocky points and the ledges with sharp drop-offs is the best bet to catch bass during the day. Plan on fighting the waves from boat traffic all day, especially on weekends.

Night fishing is best. Spinnerbaits, small crankbaits, Rat-L-Traps and an occasional topwater bait will take fish around lighted piers in the area from the sailboat club to Needle's Eye.

Cool and cold weather bass fishing

Above Pineywoods area

The narrow water above Pineywoods Marina can offer some good fall and winter fishing if the area has some water depth. Hillabee Creek is full of structure including downed trees, big rocks and brush. Spinnerbaits, topwater, small crankbaits and worms will catch both fall and winter fish around this structure.

Pineywoods area to Lake Martin Dam

Three baits tell the story of fishing this area of the lake during the cool and cold months. Bring a few white spinnerbaits, a few topwater baits such as a Crazy Shad or some type of chugger and a Carolina-rigged lizard and leave the rest of your tackle box at home.

Any of those baits can produce fish this time of the year. A Crazy Shad fished around shallow rocky points, brush piles and piers offers the most exciting fishing during the fall, but spinnerbaits are more effective if you can find water with a little stain.

Madwin Creek, Manoy Creek, Sandy Creek and Wind Creek are key creeks for topwater this time of the year, but can also be productive with spinnerbaits if they have a little water color due to heavy rains. Elkahatchee offers the best spinnerbait fishing. Elkahatchee Creek has some funky-colored water at times, depending on what color dye a local manufacturer in the area is using at the moment.

All of the above creeks, except Elkahatchee, can offer some good late fall fishing with Carolina-rigged lizards fished on deeper, rocky points.

Another pattern that works in this area during the cold-weather months is pitching. Pitching with a jig-and-pig is basically the flipping pattern with longer casts. The clear water makes flipping tough. Look for man-made brush piles on deep points to pitch in.

Many anglers believe this is the state's best lake for December and January fishing.

Kowaliga Creek

If you fish for sheer enjoyment and would rather feel a tug on your line rather than struggling for hours in an attempt to catch the big one, Kowaliga is an excellent place to fish during the fall.

Prop baits and buzzbaits fished before the sun gets over the trees is the top producer. Points and islands next to the main creek channel begin producing topwater bass (mostly spots) as early as mid-September. These fish will gather in these areas for several weeks before moving up on the flats and into the back of sloughs and creeks.

The area behind Pine Point Marina (formerly Veasey's) is a prime area during this time. Pitchford Hollow, Chapman Creek above Needle's Eye and the open water above The Narrows are prime locations for fall fishing. All three of these areas will have a limited amount of stain during this time and shad gather there. Once the sun gets over the trees, it's strictly worm fishing time.

These areas are full of crawfish at this time and bass feed heavily on them. The crawfish will be brown and their pincers will have a lime color. Due to that, a motor oil worm with a chartreuse firetail has withstood the test of time in this area.

The dead of winter means jig-and-pig time. The fish caught on that bait will usually be larger than any fish you catch in this area any other time of the year. Deep points with brush tend to hold these bass during January and February.

Other types of fishing

Hybrid and stripe fishing

Lake Martin received its first stocking of striped bass in 1965 and pretty much held the title of stripe capital of Alabama until Smith Lake began to challenge that crown in the late 1980s. Smith Lake, although it is now producing numbers of fish in the 30- to 40-pound-plus range, isn't ready to challenge Lake Martin in size of fish, however.

Lake Martin has produced hundreds of 30-pound fish, several 40-pound fish and at least one 51-pounder.

Saltwater stripes, hybrid bass and white bass all start the year off in early spring by making a spawning run (stripes and hybrids can't spawn, but do try) into the area above the U.S. 280 bridge and, over on Kowaliga Creek, into the area above the narrows.

These fish can be caught by several methods. The most exciting, but least productive, method is fishing a Redfin or Woodchopper-type lure on topwater. A 20-pound fish hitting a top water bait is unlike anything most anglers have ever seen. These topwater baits should be fished early on long sloping points and in narrow pockets.

The most productive method, but probably the most boring, is fishing a live shad on a free line. Twenty-pound-test line outfitted with nothing but a hook should be used with a live shad hooked through the nostrils. With the boat floating in the mouth of pockets or creeks, the shad should be given plenty of line and be allowed to roam free. It's not uncommon to fish three or four hours without even a strike, then catch a 30-pounder using this method.

Another spring pattern, especially early in the morning when the stripes and hybrids are shallow, is to use a small balloon as a float with a live shad. This allows the shad to run into shallow water without the boat scaring the fish off.

After Memorial Day, the stripes and hybrids will move back into the area around Pineywoods, Bay Pines and Young's Island.

The fish will hold on points and in the main river channel until approximately the first of June. Tight-lining live shad at the depth where the fish are is the only viable way to catch these fish.

The first of June will see these fish move further down river, gathering below the sailboat club, Blue Creek and Kowaliga Creek.

These fish will normally stay in these areas into the fall. Once again, tight-lining live shad is the best method of catching these fish.

The better stripe and hybrid anglers locate these fish with their depthfinders before daylight and fish them until approximately 8 a.m. After the sun gets high in the sky, the fish disappear into the depths and are almost impossible to locate.

Lake Martin's best fishing spots

1. Big boulders in main river channel. Good fishing in spring. 5-10 feet deep.
2. Island. Good in spring when water is up. Topwater.
3. Mouth of Hillabee Creek. Good when water is up in spring.
4. Island in mouth of Hillabee Creek. Stumps on downstream side of the island. Good spring fishing with topwater.
5. Excellent point when water is up. Good spring and winter.
6. Underwater point leading to main channel. Good behind point when water is moving.
7. Excellent point adjacent to island. Good funnel area in spring.
8. Underwater island. Good for schooling bass and hybrids in spring, summer and fall.
9. Railroad trestle. Good night fishing for crappie under bridge with lights.
10. Underwater road bed.
11. Good point. Old building foundation out from point.
12. Underwater railroad bed in Sturdivant Creek. Good spring and summer. Locate by finding where railroad ties run into sandy bank on left side of creek.
13. Lower point in Sturdivant Creek. Excellent fishing year round.
14. Piers filled with brush in Sturdivant Creek. Excellent year-round fishing.
15. Excellent point for hybrid fishing early in year.
16. Excellent slough below U.S. 280 bridge. Many piers there surrounded by Christmas trees and other brush. Good in spring and fall.
17. Pockets north of Pineywoods Marina. Downed trees, brush around piers, rocky secondary points. Good year round.
18. Dennis Creek. Lot of red clay points and man-made brush piles.
19. Elkahatchee Creek. Good spring and fall with topwater baits, spinnerbaits. Good color.
20. Underwater road bed.
21. Underwater road bed.
22. The Narrows. West bank very deep. Great winter fishing with hair jig or grub. Some good fishing in February.
23. Underwater road bed.
24. Excellent slough full of piers with brushpiles.
25. Underwater road bed.
26. Madwin Creek. Full of rocky points, gravel flats and brush-filled piers. Good year round.
27. Manoy Creek. Spring and fall. Deep creek, but full of points.
28. Logjams and blowdown timber. Willowleaf spinnerbaits in spring and fall.
29. Underwater road bed.
30. Underwater road bed.
31. Excellent spring fishing in back of Wind Creek. Good shallows with some weeds. Small crankbaits and spinnerbaits in very back.
32. Wind Creek. Good spring fishing for crappie in back slough.
33. Gas pumps island at Wind Creek State Park. Long point. Fish released from tournaments gather there.
34. Young's Island. Southwest point is long, deep, rocky. Holds bass on and off year round.
35. Woods' Island area. Lots of good sloughs with blowdown brush. Roadbed runs from bank to Woods' Island to Young's Island on south points.
36. Underwater island. Key area. Top of island is 8 feet deep at winter pool and covered with brush. Good spotted bass area year round. Also, some stripe school there.
37. New Hope Church area. Sloughs behind church full of man-made brush piles. Excellent spring fishing. Gravel bottoms good for spawning bass.
38. Underwater road bed.
39. Island with stumps on west bank.
40. Underwater road bed.
41. Underwater island with stumps on west bank.
42. Underwater islands with stumps on west bank.
43. Underwater road bed.
44. Underwater road bed.

45. Big Sandy Creek. Everything you need for all day fishing in spring and fall. Blowdowns, brush, rocky points.

46. Slough at mouth of Sandy Creek. Long, gravel-covered points and brushpiles. Excellent spring fishing. Brushpiles good in fall.

47. Wicker Slough. Excellent spring fishing. Gravel bottom with blowdown timber. Gravel points have stumps. First long slough on right is full of bass during pre-spawn.

48. Blowdown timber in back of slough. Good spring fishing.

49. Stumps on southwest side of island. Good spring fishing.

50. Underwater boulders on northwest side of small island. Good year-round fishing.

51. Underwater road bed from bank through middle of channel and back up other side. Key area. Good year round.

52. Stumpy bank on east side of slough. Good spring fishing.

53. Underwater road bed completely across Blue Creek. Good year round.

54. Underwater road bed.

55. Underwater road bed.

56. Underwater road bed.

57. Stump-filled sloughs. Excellent spring fishing with buzzbaits and spinnerbaits.

58. Underwater road bed.

59. Underwater road bed.

60. Underwater driveway. Home foundations on both sides. Good spring fishing.

61. Underwater road bed.

62. Underwater road bed. Runs parallel to old creek channel. Good spring fishing area.

63. Underwater road bed. Adjacent to old creek channel. Good spring fishing.

64. Underwater road bed across Cooper Creek. Both deep and shallow water. Good year round.

65. Underwater driveway.

66. Underwater road bed across mouth of Blue Creek.

67. Mouth of Kowaliga Creek at Dixie Sailing Club. Tremendous schooling area for hybrids in summer before 8 a.m.

68. Underwater road bed.

69. Underwater road bed. Surrounded by old home foundations.

70. Underwater road bed.

71. Underwater road bed.

72. Underwater road bed in slough.

73. Highway 63 bridge pilings. Good crappie fishing area at night with lights.

74. Underwater islands. Good spring fishing.

75. Good ledges next to shallow island. Good year round.

76. Excellent fishing point in spring.

77. Stump-filled slough. Good spawning area. Buzzbaits and spinnerbaits.

78. Good pockets for spring bass fishing.

79. Island in shallow water next to deep water. Good spring and fall area.

80. Mouth of pocket with good ledges. Good year round.

81. Excellent point for spring fishing. Out from Real Island Marina.

82. Excellent spring fishing point.

83. Shallow water island. Good spring fishing area.

84. Good spawning pockets.

85. Underwater rock point. Good early in fall and spring. Some good year-round fishing.

86. Underwater humps covered with brush piles. Marked fish attractor.

87. Man-made brush piles on banks and under piers in slough.

88. Big rocks off point out from Castaway Marina. Marked with danger sign. Good spotted bass fishing.

89. Good rocky point with brush.

90. Underwater rock piles. Marked with danger sign. Nearby pockets have stumps on points. Some brush in back of pockets.

91. Excellent point. Deep rock piles.

92. Long gravel-covered underwater ledge.

93. Steep ledge out from rock wall. Good year-round fishing.

94. Good trash-filled pocket.

95. Excellent point. Rocky with quick drop.

96. Point with rocks. Out from house with sliding board.

97. Islands next to deep water. Good spring and fall fishing area before sunup with buzzbaits and jerkbaits.

98. Back of Jack's Hollow. Lots of man-made brush piles.

99. Good stump-covered point. Out from big house. Good fishing in sloughs on both sides. Good fishing around islands in spring and fall. Good ledge fishing in area, too.

100. Good point with underwater rock pile.

101. Underwater rock pile. Good year-round fishing. Visible when water is down 8-10 feet.

102. Slough in back of Kowaliga Marina. Filled with rocky points and brush. Big iron pipe on left bank with

automobiles sunk beside it. Lot of trash around autos.

103. Big slough. Blowdown timber and rocks. Piers on right side have brush.

104. Block wall on point. Falls into deep water. Good topwater area in spring and fall.

105. Long point. Good spotted bass year round.

106. Excellent point above islands. Holds fish year round.

107. Good pockets. Filled with trash. Good spring and fall.

108. First point on left in Needle's Eye holds lots of fish. Back of pocket also good.

109. Excellent point right before the Narrows.

110. Pleasure Island. Pockets behind island contain pea gravel. Good spawning area for bass.

111. Back of Porter Branch. Ski course area. Entire pocket good. Full of brush.

112. Excellent series of points and slough; many brush piles.

113. Good points and pockets. Brush usually visible.

114. Big rocky point. Good year round.

115. Good pockets. Lots of small points. Full of brush.

116. Back of Little Kowaliga Creek. Good flats. Good spawning area with brush.

Mike Bolton

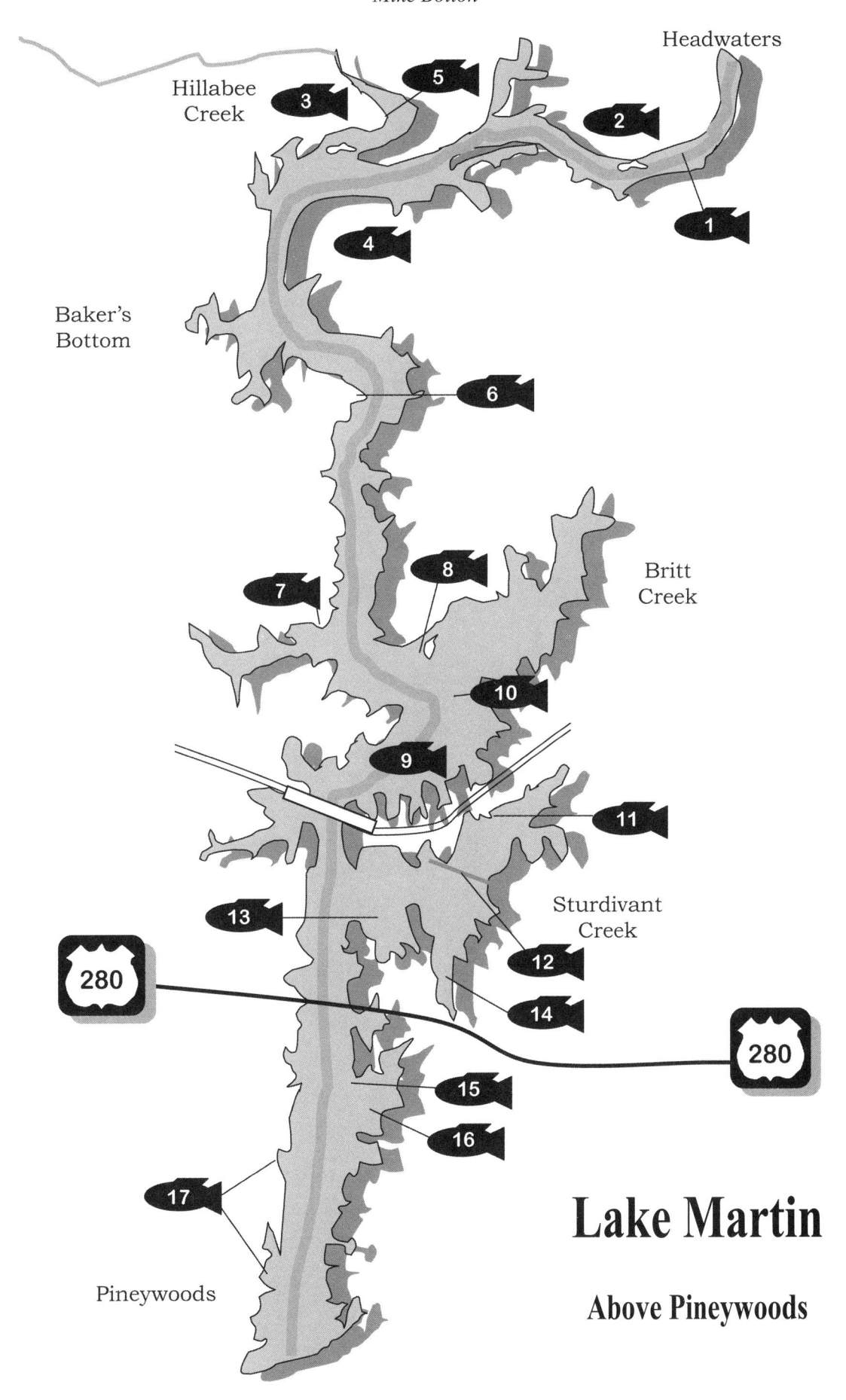

Lake Martin

Above Pineywoods

193

Lake Maartin

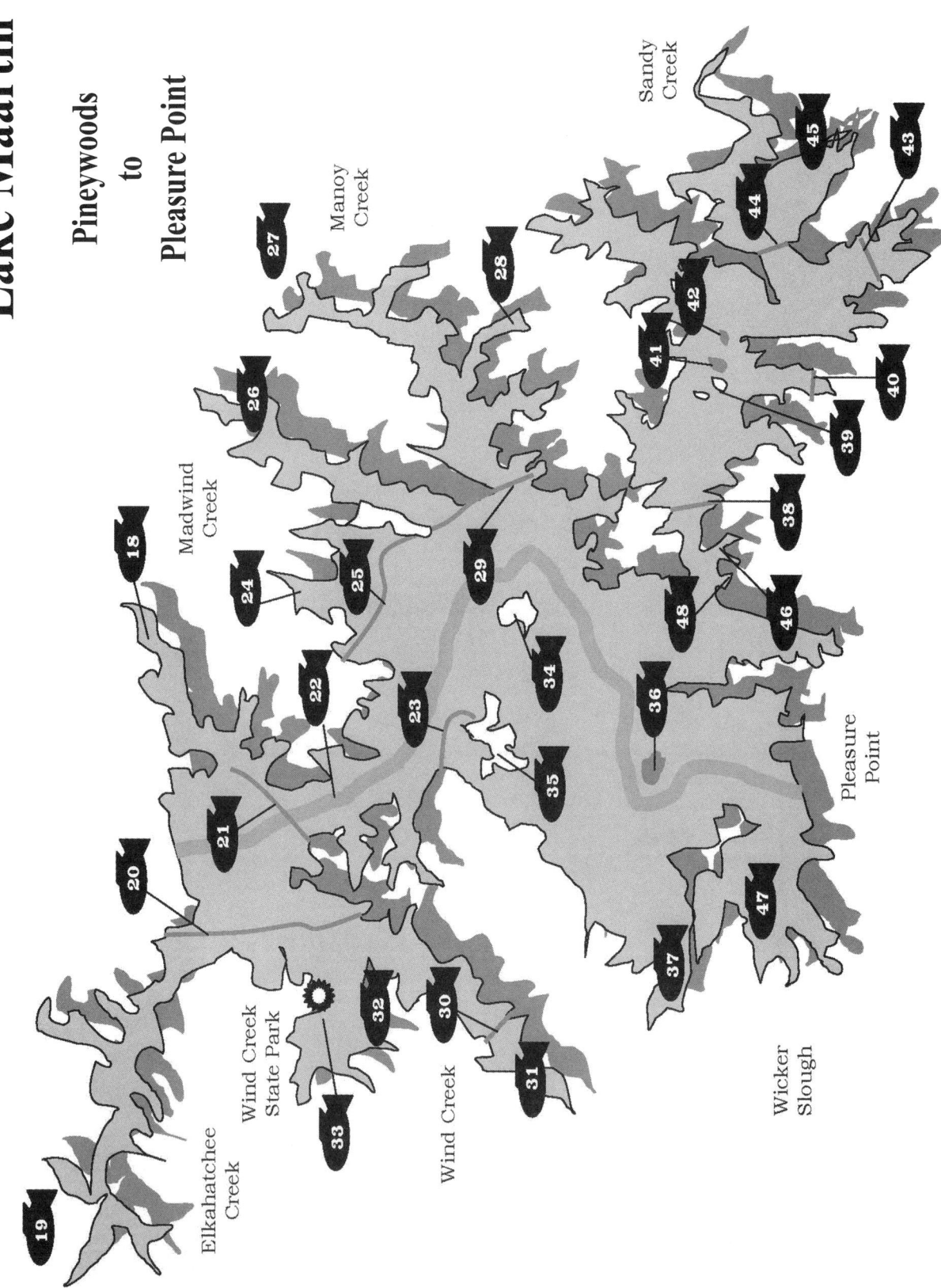

Pineywoods to Pleasure Point

The Ultimate Guide to Alabama Fishing

194

Lake Martin

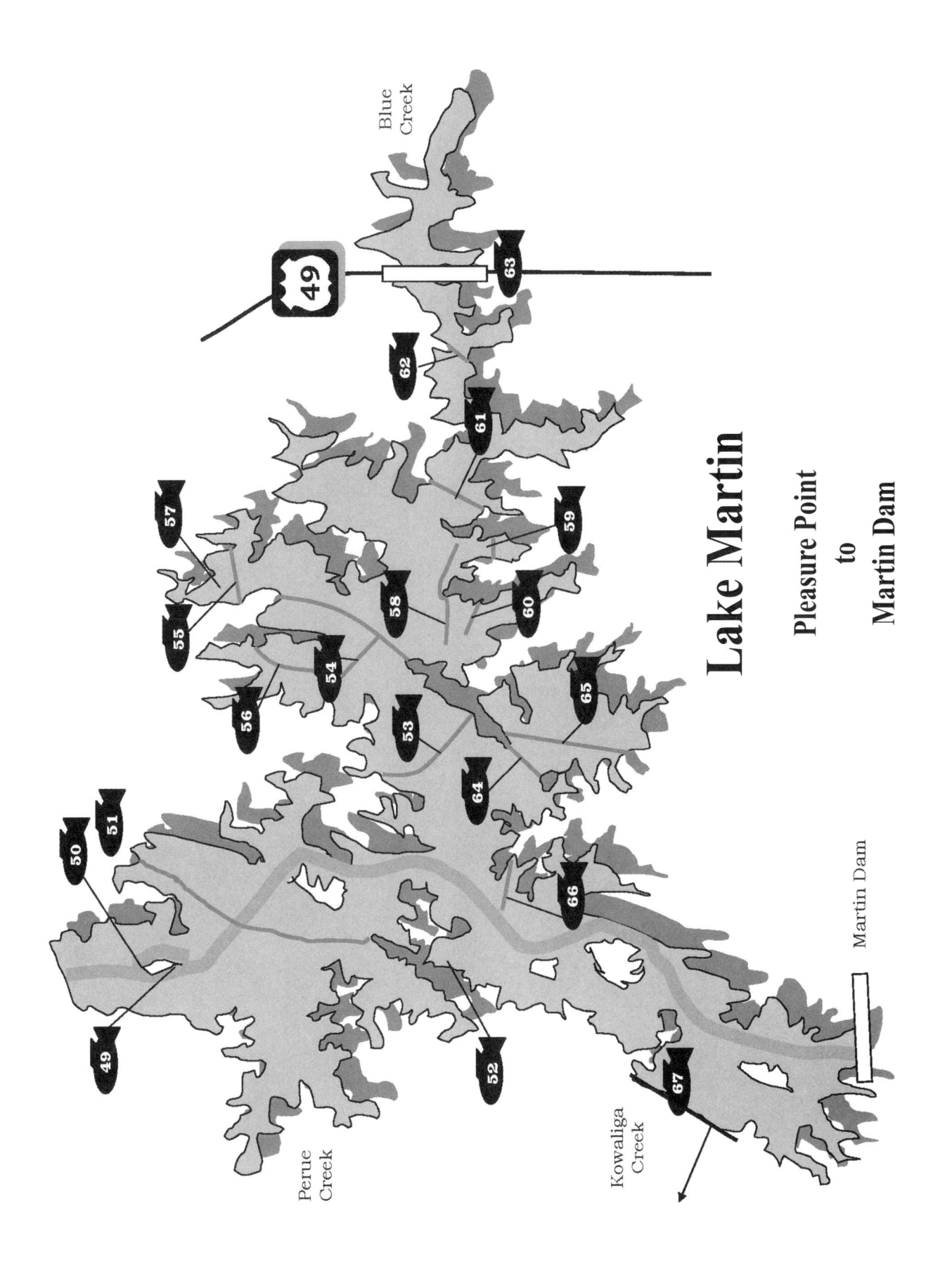

Pleasure Point to Martin Dam

Lake Martin
Kowaliga Creek

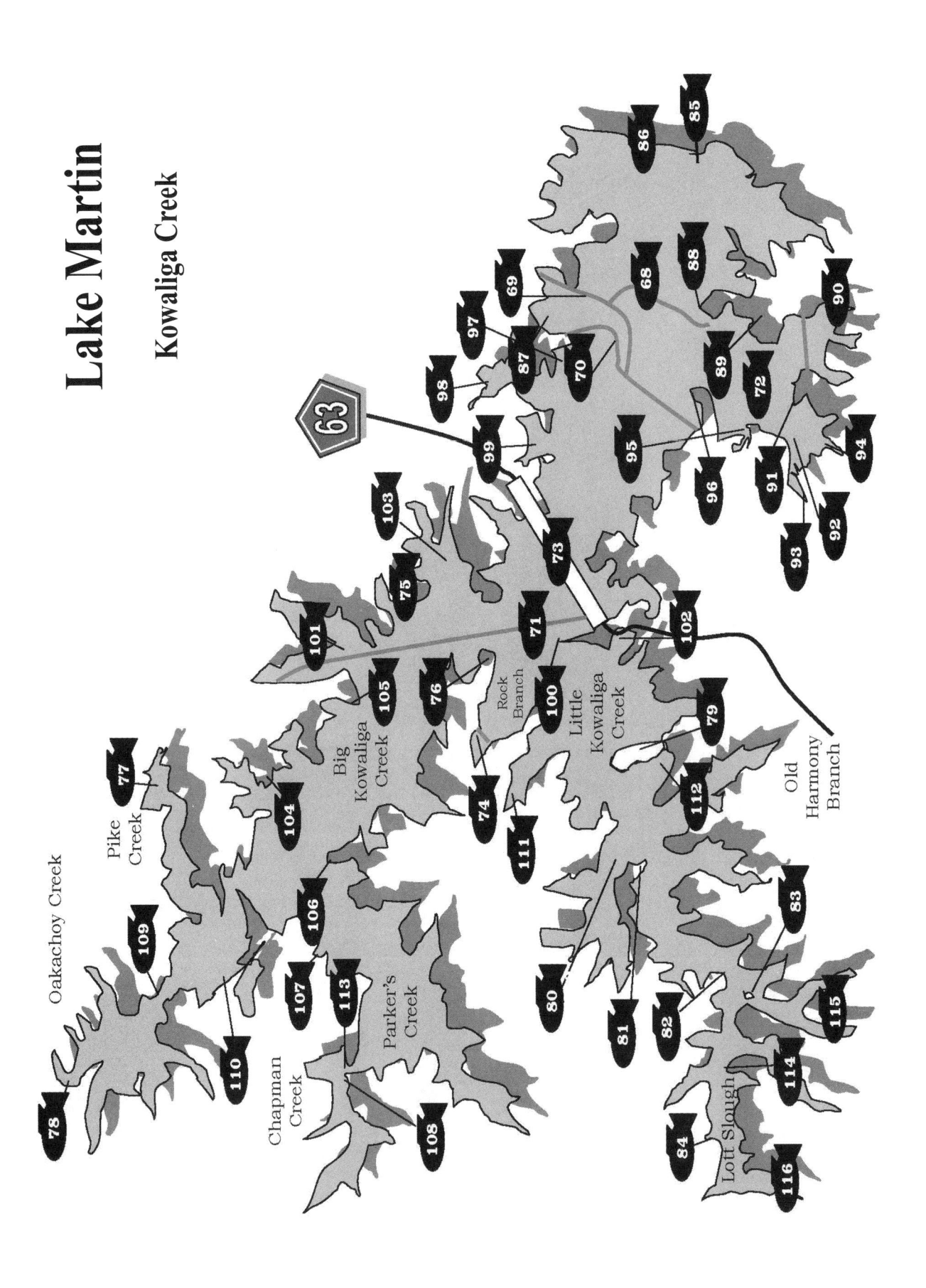

Mike Bolton

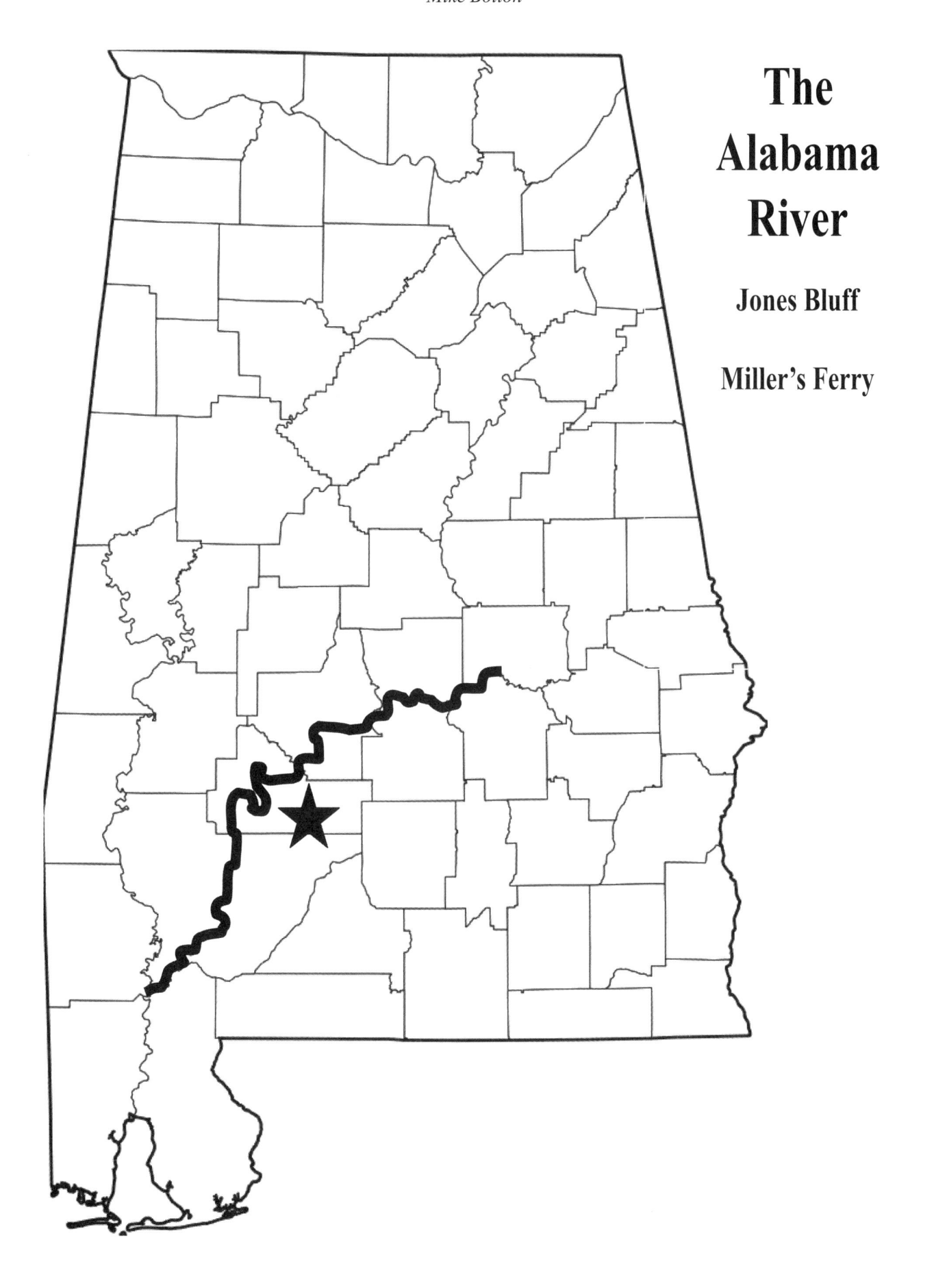

The Alabama River

Jones Bluff

Miller's Ferry

The Ultimate Guide to Alabama Fishing

Jones Bluff Reservoir (R.E. "Bob" Woodruff Lake)

Jones Bluff Reservoir is more river-like than most other reservoirs in this state but its close proximity to one of Alabama's largest population areas makes it one of the most fished reservoirs in the state.

It is a popular stop for central Alabama bass clubs and it has also become an occasional stop in recent years for pro tours including BASS. The reservoir flows along the northern edge of Montgomery's downtown and beautiful Riverwalk Park gives fishing in that area a carnival-like feel.

The reservoir got its first real publicity in 1981 and 1982 when BASSMASTER Classics were held there. Fitzgerald, Ga.'s Stanley Mitchell won the 1981 Classic with 15 fish weighing 35.2 pounds and Paul Elias, of Laurel, Miss., won the 1982 Classic with 15 fish weighing 32.8 pounds.

In recent years BASS has held the Evan Williams Bourbon All-Star Championship on the reservoir with the top pros producing some good numbers fishing a drop-shot.

The upper end of the reservoir is full of spotted bass while largemouth are more common on the lower end.

The lake also offers good fishing for crappie, bream, catfish and hybrid bass.

Spring bass fishing

Dead River to North Bypass

The section of the Alabama River that winds its way through Montgomery is little more than straight river banks. Because of a lack of creeks for the water to back up in, it's prone to flooding conditions during the spring and is often extremely muddy. For those reasons, the upper end isn't normally the place to plan a spring fishing outing.

During those years when spring flooding isn't a problem, a few fish can be caught in the upriver area, however. The area around Parker's Island and the Dead River area is full of sand bars, pea gravel bars and logjams. A 1/2-ounce chartreuse/white spinnerbait with gold blade and a No. 11 chartreuse pork frog trailer is the best bait in this area.

Anywhere the river bends is worth a look in the spring. The outside of the bend will have deep water, but the inside bend will be shallow bars, many times with trash and logjams. The same spinnerbait should be fished in these bends.

Never pass up the mouth of Coosada Creek. This small creek offers one of the few areas where largemouth can escape the swift current and fish gather in the mouth, muddy water or not. Several days without rain is usually needed to suck the mud out of the creek during spring floods, but don't hesitate to go up in the creek and fish the line where the muddy water and clear water meet. Fish are sometimes gathered there in numbers.

If you enjoy fishing escape areas in the spring, give the backside of the islands a try, too.

Further downstream, the Jackson Lake area is worth a look. This oxbow lake is full of cypress trees and holds spring bass. The gravel pits (No. 9 on the map) is a good spawning area, but high water is needed to get into the area and high water usually means muddy water. If you can find a happy medium, give the pits a try.

Don't mistake the listing of several areas in this upper section as an endorsement for fishing the area during the spring. Rarely will you ever find the lake's better fishermen in this area during spring tournament time.

North Bypass to Tallawassee Creek

Once you pass downstream of the North Bypass bridge, you will discover two things you will like—wider water and water with less stain.

Unlike the upper end of the lake, this area has many wide creeks which will back up and lessen water levels during heavy spring rains. The water still will be high at times, but it's less muddy and fishing isn't affected nearly as much.

Cooter's Pond, Catoma Creek, Pintlala Creek, Noland Creek and Tallawassee Creek help alleviate the spring flooding problem.

The key to fishing this area in the early spring is determining exactly where the fish are. These fish will be leaving the deep water of the main channel and staging themselves up the ledges on the way to the creeks.

Deep-running crankbaits and Texas-rigged lizards are key baits at this time. Let the water clarity determine the bait color. Heavily stained water will call for a lime or chartreuse crankbait or a dark worm with a chartreuse or orange tail. Less stained water will require more subtle colors, such as a shad-colored crankbait or a pumpkinseed-colored 7-inch Zoom lizard.

All of the upstream and downstream points leading into the creek mouths are surrounded by 2- to 4-foot sycamore stumps, and the dirt has washed away from the tremendous root systems of these stumps. It is excellent structure to hold staging fish. Bring lots of baits when you come, however. These roots keep a lot of crankbaits.

Once the water hits the 62- to 65-degree range, bass will have spawning on their mind and will move into the creeks using the channels and ditches as their roadways.

These fish will stack up in bends of the original creek channel and hold until the water temperature allows them to move to the ledges. Fishing can be confusing during this time because the fish will flip-flop from the channel to the ledges depending on the weather. Worms and Sassy Shads are good baits when the fish are holding in the channel, but buzzbaits and 3/8-ounce spinnerbaits with big willow-leaf blades are best when the fish move to the ledges.

One of the better secrets of the lake's spring tournament fishermen is using a blue/chartreuse, straight blue or straight red spinnerbait.

These spinnerbaits tend to catch sluggish fish that are tired of seeing white or chartreuse baits thrown at them day-in and day-out.

Go with a gold blade in light stain conditions and a copper blade or a painted orange or chartreuse blade when the water is heavily stained. Use a No. 11 pork frog trailer, a twin-tail frog or a 3-inch chartreuse grub to dress up the baits.

One of the strangest things about this lake is that the pre-spawn and post-spawn bass on the ledges in the spring often roam open water, rather than sticking close to structure. It's not unusual to throw a spinnerbait or buzzbait onto a muddy ledge void of structure and catch fish. Why are they there? Who knows?

There are several creeks to concentrate on during the spring. Cooter's Pond has lots of shallow water conducive to spring bass fishing. Catoma Creek is an excellent spring fishing area. It has the structure, flats and bends that concentrate fish and make locating them easier. Several junkyards were covered when the lake was impounded. There are also a lot of stumpy points in shallow water.

Noland Creek is a deepwater creek, but it has some flats and shallow points that hold spring fish. Pintlala Creek is narrow, but it has a lot of hard bends, blowdown timber and stumps in shallow water that make excellent spawning areas.

Tallawassee Creek is one of the better creeks for spring fishing. The Corps of Engineers left several stands of timber in the creek bends to attract fish. Much of it has now rotted and fallen, leaving stumps at the water level. This was also one of the original stocking areas for the lake and many anglers believe a large number of those fish have stayed in the area.

Tallawassee Creek to Brown Branch

Anglers will notice several distinct changes in the lake once they go downstream of Tallawassee Creek. For starters, the water clears noticeably. The water is also wider and has more grass.

The patterns used in fishing this area in the spring is virtually the same as the stretch of lake above it. You may want to go to a lighter spinnerbait and switch to a silver blade, but pattern-wise, very few changes are in order.

Crankbaits and worms should be fished in the mouth of the creeks early, and spinnerbaits and buzzbaits are needed when the fish move up into the creeks.

If there is a pattern change on this lake, it's paying more attention to the grass. The grass that can be found in 5 feet of water will hold fish and the fish will be there before, during and after the spawn. This grass provides all the basic necessities including food, cover and oxygenated water. Artificial crawfish and jig-and-pig flipped into this grass can produce some trophy fish this time of year.

There are numerous areas to concentrate on in this end of the lake in the spring. Newport Landing Slough is full of standing timber in the back and is an excellent spawning area. Tensaw Creek has both deep and shallow water and is full of good points that hold bass pre- and post-spawn.

Swift Creek shouldn't be overlooked. There is much farming in the area and animal wastes and fertilizer wash in and keep it stained.

It is one of the earliest creeks on the lake to warm and the first place on the lake you can catch a fish with a spinnerbait each year. This creek is virtually a private pond.

Cypress Creek has many humps and ledges, as well as grass the first few hundred yards inside its mouth. The west bank is deep and the right bank somewhat shallower. The back end is full of cypress trees, hence its name. This creek can be rated good for spring fishing.

House Creek is the camp-out spot on the lake. Tent-pitchers and private campers frequent the area. House Creek is narrow and over-fished. Pass it up.

The Three Fingers area, officially known as Cooper Howard Creek, is an excellent spring spawning area. The east fork stains good from farming and warms much earlier than the middle two forks, which are usually clear. Weeds, standing timber and points can be found on all three fingers and fishing can be excellent, especially on the west fork in March.

Brown Branch is a narrow creek surrounded by extremely high bluffs. Its water is much too deep for spring fishing.

Beaver Creek is a wide creek that is good in early spring. Its mouth is full of trash, floating debris, logjams and blow-down timber. There is some grass up in the creek. This creek has a lot of shallow water and is buzzbait territory.

Hot weather bass fishing

Dead River to North Bypass

If there's a good time to fish this end of Jones Bluff Reservoir, it's during the hot months of summer. Unlike most lakes in Alabama, this lake gets a good amount of water flow during hot weather and this water movement triggers the bass into feeding.

Water will move in this area nearly every weekday during the summer and the lack of creeks makes the water flow more forceful here than any other area on the lake.

The outside bends of the main river channel are the areas to concentrate on during this time. Lots of bait fish gather in these bends when water is being pulled. Rat-L-Traps, spoons, Little Georges and plastic worms will take the bass gathered in these bends.

The Big Curve area has deep banks with blowdown timber and day-in and day-out is the top producing bend on the river's upper end.

If water isn't being pulled, don't waste your time in this end of the lake during summer.

North Bypass to Tallawassee Creek

Before Jones Bluff Reservoir was impounded, chainsaws took down all the timber that would eventually be underwater. This left ledges full of standing stumps.

Anywhere along these stump-filled dropoffs is a threat to hold schooling summer bass, but they gather best in the river bends because that's where the bait gathers. These bends also have more blow-down timber because the moving water erodes these banks more quickly.

It's not unusual to find smaller bass up in this blow-down timber from dark to 7 a.m. during summer. Worms and topwater baits will normally take these fish.

The bigger fish will stay in the deep water, however, electing to suspend themselves along the ledge. They will make occasional runs onto the ledge to feed. The best method to locate these fish is to watch for shad. Shad will ripple the water and actually jump or get knocked out of the water while the bass are feeding on them.

The angler can use several baits at this time. A worm is probably the best bait, but a swimbait fished on light line is close behind.

Some opt for small, medium-running crankbaits, but the number of baits lost doesn't justify their use.

During the night or during the wee hours of the morning, don't hesitate to throw a topwater lure where the ledge meets the deep water.

Once again, moving water is the key in this area. With the water off, it's possible to fish your heart out at daylight and not get a bite, then when the sun gets high, have the water start moving and load the boat. Having the water move is often like flipping a light switch or ringing a dinner bell.

If you're a creek fishermen, several creeks in this area have water deep enough to support summer bass. Catoma Creek and Pintlala Creek have such water.

Pintlala Creek is a creek you should especially keep an eye on. This narrow creek has shade trees on both banks and the amount of shade these trees provide results in a creek with a lower water temperature than the creeks around it.

Pintlala Creek also has a lot of blow-down timber and brush which hold fish on the banks during the summer.

Tallawassee Creek to Brown Branch

Because of the amount of water flow this lake gets during the summer months, the water flow into the creeks —especially the gaps—is well pronounced. The bass like gathering in these areas.

One of the top baits during the summer is a 1/4-ounce hair (bucktail) jig. White/yellow or white/chartreuse colors seem to work best. This bait apparently matches the size of the shad hatch well because it is dynamite this time of year and not much use the rest of the year.

The bucktail jigs should be cast into the creek mouths when the water is moving and hopped back to the boat. The hair jig rarely produces big fish, but it produces numbers of fish.

Worms and spoons also will work in the creek mouths, and on the ledges as fished in the lake's middle section.

The flats and creek mouths above the dam are some of the best areas for this type of fishing.

Cool and cold weather bass fishing

Dead River to North Bypass

Once again, moving water is the key to good fishing on Jones Bluff Reservoir in the fall. Falls rains and the pulling down of Lake Martin to winter pool means there will be water movement almost every day during September, October and November.

There's a lot of visible topwater activity with bass busting shad on top in the outer bends, behind islands and in the mouths of creeks—the traditional escape areas for shad.

The smart angler keeps two loaded rods in the boat. A jerkbait should be on one rod, while a Rat-L-Trap should be on the other.

Use the topwater baits when the bass are knocking shad out of the water, being careful to position the boat as far away as possible and still cast well past the schooling activity

When the school of bass suddenly go under, switch to the Rat-L-Trap and and throw past the school and rip the lure back to the boat.

One of the tricks of the better anglers is that when a school suddenly goes under, they rush to the scene to get a look at wounded shad that may be left on top. This allows them to see the size of the baitfish and match their lures to that size. That matching is of utmost importance.

The bass will often be shallow when the water isn't running in this area—early in the morning, late in the evening or on overcast days.

Key areas to concentrate on are the ditch across Parker's Island, the backside of Mud Island, the Big Curve area, the Jackson Lake area, the mouth of Coosada Creek, and the primary points on flats and the mouths of the strip mine areas.

North Bypass to Tallawassee Creek

Fall fishing in this area is exciting because it offers so much opportunity for topwater fishing. If you purchase no other bait this time of the year, make sure your tackle box has a selection of buzzbaits and chatterbaits. These tried-and-true lures are strong on this area of the lake in the fall.

As mentioned before, stumps can be found throughout the flats in this area and are especially concentrated on the main points. A buzzbait fished on the points early on cool mornings often produces 30 to 40 bass a day when you catch it right.

Other baits that are productive include chatterbaits and worms. When water is being pulled, and that's often, fishing can be good all day long.

A good number of fish migrate up into the creeks during this time, too. Cooter's Pond, a hotbed for recreational boating and skiing, can offer good fishing in the back where the grass and lily pads are.

This is an excellent flippin' and buzzbait area. Both Catoma and Pintlala creeks are excellent fall fishing areas to concentrate on, too. Both creeks have good concentrations of shad.

Tallawassee Creek to Brown Branch

The cool and cold weather patterns change very little when you enter the lake's lower end. Just as in the middle section, you'll find some fish in the mouth of creeks, some in the creek channels and others up on the ledges of both the main river and the creeks.

Deep-diving crankbaits fished along the underwater bends in the creek channels is one of the best ways to take quality fish and is a little more effective on this end than in the middle or upper section.

Tensaw Creek is a good fall creek, but it is small and receives a lot of pressure. Cypress Creek is another area to key on, but it's often filled with crappie fishermen. The Three Fingers area is a good fall fishing area, with the western finger usually holding the better fish.

Brown Branch is one of the lake's few creeks that enter in a bend, and as a result, current forces large numbers of shad into it. Although it's a deep creek, it offers excellent fall fishing.

Jones Bluff Reservoir's best fishing spots

1. Dead River. Connects the Tallapoosa and Coosa River with the Alabama River and Jones Bluff Reservoir. Some decent fishing when current is flowing.
2. Intersection of Alabama and Tallapoosa Rivers. Excellent sandbar for summer and fall fishing.
3. Flats meet river channel. Good summer schooling area.
4. Mud Island. Good area in spring.
5. The "Big Curve" area. Good summer area. Baitfish load up in curve's upper end.
6. Gun Island. Underwater island. Downstream point good when current is moving.
7. Flats adjacent to main river channel. Good in spring.
8. Jackson Lake. Oxbow lake. Natural lake with cypress trees. Good in spring.
9. Good escape area when current is moving. Bass school on this point.
10. Gravel pits. Good in spring when water level allows.
11. Location where treated sewage from Montgomery enters the reservoir. Known as the "Shad Hole." Good year round fishing in area.
12. Gravel pits. Good spring fishing.
13. Bridge pilings. Good bass fishing when current is off.
14. County line bend. Excellent summer fishing as baitfish gather in bend.
15. Tyler Goodwin Bridge debris. Underwater debris. Good summer fishing with worm.
16. Bridge pilings and riprap on both sides. Good summer fishing area.
17. City storm sewers entrance. Spring rains bring warm water off of nearby pavement. Key spring area during rains.
18. Pocket. Good escape area.
19. Bridge pilings and riprap.
20. Flats. Excellent crappie fishing in spring.
21. Good flats. Same as No. 20.
22. Island. Falls directly into main river channel on west side. Shallow water on east side. Good fishing on both upriver and downriver points.
23. Good spring fishing in what was a private lake before river was impounded.
24. Mouth of ditch. Good year round.
25. Good bend in river. Good summer fishing.
26. Small ditch going to Cooter's Pond. Good fishing in mouth in spring.
27. Entrance to Cooter's Pond. Good ledge fishing in spring and fall.
28. Good spring spawning area around island.
29. Good spring fishing around island.
30. Standing timber patch. Good spring fishing.
31. Lily pads in back of Cooter's Pond. Good spring and fall.
32. Bridge pilings. Good spring fishing in the narrows.
33. Crescent Lake. Can get an aluminum boat in area. Good spring fishing.
34. Mouth of Pine Creek. Excellent fishing in mouth during summer.
35. Good points in mouth of Pine Creek.
36. Bridge pilings. Pilings on both sides in shallow water good for spring crappie.
37. Bridge pilings under railroad trestle. Good crappie fishing.
38. Excellent cuts.
39. Good sandbar points.

40. Autauga Creek. Lot of moving water in creek. Good summer area.
41. Good sandbar points. Good in spring, summer and fall.
42. Mouth of Catoma Creek. Excellent year-round.
43. Grass filled cuts.
44. Excellent bank. Blow-down timber.
45. Islands. Good spring fishing around island with spinnerbaits and topwater in spring.
46. Antioch Creek. Mouth offers good year-round fishing.
47. Backside of island. Good area when current is moving.
48. Standing timber.
49. Mouth of Noland Creek. Good year round.
50. Noland Creek. Good spring and fall.
51. Possum Bright Bar. Underwater island. Good summer area.
52. Sheer dropoffs. Good summer fishing on ledges.
53. Mouth of PintIala Creek. Good year round.
54. Bridge pilings. Good spring area for bass. Good in summer for crappie at night.
55. Railroad trestle. Good spring area for bass and crappie.
56. Mouth of cuts full of standing timber. Excellent spring crappie area.
57. Tallawassee Creek. Good in spring and fall for bass and crappie.
58. Downstream point of Tallawassee Creek. Excellent year-round fishing.
59. Standing timber filled slough.
60. Mouth of Billdee Creek. Good spring staging area.
61. Mouth of Rocky Branch. Points good in early spring and summer.
62. Standing-timber-filled slough.
63. Bear Creek. Mouth of creek is good fall area.
64. Tensaw Creek. Good spring and fall.
65. Pockets. Good spring and fall.
66. Flats. Good spring fishing.
67. Mouth of Swift Creek. Good year-round.
68. Creek channel. Well-defined. Good year round.
69. Dutch Island. Stumps on upper end. Good spring and fall.
70. Grass-filled flats.
71. Cut-off. Excellent when water is moving.
72. Stump-filled slough.
73. Mouth of Cypress Creek. Good spring, fall and summer.
74. Cypress Creek. Lots of grass. Cypress trees in back.
75. House Creek. Good fall creek. Typically a lot of shad.
76. Grass flats.
77. Mouth of cut. Good year round for bass.
78. Three-fingers area eastern finger. Good color from farm runoff.
79. Middle finger. Deep water.
80. West finger. 5-6 feet of water.
81. Brown Branch. Deep water. Excellent in spring.

The Ultimate Guide to Alabama Fishing

Coosa River

Tallapoosa River

Alabama River

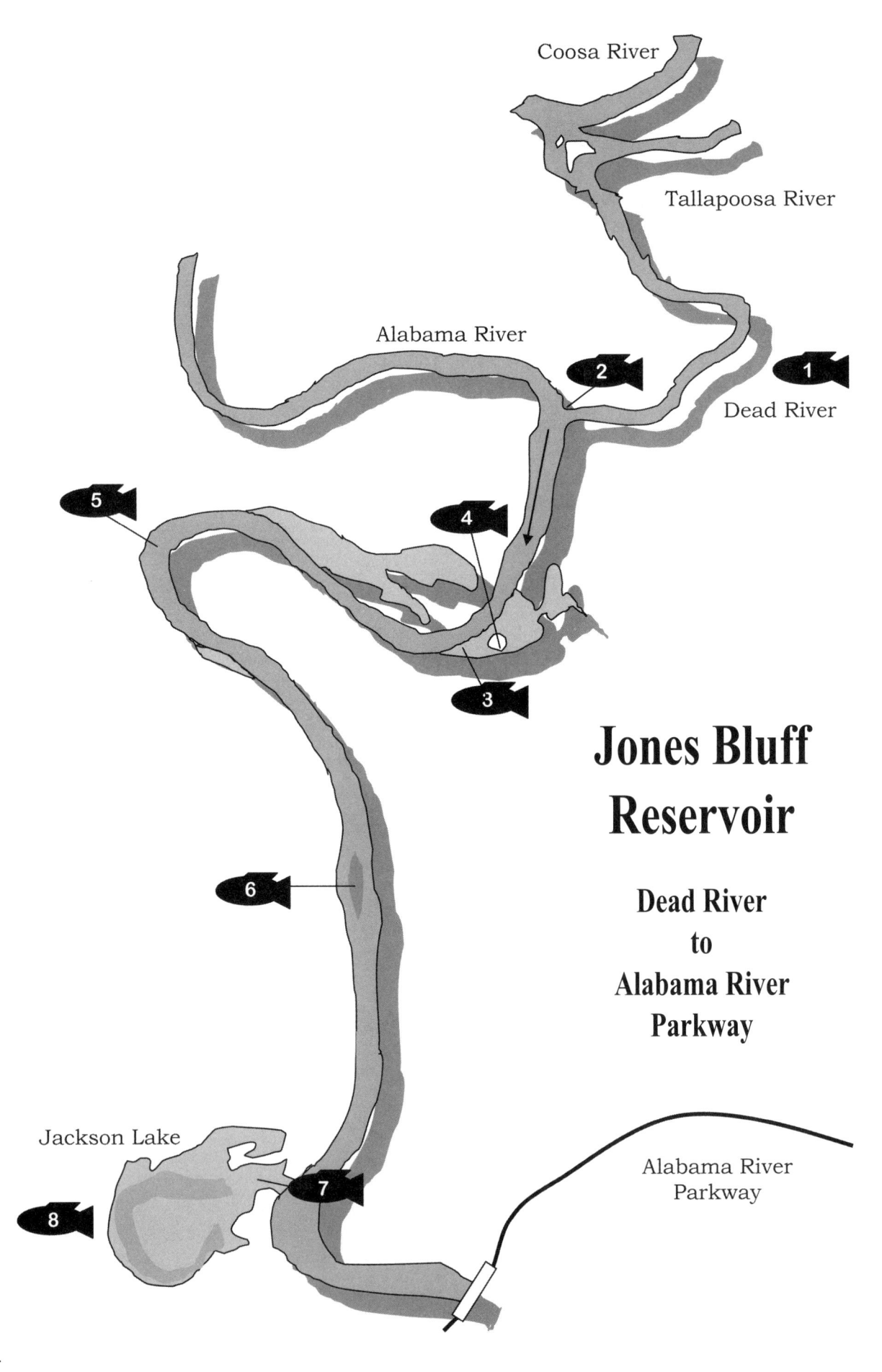

Dead River

Jones Bluff Reservoir

Dead River to Alabama River Parkway

Jackson Lake

Alabama River Parkway

204

Mike Bolton

Jones Bluff Reservoir

Alabama River Parkway to North Bypass

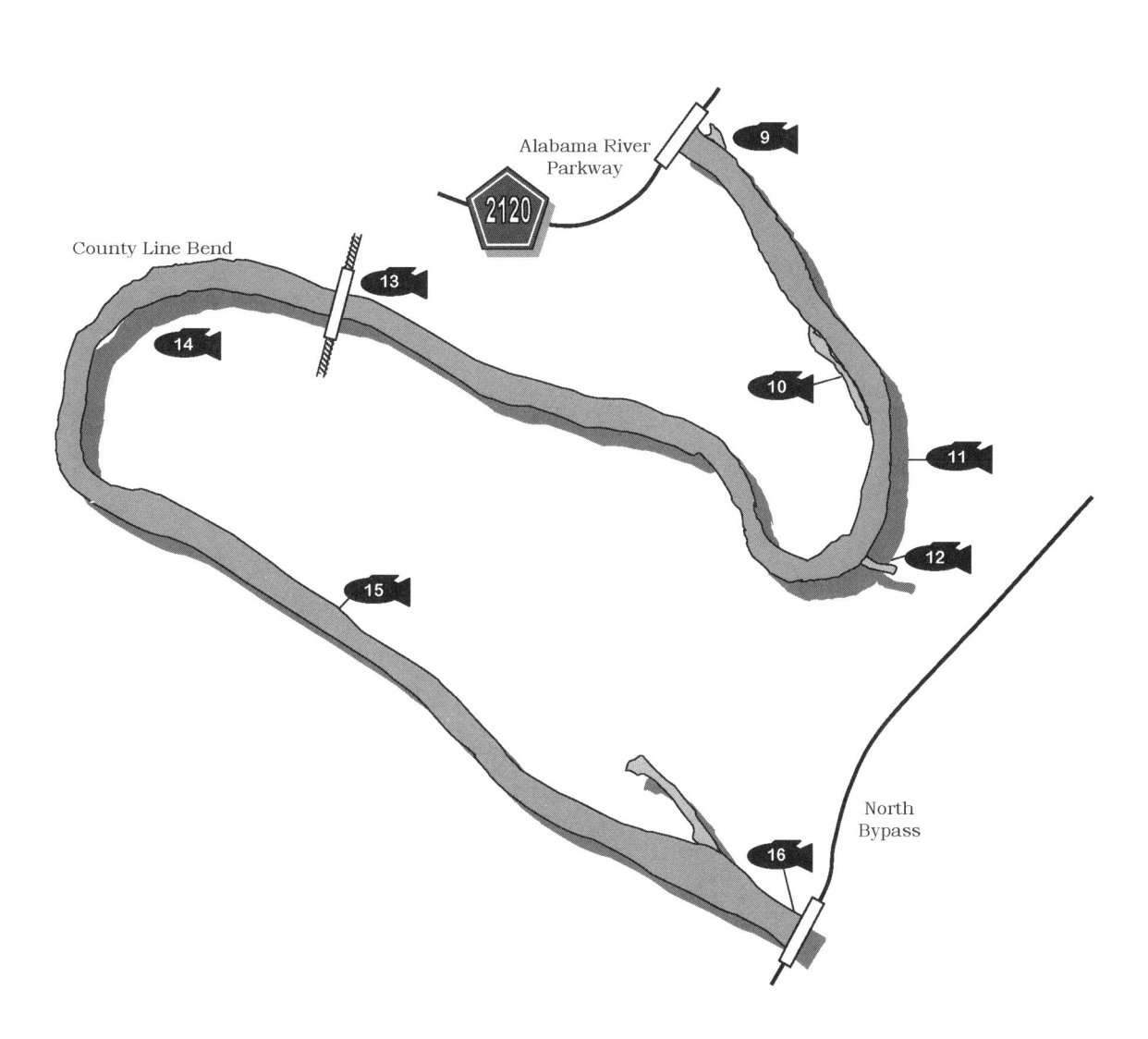

The Ultimate Guide to Alabama Fishing

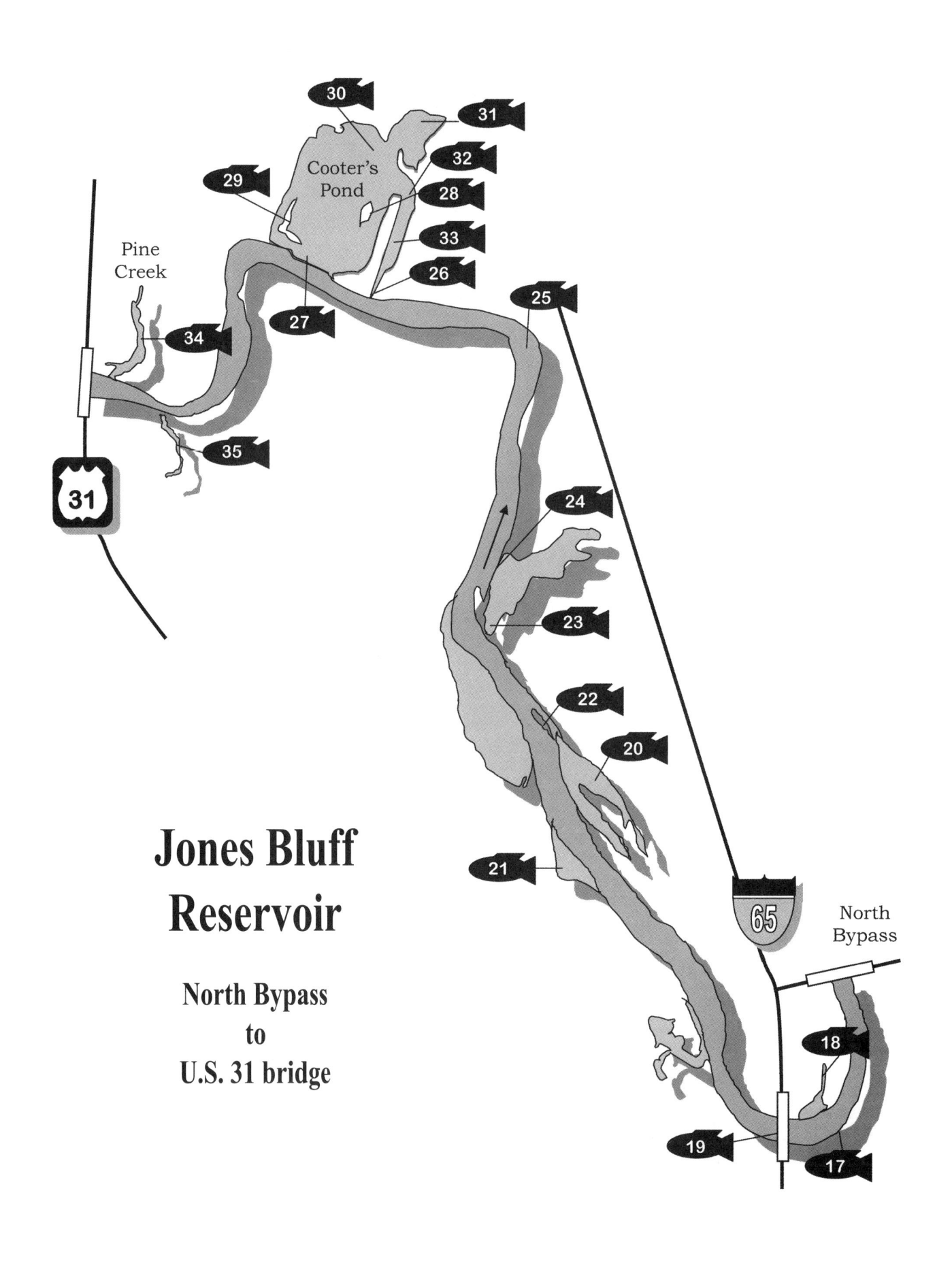

Jones Bluff Reservoir

North Bypass to U.S. 31 bridge

206

Mike Bolton

Jones Bluff Reservior

U.S. 31 bridge
to
Catoma Creek

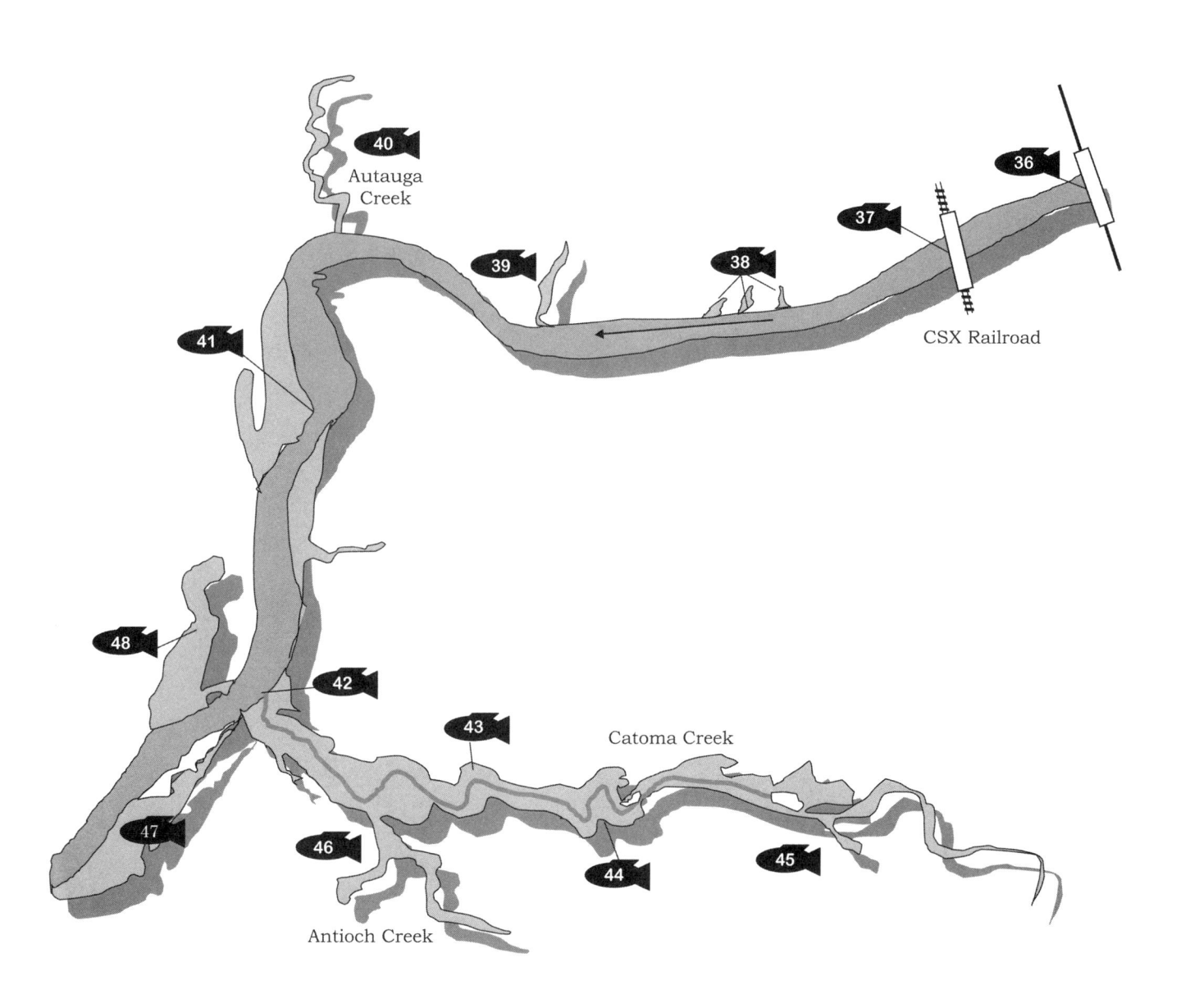

The Ultimate Guide to Alabama Fishing

Jones Bluff Reservoir

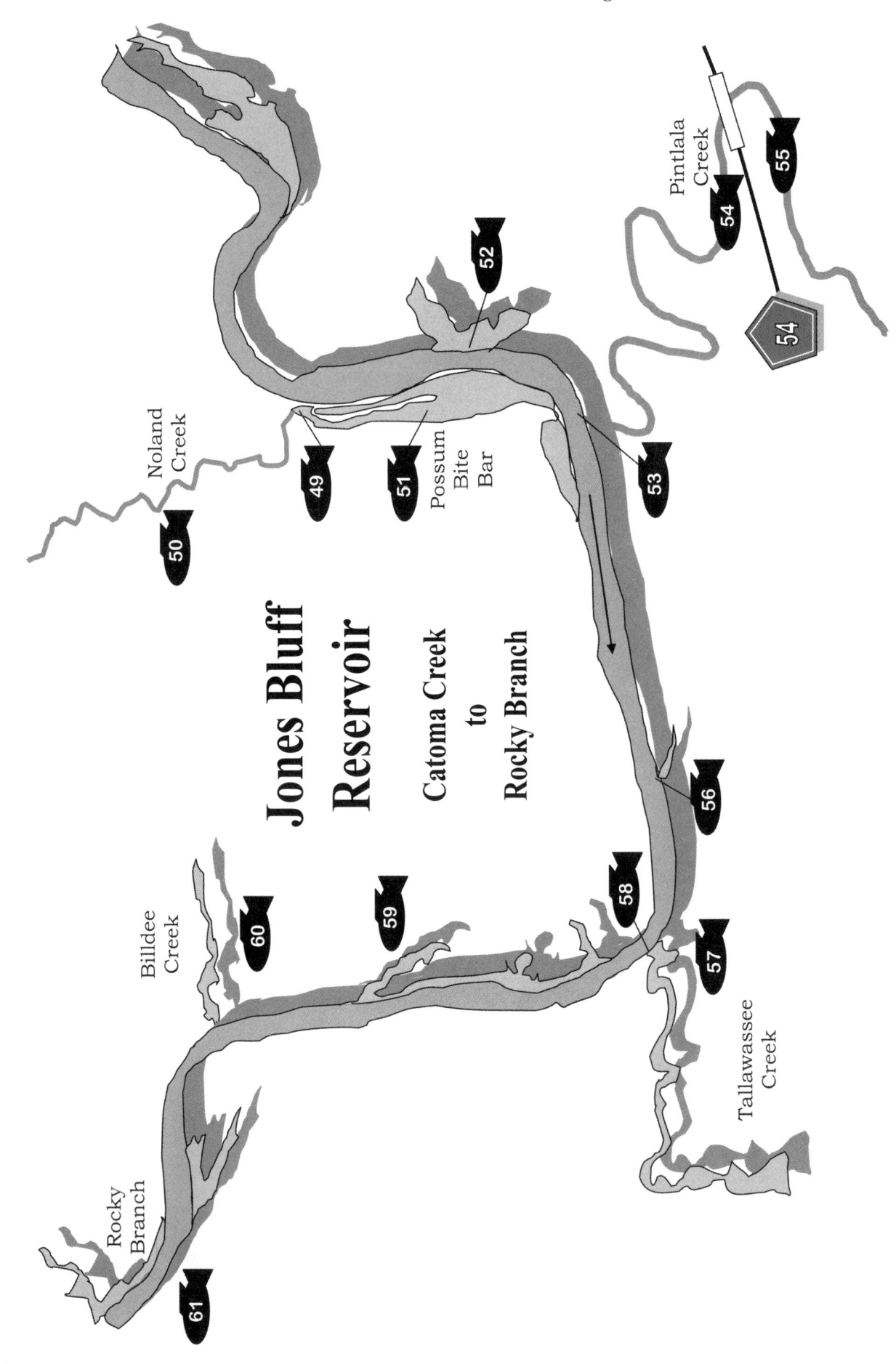

Catoma Creek to Rocky Branch

Pintlala Creek

Noland Creek

Possum Bite Bar

Billdee Creek

Tallawassee Creek

Rocky Branch

208

Jones Bluff Reservoir

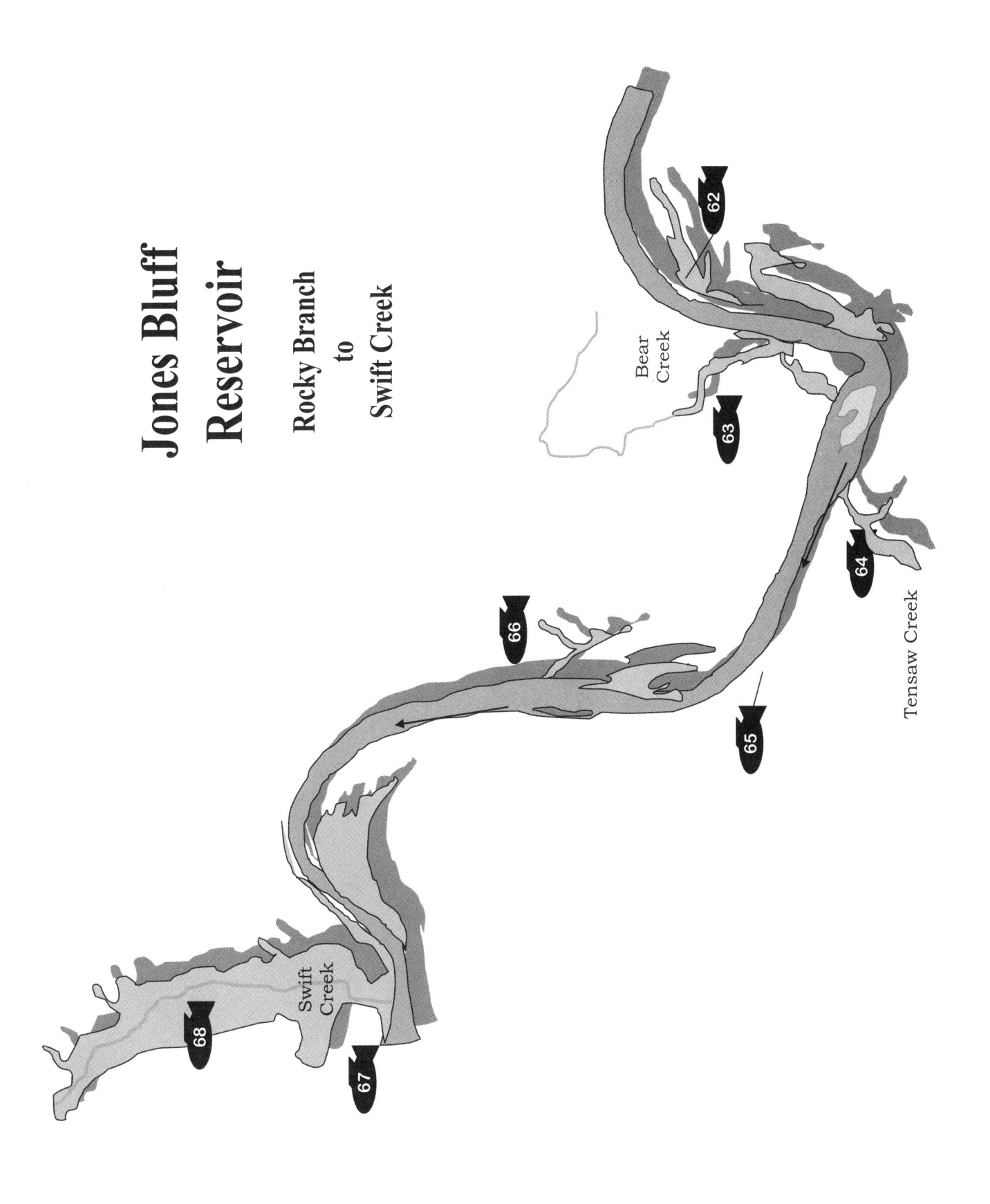

Rocky Branch to Swift Creek

Miller's Ferry

The U.S. Army Corps of Engineers' impoundment of the Alabama River in 1969 to form Miller's Ferry Reservoir flooded thousands of acres of timber to create a remote, 17,200-acre lake perfect for bass and crappie.

According to data compiled annually by the state from the Bass Anglers Information Team (BAIT) surveys, Miller's Ferry is always near the top of the list of Alabama's finest bass lakes. Creel surveys also show it is normally one of the state's top crappie producers.

You'll find something rare on this lake, too. From Gee's Bend to Canton Bend, a distance of approximately 9 miles, the river flows north. That's something seen on only one other lake (Pickwick) in Alabama.

Spring bass fishing

Miller's Ferry Lock and Dam to Gee's Bend

Probably nowhere else in Alabama is a water surface temperature gauge as valuable as on Miller's Ferry Reservoir in the spring. The lake's combination of spring-fed and non-spring-fed creeks and sloughs, as well as its long, thin sloughs and pockets protected by shade trees, results in widely varying water temperatures throughout the lake early in the year. It's not unusual to find two sloughs side-by-side, one being 5 degrees warmer than the other.

There are two other key factors in fishing this lake and they apply year round—water clarity and water movement.

This is a stain lake. Almost never will you find clear water. Even in the dead of winter, rarely will you see water with three-foot visibility.

This lake is sandwiched between two hydro-electric plants and current is common. During the spring, you'll often find water being pulled or pushed 24 hours per day. An angler can call 334-682-4896 to find load schedules. Keep tabs on the water schedules. Every trip to the lake should be preceded by a call to that number. Water movement during the spring, especially when the water level is dropping rapidly, hurts fishing.

The quality bass in this lake spend the winter deep in the main river channel, but 55-degree water temperature will snap them out of the doldrums and start them toward their spring activity. They'll begin staging out of deep water by moving up the ledges and points. Their food needs increase with this movement.

Jig-and-pigs and Carolina-rigged plastic worms and lizards, and to a lesser degree deep-diving crankbaits, are key baits during this time of year. This is the time for fishing patiently, and both jig-and-pigs and lizards can be fished deep and slow. Many of the better anglers report success by throwing a Carolina-rigged lizard on the deep points showing fish on the depthfinder and literally allowing it to sit there 5 minutes or more, the only movement being the lizard moving slightly in the current.

For fish suspended on the ledges, a deep-diving crankbait with extreme action should be used. The strong action of this bait causes the lure to move slowly just as the fish like it.

Once the water temperature jumps to the 62 to 64 degree range, the bass will be on their pre-spawn feeding binge and periodically will be up on the ledges and in the creeks and sloughs chasing shad.

Shell Creek, Mill Creek, Alligator Slough, Houseboat Slough and any of the structure-filled sloughs and flats next to the main river channel will hold these pre-spawn bass.

The increased water temperature will have increased the metabolism of the fish and they will be much quicker. A white or chartreuse/white 1/4-ounce spinnerbait with gold blades and a pork-frog trailer or a buzzbait should be the angler's top choices.

Both baits can be fished around the blowdown timber, standing timber, stumps and brush piles found throughout this end of the Miller's Ferry.

The spinnerbait and buzzbait works more effectively early in the spawning period when the fish are feeding heavy, but a 7-inch Carolina-rigged spring lizard—red-shad or pumpkinseed are the top colors—may be needed when the fish are beginning to feel the stress of the spawn and are holding closer to timber. This lake has a short spawning period for reasons no one can seem to explain.

It boils down to the fact that most of the fish here spawn all at once rather than at staggered intervals as they do many other places.

The fish on Miller's Ferry come off the bed looking for action and heavy feeding is the norm the first two weeks following the spawn. These fish can be found in the same areas you found them in the pre-spawn period, but keep in mind the water temperature will now be 70 degrees or warmer compared to the 64 to 68 degree range when they were here earlier. Increase the speed of the bait accordingly. Smaller blades on spinnerbaits will speed up that bait.

Lipless crankbaits such as the Rat-L-Trap can be effective, as are buzzbaits and other topwater lures. And of course, worms will take fish, too.

Gee's Bend to Bogue Chitto Creek

The only difference in fishing this area in the spring is the name of the creeks. Buzzard Roost is a shallow, timber-filled slough perfect for spring spawning. Pine Barren Creek is a timber-filled creek that offers excellent spawning grounds throughout much of its length. Foster Creek has a lot of deep water, but does have some shallow ledges that will hold spawning fish. The Gold Mine is a shallow water slough that will hold an occasional spawning fish. Chilatchee Creek is a wide creek that has a deep, well-defined creek channel with plenty of flats and islands and is perfect for spring spawning. Bogue Chitto Creek is similar to Chilatchee Creek, but has more structure.

Hot weather bass fishing

Miller's Ferry Lock and Dam to Gee's Bend

During years of normal rainfall, this lake will have at least some current moving some time during the day. If possible, aim for those times when there is moving water.

It's possible to find some bass in shallow water year round on this end of the lake, but as on other lakes in the state, these are almost always small fish. These shallow water fish can offer excitement for those who want to feel a tug on the line. Buzzbaits and other topwater will catch some fish early in the morning and late in the evening, especially on the shady banks in Mill Creek and Shell Creek.

Plastic worms fished in or near the ledges on the creek channel will produce some fish up in the day. Rarely will these fish be larger than two or three pounds, however.

If you want big fish during the summer months, you have to work for them. The quality fish have more stringent temperature and oxygen requirements and the larger fish will be in deeper water, namely on deep humps and points on the main river channel or in the mouth of creeks.

The deep points in the mouth of Shell Creek, Mill Creek, the humps and brush piles out from the mouth of Rex's Hole Slough, the underwater sandbar in Canton Bend (No. 28 on the map) and the deep points in the original creek channels in the mouth of Cottonhouse

Slough, Alligator Slough and River Bluff Slough are prime locations to search out with the fish finder.

What the depthfinder shows you should determine how you fish for these fish. A few scattered fish should be approached with a deep-diving crankbait, jig-and-pig, a Texas-rigged worm or Carolina-rigged worm. Deep, schooling fish should be attacked with a jigging spoon or Rat-L-Trap.

A day of riding on the river with the depthfinder won't be wasted, especially if the current is off and fishing would be poor. Locate fish and come back to them when the current is moving and they are feeding.

Gee's Bend to Bogue Chitto Creek

Once again, the fishing pattern doesn't change when you move up into this area. It's possible to catch fish year round in some of the creeks with deep channels such as Pine Barren Creek, Bogue Chitto Creek and Chilatchee Creek, but don't expect big fish. The fishing for bass in the main river channel is so difficult for many anglers that most are willing to settle for small fish, however.

This portion of the lake does have several easy-to-locate sharp river bends and creek mouths that hold fish. Because of the large amount of timber in this lake and the current, sharp bends in the river are normally full of logjams and this is prime habitat for summer bass. Gee's Bend, the bend directly across from Camden State Park, the bend one mile above Foster Creek and the bend before you reach Bogue Chitto Creek are prime areas for this type of fishing. A jigging spoon works wonderfully in this area, but plan to use a heavy spoon because they hang up easily in the timber. The weight of the lures are often enough to pull the hook out of wood. Another excellent bait is a jig-and-pig. Some anglers try to use worms, but it's hard to detect a strike on a worm at these depths.

When you locate one of the logjams, throw a marker buoy and anchor. You must be directly on top of one of these spots to be effective. Also, don't hesitate to fish the mouths and secondary points of Chilatchee Creek, Pine Barren Creek and Bogue Chitto Creek with deep-diving crankbaits.

Cool and cold weather bass fishing

Miller's Ferry Lock and Dam to Gee's Bend

The cooling temperatures of October will see a lot of fish movement in this end of Miller's Ferry Reservoir.

The bass will leave the deep water they inhabited during the summer and seek out thermal refuges such as Mill Creek, Shell Creek and Marina Slough. Not only will the water be cooler, they'll also be full of shad.

The bass this time of the year won't be steadfast on any structure, but rather will be mobile. Large schools of bass will follow large schools of shad up and down the length of the creeks.

Any shad imitation bait such as Rat-L-Trap, Sassy Shad or spoon will catch these fish. The easiest method of locating these fish is simply watching for shad trying to jump out of the water and bass hitting them on the surface.

It's a good idea to have at least one topwater bait tied on a rod during this period, too. This can result in some explosive topwater action. When the school submerges, go back to the previously mentioned baits.

During the dead of winter, the bass return to the same areas you found them in the summer, possibly a little deeper. It becomes finesse fishing time again then. A jig-and-pig is a bait you can't beat during this time.

Gee's Bend to Bogue Chitto Creek

In October, the bass head back to shallow water and there's plenty of it on this portion of the lake. Concentrate on Buzzard Roost and Pine Barren Creek, for it's there that shad migrate. The same can be said for Chilatchee Creek, Foster Creek, Lidell's Slough and Bogue Chitto Creek.

Other types of fishing

Crappie fishing

The thousands of acres of wide, structure-filled water on Miller's Ferry Reservoir makes for a tremendous crappie fishery. Locals here jest if that Weiss Lake is the crappie capital of the world, this must be the crappie fishing capital of the universe.

Many of the wide creeks not only have underwater stumps, standing timber, blow-down timber and brush piles, they are also blessed with deep creek channels next to shallow flats. You will notice that any lake in Alabama that has that combination – and that includes well-known crappie hotspots Weiss Lake, Lake Eufaula, Aliceville Lake, Gainesville Lake, Demopolis Lake and others– offers tremendous crappie fishing.

As you may expect, this structure-filled shallow water means excellent crappie fishing in the spring. Arguably, it's the best spring crappie fishing in the state.

When the water temperature reaches the 65-degree

mark, the crappie begin to ease out of the deep water that held them during the winter and begin to seek out the flats next to their eventual spawning beds. These spawning beds will have a water depth of 1 to 3 feet (18 inches is preferred) and some form of structure.

On this lake, water that fits that description is almost everywhere you look, so the smart angler has several factors he uses in determining where to find the prime spawning water.

First, he looks for spawning flats near deep water. Not only does this allow crappie access to flats during the staging process, it also tends to allow the area to stay somewhat clearer during the muddy water conditions that this lake often encounters during spring spawn time.

Second, he looks for banks receiving long periods of sunlight and areas that aren't next to large shady trees.

Third, he looks for underwater structure that isn't blatantly obvious. These areas receive extreme fishing pressure and cause fish to be spooky during what is a stressful period for them.

During the pre-spawn period, crappie move up on the flats when the water temperature hits 65 degrees, but will usually not go to the 18-inch water until the temperature hits the 68-degree mark. A good rule-of-thumb is that the 65-degree water will make them think spawn, but it takes 68-degree water to make them act it out.

During the pre-spawn period, one of the best patterns is to use 4- or 6-pound test line on a spincast or spinning reel outfitted with a 1/32-ounce lead-head chartreuse jig and popping cork. This rig allows the angler to cover vast amount of water during the period the crappie are roaming and feeding heavily right before the spawn.

These fish will be in open water and not holding tight to structure. The popping cork not only allows the jig to be set at different depths, it also allows some noise to be made to attract attention. The angler should cast the rig parallel to the bank away from structure and pop the jig back to the boat. If you think you're in an area that is holding fish, don't hesitate to experiment with varying depths and jig colors.

Once the fish move in tight on structure, leave the spinning and spincast tackle at home and finesse-fish with a cane pole or fiberglass pole, the longer the better. Outfit this rig with 4- or 6-pound test, a small foam cigar float and either a jig or live minnow. This rig allows the angler to swing the bait exactly to a spot with a gentle splash. It also eliminates many of the inaccurate casts that result in tangles, and eventually scare fish. The jig or minnow should be placed as tight to the structure as possible. The better fish will usually be buried in the thickest roots, limbs and brush.

Miller's Ferry's widest water—Pine Barren Creek, Mill Creek, Shell Creek, Gee's Bend and other such areas—is the most popular for spring fishing and for good reason, but the smaller sloughs that sit next to the main river channel also hold their share of crappie in the spring.

There's no question why crappie fishing is more popular then than any other time of the year—they are easier to get to. Once you begin pursuit of crappie the rest of the year, plan for a little work.

Deep-water summer fishing takes a little effort, effort many anglers aren't willing to do. Following the spawn, many of the better crappie slide out of the flats into the deep river and creek channels, These summer crappie will hold on deep points, ledges and humps—the same areas you will find summer bass.

Some anglers claim that these crappie are distinguishable from bass on the fish finder, however. The crappie travel in tight schools and will often appear as a large school of bait fish. A school of 20 or 25 crappie can be in an area no larger than the size of a bathtub.

A spincast or spinning reel outfitted with 4- or 6-pound line should be used for these fish with either a jig or live minnow as bait.

Since you will be fishing deeper than you did in the spring, go with a heavier jig than you did in the spring if you plan to jig fish. A different color of jig likely will be needed, too. A dark green, white or smoke grub may be needed if the water is clear, while a chartreuse/white or chartreuse/black may be needed if the water is dingy. Do a little experimenting.

If you plan to use a live minnow, use a bb shot 18 inches above the bait. This will keep the minnow down, but will also give it some freedom to move around.

In either case, fish with a tight line rather than a float. Locate the crappie with a depthfinder, or fish in areas you know hold fish, and drift the bait over them.

Fall crappie fishing varies every year on this lake. During the fall of many years, there will be a two- or three-week period where they will move shallow and can be caught near where you caught them in the spring. Other years, they never make this shallow run. During those times, you will find them where you found them in summer.

The dead of winter—January and February—is prime time for deep crappie fishing. Structure in 20 feet of water next to the river bank holds the fish. When water is being pulled, these fish feed heavily. Because of the water flow, a heavier 1/4-ounce jig is needed to put the bait where you want it. Buzzard Roost and the mouth of Mill Creek and Shell Creek are prime areas for this type of fishing.

Miller's Ferry's best fishing spots

1. Beaver Dam in Marina Slough. Good spring fishing.
2. Beaver Dam. Good spring fishing.
3. Standing timber. Good spring area with buzzbaits and spinnerbaits. Good early spring with jig-and-pig.
4. Boat docks. Good all year. Lots of brush.
5. Underwater road bed.
6. Marker buoy. Good crappie fishing at night.
7. Standing timber.
8. Slave ditch. Narrow channel dug by slaves to irrigate adjacent fields. Good path for bass to move back-and-forth from shallow water to deep water. Good spring and fall.
9. Slave ditch.
10. Shell Creek. Good year round fishing.
11. Man-made fish reef. Good year round for a variety of species.
12. Snags. Good year round.
13. Good trash-filled slough. Good spring fishing area.
14. Slave ditch.
15. Slave ditch.
16. Slave ditch.
17. Underwater hump in Mill Creek. Good summer, fall and winter.
18. Underwater hump in Mill Creek. Good summer, fall and winter.
19. Standing timber. Good year round.
20. Mill Creek. Good year round.
21. Underwater stumps.
22. Underwater brushpiles.
23. Underwater road bed.
24. Underwater brushpiles.
25. Rex's Hole Slough. Good spring fishing in back. Good spawning area.
26. Standing timber area. Good year round.
27. Location of farm pond before lake was impounded. Good summer schooling area.
28. Underwater sandbar. Good summer schooling area.
29. Cuts. Good spring fishing area for bass and crappie.
30. Slave ditch.
31. Slave ditch.
32. Slave ditch.
33. Slave ditch.
34. Stumps next to island. Excellent spring buzzbait area. Spinnerbaits and worms early summer.
35. Stumps next to island. Deeper than No. 34. Jig-and-pig all year. Topwater in spring.
36. Underwater road bed.
37. Slave ditch.
38. Slave ditch.
39. Slave ditch.
40. Alligator Slough. Full of standing timber. Good fishing year round with jig-and-pig. Deep slough.
41. Slave ditch.
42. Man-made fish attractor.
43. Location of two farm ponds before lake was impounded. Adjacent to creek channel. Good worm fishing area in summer.
44. Swimming area. Good fishing on point year round.
45. Legion Hut Slough. Some good spring fishing around brush piles.
46. Log Barge Slough. Good crappie fishing all spring.
47. Bamboo Slough. Good bass and crappie fishing in spring.
48. Ellis Landing area. Well-defined creek channel. Good crappie fishing in spring.
49. Underwater road bed.
50. Cypress trees. Good in spring for crappie around trees. Bass fishing good all year.
51. Underwater road bed.
52. Slave ditch.
53. Mouth of Gee's Bend. Good year round fishing. Deep channel in mouth. Good spring fishing in shallows.
54. Man-made fish attractor in Houseboat Slough.
55. Man-made fish attractor in River Bluff area.
56. Standing timber and stumps. Good year round fishing.
57. Slave ditch.
58. Stump flats around slave ditch.
59. Gold Mine or gravel pit area. Good year round.
60. Liddell's Slough. Good bass and crappie year round.
61. Standing timber in Liddell's Slough. Good year round.
62. Location of farm pond before lake was impounded. Good worm area in summer.
63. Standing timber throughout Buzzard Roost area. Good year round.
64. Good bass fishing around island in Foster Creek during spring.
65. Foster Creek. Good year round.

66. Man-made fish attractor in mouth Pine Barren Creek.

67. Standing timber throughout Pine Barren Creek.

68. Man-made fish attractor.

69. Man-made fish attractor. Tires and brush. Big pile.

70. Chilahatchee Creek. Good year-round fishing for bass and crappie.

71. Location of pond before lake was impounded. Excellent spring fishing area.

72. Man-made fish attractor in mouth of slough.

73. Man-made fish attractor in mouth of Bogue Chitto Creek.

74. Island adjacent to creek channel. Good year round fishing.

75. Underwater stumps. Good in spring and summer.

76. Standing timber.

77. Creek bend. Good spring fishing.

78. Location of pond before lake was impounded. Good spring topwater area.

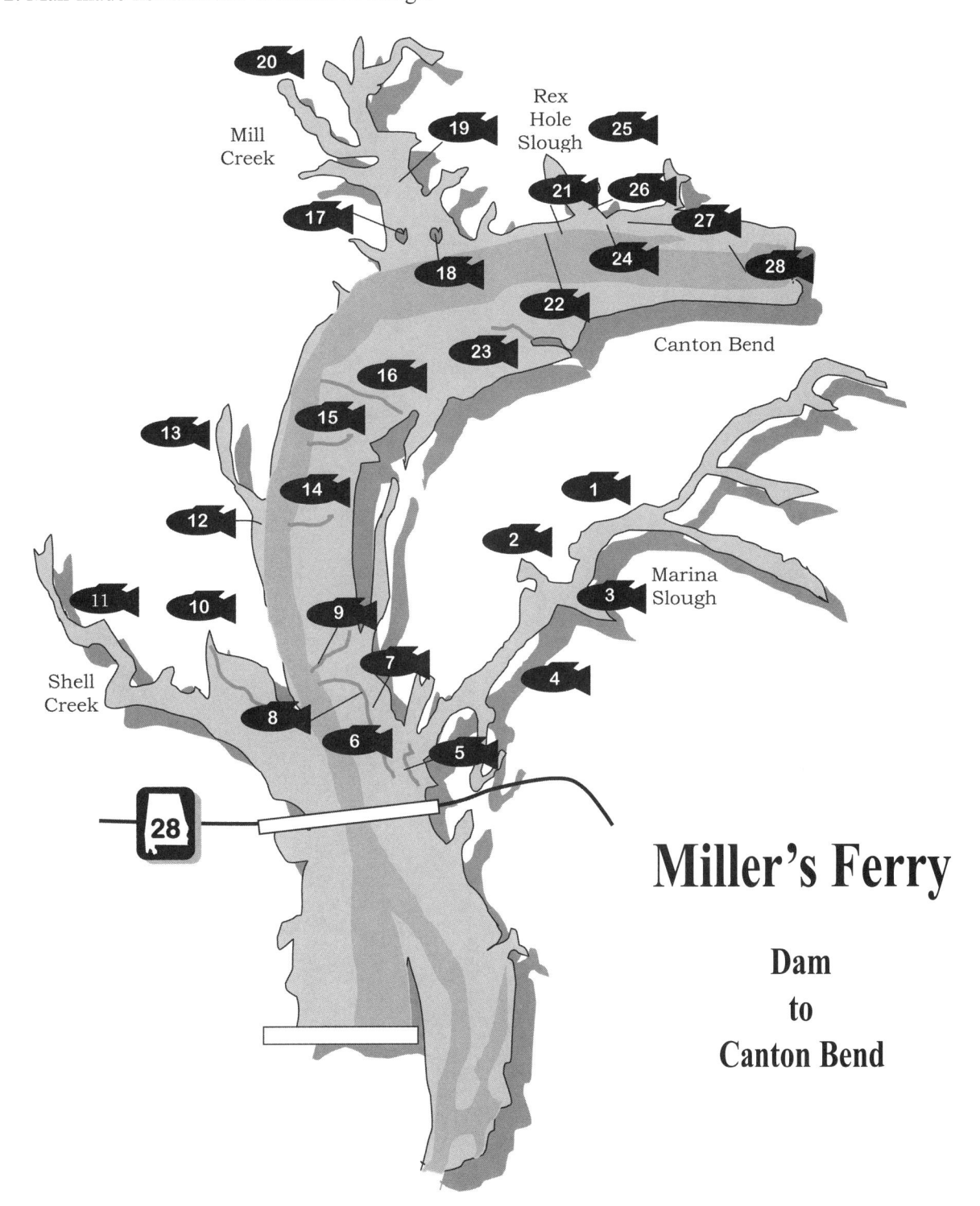

Miller's Ferry

Dam
to
Canton Bend

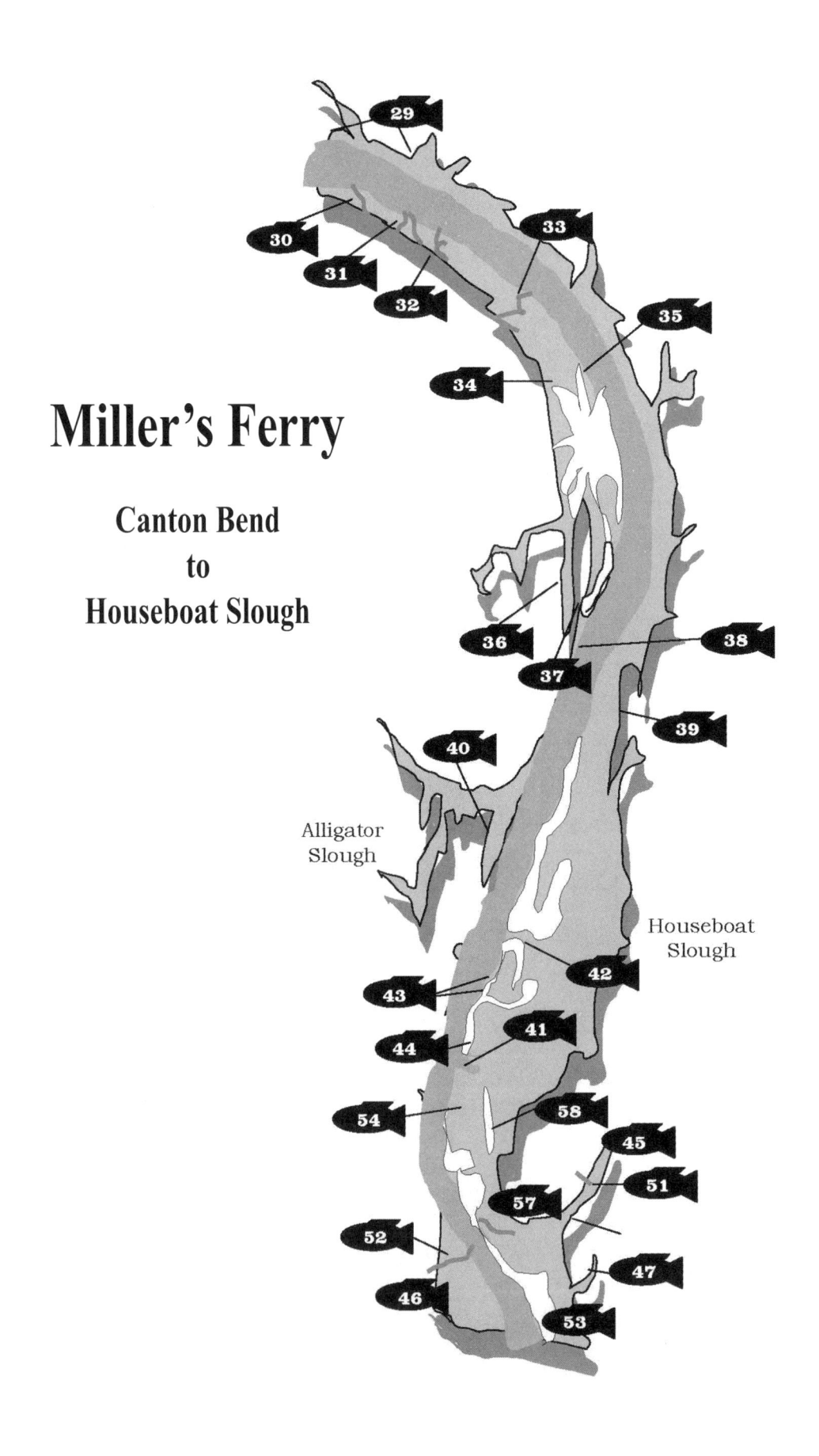

Miller's Ferry

Canton Bend to Houseboat Slough

Alligator Slough

Houseboat Slough

Miller's Ferry

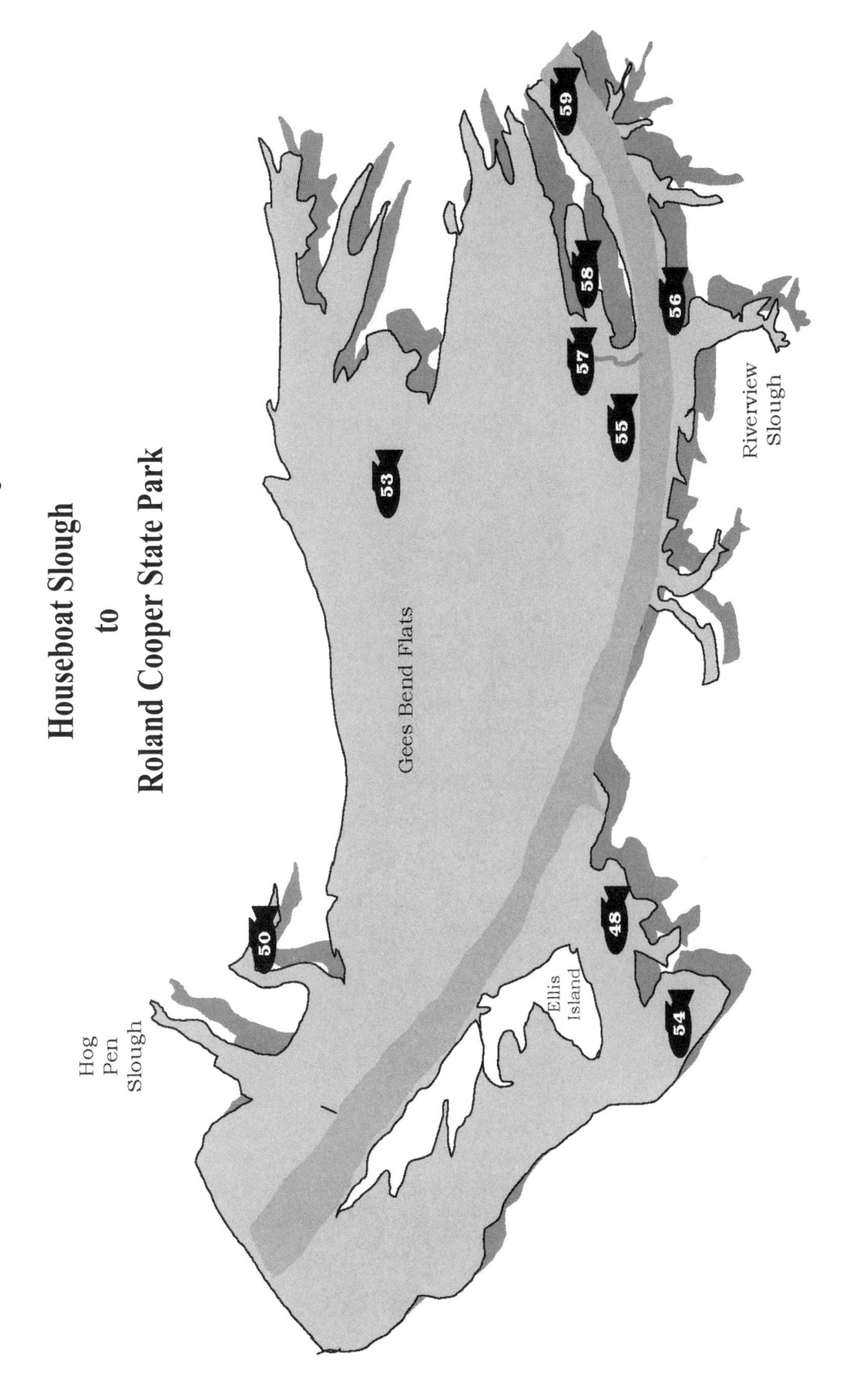

Houseboat Slough
to
Roland Cooper State Park

Riverview Slough

Gees Bend Flats

Hog Pen Slough

Ellis Island

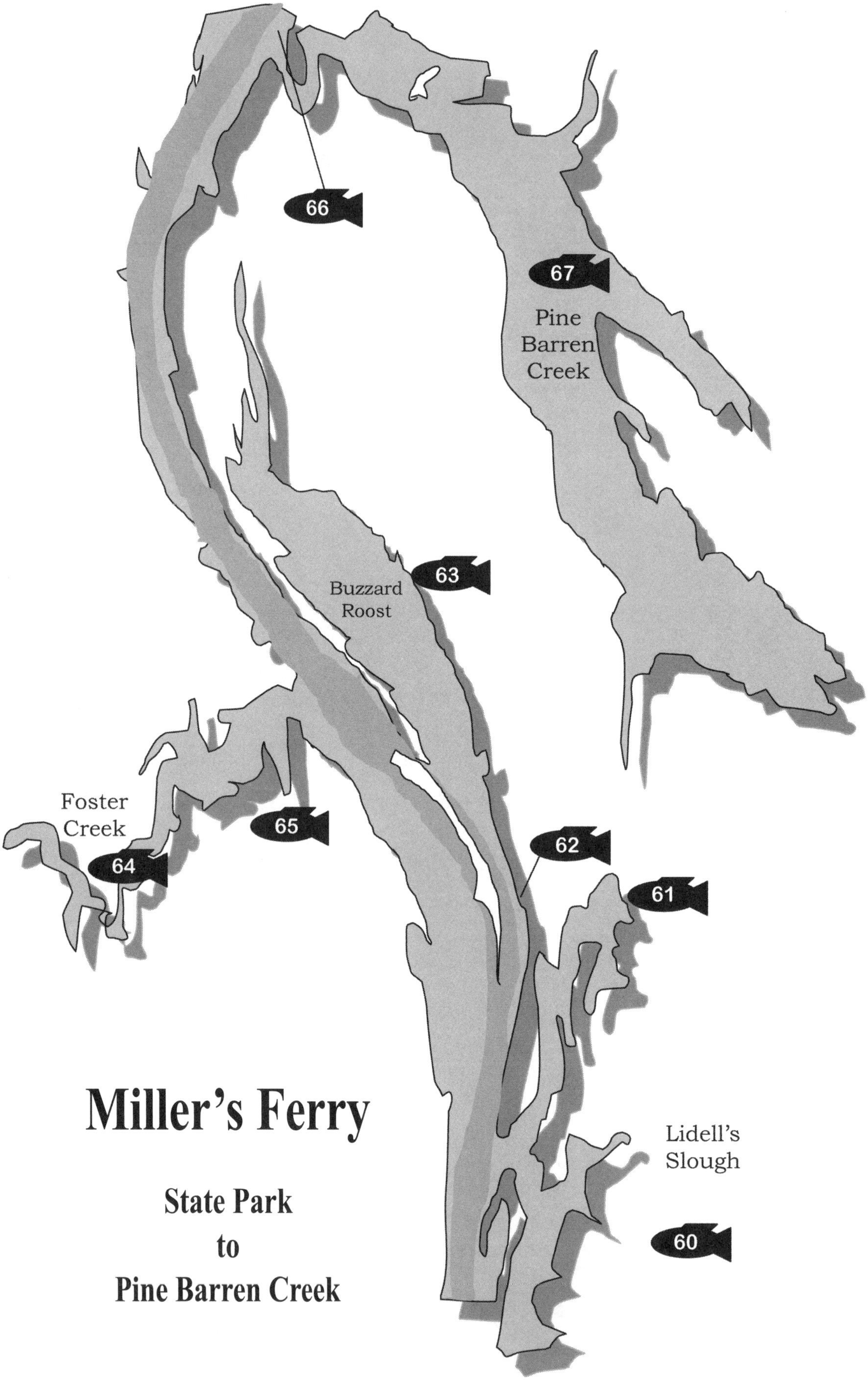

The Ultimate Guide to Alabama Fishing

Pine Barren Creek

Buzzard Roost

Foster Creek

Lidell's Slough

Miller's Ferry

**State Park
to
Pine Barren Creek**

Mike Bolton

Miller's Ferry

Pine Barren Creek to Bogue Chitto Creek

Bogue Chitto Creek

Chilatchee Creek

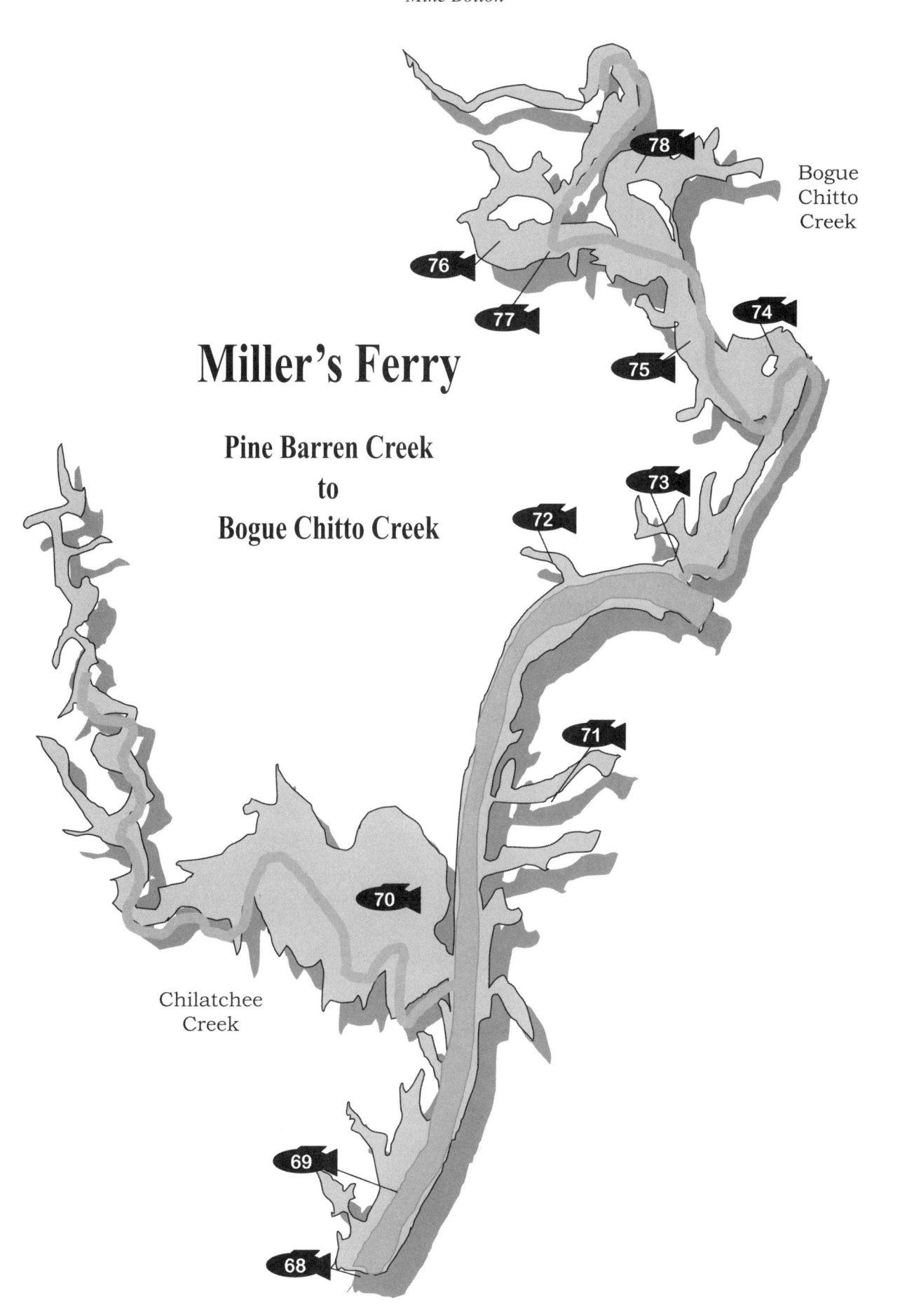

219

The Ultimate Guide to Alabama Fishing

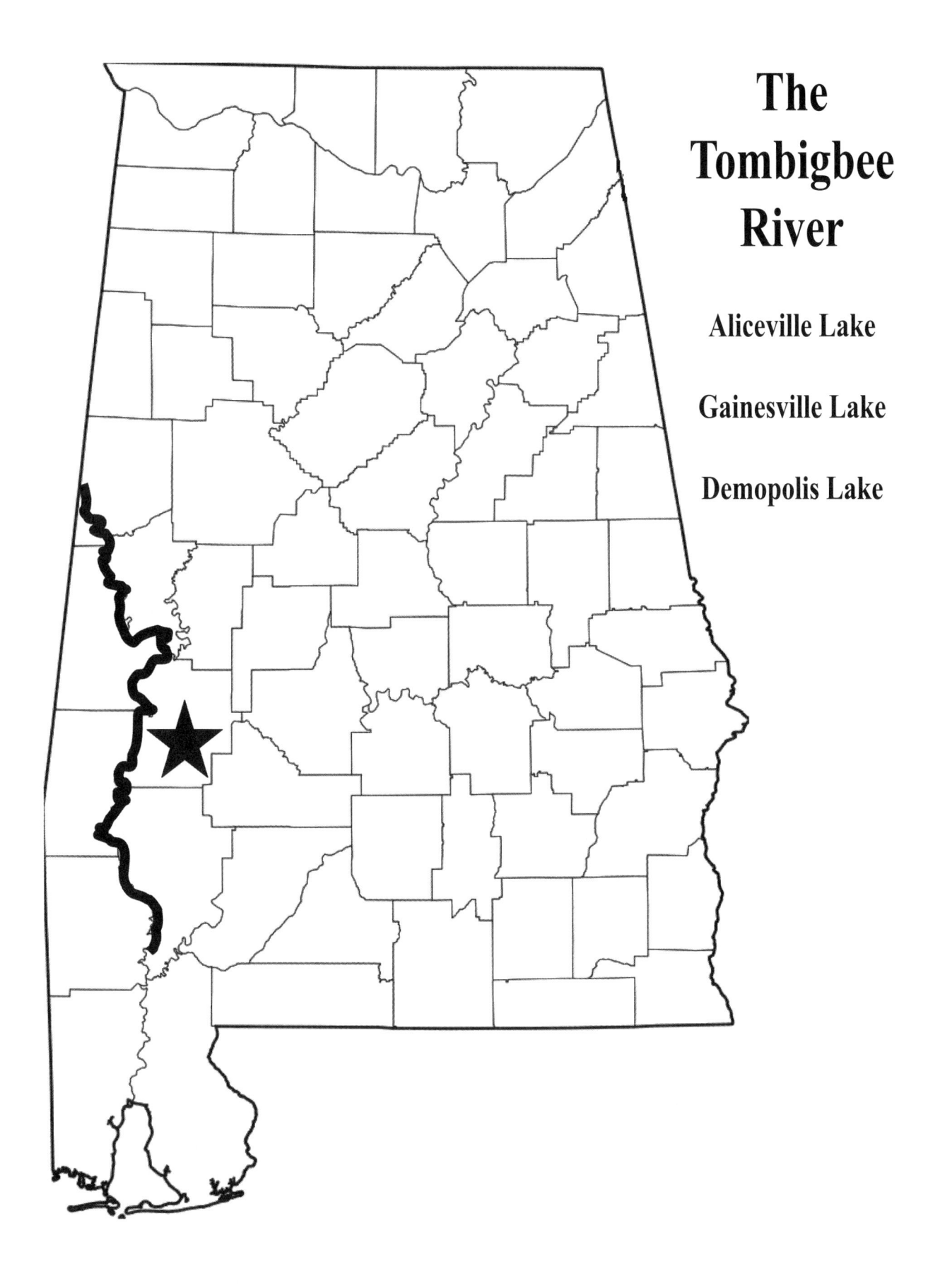

The Tombigbee River

Aliceville Lake

Gainesville Lake

Demopolis Lake

220

Aliceville Lake (Pickensville Lake)

Aliceville Lake, impounded in 1979 as part of the Tennessee-Tombigbee Waterway project, is a structure fisherman's dream. Standing timber, cypress trees and well-defined creek and river channels make for the type fishing even the most basic of fishermen find easy.

The straightening of the main channel by the dredging of Hairston Bend Cut-Off, Coal Fire Creek Cut-Off and Pickensville Cut-Off resulted in flooded lowlands left in their pristine state. The original river run in these areas is void of big boat traffic which damages the banks and surrounding standing timber.

The stocking of 100,000 Florida-strain bass into the lake in 1980 resulted in terrific fishing almost immediately. The lake peaked in the late 1980s as all new reservoirs do. Two bass larger than 14 pounds were recorded in those days and 10-pound bass were caught on a somewhat regular basis. The lake suffered a major setback in the summer of 1989 when a decision was made to spray the milfoil and hydrilla with chemicals in an attempt to eradicate them. This process by the U.S. Army Corps of Engineers continues today and it has left the lake as just a shell of what it once was and still could be.

Today Lake Aliceville is just a good bass fishery.

Spring bass fishing

Tom Bevill Lock and Dam to Lindsey Creek

Once winter begins to loosen its hold in early February, the bass on the lower end of Lake Aliceville start thinking spawn. Most bass begin a staging pattern that sees them stair-step up out of deep water, never hesitating to dive back deep with the slightest change in weather conditions. These bass move up on points and begin a feeding ritual that prepares them for the difficult spawning time which lies ahead.

Long, sloping points leading directly from deep water to flats act like an interstate and offer a route that not only has the structure the bass desire, but food as well. Both shad and bream use these points as their route to shallow water.

During this transition period, several baits will produce fish. Keep in mind that the cold-blooded bass are still sluggish from a long winter and water temperatures still on the cold side. The baitfish they are after are suffering from the same conditions and are slow themselves. When choosing a bait, remember that in early spring, the bait must be fished slowly. As the water temperature warms, allow the speed of the bait to increase.

During this early spring period, relatively heavy slow baits like jig-and-pigs, 3/4-ounce spinnerbaits, large deep-diving crankbaits or Carolina-rigged worms or lizards work best. Weight is needed because many fish are still deep and you want a bait that falls quickly into deep water.

As an early spring bait, jig-and-pigs are hard to beat in this area. If the water is heavily stained, and it usually is this time of the year, a 1/2-ounce or 3/4-ounce brown jig with brown/orange skirt and a brown pork frog trailer works wonderfully. If you like spinnerbaits, a 3/4-ounce white/chartreuse with a chartreuse pork trailer and nickel willow leaf blade work well. This color can be seen well in dingy water, the weight will drop it quickly and the big willow leaf blade will force you to fish it slowly.

Early in the spring, search for long, sloping points leading to flats that receive a lot of sunlight. These flats will often have a surface temperature as much as three degrees warmer than other areas and fish will concentrate in the slightly deeper water next to these flats.

Keep in mind that this time of year, any small body of water even one degree warmer than adjacent water will often hold fish. Once the water warms to the point that bass can stay in the shallow water comfortably, spinnerbaits, Rat-L-Traps, shallow-running crankbaits, jig-and-pigs and worms will take fish on these flats. Rat-L-Traps are the easiest bait to fish and produce the most fish, but jig-and-pigs tend to take the larger fish.

Once the water temperature hits the right mark, usually around 65 degrees, these fish move into the flats to spawn. At that point, spinnerbaits, buzzbaits and plastic worms and lizards become the deadly baits.

Noisy buzzbaits make for the most exciting fishing, but a floating lizard fished patiently will take the bigger fish.

The Big Bee Valley Landing area, Lindsey Creek, Pumpkin Creek, Little Coal Fire Creek and Big Coal Fire Creek, and the sawdust pile that connects Little Coal Fire and Big Coal Fire are key bass fishing areas in the spring.

There are also generous stands of timber in this area of the lake and you can find spring bass in these areas. Standing timber also tends to warm the water quicker than vacant water. Most of the standing timber in this end of the lake is in seven feet of water, however, and normally only big bass spawn in the timber. That sounds good, but the patches of standing timber are so thick that it may take all day to fish a quarter-mile-long stand of it.

It should be stressed that navigating this area of Aliceville Lake is treacherous. Shallow water and stumps are everywhere. Many rookie anglers have left early with their lower units separated from the rest of the outboard motor. Proceed with extreme caution.

Lindsey Creek to Hairston Bend Cut-off

Once the angler moves into this section of the lake, he's faced with a much different fishing situation than he finds on Aliceville Lake's lower end.

Acres of standing timber is the dominant feature and these trees hold the majority of spawning bass.

There's little argument about what the most productive spring bait is in this area. Because of the vast acreage of standing timber, spinnerbaits, with their ability to cover a lot of water quickly, are a must. A 1/4-ounce mixed-color spinnerbait such as a chartreuse/black, chartreuse/blue or yellow/black tipped with a black pork frog are perfect.

The spinnerbait should be tipped with two No. 4 willow leaf blades. You'll find yourself in a Catch 22 situation here when it comes to line selection. Lighter lines result in more strikes, but heavier lines make for better odds of getting a fish out of heavy timber. The best choice is a medium-size line such as 14-pound-test and a prayer that the water will have a good amount of tint to conceal the line.

Plastic lizards and jig-and-pigs are two more key baits anglers use with great success here. Their only drawbacks is inability to cover water quickly. This area's biggest spring fish are usually caught flipping jig-and-pigs, but due to the amount of standing timber, it would literally be possible to go a 1/4-mile and flip 500 trees in that distance. If you're looking for that fish of a lifetime, and have time and patience to flip, that is the pattern and this is the place you need to be.

This area is so full of standing timber that holds fish that narrowing down a few good fishing locations is difficult. The standing timber near the mouth of Hairston Bend and the mouth of Arkansas Creek are good places to start, however.

Summer bass fishing

Tom Bevill Lock and Dam to Lindsey Creek

Once the spawn is over, the bass direct their attention to their basic needs of cover, oxygenated water and food. Unlike in most Alabama reservoirs, the big bass on this lake don't head directly to the original channel to suspend themselves on the ledges for the summer.

A small percentage of fish will make their way to the main channel, but these are almost always smaller, yearling bass.

Lindsey Creek to Hairston Bend Cut-off

Summer fishing in Alabama is normally a less-than-thrilling experience. Deep fish, as well as weather that forces robins to pick up worms with potholders, combine to make many anglers stay at home and watch Bill Dance re-runs on television.

This area of Aliceville Lake is a rare exception. Being able to fish standing timber does something to the angler's inner mind. Most anglers agree that being able to fish visible structure that the mind believes holds a fish is much better rather than fishing an underwater hump.

This area of the lake is overwhelmed with standing timber. It also provides for the type of fishing most anglers enjoy most—worm and topwater.

The construction of Hairston Bend cutoff during the building of the Tenn-Tom Waterway resulted in Hairston Bend becoming a fisherman's dream. Due to the bypass of Hairston Bend, no longer does traffic from big boats disturb the area and its standing timber and other structure is left unharmed. This area, as well as Arkansas Creek, are summertime fishing delights.

For the top-water fisherman, the standing timber offers the chance for a strike behind every tree. Buzzbaits, jerkbaits, twitch baits and other topwater lures offer terrific fishing in the early-morning hours, after sunset and all day on overcast days.

Plastic worms or lizards are the best weapons during daylight hours. Fished around the deep standing timber in the mouths of Hairston Bend or Arkansas Creek, or in the bends of the original river channel in Hairston Bend, these baits produce the majority of the summertime bass.

Cool and cold weather bass fishing

Tom Bevill Lock and Dam to Lindsey Creek

A lot of anglers don't enjoy fall fishing on Aliceville Lake. Locating fall bass is most often a difficult chore. But when bass are located, they'll often be schooling and it's not unusual to catch a number of them.

More often than not, in early fall, the fish end up in shallow-water flats (3 to 7 feet of water) adjacent to deep water.

Small, sinking crankbaits such as Rat-L-Traps are the lures to use this time of year. But instead of quickly retrieving that lure across the shallow water, it is best to hop it across the bottom as if jigging a spoon in shallow water. Cast the lure and allow it to settle to the bottom, then rip it off the bottom a few feet before allowing it to settle again. When the lure is falling, pay close attention because it is often when the fish strikes.

Big Coal Fire Creek is an excellent fall fishing area. An island-like flat sits adjacent to the main river channel there. This flat sits three feet under the water and falls off into six feet of water on all sides.

Pumpkin Creek is another excellent fall fishing area. The fish will stay in these flats and creeks until January.

When extremely cold weather comes in January and February, you have to pick your days to fish. When the temperatures have been in the 20s and 30s for a week or more and suddenly you have a 42-degree day with another such day forecast for the following day, go fishing.

Aliceville Lake warms quickly and a quick warming will send winter bass moving in search of food. Jig-and-pigs, worms and slow-moving crankbaits will be needed as fish head for the shallow, flat water.

Lindsey Creek to Hairston Bend Cut-off

Having a temperature gauge on your boat is the key to fishing this area during the cool and cold weather months. During early fall, the influx of any cool water from incoming creeks can trigger bass into their fall feeding frenzy. The mouths of McCower's Creek and Cedar Creek where they dump into Hairston Bend are areas to watch for early fall feeding activity. Also keep an eye on the mouth of Arkansas Creek where it dumps into the flooded flats.

The fall bass are coming out of their deep-water summer pattern from the main river channel which runs through Hairston Bend, Coal Fire Creek and the Pickensville Cut-Off area. The deep water in these areas is adjacent to structure-filled flats and the bass will come right out of the deep water to feed in these flats when the water cools. The key to finding these fish is a temperature gauge. A stretch of cooler water, which may be caused by an incoming creek, underground spring or just being in the shade, will hold numbers of fish.

One of the best baits this time of the year is a shad-colored, medium-running crankbait, but be forewarned that it is a difficult bait to fish in these waters. Hangups are common, so bring a plug-knocker and a handful of crankbaits.

Once the water cools to the point that the bass are too sluggish to chase crankbaits, a jig-and-pig becomes the bait to use. This bait is much more weedless than a multi-hook crankbait and hangups are less likely. A 1/2-ounce black/brown jig with a black pork frog trailer works best as the pork frog gives the bait added buoyancy and further prevents hang-ups.

Casting into standing timber is difficult, and casts longer than 20 feet are tough and not advised. Any fish further than 20 feet out in the timber is an unlikely candidate for the net.

Other types of fishing

Crappie

The massive amount of standing timber in Aliceville Lake makes it an excellent lake for crappie fishing. Many of the lake's better crappie fishermen believe crappie fishing here is the best-kept secret in Alabama.

Crappie fishing is best during the spring when the fish move into the shallow water to spawn. These fish can be found in two locations during this time.

The better fish leave the original river channels in the Coal-Fire Creek Cut-Off area and the Pickensville Cut-Off area and move into the weed beds adjacent to the deep channels.

Naturally, the better fish are the toughest to get at during this time. Better fishermen use a long, telescoping fiberglass pole outfitted with 12-pound-test line and a curly-tail chartreuse jig with a white lead head. This setup can be used to gingerly reach over and drop the jig right into the fallen holes. This produces the true slabs.

An easier, but less productive method is to line up parallel to the blow-down timber and, using a spinning or spincast reel, cast a jig hanging 18 inches below a cork as far as possible and as close to the structure as possible.

This rig should then be "popped" all the way back to the boat. This popping draws attention and the jig right on the weed line is often too much for the fish to resist.

These fish can also be targeted in the traditional method, and that's with a minnow and cork fished near the weed line. This method is good right before and right after the spawn when the fish are holding in 5 or 6 feet of water.

Although the cut-off areas offer the best fishing, any timber- or weed-filled slough or pocket adjacent to the main river is likely to hold fish during this time. The same patterns will work there.

Following the spawn, the crappie will hold on the flats in about 5 to 10 feet of water until June. Stumps and blow down timber will concentrate these fish. Live minnows fished under a cork is the best bait.

Once the hot part of the summer arrives, the crappie in Aliceville Lake head for deep water. The fish that head to main river channel are often difficult to locate, so the better fishermen head to the cut-off areas. Crappie gather here in the original river channel and stay until winter. The cut-off areas are by-passed by the big boat traffic, thus the crappie find an undisturbed summer.

One of the better patterns used by the lake's best crappie fishermen at this time is trolling. A multi-rod setup with jigs set at varying depths will produce tremendous numbers of fish. Experimenting with jig colors and jig depths may be needed to fine tune this pattern.

Another summer pattern that works well is bottom-bumping with live minnows. The crappie change their pattern very little during the fall. Most of the better fish stay deep in the channels, only occasionally making a run to the ledges to feed. Trolling or bottom-bumping works best, but a live minnow fished under a cork over a ledge with structure on the lip will produce some good fish.

Winter crappie fishing is tough on Aliceville Lake. The fish get deep in the channels and are seemingly lifeless. A pattern that works to stir up these fish is first locating the fish with a fish finder, then jigging a spoon right in there faces. This will sometimes provoke the fish into striking.

Aliceville Lake's best fishing spots

1. Island in Hairston Bend. Excellent shallow water fishing in spring.
2. Standing timber in Hairston Bend. Excellent fishing year round.
3. Excellent point in Hairston Bend. Standing timber adjacent to deep water channel. Good fishing in shallow timber in spring and fall; good fishing in channel summer and winter.
4. Upper end mouth of Hairston Bend. Excellent fishing for schooling summer bass.
5. Cedar Creek. Good standing timber. Good all year except hot summer and coldest part of winter.
6. Lower end mouth of Hairston Bend. Excellent fishing for schooling summer bass.
7. Mouth of Arkansas Creek. Difficult to locate. Island surrounded by standing timber just inside mouth. Weed beds in upper crook. Excellent spring fishing in mouth and along creek. Prime fall fishing area in mouth.
8. Island in flats west of main channel. Surrounded by weed beds. Excellent in spring. Good in summer.
9. Location of private lake before lake was impounded. Excellent shallow water spring fishing.
10. Excellent standing timber in shallow water adjacent to main river channel. Good year round fishing.
11. Back of flats. Good grass beds. Good spring fishing with buzzbaits.
12. Island in shallow flats. Good spring fishing around island with topwater.
13. Standing timber patch.
14. Standing timber patch.
15. Underwater island adjacent to main river channel. Good year round fishing.
16. Island splits main river channel. Excellent summer schooling area for small bass all around island.
17. Standing timber in mouth of Kincaide Creek. Excellent spring spawning area for bass and crappie. Excellent jig-and-pig or topwater area for spring bass.
18. Standing timber patch. Good spring spawning area for bass and crappie.
19. Standing timber in back of Kincaide Creek.
20. Islands in Kincaide Creek.
21. Mouth of James Creek. Excellent area for schooling summer bass.
22. Island in James Creek. Good spring fishing.
23. Island in James Creek. Good spring fishing.
24. Standing timber in James Creek. Good year round fishing for bass and crappie.
25. Standing timber in back of James Creek. Excellent spring fishing for bass and crappie. Surprisingly good bass fishing area in fall.
26. Good ledge fishing area in summer.
27. Visible island adjacent to main river channel. Good year round fishing.
28. Underwater island adjacent to main river channel. Good summer fishing for bass with deep-running crankbaits or Rat-L-Trap. Good fall and winter fishing with jig-and-pigs.
29. Underwater island in mouth of small unnamed creek. Adjacent to main channel. Excellent area year round. Jig-and-pigs late winter, early spring. Good summer schooling area for bass.
30. Standing timber in small creek. Excellent buzzbait area. Excellent spring and fall area for crappie.
31. Mouth of Lindsey Creek. Excellent summer and fall. Excellent buzzbait area in standing timber in spring.
32. Standing timber in back of slough in Pumpkin Creek area.
33. Standing timber in back of slough.
34. Standing timber in back of slough.
35. Man-made underwater brush pile.
36. Man-made underwater brush pile.
37. Man-made underwater brush pile.
38. Man-made underwater brush pile.
39. Standing timber in back of Pumpkin Creek.
40. Long row of man-made brush piles.
41. Mouth of Pumpkin Creek. Full of underwater islands. Good year round fishing.
42. Underwater island at fork of main river channel and Coal Fire Creek Cut-Off. Good summer and fall schooling area for bass.
43. Islands adjacent to original river channel. Good spring and summer fishing for bass.
44. Islands off of point. Good spring spawning area.
45. Man-made underwater brush pile.
46. Massive man-made underwater fish attractor.
47. Man-made underwater brush pile.
48. Large slough full of standing timber.
49. Standing timber-filled slough.
50. Standing timber patch.
51. Boat lane cut in standing timber patch.
52. Island. Excellent spring spawning around island.
53. Man-made brush pile.
54. Standing timber in back of slough.
55. Large man-made brush pile.
56. Island. Good topwater area.
57. Massive man-made brush pile.
58. Man-made brush pile.
59. Standing timber.

60. Standing timber.
61. Large man-made cedar tree pile.
62. Standing timber.
63. Standing timber.
64. Island. Good spring and fall fishing area.
65. Small man-made brush pile.
66. Man-made brush pile between two stands of standing timber.
67. Man-made brush pile.
68. Man-made brush pile.
69. Large underwater cedar tree pile.
70. Standing timber adjacent to main river channel. Excellent year round fishing.
71. Standing timber adjacent to main river channel. Good year round fishing.
72. Fork where original river channel and Coal-Fire Creek Cut-Off meet. Excellent summer schooling area. Good year round fishing.
73. Standing timber in back of slough. Excellent spring spawning area for bass and crappie.
74. Point where original river channel and Coal Fire Cut-Off meet. Good year round fishing.
75. Man-made cedar tree pile.
76. Man-made brush pile.
77. Man-made brush pile.
78. Island. Good spring fishing.
79. Island. Good spring fishing.
80. Man-made brush pile on point.
81. Man-made brush pile.
82. Point of original river channel and Coal Fire Creek Cut-Off. Good summer schooling area.
83. Large islands above dam. Good spring spawning area for bass and crappie. Good year round point.
84. Island adjacent to main river channel. Good year round fishing.

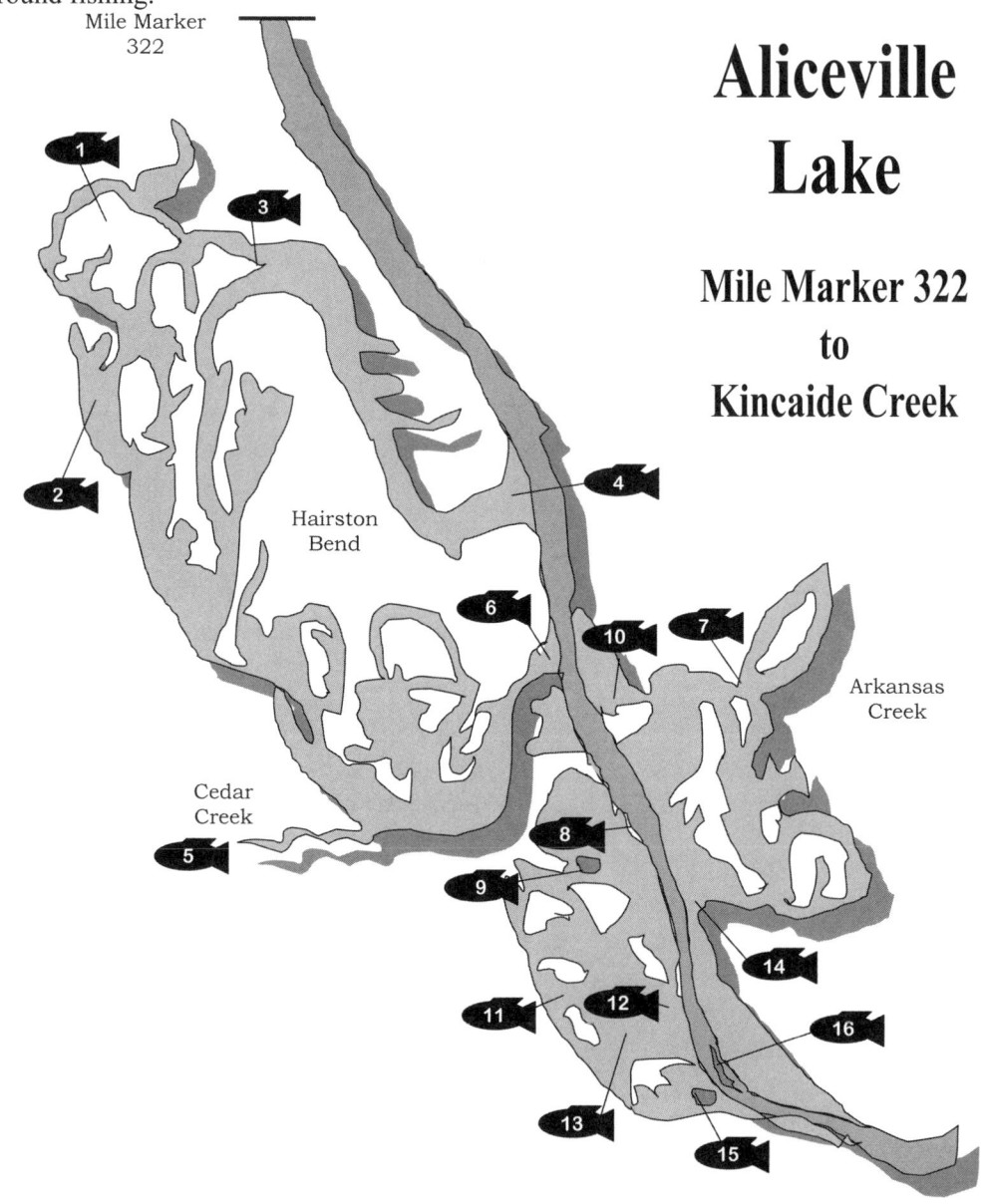

Aliceville Lake

Mile Marker 322 to Kincaide Creek

Mike Bolton

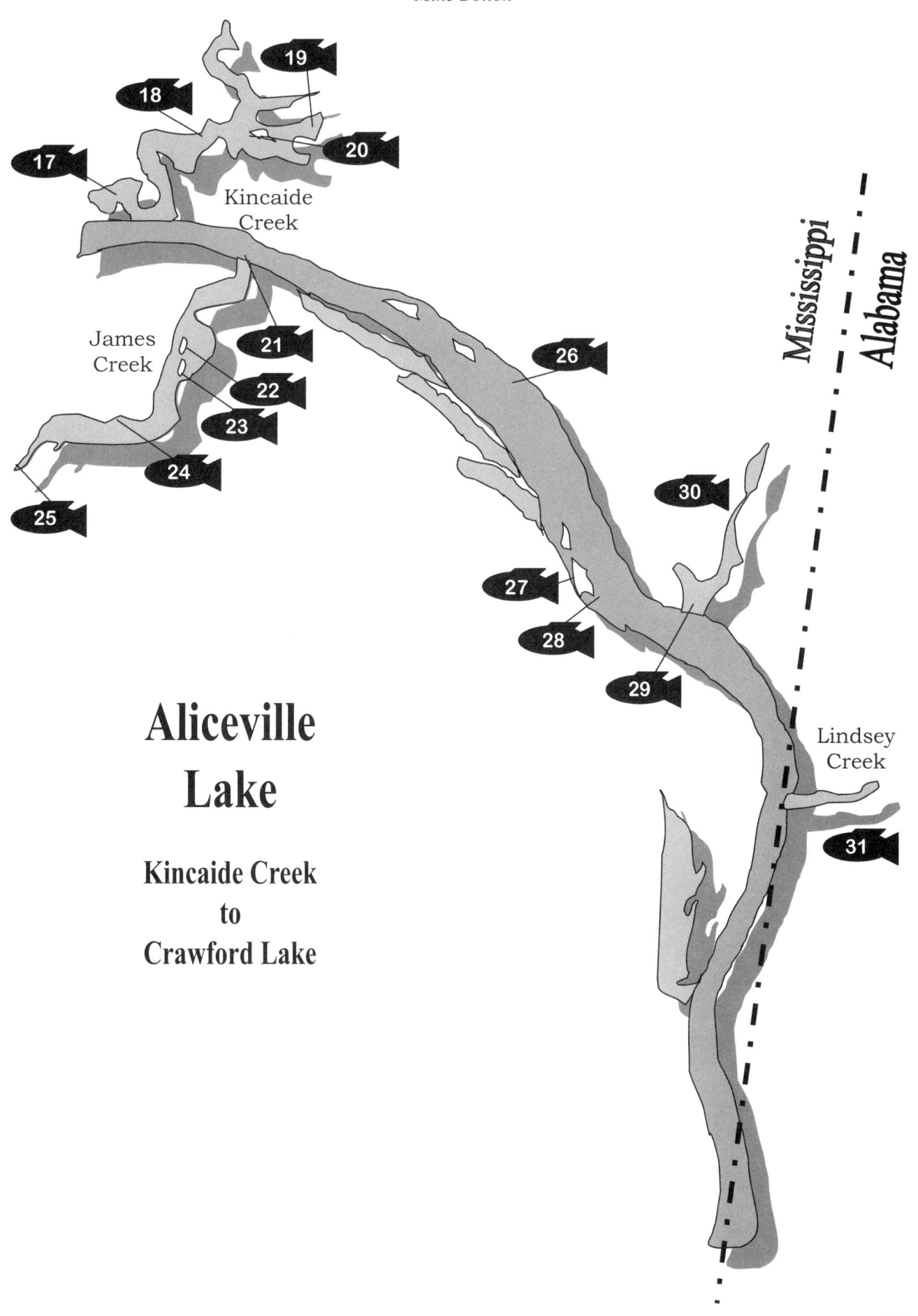

Aliceville Lake

Kincaide Creek to Crawford Lake

227

The Ultimate Guide to Alabama Fishing

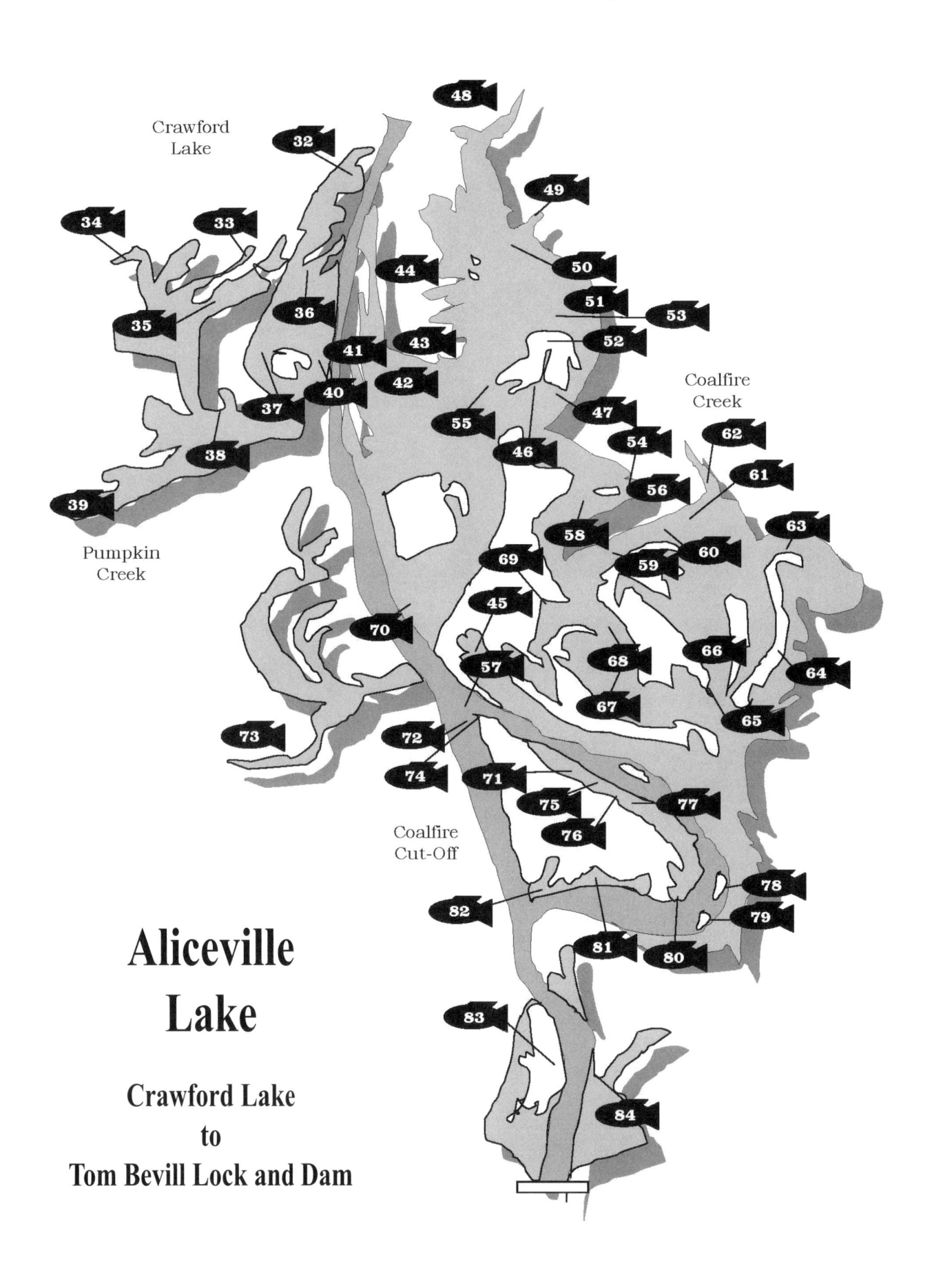

Crawford Lake

Pumpkin Creek

Coalfire Creek

Coalfire Cut-Off

Aliceville Lake

Crawford Lake
to
Tom Bevill Lock and Dam

228

Gainesville Lake

Take a major river located away from a major metropolitan area, impound it, fill it full of standing timber and grass, straighten out its curves with cut-off areas, stock it with more than 60,000 Florida-strain largemouth and what do you have? A situation like you'll probably never see again in Alabama.

The creation of lakes Gainesville (1978), Aliceville (1980) and Harris (1983) likely ended the creation of new, major lakes in Alabama. Environmental concerns, construction costs and the lack of feasible locations for impoundments has probably ended the building of major lakes in this state.

Gainesville Lake is one of the newest lakes in the state and it experienced that new-lake high in the late 1980s that all new reservoirs experience. Reports of a 16-pound and 14-pound bass caught in the lake by one angler in one day in January 1989 brought nationwide publicity to the lake.

It is still one of the great fishing lakes in Alabama even though its ability to produce monster bass has diminished with time and fishing pressure. The U.S. Army Corps of Engineers decision not to spray and kill the grass in Gainesville Lake like it did in Aliceville Lake has guaranteed that it will be one of Alabama's best bass lakes for years to come.

Spring bass fishing

Tom Bevill Lock and Dam tailrace to Aliceville Bridge

The upper end of Gainesville Lake is strictly run of the river from Tom Bevill Lock and Dam to Aliceville Bridge, and as a result spring bass fishing is limited. But take note that "limited" doesn't necessarily mean poor.

Whenever you have a situation where the upper end

of a lake is river run and the number of incoming creeks is few, the creeks are forced to concentrate bass during the spring spawn. The result is terrific fishing in areas easy to pinpoint.

One doesn't have to be a fisheries biologist to see that the Big Creek Cut-Off area (see map) has wonderful spring possibilities. By being the only incoming creek within a four-mile stretch below the dam, bass seeking shallow water to spawn or shelter from the strong muddy current caused by spring rains have no choice but to go there.

There are similar situations downriver at Owl Creek Cut-Off area, Beaver Creek, Cedar Creek, Curry Creek, Bogue Chitto Creek and the many small standing-timber-filled pockets. Keep in mind the farther upriver you go, the stronger the current becomes during times of spring flooding and the stronger the current, the more important these incoming creeks become as a place for fish to escape.

The tailrace below Tom Bevill Lock and Dam (mile marker 306) produces bass year round, but access is difficult and most anglers choose to ignore the area. There's a launch at Ringo Bluff (mile marker 303), but getting there is an experience. The two-mile dirt road that leads to the launch is more than many anglers want to put their boats and vehicles through.

There's another landing, a good one, at Cochrane (slightly above mile marker 294), but the tailrace is a 12-mile boat ride from that point. Most anglers simply choose to head downstream to wider expanses of water. The result is a year-round fishery that receives little pressure.

Under normal early spring water flow conditions, pearl-colored swim baits, chrome Rat-L-Traps and white grubs fished on the walls of the lock, in the eddy pool to the left of the dam and in the eddies behind logjams on both banks will produce both spotted and largemouth bass.

During the years of heavy spring rains, fishing in this area gets even better. The swift current forces bass to seek shelter to avoid being washed away. The diversion canal on the west bank, the logjams and eddy pools on both banks and the Big Creek Cut-Off area downstream on the east bank are the bass' only real options of escape.

The diversion canal's walls and mouth are good for only a short period of time once the swift current begins. Rat-L-Traps fished in these areas will take fish that are feeding heavily in anticipation of having difficulty in finding baitfish for awhile. The diversion canal will quickly become as muddy as the surrounding water, however, and fishing will become difficult.

The procedure in the Big Creek Cut-Off area is different. This area will quickly muddy at the first sign of spring rains, but within a few days, the swift current will suck the muddy water out. Remaining is a clear-water refuge full of concentrated fish.

It is in this same Big Creek Cut-Off area, and in the other creeks and sloughs downstream, that these bass will eventually spawn. Standing timber and grass beds are common in the flats of these creeks and the typical spring baits—buzzbaits, spinnerbaits, twitchbaits and floating worms and lizards—take the fish.

Aliceville Bridge to Big Slough

Once the angler ventures below Aliceville Bridge, he'll find a lake so much unlike the upriver portion he may find it difficult to believe he's on the same body of water.

No fewer than three cut-offs have been dredged between Aliceville Bridge and Big Slough to straighten the winding river. The result is three slough-like areas full of standing timber, fallen timber, grass beds and a deep-river channel adjacent to ledges. It's simply Alabama spring bass fishing at its best.

The key location in this area is unquestionably the Dead River, Lubbub Creek Cut-Off area. This is the area that the 16-pound and 14-pound bass were caught on Jan. 19, 1989. This double loop (see map) was once the original river channel, but dredging the cut-off during construction of the Tenn-Tom waterway left the area untraveled by big boat traffic.

This area features a well-defined river channel (in the west loop) and a well-defined creek channel through the Dead River area (in the east loop). The river channel loop is strictly run of the river and doesn't offer good fishing except at the four intersections, but the creek channel loop is a different matter.

The creek channel meanders its way through both standing and fallen timber and past the mouth of Lubbub Creek. Most of the standing timber has broken off at water level, or slightly above, and has fallen into the flats next to the banks and the channel. It's here that the angler will find some of the best bass spring spinnerbait fishing that exists in this state.

A 1/2-ounce white spinnerbait with some type of trailer—a white pork frog, a white split-tail eel, a white grub or a white worm—should be used. This area of the lake usually has good water color in spring, so tandem nickel willowleaf blades should be used to draw attention.

The only other pattern that should be considered is flipping. A black/brown jig-and-pig used with 20-pound or larger line should be used for bigger bass in the standing timber in water 4 feet or deeper. The standing timber is in water from 1 to 11 feet deep.

The fish will be concentrated in water warmer than surrounding areas, so locating the fish can be somewhat difficult. The smart angler covers a lot of water quickly with the spinnerbait to locate fish, then switches to a flipping pattern to cover that area thoroughly after the spinnerbait produces a fish.

Other good areas where you can use similar patterns include the standing-timber-filled slough known as Buzzard's Roost (a lot of late-spring tournaments are won there), the standing-timber-filled pocket in the back of Cochrane Cut-Off, the standing-timber-filled slough adjacent Catfish Bend, the standing-timber-filled Pump House Slough (also has grass), and the loop known as Lard Lake. Lard Lake has both standing timber and grass.

Although spinnerbaits and jig-and-pigs are the key baits in these areas, buzz baits can be productive, but to a lesser degree.

Big Slough to Gainesville Lock and Dam

Once the angler moves downriver of Lard Lake, he finds a river that doesn't widen greatly the rest of the way to the dam.

Despite this area's lack of wide water, it does have incoming creeks and an incoming river (the Sipsey), as well as numerous grass-filled and timber-filled sloughs. While the area from Aliceville Bridge to Big Slough is best known as the lake's spring fishing area, spring fisherman will have no problem finding fish here as well.

Two areas stick out for spring fishing in this portion of the lake—the Sipsey River and Bend Lake.

The Sipsey River is a long, navigable river which offers miles of good fishing. It's full of standing timber, a few islands, hundreds of good points and a well-defined channel adjacent to flats. This river has the structure and water depth to hold fish year round. The Sipsey River's only real problem is that it's too obvious a spot. There's easy access and as a result, it's fished hard. The lake's better bass fishermen, especially at tournament time, stay out of the area.

That doesn't mean fish can't be caught there. The concentration of spotted bass can make for a fun day of fishing. The Sipsey is well known as a spinnerbait area in spring and fish found there like a falling spinnerbait rather than one pulled at a constant speed. An occasional pause to let the lure fall produces more fish for a reason that hasn't been explained.

Bend Lake isn't really a lake, but a slough fed by a small creek in the farthest reaches of the slough. This slough doesn't appear to be anything special on the map, and many anglers pass it up for better-looking water. But the long, narrow slough is filled with timber from front to back and holds a lot of spring fish.

A Rat-L-Trap fished in the first few hundred yards of the mouth of Bend Lake is an excellent pre-spawn and post-spawn pattern, while spinnerbaits fished in the shallows around the standing timber is best during the spawn.

Several other areas are worth trying. If the angler likes grass and standing timber, Big Slough has just what the doctor ordered. The site of a large lake before Gainesville Lake was impounded, Big Slough is now connected to the main river and is an excellent spring spawning area. Spinnerbaits, jerk baits and buzzbaits are key baits in fishing this area.

Goat Slough has some standing timber in its left fork and offers excellent fishing in the spring and fall.

There is some standing timber in the first slough you come to while going south in Cook's Bend, as well as in the last slough you encounter before getting back out in the main river channel.

Phanachie Creek, which leads into Warsaw Bend, has some good standing timber above the bridge and will give up fish in the 3- to 5-pound range on buzzbaits spring, summer and fall. Bass spawn by the thousands in the creek, but where they spawn is protected by shallow water and anglers can't get there. The spawning flats, which lie out of the 10- to 12-foot channel are sometimes only a foot deep and boats can't hack it. When the spawn is on, fish elsewhere.

Turkey Paw Branch directly above the dam also has standing timber, as well as some grass scattered throughout. Turkey Paw is difficult to navigate. Anglers should stay on the right bank halfway back into the creek, then move to the left bank. Turkey Paw is outstanding for topwater fishing spring and fall. Topwater baits such as Lunker Lures and jerk baits work below Bend Lake better than they do in the areas above it.

Hot weather bass fishing

Tom Bevill Lock and Dam tailrace to Aliceville Bridge

If you were to design an area non-conducive to summer bass fishing, it would probably look a lot like the upper end of Gainesville Lake.

This area, as said before, is strictly river run. Bass are unquestionably in this main channel during the hot summer months, but the lack of ledges and the small number of points doesn't give the angler structure to develop a summer pattern.

Save your gas and fish elsewhere during the summer.

Aliceville Bridge to Big Slough

The creek channels through the cutoff areas have water deep and cool enough that bass can comfortably reside there during the difficult months of summer. This is good news for fishermen because this keeps the bass out of the main river channel where they are much more difficult to locate and catch.

The water is usually cool enough that a good bit of top water action takes place during the early-morning and late-afternoon hours. The deep channels next to ledges filled with standing timber make it easy for bass to chase shad and corner them. These ledges make excellent summer fishing.

The angler needs to keep two rods outfitted with different baits during these months. A rod with a jerk bait such as a Pop-R or Bang-O-Lure needs to be on hand when the bass are chasing shad on the ledges, and a rod with a plastic worm or lizard needs to be in reach when the bass are playing hit-and-run with the shad.

The angler needs to fish the worm or lizard on top of the ledges where they fall off into the channels, concentrating on areas with fallen timber, stumps and other irregularities. The mouth of Lubbub Creek and the fingers off the channel in Dead River are key areas.

Whenever the angler spots bass hitting shad on top of the ledges, he should quickly switch to the topwater bait. When the topwater ceases, he should quickly go back to the worm and throw on top of the ledge where the activity took place, slowly bringing that worm back and letting it fall off the ledge. The bass will usually be suspended there two or three feet deeper than the ledge awaiting the shad to return to deep water. Be ready when the worm falls. The bass will often explode on the worm and a quick hookset is needed.

The Cochrane Cut-Off area, Catfish Bend and the mouths of Lard Lake and Pump House Slough are other areas where this pattern will work.

Big Slough to Gainesville Lock and Dam

The hot months of summer scatter fish widely in this area. The largemouths head to the points and dropoffs in the main river channel, while the spots move to the timber that has fallen into the main river.

Fishing for these spots can offer some exciting fishing. A large deep-diving crankbait fished on the fringes and between the limbs of the fallen trees takes the best fish. Spinnerbaits, worms and lizards can be fished in tighter, but they take fewer fish.

Deep-diving crankbaits can be used on points for largemouth, but this fishing is hit-and-miss. It's possible to come back from a long hot day of fishing with nothing to show but a sore arm from cranking a reel.

It's possible to catch a few summer fish in the creeks, but a good fish is rare. Phanachie Creek has a channel deep enough for summer bass, as does Warsaw Bend and Cook's Bend.

The message is this—you might catch some great fish in this area. Then again, you may not.

Cool and cold weather bass fishing

Tom Bevill Lock and Dam tailrace to Aliceville Bridge

During the fall and winter, bass fishing is a mixed bag here. The angler who finds fishing so good here in the fall often finds it dismal just a few months later.

Because the area lacks ledges and points, fall bass that are gorging themselves for the winter are forced to corner shad the only way they know how. That is by forcing them to the surface and to the banks.

The result is some spectacular topwater action in this area. The angler needs three things to be successful during the fall.

First, he needs a rod rigged with some kind of jerk bait. It should be outfitted with 10 or 12-pound test line. This will be used to take bass schooling on top water.

The second thing he needs is a rod rigged with a lipless crankbait. A chrome/black backed Rat-L-Trap will work nicely. This rod should be outfitted with 8-pound test line. It will be used to take schooling bass that have been hitting topwater and have temporarily returned to below the surface.

The larger fish are often taken by this method. The larger bass often let smaller bass attack the school of shad while they wait for the cripples that begin to fall.

The third thing the angler needs is a pair of binoculars. This will allow him to scan the water surface up to a half mile away for signs of schools feeding on topwater.

Once a school is sighted (the tailrace and the mouths of creeks are prime areas), the angler should make a dash in that direction, stopping at least 100 yards away. The trolling motor should be used to approach the schooling fish quietly. The angler should make the farthest cast possible.

Don't hesitate to throw a jerk bait in the middle of the schooling fish. Once the school submerges, pick up the other rod and throw the lipless crankbait well past the area where the fish went down and pull the bait through the area with an irregular jigging motion. The bass will often take the bait when it's falling.

Many anglers would rather catch one bass on topwater than 10 by any other method and if that's the case, this is the area and the time of the year. This exciting method of fishing is known locally as "jump fishing."

Surprisingly, this area is Dullsville when the water temperature drops into the 40s. The lack of structure doesn't concentrate the fish in any one area and they are almost impossible to locate. The other sections of the lake offer much better winter fishing.

Aliceville Bridge to Big Slough

It may be hard to believe, but the bass that bunched up so tight in this area during the spring and summer will be in even tighter schools during the fall months. That makes these fish difficult to locate, but once they are found, it's not unusual to load the boat with nice bass.

For the angler who loves to fish a Rat-L-Trap, this is it. The wobbling, noisy bait drives schooling fish crazy.

The angler should locate schools of bass or schools of baitfish (if the baitfish are there, the bass are nearby) with his depthfinder. The mouth of Lubbub Creek, the points of the cut-off areas, the mouth of Lost Slough, the mouth of the small creek in the back of Lost Slough and the mouths of Lard Lake and Pump House Slough are prime areas to find schooling fall bass.

Once the fish are located, the angler should back off the area as far as possible and cast the Rat-L-Trap well past the area. The retrieve should be just fast enough to keep the lure on a plane just over the schooling fish. An occasional pause that lets the lure fall slightly is often all that is needed to trigger the schooling fish into a feeding frenzy.

Once the water surface temperature dips below 50 degrees, bass fishing here gets difficult. The bass settle into the deep creek channels and on the points of the Cut-off areas and get sluggish. The winter angler needs to switch to slow-crawl fishing then. A jig-and-pig or a 4-inch worm fished slowly is the best shot at a good fish.

The only exception to this is on those rare occasions when several unseasonably warm days string together. This will kick the temperature in shallow water up to the 48-degree range and send baitfish scurrying to the area. Bass will follow. A worm or medium-running crankbait fished slowly in this shallow (4 to 8 feet) water will produce fish then.

Other types of fishing

Crappie

The crappie fisherman who finds success on Aliceville Lake should have no problem finding similar success on Gainesville Lake. Gainesville Lake is remarkably similar to its upstream neighbor.

Just as Aliceville Lake, the massive amount of shallow structure next to deep water makes good crappie fishing year round. Crappie fishing is best during the spring when the fish move into the shallow water to spawn. The fish can be found in several locations during this time.

The majority of the better fish leave the original river channels in the Big Creek Cut-Off area, the Owl Creel Cut-Off area, the Dead River or Lubbub Creek Cut-Off area, the Cochrane Creek Cut-Off area, the Cook's Bend Cut-Off area and Warsaw Bend Cut-Off area to move into the weed beds, logjams and standing timber next to the deep channels.

Naturally, the better fish are the toughest to get at. Better fishermen use a long, telescoping fiberglass pole outfitted with 12-pound test line and a curly-tail chartreuse jig with a white led head. This setup can be used to gingerly reach over and drop the jig right into the holes in the weeds. This produces slabs.

An easier, but less productive method is to line up parallel to the weed beds and, using a spinning or spincast reel, cast a jig hanging 18 inches below a cork as far as possible and as close to the weeds as possible. This rig should then be "popped" all the way back to the boat. This popping draws attention and the jig right on the weed line is often too much for the fish to resist.

These fish can also be targeted in the traditional method, and that's with a minnow and cork fished near the weedline. This method is also good right before and right after the spawn when the fish are holding just outside of the weeds in 5 or 6 feet of water.

Although the cut-off areas offer the best fishing, any timber- or weed-filled slough or pocket next to the main river is likely to hold fish during this time. The same patterns will work there.

Following the spawn, the crappie will hold on the flats in about 5 to 10 feet of water until June. Stumps and blow-down timber will concentrate the fish. Live minnows fished under a cork is the best bait.

Once the hot part of the summer arrives, the crappie head for deep water. The fish that head to the main river channel are often difficult to locate, so the better fishermen head to the cut-off areas. Crappie gather here in the original river channel and stay until winter. The cut-off areas are by-passed by the big boat traffic, thus the crappie find an undisturbed summer.

One of the patterns used by the lake's best crappie fishermen at this time is trolling. A multi-rod setup with jigs set at varying depths will produce numbers of fish. A little experimenting with jig colors and jig depths may be needed to fine tune the pattern.

Another summer pattern that works well is bottom-bumping with live minnows. The crappie change their pattern very little during the fall. Most of the better fish stay deep in the channels, only occasionally making a run to the ledges to feed. Trolling or bottom-bumping works best, but a live minnow fished under a cork over a ledge with structure on the lip will produce some good fish.

Winter crappie fishing is tough on Gainesville Lake. The fish get deep in the channels and seem lifeless. A pattern that works to stir up the fish is locating the fish with a fish finder then jigging a small spoon right in their faces.

Gainesville Lake's best fishing spots

1. Tom Bevill Lock and Dam tailrace. Good spring, summer and fall for bass and crappie.

2. Mouth of Big Slough. Excellent during summer with Rat-L-Traps. Good in early spring with jig-and-pigs and worms.

3. Upper end of Big Creek Slough. Standing timber, grass beds.

4. Aliceville Bridge. Good around pilings in summer. Deep-diving crankbaits.

5. Pulpwood Canal. Holds summer fish.

6. Mouth of Dead River. Good worm area in summer.

7. Point in mouth of Dead River area. Good topwater fishing in late spring, early summer.

8. Upper mouth of east loop entering Dead River. Excellent spring, summer and fall.

9. Lower mouth of east loop exiting Dead River. Excellent spring, summer and fall.

10. Fingers in Dead River. Full of standing timber. Excellent in spring.

11. Standing timber and ledges on inside bank of Dead River. Excellent spring, summer and fall.

12. Mouth of Lubbub Creek. Good summer fishing.

13. Lubbub Creek. Excellent spring, summer and fall. Spring and summer with spinnerbait on bends, fall with crankbait in bends. Also good creek for sauger and walleye in fall.

14. Buzzard Roost area. Full of standing timber. Excellent in spring with topwater. Good in fall with Rat-L-Trap.

15. Good springtime pocket full of standing timber.

16. Point of Cochrane Cut-Off. Good for schooling fish in summer.

17. Good standing timber. Good fall fishing area.

18. Standing-timber filled slough. Good in spring.

19. Point in Catfish Bend. Good summer and fall.

20. Mouth of Lost Lake. Good in summer and fall.

21. Lost Lake. Standing timber. Good topwater summer and fall. Tough area to get into, but it's possible.

22. Mouth of Lard Lake. Good in summer and fall.

23. Good standing timber and weeds in Lard Lake. Push pole territory. Good spring fishing with topwater. This area has given up several 12-pound largemouth.

24. Pump House Slough. Standing timber and weeds. Good in summer with buzzbaits.

25. Mouth of Big Slough. Good good island inside mouth. Good fishing year round first 100 yards.

26. Back of Big Slough. Area around culvert is excellent after a good rain. Water running in from culvert oxygenates water.

27. Mouth of Sipsey River. Good spring, summer and fall.

28. Moon Lake. Standing timber. Good in spring with spinnerbaits, summer with buzzbaits.

29. Sipsey River. Miles of standing timber. Weeds. Good in spring when not muddy. Good in summer and fall.

30. Mouth of Goat Slough. Good fishing spring, summer and fall.

31. Left fork in Goat Slough. Full of standing timber. Excellent spring, summer and fall.

32. Right fork of Goat Slough. Good spring fishing with lizard.

33. Back of Cook's Bend. Good year-round fishing.

34. Timber-filled slough. Good spring fishing with spinnerbait, good summer fishing with buzzbait and worm.

35. Bend Lake. All standing timber. Good year round. Bass spawn on little points just inside mouth in spring.

36. Standing-timber filled slough. Good year-round fishing.

37. Wilkes Creek. Standing timber in back. Weeds throughout. Good year-round.

38. Mouth of unnamed slough. Good in summer for schooling bass.

39. Back of slough. Grass and standing timber. Good year round.

40. Timber-filled slough. Good crankbait area in fall.

41. Mouth of Turkey Paw Branch. Left bank is standing timber. Excellent with spinnerbait and jig-and-pig in spring.

42. Back of Turkey Paw. Points of fingers offer good year-round fishing.

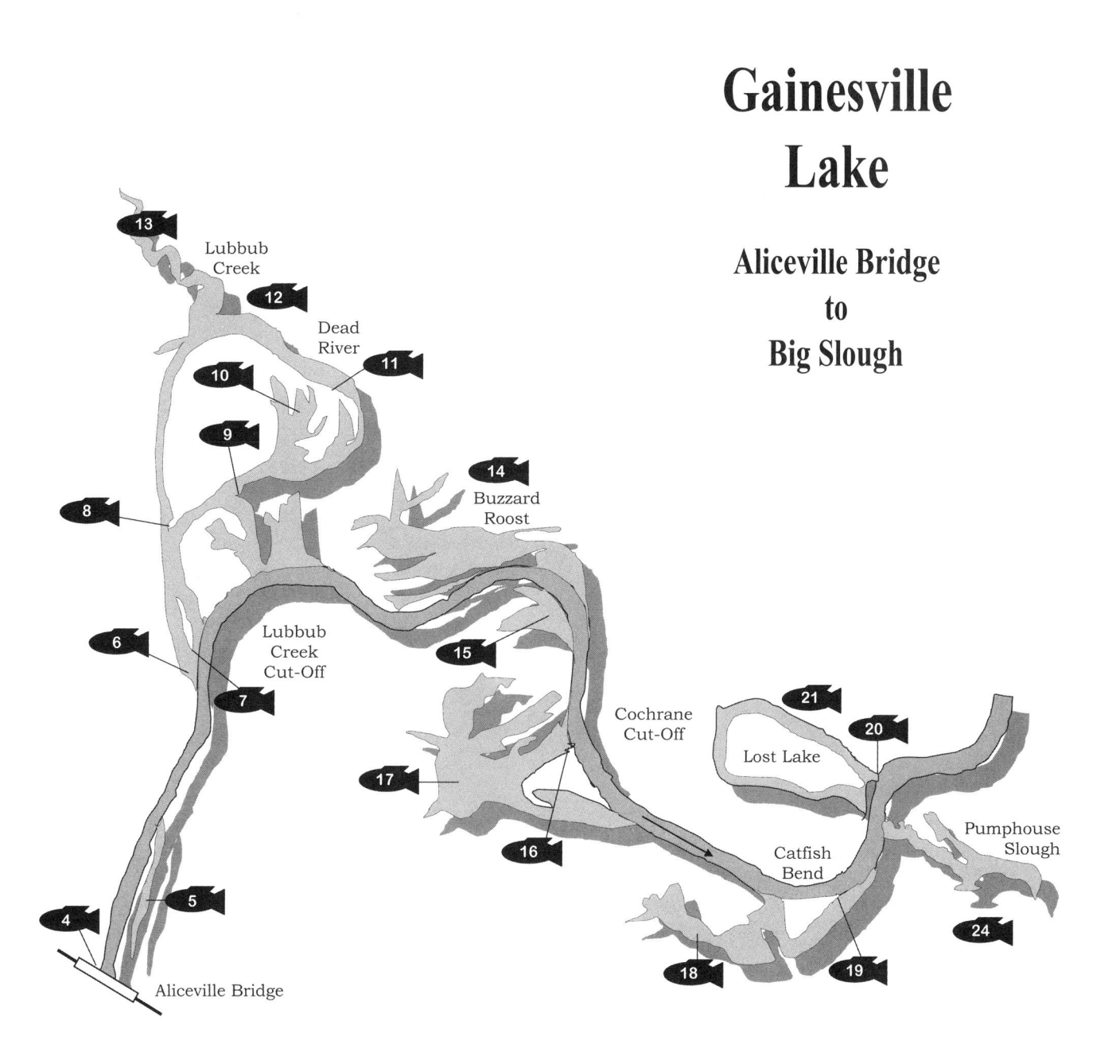

Gainesville Lake

Aliceville Bridge to Big Slough

The Ultimate Guide to Alabama Fishing

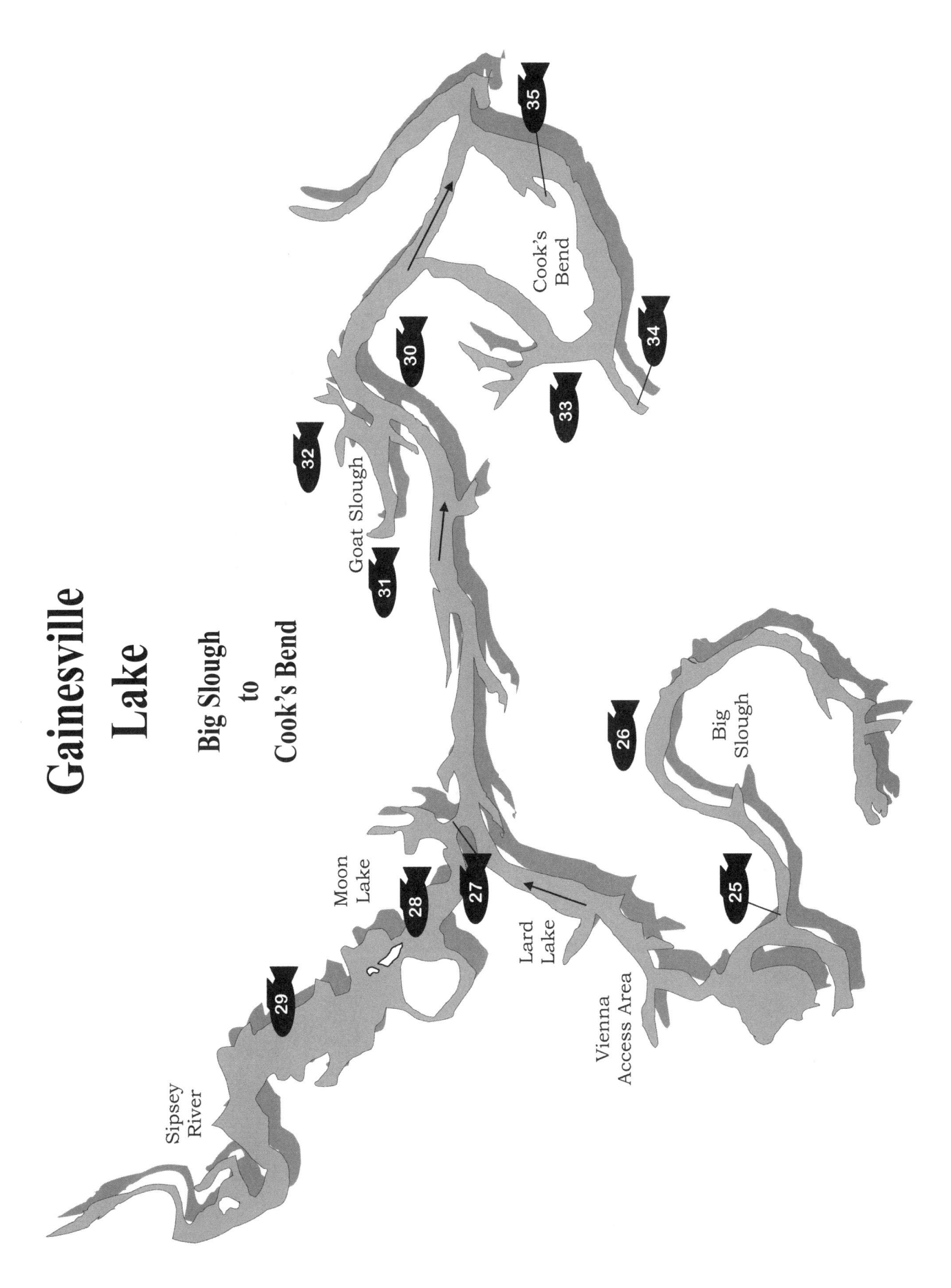

Gainesville Lake

Big Slough
to
Cook's Bend

Cook's Bend

Goat Slough

Moon Lake

Sipsey River

Lard Lake

Vienna Access Area

Big Slough

236

Gainesville Lake

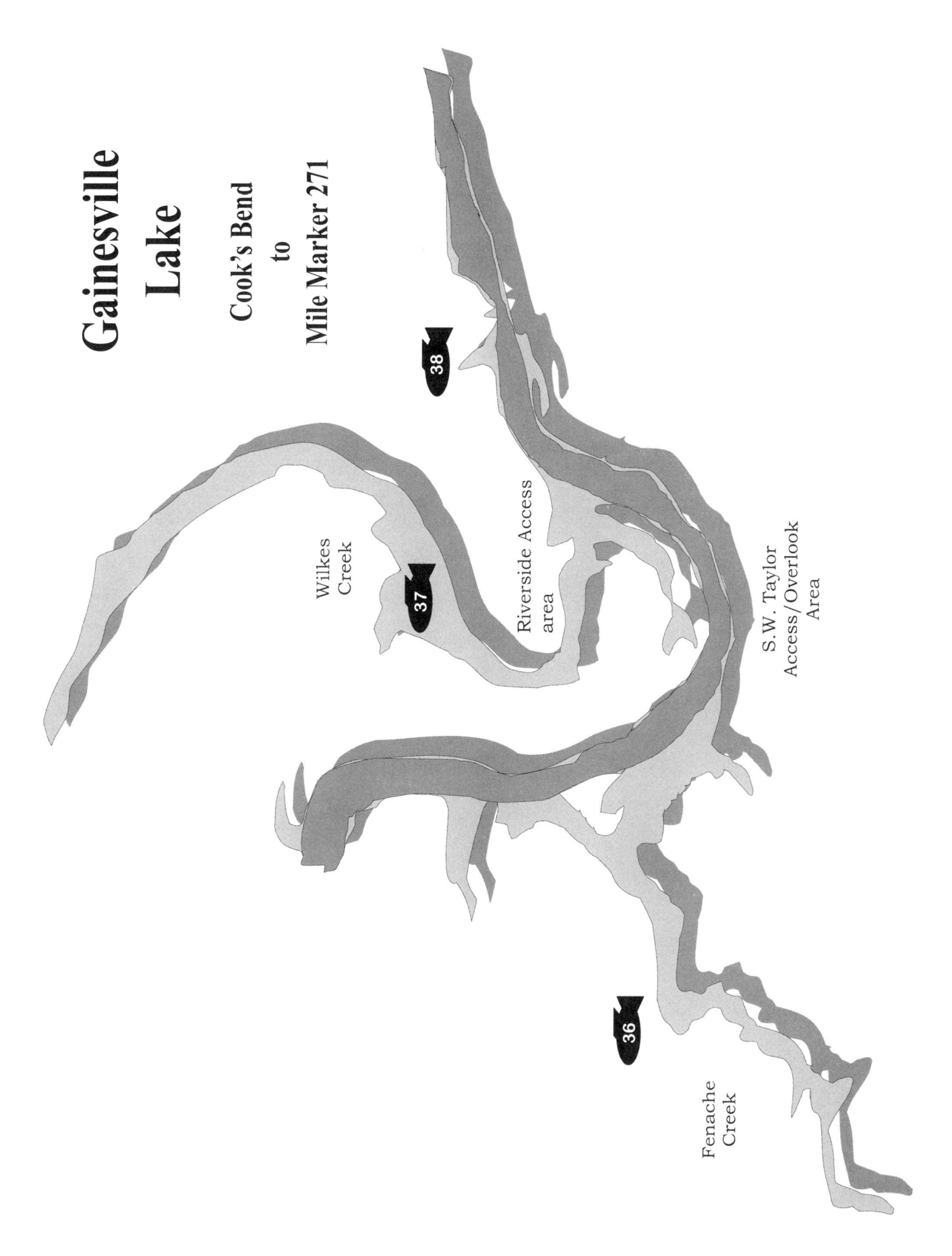

Cook's Bend to Mile Marker 271

Mike Bolton

Wilkes Creek

Riverside Access area

S.W. Taylor Access/Overlook Area

Fenache Creek

The Ultimate Guide to Alabama Fishing

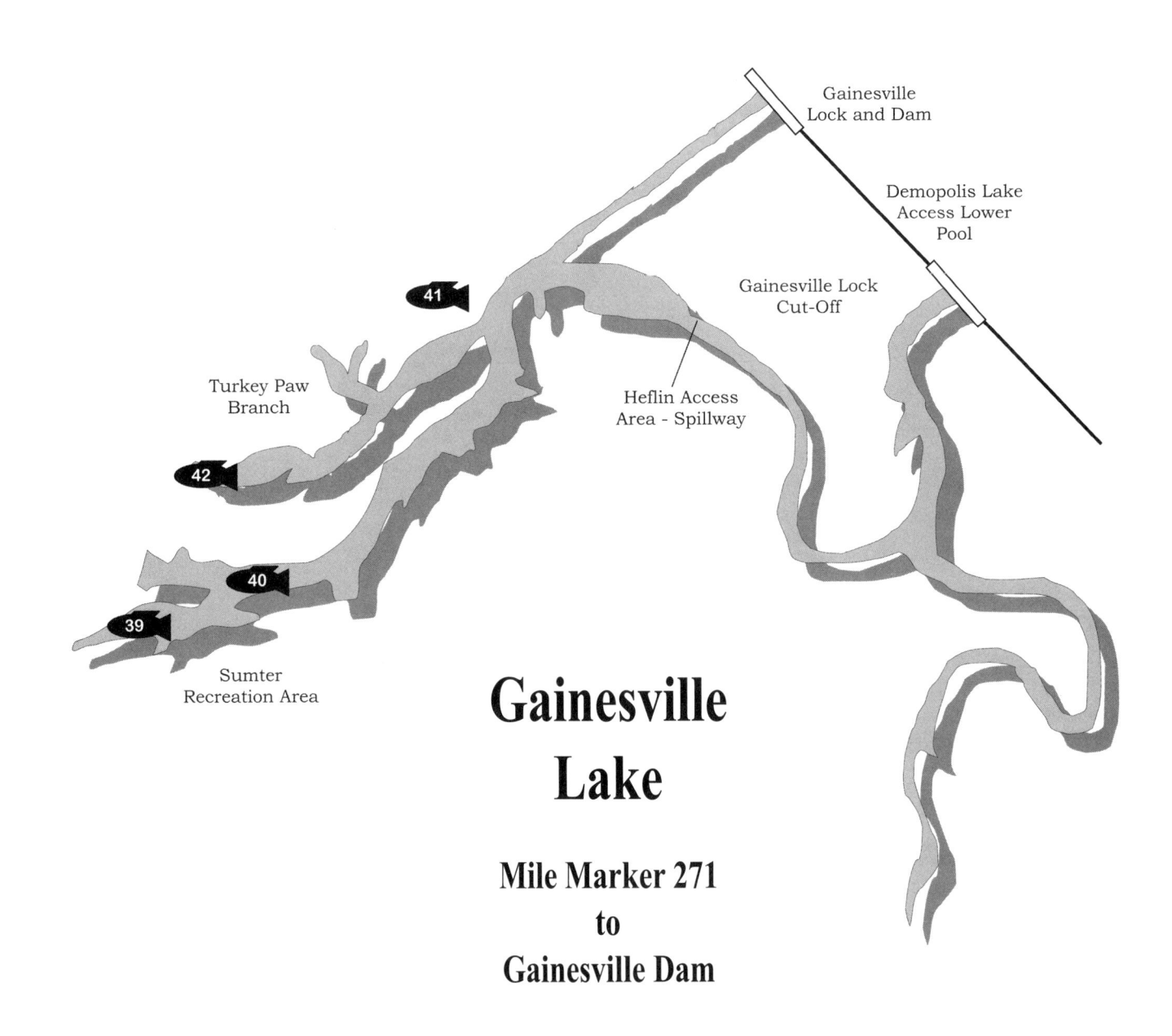

Gainesville Lock and Dam

Demopolis Lake Access Lower Pool

Gainesville Lock Cut-Off

Heflin Access Area - Spillway

Turkey Paw Branch

Sumter Recreation Area

Gainesville Lake

Mile Marker 271 to Gainesville Dam

238

Mike Bolton

Demopolis Lake

Spring bass fishing

Tombigbee River above the Black Warrior intersection

By no stretch of the imagination is Demopolis Lake a lake, at least by the standards of appearance set by many other Alabama lakes. The damming of these rivers did little to change the original river channels. There isn't a wide spot anywhere on the main rivers that would give the appearance of most lakes in Alabama.

Demopolis Lake is also unusual in another way. While most lakes are the result of the damming of a river, Demopolis Lake was formed by backing up two rivers—the Tombigbee and Black Warrior. It is possible for an angler—in a space of just a few feet—to fish different rivers with different water colors.

Although Demopolis Lake is unusual by most standards, many an angler would swap his best rod and reel for his home lake to have the structure Demopolis Lake has. Standing timber and cypress stumps can be found tucked away in a large number of sloughs, swamps and creeks. The shallow standing timber and cypress stumps means shallow-water fish easily accessible to most anglers.

The bad news is that the U.S. Army Corps of Engineers has made the lake just a shell of what it once was by making the decision to eradicate the grass throughout the lake. Most fishermen said good riddance to the water hyacinth that choked the lake and held few fish, but there was no select spraying. Also eradicated was the alligator grass that held so many fish and made Lake Demopolis one of the state's great flipping lakes.

Anytime a lake is blessed with shallow standing timber and shallow cypress trees it's a foregone conclusion that tremendous springtime fishing awaits bass anglers. The Tombigbee River section is a perfect model of water that offers spectacular springtime fishing.

Bass in this section of Lake Demopolis have very few winter options. The main river has the only deep water to offer comfort for big bass during the harshest part of winter, and it's out of this main river and into the sloughs the bass move when the first warm weather approaches. The shallow and medium depth sloughs, usually laden with good stain due to heavy spring rains, warm much quicker than the main river. The dingy water absorbs heat rather than reflecting it as clear water does, and the warm water sends bass shallow.

This warm, structure-filled water combined with the baitfish that are attracted to it, computes into excellent pre-spawn, spawn and post-spawn water for bass. These bass first move into the mouths of the sloughs and creeks and alternate between there and the main river as the weather does its usual early spring flip-flops.

Several days of sunshine will warm the mouth to the point the fish will find it comfortable enough to remain there, while the periodic cold snaps or cold nights will send them back to the deep water of the main river channel.

Crankbaits and worms are the first effective baits. The points and secondary points inside the sloughs hold the fish, and medium running crankbaits will shake them loose. Shad-colored crankbaits work well on the rare occasions the water is clear this time of year, while chartreuse and other brightly-colored baits work well when the river has its normal spring stain. Belmont Landing Slough, Spidle Lake, Logjam Slough, Taylor Lake Slough, Birdine Creek, Birdine Slough and McConico Creek are shallow-water refuges in which you'll find early spring bass.

When warm days finally begin to be followed by warm nights on a regular basis, the bass will move into these areas and stay until post-spawn. The bass will hold temporarily on points and deepwater timber and will feed heavily in anticipation of the spawn.

Buzz baits, twitch baits or some other form of topwater bait will take these fish in exciting style, but die-hard spinnerbait fishermen can also be successful with their favorite lures. A chartreuse/white spinnerbait tipped with a chartreuse pork frog and retrieved through the flats with a slow roll will provide vicious attacks. Odds are that any of the fish taken by any method this time of the year will be quality fish.

Every slough or creek between the Tombigbee River-Black Warrior River intersection and the upper end of Rattlesnake Bend cutoff is filled with standing timber, stumps or cypress trees, and most have a combination of of structure. It's these good hiding places bass seek out to build their beds. Most of the beds in these sloughs will be in 3 to 5 feet of water.

The big females lay their eggs and vacate the premises in short order. The males are left behind to guard the eggs. The males are always smaller fish.

Once the spawn is over, Demopolis Lake goes through a brief period that is unquestionably its best fishing time of the year. For two, three or four weeks, depending on the weather, the bass, reeling from the hardship of the spawn, begin to revitalize themselves by going into a feeding frenzy.

Both the males and females are hungry and the relatively cool water allows them to stay in the shallow structure to feed. Topwater action can be tremendous. Buzzbaits, twitch baits, pop baits, chug baits, floating worms and lizards – just about anything that floats, will produce fish.

Tombigbee River/Black Warrior intersection to Yellow Creek

The patterns that lead to successful spring fishing on the Tombigbee River on Demopolis Lake work just as well on the Black Warrior River fork. The only real difference between the two rivers during the spring is that following heavy rains, the Black Warrior normally will have somewhat less stain and will clear more quickly.

Medium- or deep-running crankbaits or lizards rigged Carolina style are effective when the fish first move out of the main river channel to the mouths of sloughs and swamps. The angler should use a 7-inch or larger lizard such as Zoom makes because the larger the lizard, the better it floats. A 7-inch pumpkin Zoom worked slowly in the mouth and secondary points can be deadly those first few days that the bass move shallow.

Once the bass begin their transition from the mouth to the shallow water back in the creeks, sloughs and swamps, spinnerbaits work wonderfully. You're going to find stained water most of the time, so use a chartreuse or chartreuse/white spinnerbait tipped with a chartreuse pork frog and a nickel willow leaf blade.

It should be noted that some creeks are clearer than others this time of year for one of two reasons. Either the slough or creek is deep, or it has a good creek running through it and pumping muddy water out of it. Pay close attention to that and the water temperature. The creeks that have moving water have good oxygen content and, thus, more active fish, but only in middle or late spring. Moving water is more difficult to heat and the creek with good moving water, in early spring, may be actually pumping cooler water on top of warm water.

It's such little tricks that experts use to make the difference between good and great fishing.

Once in shallow water, topwater baits such as buzzbaits, floating lizards (once again, use the largest lizards you can find so they will float a hook) and twitch baits are the key. When hard on spawn and in the weeds, buzzbaits will work, but 3/4-ounce jig-and-pigs are sometimes needed to punch through the heavy greenery to get where the fish are.

Just as on the Tombigbee fork, most spring fishing in the Black Warrior fork takes place in the creeks and sloughs. Having a good knowledge of the areas and the type structure they contain is important.

Dobbs Swamp is a shallow, multi-fingered swamp that lives and dies with spring floods and summer drought. It has a narrow opening, but opens to a wide expanse of water. The swamp is full of fallen trees.

The shallow water is an excellent spawning area.

Buzzbaits and floating plastic lizards work well during spring.

Runaway Branch is a moderately deep slough that has scattered standing timber and much blow-down timber. Camp houses can be found on both sides of the branch and many have piers and docks laced with underwater brush piles. Worms, jigs and swim baits work well in that brush during spring. There also is a landing accessible from U.S. 43.

Kelly's Slough is the closest slough to the old U.S. 43 bridge. The slough is void of points, and is fairly shallow. Camp houses and piers can be found in the slough and many areas surrounding piers are loaded with brush. Cypress trees can be found in the back. It's a natural location for the bass spawn. Buzzbaits, worms, spinnerbaits and topwater twitch baits work well.

The Citadel Cement Slough has a small opening laced with old bridge pilings, but it opens into a fairly decent-size fishing hole. Throughout the slough are steep lime-rock banks. It offers only average spring fishing.

French Creek is one of the three largest creeks on Demopolis Lake. It is the deepest creek in the lake and the last creek on the lake to get muddy during spring rains. This wide creek offers islands, a multitude of points, rock banks and a well-defined creek channel. For the average angler, its spring fishing is only fair, but top anglers do quite well there in the spring. The multiple offerings of other creeks with standing timber and the resulting shallow spring fish make other creeks a better bet for the average angler, however.

Powerline Slough is a slough that offers deep water in the front and shallow water in the back. It has its share of standing timber in the back. It also has three major points. The shallow water in the rear offers excellent bass and crappie spawning. This slough is better known as a crappie producer, but nice bass can occasionally be found during the spring. The spring bass baits mentioned for other creeks and sloughs work well.

Backbone Creek is a long, narrow creek with moderately deep water. There is a launch site. It has a good stand of cypress trees in the back which offer a safe spawning area to bass in the spring. Spinnerbaits and topwater are best.

Yellow Creek appears on the map as a long, navigable creek, but anglers find difficulty in fishing more than a half-mile mile back. By being in a major river bend, the mouth of this creek has silted in and passage can be difficult. For the angler who finds water deep enough to get through, the shallow-water creek offers good points and grass beds. This is primarily a spring fishing area because passage is limited other times of the year.

Yellow Creek to Candy Landing

Big Prairie Creek is the largest water inlet on the Black Warrior side of Demopolis Lake and offers excellent fishing year round. It is possible to go as far as four miles back in Big Prairie and still find 17 feet of water. Yet, there is plenty of shallow water throughout to make for excellent spring fishing. There are plenty of shallow water ledges that fall directly into deep water and these are excellent areas to find spring bass.

In early spring, the bass will inhabit the deep water points and use these points as a highway to travel back and forth for brief feeding runs in shallow water. With your boat positioned in deep water, shad-colored deep-running crankbaits cast into shallow water and reeled down the deep point produce nice catches. Large (3/4- or 1-ounce) spinnerbaits with big blades also work well here. The heavy spinnerbait falls quickly down the steep points and the larger blades force the angler to fish the bait slowly at the speed these sluggish fish need.

The water normally isn't heavily stained here, so use spinnerbait colors such as white or shad and blade colors such as copper or gold that work well in clearer water.

Other baits that work well here in the early spring are jig-and-pigs, plastic worms and Carolina-rigged lizards.

Once the water warms to 65 degrees, the bass are thinking spawn and shallow water. Buzzbaits, spinnerbaits and other topwater baits mentioned elsewhere for fishing grass can be used.

These are not the best patterns for taking spring bass, however. If you've ever considered jiggerpole fishing, there are few places better than Big Prairie Creek for you to witness it at its best. It should be noted that fishing with jiggerpoles is not allowed in tournaments.

For those who haven't heard of this art of fishing (a method your grandaddy might have used if he fished), all that is needed is a 15- to 18-foot cane or fiberglass pole, a large broken-back minnow bait and some braided 60-pound test line.

The lure should be tied to the pole with 6 inches of braided line. Using the trolling motor, simply run the bait inches away from the structure, all the while shaking the pole so the tip of the pole barely ticks the water with a steady rhythm. This may seem unnatural at first, but it gives the appearance of a minnow chasing a small bug and the result is an offering that the bass, hiding in the structure awaiting an easy meal, can't turn down. This method of fishing is thorough. It allows the angler to cover every inch of the structure instead of casting every 15 feet or so.

When a fish decides he wants the bait, the explosions are tremendous. The angler must then swing the fish around to the opposite side of the boat into deep water and pull the pole in hand-over-hand to retrieve the fish.

Big fish are common, and a 5-pound bass on 6 inches of line is a thrill unlike any other you'll find in fishing.

If you're looking for water that has less fishing pressure than Big Prairie, Bream Slough is a moderately deep slough. Its mouth silts in often which makes it difficult to get into, but getting inside with a small aluminum boat can have its rewards. Bream Slough is better known as a spring bream producer, but it does hold some bass, especially around the lone stump that can be found there.

Baker's Slough is another such place. It has a tiny mouth, but it is filled with structure.

Hot weather bass fishing

Tombigbee River above the Black Warrior intersection

Because the grass has been eradicated in Lake Demopolis, No longer do the bass stay shallow in the summer months.

Several areas off the main river in this portion of the lake contain water deep enough to hold summer bass. It's not unusual to find bass holding on points in Rattlesnake Bend, Twelve-Mile Bend, Taylor Lake Slough, Spidle Lake and McConico Creek even when the surface temperature passes the 90 degree mark. The deep south bank in Spidle Lake is a solid summer place to go. Twelve-Mile Bend is virtually nothing but river run filled with enticing structure, and the dredging of the Twelve-Mile Bend Cut-Off to divert barge traffic during construction of the Tenn-Tom Waterway turned this into an excellent summer fishing area, and other times of the year as well. Twelve-Mile Bend is virtually a half-mile-long deep slough.

Other key holding areas during this time are the wide mouths of Taylor Lake Slough and Spidle Lake. The fish suspended in the main channel dash into the wide areas to feed, but are always just seconds away from deep water again.

Deep-running crankbaits, Rat-L-Traps, plastic worms and Carolina-rigged lizards are the key baits to fish these points with during the dog days of summer.

Both Taylor Lake Slough and Spidle Lake have gradual slopes leading out of their wide mouths into the main river channel and a Carolina-rigged lizard can be effective.

If you're willing to swat mosquitoes and fish at daybreak or dusk, don't forget to bring your topwater baits along, even if common sense tells you it's not really topwater season.

Spidle Lake, Taylor Lake Slough and McConico Creek are areas where shad surface near the mouth in the early morning and late evening hours and are attacked by small, schooling bass. Twitch baits and other injured shad or minnow imitations often result in a limit of bass in minutes. Don't expect them to be quality bass, however. They're usually just big enough to stink up a frying pan.

Barge and yacht traffic on the Tombigbee River is cursed by many fishermen, but the disturbances do result is some good summertime fishing. The waves from the vessels have washed the bank from under a large number of trees and these trees have fallen into the river. The trees offer excellent summer areas for flipping.

Tombigbee River/Black Warrior intersection to Yellow Creek

Once the spawn is over, the fish then seek deeper water with milder temperatures.

The mouth of Runaway Branch is a summer schooling area on this portion of the lake. Chrome/blue-back Rat-L-Traps and similarly colored medium-running crankbaits are excellent when fish are schooling hard in open water. When the fish are running up on the ledges to feed, 4-inch Slider worms or swim baits cast from deep water onto the ledges work well.

Summer bass also school in the mouth of French Creek. Lesser fish can be found in both creeks, especially in the channel in French Creek. Another good possibility is fishing the piers and docks in both creeks. The fish find shade under the structures during the hottest part of the day, and for night fishermen, feed under them in the brush at night.

Yellow Creek to Candy Landing

Summer fishing is rarely tremendous on any lake in Alabama, but the wide expanses of deep water found in Big Prairie Creek make it a good summer location by summer standards. Bass here can sometimes find enough deep, cool water that they can find refuge, at least in the early morning hours, in the late afternoon or at night.

These fish won't be aggressive, so a slow bait like a heavily weighted worm is needed. Other fish here hold on the points in the deep creek channel. Worms, Carolina-rigged lizards or jigging spoons will take these fish.

A large number of bass head to the main river channel during summer and the fallen trees and logjams offer cover they need. Flipping a jig-and-pig at the base of trees that are in 20 feet or water can catch the fish. The water is fairly clear this time of the year, so a brown/black jig-and-pig tipped with a brown or black pork frog works well.

Cool and cold weather bass fishing

Tombigbee River above the Black Warrior intersection

Cooling water temperatures in late September send bass fleeing from the dull, stale, deep water of summer. Just as for humans, this must be an enjoyable time of the year for bass. They find temperatures much more bearable. And it's the time of the year for the bass to harvest and store food for winter. The cooler water sends the baitfish into shallower water and the bass readily follow.

If any good has come from the eradication of the grass in Lake Demopolis is that it has not hurt fall fishing as bad as it was hurt in late spring and summer. The shallow grass that was once here is no longer dying and robbing the water of oxygen. Spinnerbaits, worms and, to a lesser degree, topwater baits and crankbaits, are effective. Sloughs with shallow water falling into deeper water close to the mouth will hold the fish and they'll be actively feeding in anticipation of winter.

Placing a boat in the mouth of Taylor Lake Slough and casting a Carolina-rigged lizard or worm into the mouth of the creek and into the main river channel is an excellent fall pattern. Great numbers of bass coming into the shallow water to feed use this route. Once the water temperature plunges, full bellies and lower metabolisms slow the fishes' feeding habits. The shallow water that offered the best pH and water temperature flip-flops and offers the worst conditions for fish, so they move to deeper water to compensate. There, they become complacent and uninterested in eating.

The same areas that offered refuge from the unbearable summer offer the same refuge for winter bass. A slow bait such as a jig-and-pig, a Carolina-rig or a heavy spinnerbait slow-rolled across the bottom right in the fish's face is needed to get a response. These baits should be worked on deep points and in the main river channel around any type of deep structure such as brush piles or logjams.

Tombigbee River/Black Warrior intersection to Yellow Creek

Fall fishing can be quite good in this section of the lake. The logjams and fallen trees along the main river channel hold active bass, as do the mouths of Runaway Branch, Kelly's Slough, Citadel Cement Slough, French Creek, Powerline Slough, Backbone Creek and Yellow Creek. Bass will school and feed heavily in the main river channel near sharp bends such as those found near Hines Landing, Klies Bend and at the mouths of Backbone Creek and Yellow Creek.

When the bass are schooling and chasing shad on the top of the water, Rat-L-Traps, medium-running crankbaits, spoons or swim baits thrown into the middle of visible schools will result in catches of small to medium bass.

At times, bass will leave the main river channel to feed on the narrow ledges found throughout the length of the river. Locating stumps and brush on these ledges is the key to finding fish when they're not schooling on top.

Because of the moderately deep water, both Runaway Branch and French Creek hold fall bass. The boathouses and piers hold quantities of small bass until the water temperature drops below 60 degrees.

Once winter sets in, fishing gets tough. Jig-and-pigs hopped along the bottom of points in the creek channels in Runaway Creek and French Creek will result in a few fish. The best route to go, however, is flipping the deep fallen timber in the main river.

Yellow Creek to Candy Landing

Fall fishing is usually on-again, off-again in this section of Demopolis Lake. The best bet is to watch for schooling activity in the mouth and on the secondary points in Big Prairie Creek, but that isn't much to depend on. Those fish will normally be there, however, even if they can't be seen. Worms, swim baits and Carolina-rigged lizards work well on the points, but keep a Rat-L-Trap or twitch bait tied on another rod to be ready for any visible schooling activity.

The mouth or secondary points of any opening off the main river channel is a threat to erupt with schooling bass.

After the water cools somewhat, the mouth and banks of Power Plant Slough will hold some bass. Rat-L-Traps and medium-running crankbaits will work there.

The coldest part of the winter means jig-and-pig fishing again, with timber on the main river and deep points in Big Prairie holding the better fish.

Other types of fishing

Crappie fishing

The amount of standing and blow-down timber in the wide creeks in Demopolis Lake make it a natural for crappie. Although this lake doesn't produce the numbers of 3-pound-plus crappie some waters in this end of the state like Miller's Ferry produce, it can more than compete numbers-wise. This lake is a tremendous crappie-producer year round.

If this lake has a crappie fishing secret, it's the tremendous numbers of big fish it produces in November, December and January in the Demopolis Lock and Dam tailrace. Large numbers of large crappie move into the tailrace area during these months and position themselves in the deep structure approximately 300 yards below the dam. An angler should use a depthfinder to locate this structure, then anchor right on top of it. The structure includes rocks, trees and some steel beams.

This is a pattern that works best when there's no current or little current. Using ultra-light tackle and 6-pound test line, drop live minnows hooked through the lips into the structure. This often produces numbers of crappie in the 2-pound range.

Fishing up in the lake isn't bad either. During the spring, numbers of crappie move into French Creek and Powerline Slough, as well as other similar areas. The fish will hold on stickups and stumps in 5 or 6 feet of water until the surface temperature reaches 68 to 70 degrees, then will go to the banks to spawn.

While off the banks, jigs are an excellent way to take numbers of fish. Once the crappie are on the banks, a long, fiberglass pole and minnows are the way to go. The pole allows you to drop the bait right in the crappies' faces when they're holding tight to the grass.

Following the spawn, the crappie return briefly to the areas they held in before the spawn, then head to the deep creek and river channels. A double-hook rig with minnows fished in French Creek, Taylor Lake and Powerline Slough is an excellent way to take these fish.

In September and October, the cooling water temperature sends the crappie shallow again. Tree tops, stumps and shallow humps in 4 to 6 feet of water will hold fish. Live minnows or jigs can be used.

Demopolis Lake's best fishing spots

1. Mouth of McConico Creek. Excellent spring fishing, good other times of the year. Full of standing timber in deep water.
2. Point in McConnico Creek. Excellent year-round fishing.
3. Underwater island in McConnico Creek. Excellent spring and summer fishing.
4. Small sloughs on north side of the bank. Excellent spring fishing.
5. Birdine Slough. Shallow slough offers excellent spring fishing.
6. Birdine Creek. Small creel muddies easily, but good late-spring fishing on structure on the banks.
7. Small, structure-filled slough. Excellent spring fishing. Good summer fishing in the mouth.
8. Small slough. Excellent spring fishing. Good summer fishing in the mouth. This slough and No. 7 are remotely located, so they attract large numbers of fish corning out of the channel in spring and late summer.
9. Rattlesnake Bend Cut-Off south intersection. Excellent worm fishing in early morning hours in spring and summer.
10. Rattlesnake Bend Cut-Off north intersection. Same as above.
11. Taylor Lake. Stays clear in spring. Good mouth for worm fishing. Islands, standing timber and points.
12. Logjam Slough. Cypress trees in the back corners of both forks. Deep slough holds bass year round. Several good points.
13. Unnamed slough. Excellent spring spawning area. Normally impossible to get into during summer.
14. Mouth of Spidle Lake. Good fishing most of the year.
15. Back of Spidle Lake. Structure and excellent spring spawning area.
16. Deep, south bank of Spidle Lake. Stays shady in summer. Good structure.
17. Belmont Landing Slough. Full of cypress trees. Excellent spring fishing.
18. Culpepper Slough. Good worm fishing in mouth during summer. Standing timber in back. A few good points. Good spring fishing.
19. Demoplis Yacht Basin Slough. Wide area directly off river. Boat slips are good for crankbait fishing year round.
20. Island in Yacht Basin. Good topwater fishing in spring.
21. Webb's Bend. East side is lime-rock banks good

for crankbaiting in summer.

22. Whitfield Canal. Points, boat ramps and deep water. Some good fishing year round. Good area for worm fishing.

23. Shallow water in back of Whitfield Canal. Grass beds. Good spring fishing.

24. Grain Elevator Slough. Same type of fishing as Whitfield Canal.

25. Lock and Dam Slough. North of lock and dam. Very shallow. Excellent spring spawning area. Good buzzbait area.

26. Foscue Creek. Deep slough with good year-round fishing. Points, shallow water and islands.

27. Dobb's Swamp. Shallow water with standing timber. Topwater fishing. Excellent spawning area.

28. Runaway Branch. Excellent year-round fishing. Deep water. Standing timber, boat docks, points and cypress trees. Clear slough most of the year.

29. Kelly's Slough. Long, straight slough. Standing timber. Worms, spinnerbaits and buzzbaits spring and early summer.

30. Unnamed slough. Connects with Demopolis Yacht Basin Slough in high water conditions. Standing timber. Good worm fishing in mouth.

31. Citadel Cement Slough. Both deep and shallow water. Good year-round fishing.

32. Mouth of French Creek. Good year-round fishing.

33. French Creek. Largest water inlet on Black Warrior side. Good standing timber. Deep water throughout most of creek.

34. Island in French Creek. Good spring fishing.

35. Roadbed in mouth of French Creek.

36. Powerline Slough. Very narrow opening. Standing timber and cypress trees. Good points. Shallow and deep water. Good year-round fishing.

37. Backbone Creek. Narrow slough with cypress trees in back half. Clear slough. Good year-round fishing.

The Ultimate Guide to Alabama Fishing

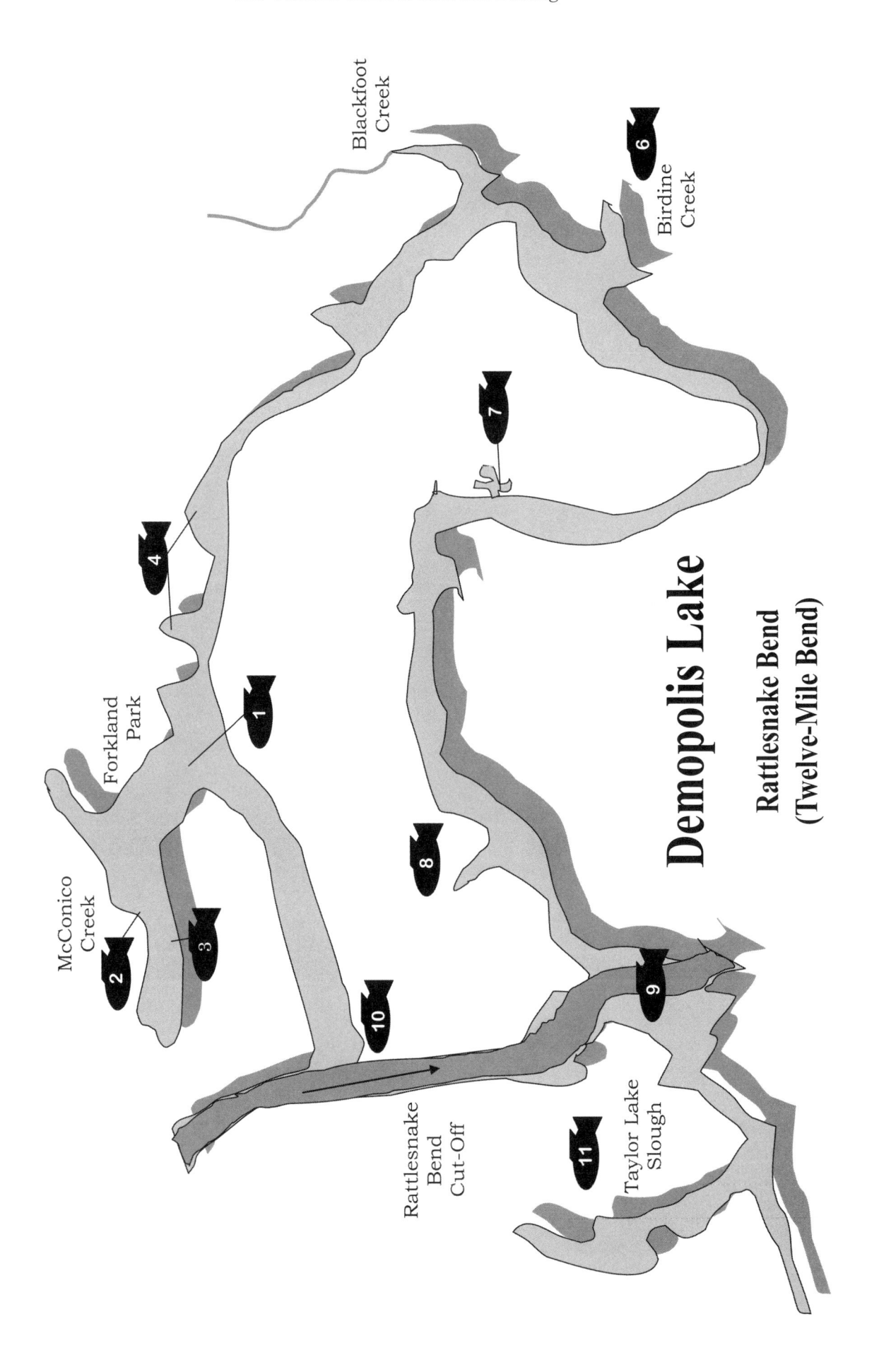

Demopolis Lake

**Rattlesnake Bend
(Twelve-Mile Bend)**

246

Mike Bolton

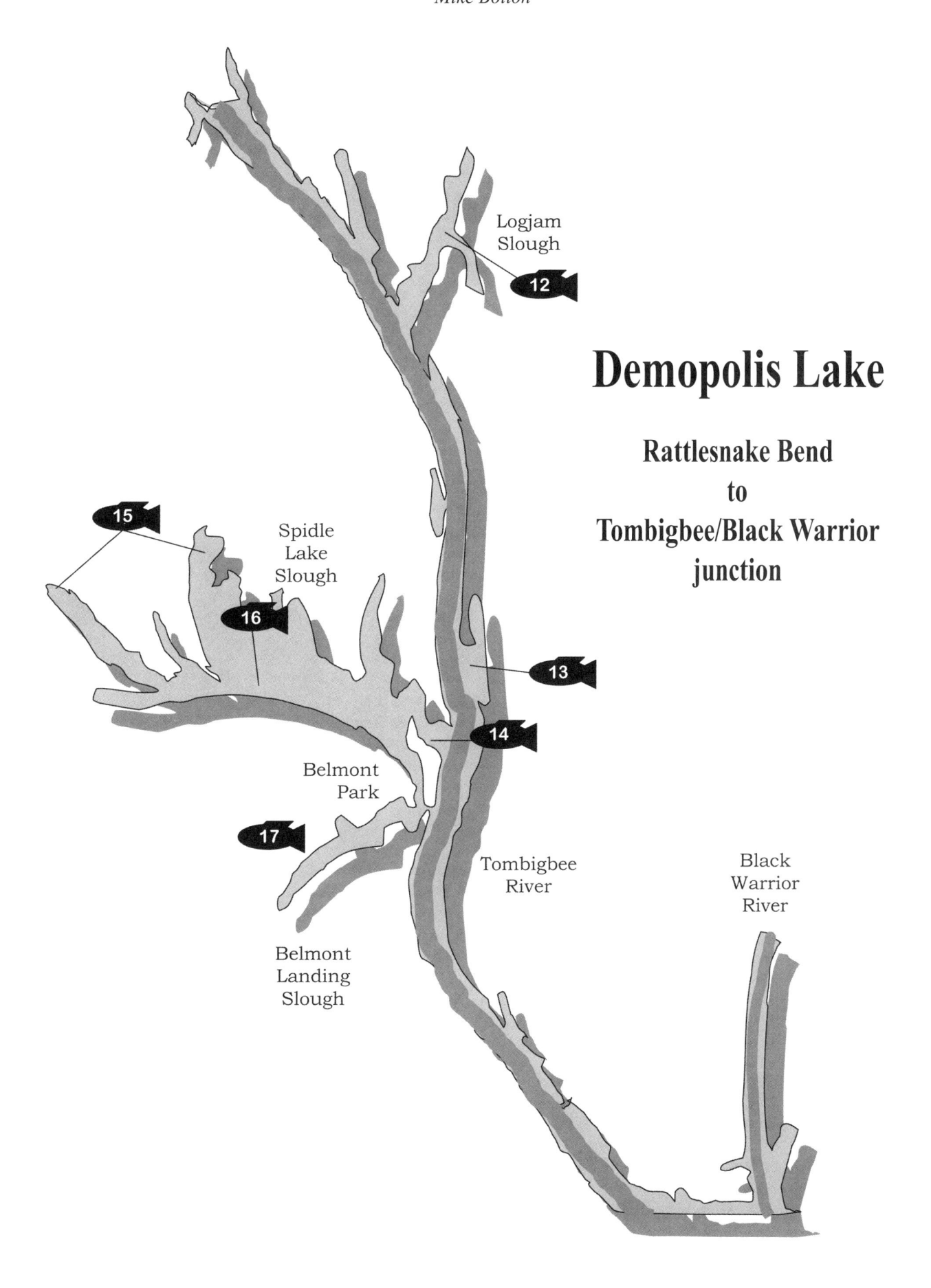

Demopolis Lake

Rattlesnake Bend to Tombigbee/Black Warrior junction

Logjam Slough

Spidle Lake Slough

Belmont Park

Tombigbee River

Black Warrior River

Belmont Landing Slough

The Ultimate Guide to Alabama Fishing

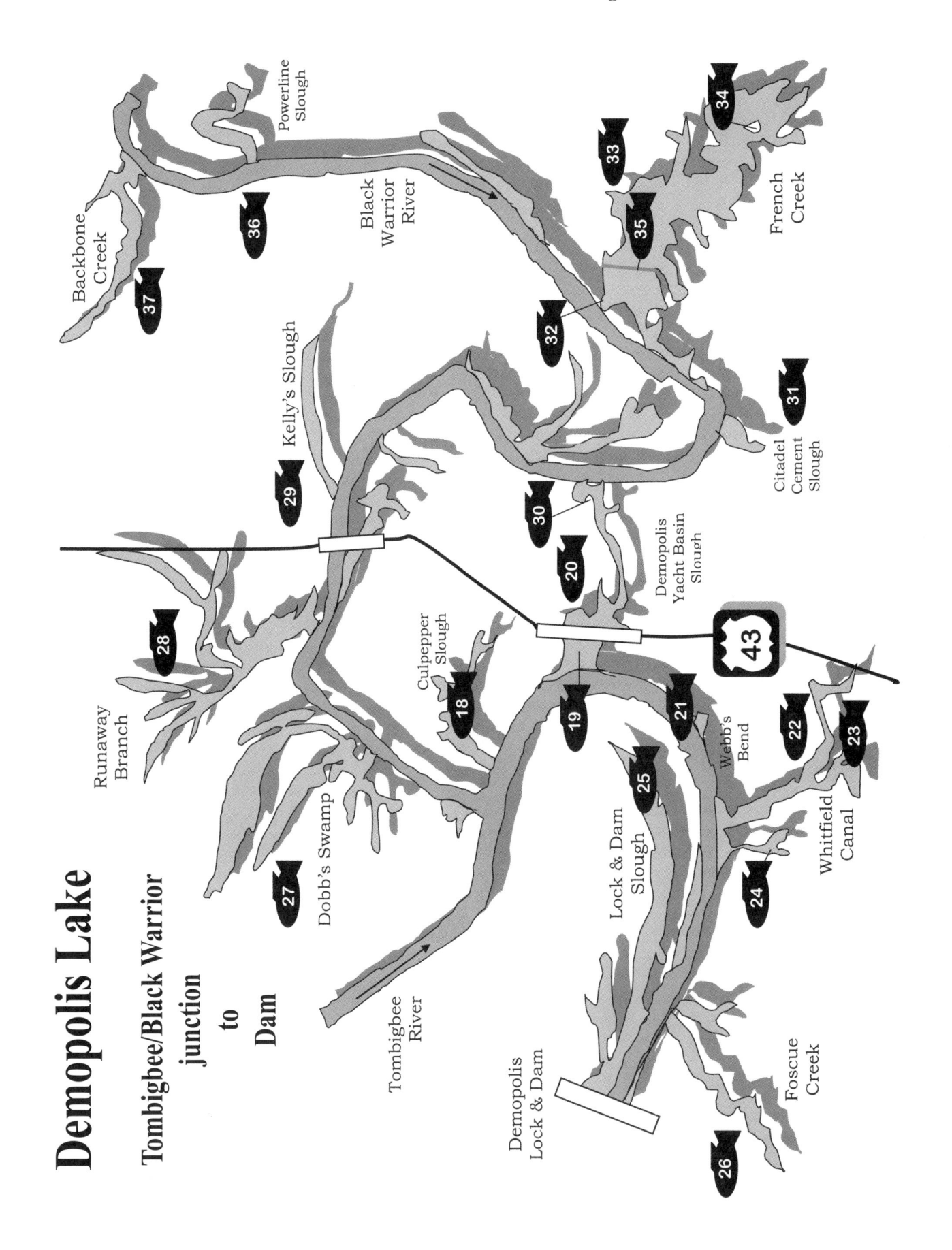

Demopolis Lake

Tombigbee/Black Warrior junction to Dam

248

Mike Bolton

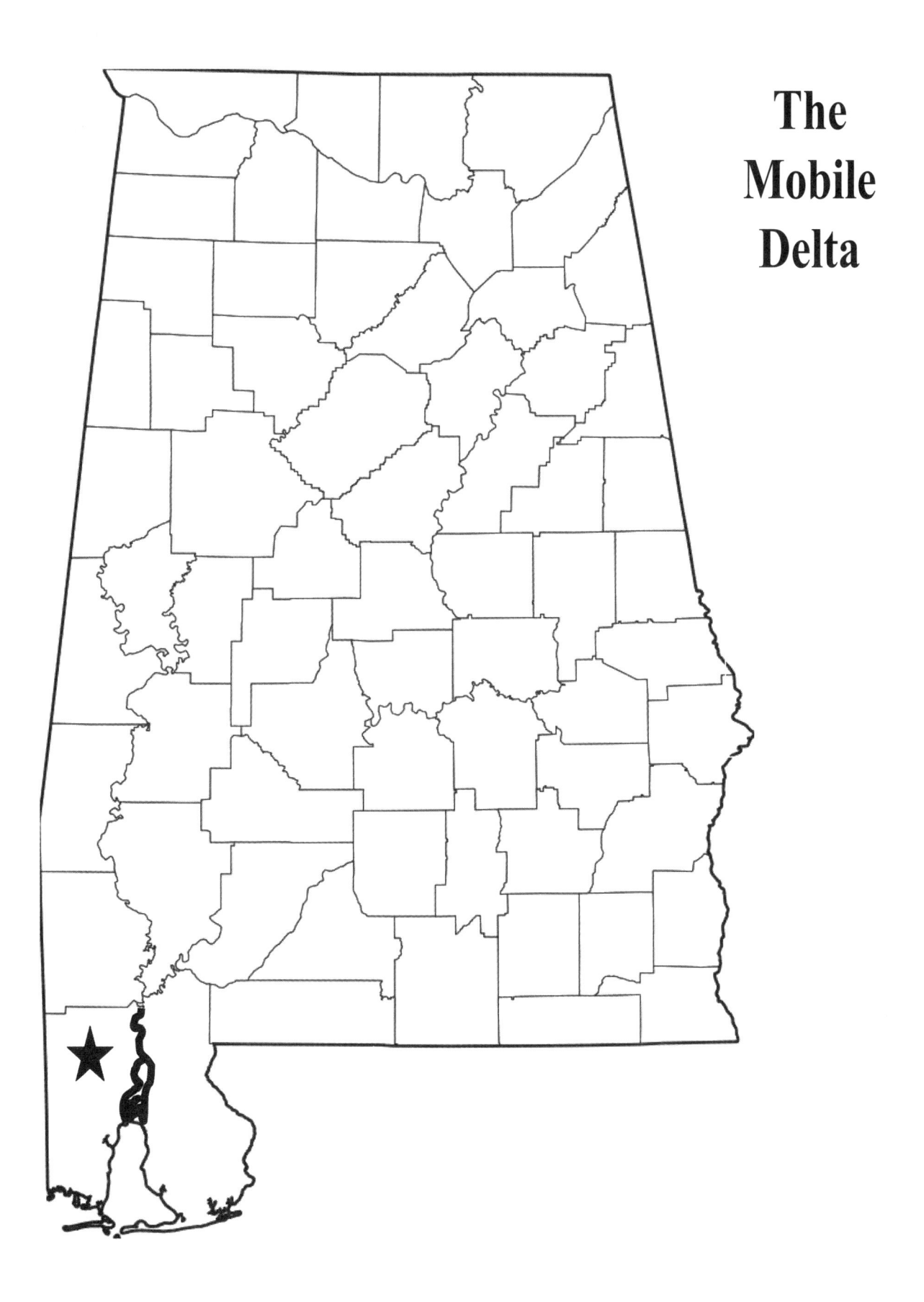

The Mobile Delta

Mobile Delta

To the angler accustomed to fishing Alabama's reservoirs, the Mobile Delta can be a perplexing and daunting puzzle. Not only does the fisherman in the Mobile Delta have to worry about the myriad of factors that affect reservoir fishing—factors such as fronts; water color and water temperature—he must also worry such things as ebb tides, salinity, wind direction and salt lines.

The angler adept at fishing the Mobile Delta is a well-rounded fisherman capable of fishing anywhere.

The Delta is made up of several major rivers and numerous small creeks and shallow bays which offer widely- and quickly-changing fishing conditions.

For instance: The shallow bays offer excellent spring fishing, but the bays closest to the Mobile River muddy quickly and often become unfishable. It's also possible for spring fishing to be terrific, then suddenly have a north wind kick up and literally push the water out of the bay. To make matters worse, these same shallow bays and the creeks leading to them are often choked off with weeds as the spring fishing gets good. These same areas offer some good fall fishing, but by winter, many are too shallow to fish.

And then there's the perplexing problem of the fish themselves. During certain times of the year, an angler doesn't merely go bass fishing. A fish on the end of the line may be anything from a bass to a flounder. If you're looking for a variety of fish, this is your best bet in Alabama. Largemouth bass, striped bass, crappie, bream and several species of catfish are abundant here as they are in most Alabama waters, but a cast at the right time of the year in the Mobile Delta will also produce saltwater fish. Speckled trout, redfish, flounder, sheepshead, croaker, drum and jack crevalle are common at certain times.

Spring bass fishing

Below I-65

By the third week of February, water temperatures in the Mobile Delta usually have risen to the point that bass are thinking about the spawn. The bass leave the wide, deep waters of the main rivers where they spent the winter and head for the shallow bays.

These shallow weed-filled waters heat up quicker than surrounding water and it here that the bass will lay their eggs and raise their young.

There are several key areas to concentrate on during the spring months. Chocolata Bay (also known as Chacalloochee Bay) is one of the first bays to hold spring fish and the bay that holds bass the longest. The area near the grass beds in the back offers excellent spring top water fishing. Conway Creek, which is near Chocolata Bay, is a key area when a north wind is forcing the water out of the bay. Delvan Bay is a wide bay with a somewhat narrow mouth that offers good spring fishing. Grand Bay and Chukfee Bay are also key spring areas.

Little Bateau is a shallow bay that is only a foot deep and is rarely fished, but a flood tide causes fish to stack up in the mouth and it is worth a cast or two when those conditions are present.

Farther north, The Basin Negro, better known as Negro Lake, and Big Lizard Creek are also worth taking a look at. Mifflin Lake, which has its mouth below 1-65 but extends well north of the interstate, is another key area.

Keep in mind that almost every raindrop that isn't soaked into the ground when it falls in this state ends up in the Mobile Delta. Much of the Tennessee River, the Tombigbee, the Warrior, the Black Warrior, the Coosa, the Tallapoosa and the Alabama River eventually flow into the Mobile Delta. Thus, spring rains all over the state have an effect on fishing here. Finding water not too muddy to fish is a key to good spring fishing.

It's not unusual to find a shallow bay full of muddy water in its mouth, but don't concede the fishing there without further investigative work. It is often possible to locate a distinct stain line separating muddy water from the clear somewhere in the bay. The northern half of these bays often clear up quickly after huge influxes of muddy water. The stain lines often offer tremendous fishing with most strikes coming by casting into the clear water and bringing it to the stain.

The shallow bays have a variety of grasses including Coontail and ribbon grass. All will concentrate fish. Another key form of structure to concentrate on in the spring is the hundreds of duck blinds scattered throughout the Delta. When these duck blinds are located in shallow water, bass will almost always use them to spawn and they are hot both pre- and post-spawn.

A wide array of lures will take spring bass from the Delta, none really much better than the other. A 1/4-ounce or 1/2-ounce spinnerbait, either white or chartreuse depending on the water clarity, is probably the most popular bait. It should be equipped with a single, small nickel blade and a pork frog trailer the same color as the skirt.

The pork frog trailer is important because it keeps the spinnerbait buoyant and allows it to slip over the grass.

Topwater baits also are deadly. Buzzbaits with a bead placed in front of the propeller to discourage the gathering of grass is probably the most fun bait to use. It also covers a lot of ground quickly. Before the grass gets too thick, stick baits work well.

Another popular bait is the swimming jig. It is a deadly spring bait. Bass larger than 5 pounds are rare in the Delta, so an angler can survive with 12-pound test line.

The Delta's largest bass are often caught in spring on live shiners fished around the duck blinds. Chocolata Bay usually gives up the Delta's largest spring bass. On occasions when a good, hard, north wind pushes the water and fish out of the bays in the spring, th deep creeks near the bays become the best locations to fish. The water will be churned up and spinnerbaits will be useless, but plastic worms rigged to fish on bottom will produce numbers of good fish. The small creeks near Chocolata Bay are especially deadly when the north wind pushes the water out. One of the key areas is the small passes in the area. Bass stack up heavily in these passes at this time and its possible to catch numbers of good fish.

Above I-65

Fishermen who bass fish in the Mobile Delta above Interstate 65 can witness a wide spectrum of circumstances during the spring. A south wind and a high tide results in high water that jumps the banks and fills the nearby woods. The very next day, the angler may find a north wind and a falling tide that pulls the water out as if someone had pulled the bathtub stopper. Fortunately, low tides and north winds are not common when the fishing gets its best.

On the occasions that the rivers and creeks jump their banks and flood the timber, bass will follow and take hold in the new-found structure. They will locate at the base of trees and stumps as far back as the water goes. They will stay there until the water recedes.

When an angler finds this situation, a spinnerbait fished in the thick stuff is difficult to beat. A 1/2-ounce or 3/8-ounce chartreuse/white spinnerbait with a single copper blade is best when the water has good stain. A white spinnerbait with white blades or a chartreuse bait with chartreuse blades is needed when the water muddies.

The spinnerbait should be cast as far up in the trees as the water goes and slow-rolled through the stickups and trees, bumping every piece of structure possible.

There are several key areas which an angler can concentrate on during the spring months. Mifflin Lake is the best and all fishing should begin there. The I-65 bridge pilings in Mifflin offer excellent structure, especially pre- and post-spawn. Bottle Creek is another key spawning area, as is Tensaw Lake. All three locations offer Mobile Delta spring bass fishing at its best.

Summer Bass Fishing

Below I-65

Unfortunately for the spring angler, just about the time the shallow-water bays hits 65 degrees and it gets time for great fishing, the grass is often so thick it is impossible to fish the area.

The bass are on their post-spawn cycle by then and their instincts tell them to clear the area while they can still get out. These bass end up in the rivers.

This often happens by the end of May or the first of June. Mobile Delta bass are forced into a summer pattern earlier than the bass in many other areas in the state.

The rivers in the Delta lack blowdown timber and other similar structure which typically holds summer bass elsewhere in the state, but these rivers have their own brand of fish-concentrating structure, namely barge pilings and feeder creeks. Barge pilings, which are large creosoted poles, are abundant on the rivers. They are usually found in 15 to 18 feet of water all up and down the rivers in the Delta, and all are a threat to hold bass. Chrome-colored Rat-L-Traps with blue or black backs, and crawfish-colored lipped crankbaits are the key baits in fishing these pilings.

All along these rivers you will find small feeder "creeks" which concentrate bass. These "creeks" may be only a foot wide in some places and are nothing more than a trickle of water, but they supply the river with a steady supply of seed shrimp, a tiny shrimp that is a major part of the diet of summer Delta bass. These feeder creeks are easily located by looking for a break in the heavy cane that lines the banks.

Seed shrimp are available at many bait shops in the Delta, or anglers may catch their own by using a fine mesh net to shake underneath tree roots on washed-out banks. Two or three of the small shrimp laced on a fine wire hook and fished under a small float and fished near the creek mouths will produce fish.

Quarter-size crab are another of the Delta bass' main foods during the summer months. Anything with crab coloring can evoke a strike. A greenish/brown crankbait with a light-colored belly, or a motor oil worm with chartreuse tail, will also catch bass.

Traditionally, the farther you go up in the Delta, the better the summer fishing. Both the Tensaw and Mobile Rivers offer good summer fishing near I-65. There's also some good fishing in Negro Lake, Mifflin Lake and any of the creeks or cuts that hold more than nine feet of water.

There's limited summer bass fishing near Interstate 10 and the bay, but what is there is excellent. A deep channel was cut through John's Bend years ago to allow barges to transport materials to build the interstate bridge and that deep water is excellent in June and July. Pay special attention to the hook in the bay and any grass that may be there.

Also, the mouth of the Blakely River where it dumps into the bay will concentrate bass in August when they gather to slaughter the shrimp.

Above I-65

Summer fishing isn't great in the Mobile Delta, but nowhere else in Alabama is summer fishing great, either.

Tides are the key to the best fishing above I-65 in the summer. Bass gather on points when in-going or out-going tides are moving water, much like the do on reservoirs when water is being pulled.

Plastic worms become the Delta's top bait during the hot summer months. A Pumpkin-colored Carolina-rigged worm fished around blow-downs and pilings on the rivers and lakes is tough to beat.

When an angler knows the location of deep-water underwater structure, flipping can be deadly. A 1/4-ounce jig-and-pig with a piece of plastic worm tail is the best bait. Orange and black colors are strong in summer.

The many lakes in the area can offer good fishing in the daylight and sunset hours, but they close down completely when the sun clears the trees. Topwater baits and worms both work well. Mifflin Lake and Tensaw Lake are key areas for this type of fishing.

The best summer fishing is definitely in the rivers, however. Visible, deep structure in the Middle River, Tensaw River and Mobile River should draw most of your attention: The north winds that occasionally hit the Delta during the summer shouldn't be a signal to not go fishing.

A north wind that pushes the water out the rivers and creeks force bass into the mouth of creeks and this is a good time to go fishing. The bass school heavily when this happens and this is one of your best chances to catch numbers of bass.

Cool and cold weather fishing

Below I-65

The Mobile Delta gets its least amount of rainfall in October and this enables the saltwater to make a surge into the rivers and creeks far up in the Delta. This is beneficial to the fisherman because this influx of saltwater normally kills the thick grass in the shallow bays.

When rains finally come, it is not unusual to see this dead grass floating away in large portions.

A bass angler must understand that the water in these creeks and bays literally separates, with the saltwater moving to the bottom and the freshwater to the top. Bass have no difficulty living in this brackish water, despite its salt content.

The bass will gang up in the creeks near these shallow bays before they go back into a spring-like pattern. They stage in this deeper water much like they did in spring and in the same areas. Shrimp—which bass love—make a run as far as the saltwater goes in the creeks and the largemouth follow and have no trouble finding a quality meal.

Bass will also gather in the rivers near the mouths of the feeder creeks at this time. These areas are likely to produce redfish at this time. Both white and chartreuse spinnerbaits and shallow-running lipped crankbaits in lighter colors are deadly during this time. Creek bends or stick ups will hold bass, as well as redfish, speckled trout and an occasional sheepshead. Those who fish plastic worms in these areas often catch flounder, too.

Colder weather naturally sends bass back into deeper water, and the deeper creeks that held bass in the post-spawn pattern in the spring will hold winter bass. Other key areas to concentrate on are the bridge pilings under 1-10, U.S. 90 and 1-65. Jigging spoons, grubs and plastic worms will catch bass in these areas.

Many of the better bass anglers also live bait fish for bass during the winter months. Shrimp are productive when shrimp are present, but when the cold drives the shrimp out, most live bait anglers switch to shiners.

Many of the better anglers ride the river on low tide situations and take notes of structure that is showing and then fish those areas when the water is back up. These areas are particularly productive during winter months and can be beneficial year round.

Above I-65

Water levels are usually low during the cool and cold-weather months, so deep banks with cover on both the lakes and rivers hold fall and winter bass. Sloping banks, even if they have cover, rarely hold fish during this time.

The better fall and winter fishing is below 1-65.

Saltwater fishing

Summer

Above I-65

The shortage of rainfall that this area traditionally sees during summer months allows heavy incoming tides to push saltwater up into the Delta. The lack of rainfall means there is no freshwater to push the saltwater back out. The brackish water eventually creeps well up into the Delta.

Redfish are usually the first saltwater fish to make the move into the Delta, followed by flounder, sheepshead and croaker. Bass anglers fishing white spinnerbaits are usually the first to discover the redfish have arrived.

Anglers wishing to catch redfish during the summer months should concentrate on areas closest to the mouth of the bay since this water will have the most salt content.

Any of the small islands near the mouth of the bay is good with the north point usually holding the fish for some reason. Little Sand Island is probably the best of all these islands. Pinto Pass is another key area. Many a tourist visiting the USS Alabama has leaned over the rail and watched anglers load the boat with redfish and flounder in Pinto Pass.

The submerged pilings east of Little Sand Island also offer good fishing. The piers all along the east bank of Mobile Bay offer good redfish, speckled trout and flounder during the summer.

One of first areas above 1-65 to find saltwater fish in the summer months is the mouth of Grand Bay on the Spanish River. The point of Polecat Bay in the Spanish River is another excellent spot.

Underwater stumps near the cypress trees across from the mouth of Bay Grass hold a lot of redfish during this time. The point of the Tensaw River where it meets Delvan Bay is another key area for redfish and flounder.

The pilings (called the "Old Battery") at the split of the Apalachee and Blakely River hold lots of sheepshead, croaker, and flounder. This "V"-shaped series of pilings once extended across both rivers to block boat traffic during the Civil War, some locals claim.

The upper point of the Apalachee and Tensaw rivers is full of sandbars and offers good fishing for redfish on live or dead shrimp. Any sandy point is usually good because one side usually drops off into the river and the other leads to three to four feet of water. These sandy points are excellent on a falling tide.

For those wishing to strictly fish for flounder, the Blakely River where it passes under the 1-10 bridge is strong. The interstate bridge pilings 100 yards each way of the river hold lots, especially in June. Natural baits are popular, with live or dead shrimp the best for redfish and live shrimp, bull minnows or shiners good for flounder.

The "fishfinder rig" is the best tackle setup. An 18-inch length of 20-pound mono should be topped with a swivel, followed by a barrel weight on the main line. This should be finished with a No. 6 treble hook. This rig allows a fish to pick up the bait and run with it while pulling the mono through the weight. The treble hook allows for a good hook-set. If fishing strictly for flounder, opt for a 1/0 single hook rather than a treble hook.

Cool and cold weather fishing

There's no question that once the cool breezes of fall arrive, the saltwater fishing in the Mobile Delta heats up. The redfish and flounder will move as far up in the Delta as the shrimp and saltwater go. Catches of reds and flounder near the 1-65 bridge are not uncommon. Speckled trout will not normally move that far.

The small island in Grand Bay and the small island nearby in the Raft River holds both reds and flounder, as do the submerged pilings in the Mobile River. The dock pilings just above Grand Bay where the Spanish River and Mobile River meet is another key area. The angler who throws a Mann's pink and white Stingray grub can catch reds, flounder, bass and an occasional sheepshead.

When September arrives, the speckled trout will be at the head of the bay just south of the 1-10 bridge. They will be staging in the cuts and channels feeding heavily on shrimp. The terns and seagulls will give the specks' location away as the dive for the shrimp injured by the attacking specks. This will usually last a month. Unfortunately, these easy-to-locate specks are usually small.

The middle of November usually signals the start of speckled trout season up in the Delta. This is when the quality specks move in. The mouth of Delvan Bay at the point of the Tensaw River is usually one of the first spots where the specks gather in numbers. Any sandy point below Chuckfee Bay is a threat to hold specks then. Cocahoe Grubs—chartreuse/black back, or smoke color—are normally the best bait for specks at this time, but unseasonably cold water will require white or pink Stingray grubs. If the specks are finicky, a small curly tail grub like you would use for crappie fishing may entice them to bite. Either white or pink colors can be used.

A 1/4-ounce lead head is needed when the fish are off the bottom and a 3/8-ounce head is needed when the fish are on bottom. Most of the better anglers opt for 10-pound test line.

When the water is warm, a steady spinnerbait-like retrieve can be used. When the water gets colder, a jigging retrieve that allows the bait to hit bottom is needed.

Another pattern that works well is fishing a live shrimp under a popping cork. The sound of the float being popped in the water often draws the attention of fish. Once again, terns and seagulls will give away the specks' location.

One of the best-kept secrets of the better anglers is paying close attention to the large pelicans that often accompany these birds. The pelicans aren't as quick and agile as terns and seagulls, so they'll actually swim along with the shrimp that the specks are feeding on rather than trying to fly and dive at the bait.

When the weather really turns cold, usually in mid-December or early January, rains have pushed much of the saltwater back out of the Delta. This forces the saltwater species back to the bay's head. The Tensaw River, where it is 18 feet deep, and the Apalachee where it's 17 feet deep just below U.S. 90, are key areas at this time. Grubs or live shrimp fished on the bottom will catch these fish.

Mobile Delta's best fishing spots

1. Lake Forest Yacht Club. D'Olive Bay good for speckled trout, flounder and redfish September through December.
2. I-10 bridge pilings. Blakely River. Flounder fishing in June, specks September, October and November.
3. I-10 bridge pilings in Ducker Bay. Bull minnows for flounder in May and June. Dead minnows for sheepshead.
4. John's Bend. Good spring bass fishing in shallow water, good summer fishing in deep cut in summer.
5. Devil's Ditch. Deep pass. Good for specks in fall, flounder on the points.
6. Seawall across from Mack's Bait and Tackle. Redfish in fall. Dead shrimp on bottom.
7. Pete's Island. Flounder and sheepshead on dropoffs and submerged pilings.
8. Underwater pilings south of Pinto Pass. Specks and reds.
9. Oyster shell reef. Fall fishing for specks, reds and flounder, good in summer when there is little rain.
10. Pinto Pass. Live or dead shrimp in summer for big specks, good in winter on artificial baits.
11. The Cut. Redfish in small channel.
12. Point of Polecat Bay and Delvan Bay. Fall fishing for reds and specks.
13. Point of Tensaw and Delvan good in August-December for reds, specks and flounder.
14. Cypress trees and stumps. Fall and winter for redfish.
15. Delvan Bay. Goodin spring for bass. In May, good for bream around duck blinds.
16. Bay Grass. Spring bass area.
17. Chocolata Bay. Good end of February for spawning bass.
18. Mouth of Conway Creek. Good when north wind pushes water out of Chocolata Bay.
19. Big Bateau. Lots of grass. Shallow. Fair in March and April for bass. Good April and May for bream around banks and duck blinds.
20. Deep water creeks around Little Bateau. Good bream fishing in spring.
21. The Old Battery. Specks, reds, flounder around pilings.
22. Sardine Pass, also known as Game Warden's Ditch. Fair spring fishing.
23. Bay Minette Basin. Good spring bass fishing. Second best spit next to Chocolata.
24. Bay Minette Creek. Submerged timber. Lots of white perch.
25. Mouth of Mud hole Creek. Big sandbar on north side. Speckled trout and redfish in the fall.
26. Tensaw-Apalachee fork. Good sandbars. Specks. Very few reds. Live shrimp under popping corks in fall.
27. Crab Creek. Good worm fishing for bass around blowdowns in spring.
28. Raft River. North points of islands good in fall.
29. Grand Bay and Spanish River point. Specks and reds in fall.
30. Grand Bay. Bass fishing in spring before grass takes it over in May. Also good in fall.
31. Visible pilings where Spanish River and Mobile River meet.
32. Feeder creeks good in fall for bass. Shallow with lots of blowdowns.
33. Chuckfee Bay. March and April for bass. Good bream fishing around duck blinds.
34. Good feeder creeks for spring bass fishing.
35. Oak Bayou. Good fall bass fishing.
36. Gravine Island. Reds, specks and flounder fishing around sandbars.
37. Negro Lake. Good spring fishing for bass.
38. Ship Canal. Man-made cut. Good point for bass fishing.
39. Good bass fishing point.
40. Good stretch of bank on Tensaw River. Cypress stumps. Good in fall for bass, specks and reds.
41. Good bass fishing in mouth of cut.
42. Little Briar Creek. Full of weeds. Good spring bass fishing.
43. Big Briar Creek. Lots of weeds. Good spring fishing for bass.
44. North point of Twelve-Mile Island. Good fall fishing for reds.
45. Pilings under railroad trestle. Fair bass fishing year round. Some reds in fall.
46. Big Bayou Canot. Some logjams. Some bass.
47. Dead Lake. Good water depth. Bream and crappie.
48. Smith Bayou. Good weeds. Good spring fishing with buzzbaits.
49. Owl Creek. Same as 48.
50. Squirrel Bayou. Same as 48.
51. Duck Bayou. Same as 48.
52. Six Bits Creek. Same as 48.
53. Bat Creek. Same as 48.
54. Bayou Avalon. Some milfoil and hydrilla.

55. Big Lizard Creek. Reeds on bank with sharp dropoffs. Dropoffs hold bass year round.

56. Not shown on map.

57. Not shown on map.

58. Little Lizard. Some good bass fishing in late spring and late summer.

59. Dennis Lake. Good spring bass fishing.

60. Mifflin Lake. Good spring bass fishing.

61. Bottle Creek. Fair bass fishing around runoffs.

62. The "T". Powerline crosses just past interstate. Deep and shallow water. Some good bass fishing year round.

63. Briar Lake. Limited spring bass fishing.

64. Tensaw Lake. Fair spring fishing when water is high. Island. North point good year-round.

65. Island. Clay point with lots of roots. Good year-round fishing.

66. Island. Clay points with lots of roots. Good year-round fishing.

67. Good stretch of river. Runoffs concentrate fish.

68. Watson Creek. A few spotted bass in mouth.

69. Douglas Lake. Fair bass fishing.

70. Chippewa Fingers good in high water in spring.

71. Good flippin' area. Lots of brush.

72. Grompau Branch. Good bass fishing in mouth when water is being pulled out.

73. Bayou Zeast. Fair in spring. No grass. Some cypress trees.

74. Dead Lake. Good year-round fishing for bass.

75. Good stretch in Stiggins Lake. Excellent in spring.

76. Bear Creek. Good summer area.

77. Sheppard Lake. Good spring fishing in back. Barges hold bass.

78. Cedar Creek. Fair spring fishing.

Mike Bolton

Mobile Delta

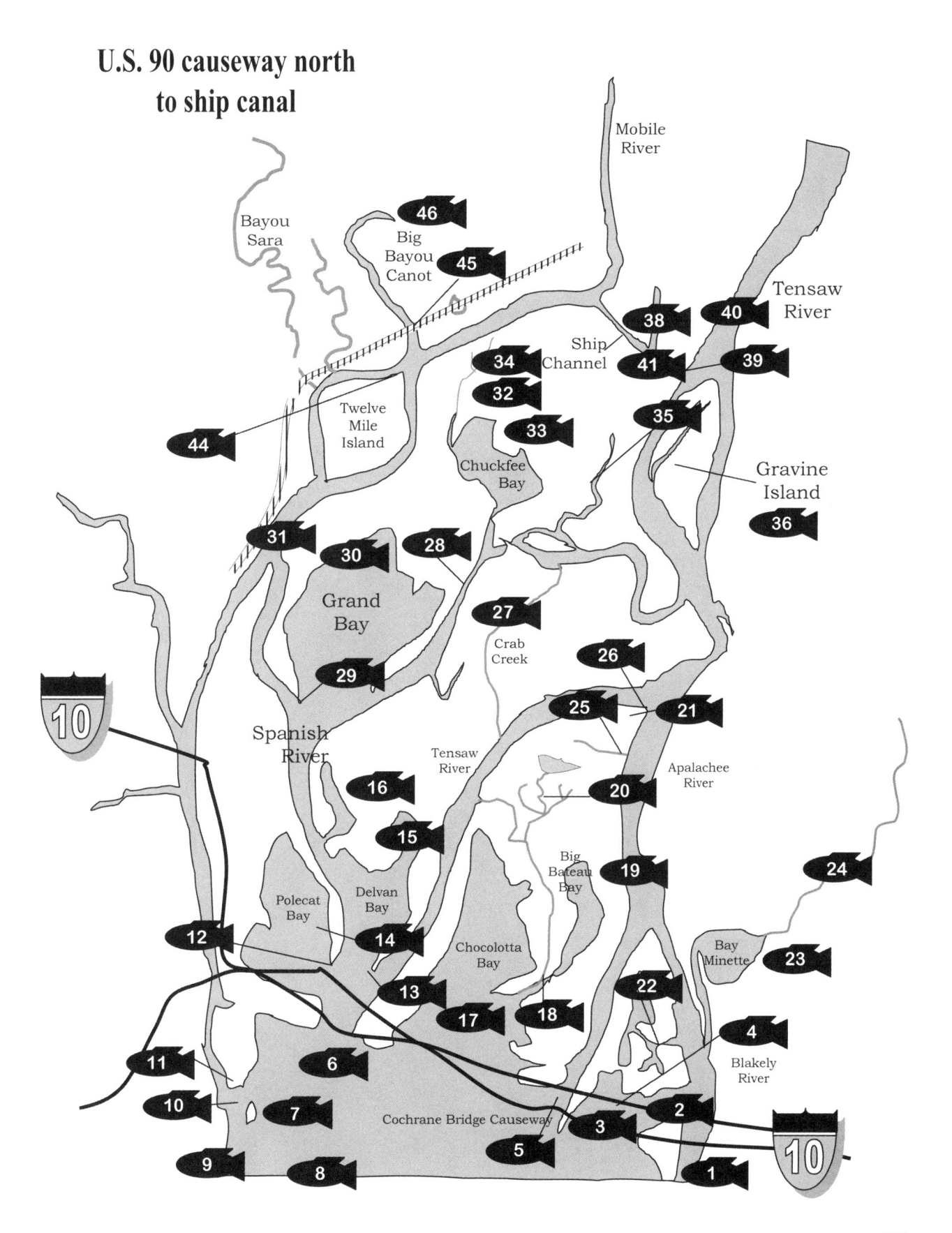

U.S. 90 causeway north to ship canal

The Ultimate Guide to Alabama Fishing

Mobile Delta

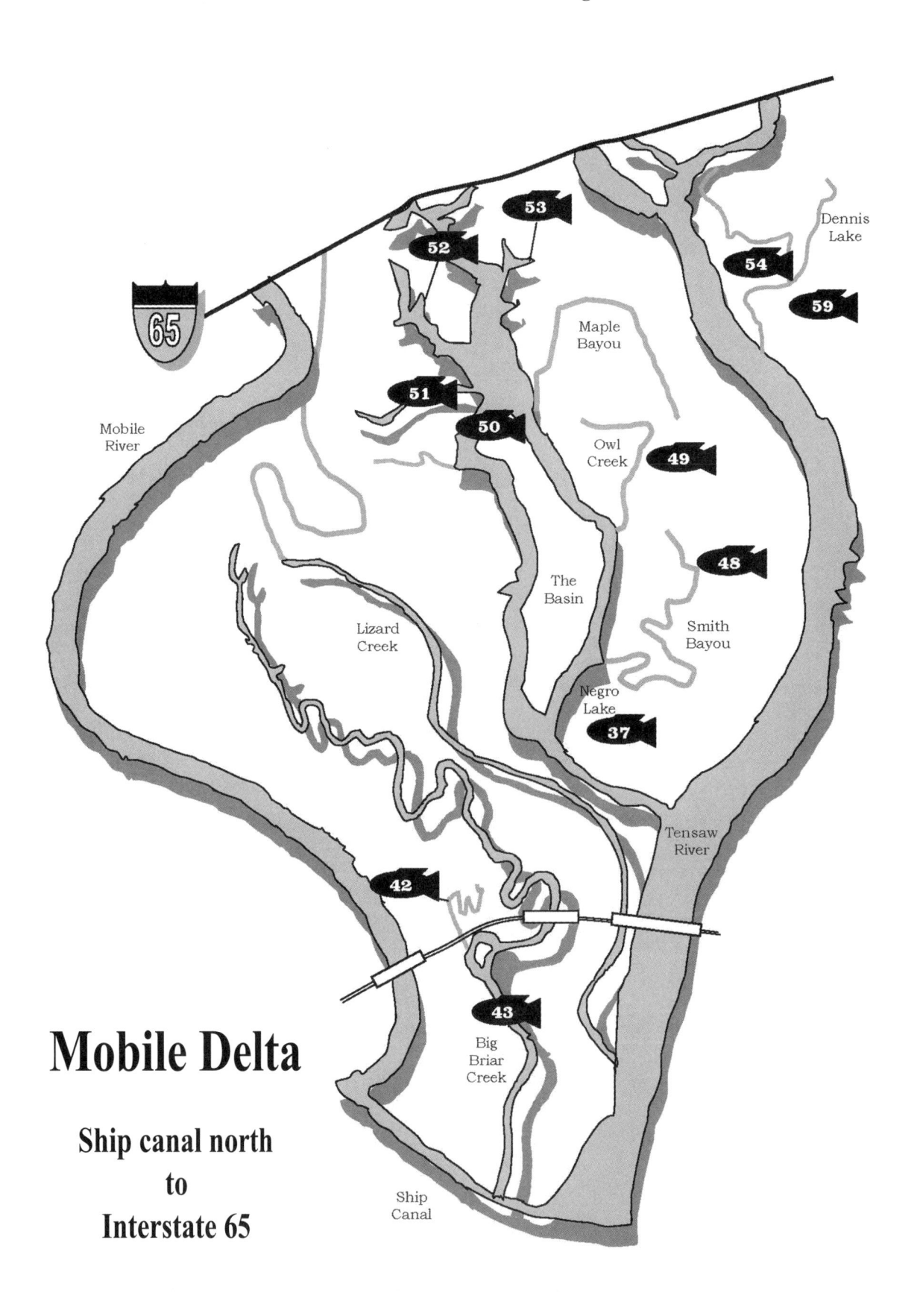

Ship canal north to Interstate 65

Mike Bolton

Mobile Delta

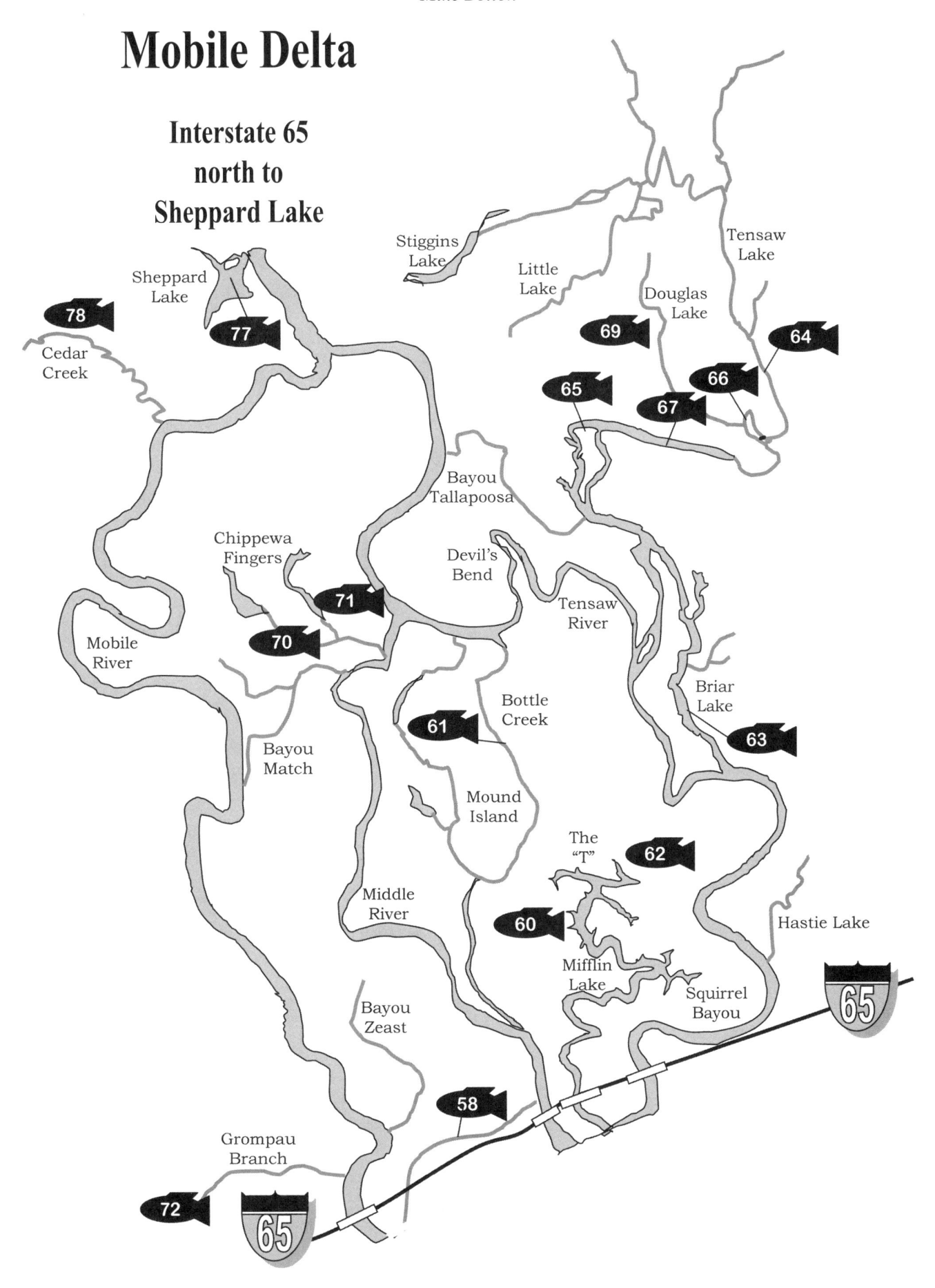

Interstate 65 north to Sheppard Lake

259

The Ultimate Guide to Alabama Fishing

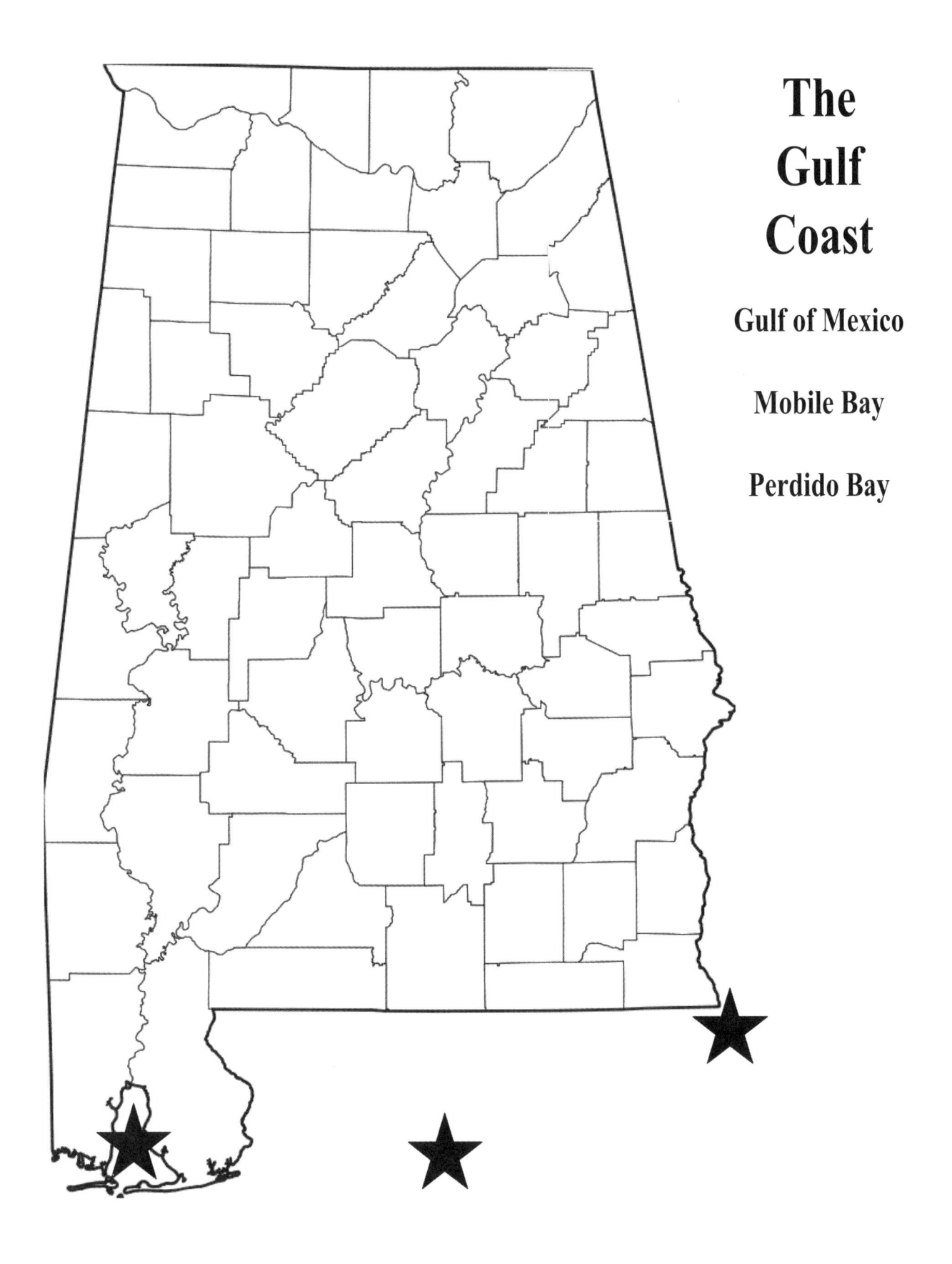

The Gulf Coast

Gulf of Mexico

Mobile Bay

Perdido Bay

The Gulf of Mexico

The gulf itself is a monstrous circular pond, bounded by Mexico's Yucatan Peninsula, the east Mexican, Texas, Louisiana, Mississippi, Alabama and Florida coasts, with an outlet into the Atlantic through the Caribbean Sea.

The Alabama portion stands out in several ways:

• The Loop Current, part of the Gulf Stream circling through the gulf, edges closest to shore along the Alabama coast. Riding that current is a rich variety of fishes; tropical species from the south that arrive in spring and leave in fall, and bottom species that range in and out as water temperatures rise and fall.

• The gulf s major geologic feature—the DeSoto Canyon—nudges closest to shore nearest the Alabama coast, beginning it slope about 30 miles due south of the Alabama-Florida state line. There the bottom drops from a depth of just over 100 feet in stair-steps of craggy cliffs. They are overgrown with marine plants and animals, home for every step in the marine food chain. The canyon is a well from which dozens of kinds of fishes ebb and flow, depending on the seasons, temperatures and currents.

• Mobile and Perdido bays play vital roles. Ringing these bays are wetlands, breeding grounds and massive incubators of sea life; safe places for fishes to be born and spend their juvenile lives.

The inland rivers push nutrients—nature's fish food—from the north into the bays and out into the gulf.

Through a million years the gulf rose and fell from ice age to ice age. At times its shoreline was many miles north with its waters covering at least part of several Southeastern states. At other times great forests grew where fishes now swim. Underwater limestone rock rises from the bottom in thin ridges, signs of one-time luxurious shallow, warm-water coral reefs that now are covered only by small soft corals, sea fans and a few deep-water hard corals. These sparse patches of rock rise out of the sand mostly in east-west strips, paralleling the coast, and are home to remarkable concentrations of sea life.

• Sea turtles, mackerel, Cobia, Amberjack and Spadefish patrol waters over the reefs for easy meals from the schools of baitfish that swarm over the rock.

• Some reefs are pocked with holes—some say

stump holes from prehistoric forests—and fractured into thousands of cracks and crevices. Lying in and schooling along the safety of these fissures are grouper, snapper, grunts, porgys and flounder. Circling the rock or hovering at the outer edge are the sharks and barracuda, ominous scavengers preying on the injured or careless, leaving the fittest to regenerate the life of the reef.

• Water temperature rises and falls with the seasons in greater extremes than any other location along the gulf rim. When summer water temperatures peak at 85 degrees or so, the mackerels, dolphin and billfish have surged northward into these waters and patrol the coast until winter cold fronts push them back toward their homes to the south.

• Bottom fishes move from the depths to the shallows, then back again as the water heats and cools.

• Great swarms of baitfish surge in and out of the bays with the heating and cooling of the water.

• As winter approaches this section of the gulf, it brings a succession of fast-moving weather fronts, packing strong winds that keep most sport-fishing boats in port through the cold months. It offers a brief respite to reef dwellers so heavily fished from spring through fall, allowing populations to rebuild as fish spawn or migrate into the structure.

In recent years, heavy fishing has taken some toll on gamefish populations. But the Alabama coast has fared better than some areas of nearby Florida where overfishing has greatly reduced fish populations.

Government regulators have put additional limits on fishes and fishermen to safeguard the region's sea-life population. The season, size and creel limited of several species have created tough times for some fishermen; it certainly has forced a discipline to take fewer fish, and a different style of fishing to ensure that future generations are able to experience the joys of a day upon the sea.

The Alabama Coast

Perhaps it is the idea that one is standing at the edge of the world that brings fishermen back again and again to venture into the gulf.

Perhaps it is the plentiful catches the area is known for, or the size of the fishes. Perhaps it is the great diversity of fishing and styles of fishing that are possible.

Whatever the reason, fishermen flock to this place by the thousands; unwind in its casual air, revel in its beauty, are awed by its mystery, respect its power and harvest its bounty.

The surf

The very thing that draws millions of visitors to the Alabama shore—the pristine white sand—is the same one that discourages many fishermen from trying their luck on the beach.

Because the surf lacks close-in reefs, rocks, vegetation or other obstructions that normally attract fish, many anglers view it as so many miles of desert when it comes to sea life. A closer look reveals something vastly different. Actually, the surf is a treasure house of marine life.

Great schools of fish cruise near the beaches, and several types of game fish patrol the sandy bottom for bits of food washed up or crustaceans and shellfish that live in the shallows.

The key is knowing how to look for the fish. Many a surf angler has spent all day without so much as a nibble, while others fishing sometimes only yards away are filling their coolers.

If you plan to bottom fish in the surf, search for potholes and cuts in the sand bars. Water moves more rapidly in the cuts, and food particles accumulate in the potholes. The food and moving water attract fish.

These cuts and potholes are easy to detect, particularly on calm days, by the color of the water. The shallower the water, the lighter color it is. The deeper it is, the darker it gets. The shallowest water appears almost clear white; when it begins getting deeper it appear light green; in the cuts and potholes it appears blue to dark blue.

There are several other ways to find fish. Watch for birds circling and diving at the surface, or a dimpling of the surface. Those are indications of schooling fish. In extremely clear water on calm days you can actually spot schools of fish along the shore.

The schooling fish move in and out from the shoreline; some days they run in thick schools; other days they disappear completely. You just have to hunt them.

One thing about surf fishing: you probably will catch something, even if it is only a sea catfish or pinfish.

Surf fishermen use tackle ranging from light spinning rigs—the same as they use back home for bream—to heavy surf rods. The only criteria is to select a rod capable of casting bait or a lure beyond the first sand bar (You don't even need to cast that far for Pompano or Whiting sometimes). Just be prepared to have your line snapped once in a while with light tackle.

Bottom fishermen use smallish hooks on one- and two-hook rigs, held on the bottom by a pyramid or barrel sinker, usually about the size of those used for crappie fishing. Popular bottom baits are frozen squid and shrimp. A small bit of bait on a small hook will do fine. Some fishermen use cut bait, sand fleas, bits of shellfish or sand crabs.

Favorite artificial lures include jigs ranging from 1/4 to 3/4 ounce. Sting Ray Grubs and silver spoons in 00 or 0 size. A small three-hook mackerel rig is popular for casting natural baits such as cigar minnows.

During summer months, casting with lures or natural bait or bottom fishing all produce fish. In winter, surf fishing is limited mostly to bottom fishing.

Expect to catch Whiting, Pompano, an occasional mackerel, ladyfish, redfish and sharks, depending on the season.

A couple of handy gadgets that make surf fishing easier include a five-gallon bucket or medium-size cooler to keep your catch in and a three-foot length of stiff plastic pipe, of sufficient width for the butt of your rod to slide down into the pipe. You stick one end of the pipe in the sand and the butt of your rod in the other to keep it out of the sand when you don't want to hold it.

A caution: Wear clothes fit for wading and shoes you don't mind getting wet. And remember the sun. A person sunburns quickly with the rays reflecting off both the water and white sand.

Piers

Some wag once said that piers were the poor man's boat. Don't believe it. Pier fishing is something different—something all its own.

The structures jut into the gulf to beyond the breakers and provide a haven for fish and a place where fishermen can cast beyond the sand bars into waters holding Bluefish, Bonito, Flounder, Pompano, King and Spanish Mackerel, Cobia, Ladyfish, Whiting, Barracuda, Shark, Redfish, Black Drum, and Blue Runners (hardtails).

Some fishermen come to the pier with light freshwater tackle, fish for such smaller species as Flounder and Whiting and do just fine. For larger fish such as the King Mackerel and Bonito, heavier gear is needed. A light to medium surf rod is recommended—one that you can get some casting distance with.

Most natural baits will draw action. For bottom fishing, either a one- or two-hook rig will do. Small to medium spoons, grubs and jigs in the 1/4- to 3/4-ounce size are commonly used. Some pier fishermen cast live or dead cigar minnows on an unweighted line for mackerel. They sometimes use a single hook through the minnow's mouth and out the eye socket, or a three-hook mackerel rig.

Piers can get crowded at times, particularly when fish are schooling. Remember to give the other fellow room at the rail and room to cast. If the fisherman next to you has a fish on, get your line out of the water. Otherwise there is a fine opportunity for a tangled line and a lost fish.

You are a good distance above the water on the pier. Large fish usually are gaffed to get them up and over the rail. Customarily the fisherman with the gaff shares it with others. Remember to put it back when you are finished using it. Pier fishermen know how tough it is to maneuver a fish from the water onto the pier, keeping it away from the huge crustacean-covered pilings that can separate a line so quickly, and then hoisting it onto the pier itself. Keep your tackle box closed and away from the pier's edge.

Don't leave sharp-finned fish for someone to step on. And keep tackle put away.

Jetties

To halt silting and erosion of channels, jetties of huge boulders and concrete have been built at the mouth of bays and in a few other areas along the Alabama coast.

Jetties are havens for an array of fish and other sea life. The cracks and crevices in the rocks draw shellfish, crustaceans, sea urchins, starfish and minnows—which draw bigger fish.

Fishermen either are addicted to the jetties or hate them. Some anglers fish jetties almost exclusively year around, and come home with heavy stringers of fish. Others swear they will never again fish jetties after leaving half their tackle tangled in the rocks. There is no question that you will use a quantity of tackle when fishing jetties.

A fish will try to dive for the rocks when hooked; and if he gets to them, odds are he'll be able to drag your line across one and cut it.

Yet, such a variety of fish—and large numbers of them at certain times of year—make jetties favorite fishing spots. Not only will you find most surf species around jetties, oftentimes such reef fishes as Triggerfish and grouper and, on rare occasions, White and Red Snapper will move into the jetties, particularly during cooler months.

While many people use light fresh-water tackle for jetty fishing, others prefer light to medium surf tackle. Also, some anglers use slightly heavier than normal sinkers because of strong tidal currents. A pyramid sinker is preferred over a barrel sinker since it will better stay put. A barrel sinker tends to roll in the current.

Open Gulf Fishing

Bottom fishing

Much of the year, your chances of catching a variety of food fish are greatest bottom fishing. The bottom fish—such as snappers, triggerfish, porgy and groupers—usually are willing biters, easily located (for those who know where to look) and caught with a minimum of expertise.

Almost every day that the gulf is calm enough during the spring, summer and fall, private, charter and party (head) boats venture from Alabama ports into the gulf for bottom fish.

If you charter a boat or get on a party boat, your bait and fishing tackle is furnished. You use heavy tackle and natural bait—either cut bait, squid or, sometimes, live or dead baitfish.

The basic bottom-fishing technique is simple, but the fine points take practice to master. Bottom fishermen usually use a rig consisting of two hooks and a heavy sinker to hold the rig near the bottom.

Once the boat captain signals that it is time to fish, the fisherman simply lowers his line over the side of the boat until he feels the sinker hit bottom. Usually he will get a bite immediately. The tough part for the novice is being able to distinguish a bite—particularly a soft nibble—from the sinker bumping bottom. The second toughest part is setting the hook. It is tough because with circle hooks that are now required for the catching of reef fish, the method is to let the fish hook himself and then just reel him in, rather than sharply raising the rod to set the hook as was the practice with "J" hooks.

If you lack your own boat, you have two choices: party boats, also known as head boats because they charge per head (fisherman), and charter boats.

Party boats are the least expensive way to spend a day on the open gulf. Depending on the boat and the day, these trips sometimes last a half day, sometimes a full day of 9 or 10 hours. Many party boats are equipped with a galley where food and drinks can be purchased.

These generally are large boats, carrying 30 or more fishermen at a time, traveling to large reefs or wrecks, usually within 30 miles of shore.

The main advantages to party boat fishing is price and the fact that all bait and equipment is furnished to the fisherman. Party boats are equipped with a variety of fish finders and navigation equipment and captains have no trouble locating fishing grounds.

Some argue that individual fishermen do not catch as many fish on party boats as they do on charter boats. That may or may not be the case, depending on the boat, the fisherman, and the day of the fishing trip. If party boats sometimes do not deliver the same per-fisherman catch as charter boats, it may not be that one is fishing better territory than the other. Because of the size of his boat and the number of fishermen aboard, a party boat captain must stop over areas capable of holding large concentrations of fish. While these spots may be larger, they also may be more heavily fished than a tiny spot that a charter boat may offer to his group of a half dozen fishermen.

Secondly, the pure economics of 50 fishermen fishing a single hole dictates that while a party boat may get as many fish from one spot as a charter boat, the number of fish per fisherman may be less. In addition, when 50 or more fishermen try to fish shoulder to shoulder there are going to be more tangled lines and more confusion. Party boats do have deck hands who help, but on party boats each deck hand often has more fishermen to watch out for than the average charter boat deck hand.

The final, and perhaps most significant, factor has to do with the skill of the fisherman. Every party boat takes out a large number of people each day who don't know anything about salt water fishing. And that factor alone keeps them from catching the numbers of fish they could catch if only they had taken some time to learn what to do.

A novice probably isn't going to have as much luck as an experienced fisherman. Most experienced fishermen often bring back heavy stringers from a day of party boat fishing. They aren't always huge fish, but there almost always are plenty of them. With some knowledge of salt water fishing techniques—even such things as how to properly bait the hook, when to improvise and put a live bait on, how to detect a bite and how to set a hook—a party boat fisherman will bring more fish over the rail than he will know what to do with.

The other option is charter boat fishing. If you don't care for the larger boat or large number of fishermen, or want some personal and special attention from a captain and deck hand, charter boat fishing should be considered. It costs more than party boat fishing.

On the whole, these are smaller boats capable of carrying from 6 to 20 fishermen and are chartered by the hour or day, with the base rate figured on 6 or more fishermen, depending on the size of the boat.

Additional anglers cost extra. Charter boat captains often will go fishing for the number of hours and on the hours that you specify. With fewer fishermen, the captain and crew have more time to offer fishing tips, usually will move from one fishing spot to another whenever you want and usually will change from one style of fishing to another whenever you want.

Charter boats also furnish all bait and tackle. Lighter tackle often is available on charter boats—making it easier to detect bites and to do more kinds of fishing, to use lighter weights or even to use an unweighted line. Most charter boats offer live-bait fishing.

Some charter boats have galley; some do not. Most allow you to bring your own lunch and beverages. Some require you to do so. Ask in advance.

Half-day and full-day trips are available. Some boats offer overnight trips for an additional fee.

While charter captains often will fish for what the fishermen demand, the smart angler will ask the captain what's biting and take his advice.

Getting a good captain

Some captains catch more fish than others, just as in life some workers are more productive than others. Following are some tips on selecting a top-notch captain:

• Before booking your trip, talk to acquaintances who have fished Alabama gulf waters recently. Get the names of captains who have provided successful trips.

• If you arrive at the coast without having already booked your trip, visit the marinas. Notice the boats. While a neat, clean, well-kept boat won't catch fish by itself, it is an indicator of the care a captain gives.

• In addition to checking boat conditions, if you visit the marinas as boats are returning from a day of fishing you can see what they catch. Visit the same marina several times in a row if possible. While the good boats will consistently come in with catches, even the best ones have a bad day once in a while.

• Talk to the captain. If you don't care for his temperament, find another one. See if he seems genuinely interested in you catching fish and having a good time.

• Ask area residents, at motels and at fish houses, and with folks you see wandering around the docks about which captains do the best job.

Throughout much of the spring, summer and fall, the better boats are booked in advance, particularly on weekends. However, they are not so fully booked on weekdays. If you want a weekend trip, call in advance or you may not be able to find a boat.

Trolling

Generally speaking, party boats do not troll. Charter boats offer two distinctively different types of trolling—inshore and offshore.

Inshore

Inshore trolling means trolling not too far from the beach—usually within 15 miles of land and often within a mile.

When the mackerel are running, this type of fishing is often so fast and furious that a fisherman will want no more than a half-day trip. Charter boats charge similarly for both bottom fishing and trolling.

Also, as with bottom fishing, boats chartered for trolling normally provide all tackle and bait.

Trolling can be much more a feast or famine proposition than bottom fishing. The migratory fish that are the bread and butter of the trolling fleets move in and out of the coastline; some years they are plentiful; other years scarce. One day a charter captain will fish hard for two or three fish, only to go out the next day and load the fish box.

Oversimplified, trolling is a matter of hooking a natural or artificial bait at the end of your line and towing it through the water, enticing a fish to strike what it thinks is its next meal.

The difference in lures generally is color and size, with smaller spoons and jigs for Spanish Mackerel, and larger dusters and natural baits more common for King Mackerel. Other trolling baits include natural and artificial squid, cigar minnows and large spoons.

Trolling boats congregate over reefs, wrecks, near cuts or channels and over any kind of obstruction or area that attracts baitfish. Often they will troll in a tight circle around buoys.

Offshore

The warm Loop Current and the blue-black depths of the DeSoto Canyon are the cruising grounds for Wahoo, Tuna, Blue and White Marlin, big Dolphins and Swordfish.

These giants migrate into the waters during the spring, and roam the deep waters until the winter chill.

You troll for these big game fish. But it is different than inshore trolling.

An offshore trip often will take at least 12 hours, and possibly require spending a night on the open gulf. Because it requires more time to get to the deeper water, it costs more to troll offshore.

The search for big offshore fish involves hours of monotony, broken by moments of spine-tingling excitement. Even after reaching offshore fishing grounds, you may troll for hours before anything rises to so much as look at your bait.

Inshore, when a fish hits it normally hooks itself.

Offshore, a billfish sometimes will follow a bait for a distance before striking; even then, it may touch the bait several times with its bill before actually taking it in its mouth. Sometimes a fisherman will fight a fish for a long time only to have it tear loose or break off.

One giant Blue Marlin is a prize catch. The success of an offshore trip is measured in the size of the fish you catch, not in numbers. It is specialized fishing. Tackle is among the heaviest available. Lures are oversize. And the length of the trip requires some stamina.

Tidbits

Seasickness

Anyone can get seasick. There are a number of remedies on the market these days. Don't wait until you are sick to take one. The smart angler checks with his druggist or physician ahead of time on the best seasickness remedy. Follow the instructions. Once you get out to sea, particularly on a party boat, there is no turning back until the captain has finished his day. To avoid seasickness, eat normally, avoid greasy foods and don't upset your normal routine. That includes not drinking alcoholic beverages heavily the night before you go out.

What to wear

Wear light-colored, comfortable clothes. In winter you will need a jacket on most days. In summer, wear light clothes and a hat, and shoes that you don't mind getting wet. You will get bait and fish slime on your hands. Bring a towel to wipe them.

Tipping

Charter boat deck hands normally are tipped 15 to 20 percent of the trip cost. That is, if they do a good job. A deck hand working hard can mean the difference in a successful trip and an unsuccessful one.

On your own

Do not challenge the open gulf in your bass boat, or any other vessel not properly equipped for such a large body of water. Piloting a small boat on the Gulf of Mexico can be an enjoyable experience. It also can be an exercise in pure terror and can end in disaster.

As seas go, the gulf is considered calm. Still, strong winter and spring winds can raise seas to more than 10 feet in a short time. Summer seas are less fickle, generally calmer, and more predictable.

Still, if you never have had experience on the open gulf, venture out cautiously—first into the bay, then a short foray into the open gulf. Build your confidence. Practice maneuvers until you are confident of the maximum seas that you and your boat are capable of handling safely.

Even before leaving port that first time, make sure your craft is rigged for open water. A ship-to-shore radio, GPS and compass are essentials. Most tackle shops have a list of equipment required for boats venturing into the gulf. It is the wise fisherman who makes sure he has everything on the list.

If you expect to catch any fish bottom fishing, a GPS and depth finder also are musts. Spend some time learning to operate them; 10 miles out is no place to learn to program a GPS.

Many anglers say they would not venture out without a backup engine, while others frequent the gulf with only one; they say their will call for a tow if the engine breaks down. The most efficient of life jackets are recommended.

An anchor designed to dig into soft bottom is suggested, along with rope approximately three times as long as the water depth you intend to fish in.

Basic salt-water tackle consists of a sturdy boat rod and a level-wind reel. Rods of solid fiberglass, with the fiberglass extending through the butt handle, in five and six foot lengths are basic. Strong, hollow rods of modern-day materials are available and are lighter than solid fiberglass. A good all-around reel is the 4/0 size. If you don't plan to venture out more than 15 miles or so, you can get by with smaller reels. If your fishing is to include heavy grouper and Amberjack, you may want to consider the slightly larger 6/0 size.

These reels should be equipped with 50 to 80 pound test line. Most fishermen now use monofilament, but braided line still is available and doesn't present the stretch problem of monofilament. Circle hooks are required for reef fishing. Three items are important in hook selection: 1. Match the size of the hook with the kind of fish you are going for (remember that grouper and Amberjack have big mouths); 2. Salt water hooks need thick shanks to keep toothy species from biting them in half; 3. If you are undecided about which size hook to use, always pick the smaller of two choices.

Tackle

Other important tackle peculiar to salt water are a gaff to hoist larger fish into the boat, a cutting board for cutting fish and squid into bait-size pieces and a marker buoy which can be purchased at tackle stores or made from an old plastic jug.

"A Big Pond"

The Alabama salt water angler with his own boat still is faced with a challenge. He can troll near shore or outward over the reefs. He can drift fish or bottom fish, use light tackle and play a fish all day long or use a heavy boat rod for big Amberjack or grouper. But he still has to find the fish.

Mostly, the shore off the Alabama coast is a gently rolling to flat sand bottom outward for 20 or more miles. This gentle sea floor is occasionally interrupted by limestone outcroppings and artificial reefs—sunken ships, car bodies, concrete bridge rubble or just about anything else that fishermen could sink to attract fish.

The Gulf of Mexico is a big pond. Reefs, artificial or natural, are tough to find. Men who make their livings fishing the gulf fiercely protect the locations of the reefs that they personally build at much effort and expense.

There are many reefs built by the government for the pleasure-boat fisherman, and the GPS numbers for those are available to the public. Many of the natural reefs are now well charted and their locations are well known.

The locations of these fishing grounds are determined with a navigation tool known as a GPS, and operates on the same principle as an automobile GPS.

Once you locates an area where you spots a concentration of fish on the depth finder, you can forever return to that spot by logging the GPS numbers displayed for that spot.

Artificial reef building off the Alabama coast.

Alabama's artificial reef program

Alabama has one of the largest artificial reef programs in the world. The natural bottom of offshore Alabama is a predominately flat sand/mud type bottom. This bottom type attracts very few fish that are either commercially or recreationally valuable.

However, it has long been known that if vertical relief is created on this bottom, many reef fish such as snappers and groupers will be produced.

In fact, artificial reefs can be created that over time will appear as natural reefs with similar communities of encrusting organisms and bait fish. As various encrusting organisms such as corals and sponges cover the artificial reef material, small animals take up residence.

Ongoing research within Alabama's artificial reef general permit areas indicates that juvenile red snapper recruit to new, uninhabited artificial reefs, aging with the reef, and recruiting to the fishery at an appropriate size. At that point the artificial reef functions as a natural reef.

Alabama's artificial reef building program started in 1953 when the Orange Beach Charter Boat Association asked for authority to place 250 car bodies off Baldwin County. This proved to be very successful and in the years since, many different types of materials have been placed offshore of Alabama. These have included additional car bodies, culverts, bridge rubble, barges, boats and planes.

In 1974, in an excellent example of State/Federal cooperation, several "ghost-fleeted" liberty ships were sunk in five locations off Mobile and Baldwin Counties in 80-93 feet of water.

In 1987, a general permit was issued by the U.S. Army Corps of Engineers creating specific areas offshore of Alabama for the creation of artificial reefs.

In 1987 the areas encompassed almost 800 square miles.

In 1993, the U.S. military in addressing the need to de-militarize obsolete battle tanks realized that immersion in sea water was an acceptable method. The idea was presented to the Marine Resources Division and development of an operation plan began. In 1994, 100 M-60 tanks were deployed as artificial reefs in depths of 70 to 110 feet within the Hugh Swingle and Don Kelley North permit areas. The conservative estimate for the life span of the tanks as artificial reefs is 50 years.

In late 1997, the U.S. Army Corps of Engineers authorized an expansion of Alabama's artificial reef construction areas to allow for greater freedom in reef placement and greater variety in depth. The combined area for all reef permit zones now encompasses approximately 1,260 square miles. At the same time, the protocol for reef construction was modified. This modification limited the types of materials that can be used to construct artificial reefs. Enforcement of the protocol and placement of reefs is a joint effort of the Marine Resources Enforcement Section, the Alabama Marine Police Division, and the U. S. Coast Guard.

GPS Coordinates for artificial reef structure off the Alabama coast

Reef Name	Material	Latitude	Longitude	Distance/ Nautical Miles Perdido Pass	Distance/ Nautical Miles Sand Island Light
105 Tug	105' Wood Hull Tug Boat	30°03.888	87°42.455	14.51	19.22
Allen	WW II Liberty Ship	30°07.835	87°31.790	8.40	27.19
Anderson	WW-II Liberty Ship	29°59.041	88°04.305	31.72	12.23
Atlantic Drydock	Wood Drydock Landinge	29°56.586	88°01.846	31.40	14.67
Atlantic Marine Pipes	1 - Concrete Pipe	30°02.002	87°49.566	19.81	14.83
Atlantic Marine Pipes	2 - Concrete Pipe	30°01.042	87°49.564	20.54	15.45
Atlantic Marine Pipes	3 - Concrete Pipe	30°00.004	87°49.561	21.32	16.15
Buffalo Barge	Wood Hopper Barge	30°05.047	87°49.938	18.08	12.86
Buffalo Barge 2	Wood Hopper Barge	30°04.770	87°50.438	18.57	12.62
Callon Jacket	Oil Platform and Jacket	29°26.000	87°58.500	54.02	45.25
D.I. Bridge 01	Bridge spans and pilings	30°02.537	88°05.796	31.13	9.02
D.I. Bridge 02	Bridge spans and pilings	30°02.727	88°04.651	30.06	8.55
D.I. Bridge 03	Bridge spans and pilings	30°02.786	88°04.565	42.22	16.75
D.I. Bridge 04	Bridge spans and pilings	30°02.810	88°04.720	30.20	8.47
D.I. Bridge 05	Bridge spans and pilings	30°02.817	88°04.740	31.03	10.18
D.I. Bridge 06	Bridge spans and pilings	30°02.916	88°04.897	30.11	8.45
D.I. Bridge 07	Bridge spans and pilings	30°03.722	88°05.134	29.97	7.71
D.I. Bridge 08	Bridge spans and pilings	30°05.545	88°07.408	31.16	6.84
D.I. Bridge 09	Bridge spans and pilings	30°01.170	88°04.846	30.84	10.15
D.I. Bridge 10	Bridge spans and pilings	30°02.852	88°04.189	29.58	8.41
Drydock	Wood Drydock Wing	30°01.347	88°07.163	32.73	10.52
Edwards	W W-II Liberty Ship	29°57.931	88°06.597	34.00	13.65
Gulf State Park Pavilion 1	Concrete Walls and Pilings	30°07.737	87°32.059	7.86	27.01
Gulf State Park Pavilion 2	Concrete Walls and Pilings	30°08.701	87°32.814	6.83	26.25
Gulf State Park Pavilion 3	Concrete Walls and Pilings	30°08.749	87°33.633	6.78	25.54
Gulf State Park Pavilion 4	Concrete Walls and Pilings	30°08.897	87°34.323	6.68	24.93
Hanson Pipe 1	Concrete pipe and pieces	29°57.544	87°58.626	28.65	14.19
Hanson Pipe 2	Concrete pipe and pieces	29°56.509	87°58.577	29.30	15.21
Hanson Pipe 3	Concrete pipe and pieces	29°55.512	87°58.577	29.98	16.17
Lillian Bridge	Bridge Spans and Pilings	30°06.881	87°32.685	9.25	29.55
Liscomb Tug	Wood Hull triple deck Tug	30°04.880	87°48.180	17.00	14.29
McMillian Barge	Metal Hopper Barge	29°58.766	88°02.700	30.72	12.46
Mobile Bay Farewell Bouy	Mobile Bay Farewell Buoy	30°07.490	88°04.130	27.92	3.86
Mobile Co. Roadbuilders 1	Concrete pipe and pieces	29°55.087	87°57.535	29.63	16.82
Mobile Co. Roadbuilders 2	Concrete pipe and pieces	29°54.087	87°57.002	30.04	17.91
Mobile Co. Roadbuilders 3	Concrete pipe and pieces	29°53.045	87°57.520	31.11	18.79
Nail Ship	Cargo barge sunk 1930s	29°57.573	88°09.692	36.48	14.84

Reef Name	Material	Latitude	Longitude	Distance/ Nautical Miles Perdido Pass	Distance/ Nautical Miles Sand Island Light
Papa Joe (John Wood)	Metal Tug Boat	29°59.979	88°07.459	33.50	11.90
Perdido Pass Bridge 7	Bridge Spans and Pilings	30°08.446	87°33.956	7.68	25.26
Perdido Pass Bridge 2	Bridge Spans and Pilings	30°08.565	87°34.270	7.58	24.97
Perdido Pass Bridge 1	Bridge Spans and Pilings	30°01.913	87°33.545	14.20	27.13
Perdido Pass Bridge 3	Bridge Spans and Pilings	30°02.309	87°33.245	13.81	27.25
Perdido Pass Bridge 4	Bridge Spans and Pilings	30°02.805	87°33.100	13.31	27.20
Perdido Pass Bridge 5	Bridge Spans and Pilings	30°03.122	87°32.873	13.00	27.29
Perdido Pass Bridge 6	Bridge Spans and Pilings	30°03.630	87°32.716	12.50	27.28
Perdido Pass Farewell Buoy	Perdido Pass Farewell Buoy	30°15.545	87°33.324	0.00	26.03
Pmid 04-001	Concrete pyramids	30°02.894	87°34.648	14.52	25.92
Pmid 04-002	Concrete pyramids	30°01.844	87°34.691	15.76	26.24
Pmid 04-003	Concrete pyramids	30°00.889	87°34.744	16.89	26.55
Pmid 04-004	Concrete pyramids	29°59.876	87°34.719	18.02	26.98
Pmid 04-005	Concrete pyramids	29°58.854	87°34.800	19.23	27.36
Pmid 04-068	Concrete pyramids	30°02.020	87°34.870	15.67	26.03
Pmid 04-069	Concrete pyramids	30°00.947	87°50.711	20.71	14.76
Pmid 04-070	Concrete pyramids	30°00.123	87°50.138	20.94	15.69
Pmid 04-071	Concrete pyramids	29°59.261	87°49.511	20.20	16.69
Pmid 04-072	Concrete pyramids	29°58.327	87°48.887	21.57	17.73
Pmid 04-073	Concrete pyramids	29°57.479	87°48.246	21.90	18.73
Pmid 04-138	Concrete pyramids	29°55.828	87°36.992	19.59	27.26
Pmid 04-139	Concrete pyramids	29°56.868	87°37.046	18.57	26.65
Pmid 04-140	Concrete pyramids	29°57.857	87°36.944	17.59	26.21
Pmid 04-141	Concrete pyramids	29°58.855	87°36.978	16.61	25.69
Pmid 04-142	Concrete pyramids	29°59.853	87°36.922	15.63	25.27
Pmid 04-143	Concrete pyramids	30°00.839	87°37.009	14.68	24.78
Pmid 04-144	Concrete pyramids	30°02.005	87°36.904	13.53	24.39
Pmid 04-145	Concrete pyramids	30°02.043	87°39.276	14.11	22.49
Pmid 04-146	Concrete pyramids	30°01.015	87°39.355	15.09	22.86
Pmid 04-147	Concrete pyramids	30°00.025	87°39.381	16.03	23.30
Pmid 04-148	Concrete pyramids	29°59.066	87°39.303	16.91	23.83
Pmid 04-149	Concrete pyramids	29°58.055	87°39.348	17.89	24.33
Pmid 04-150	Concrete pyramids	29°57.077	87°39.281	18.80	24.92
Pmid 04-151	Concrete pyramids	29°56.089	87°39.303	19.76	25.48
Pmid 04-166	Concrete pyramids	29°57.391	87°41.638	19.17	23.08
Pmid 04-167	Concrete pyramids	29°58.345	87°41.663	18.30	22.50
Pmid 04-168	Concrete pyramids	29°59.394	87°41.694	17.36	21.90
Pmid 04-169	Concrete pyramids	29°59.894	87°40.765	16.58	22.32
Pmid 04-170	Concrete pyramids	30°00.482	87°41.537	16.32	21.45
Pmid 04-171	Concrete pyramids	30°01.585	87°41.610	15.37	20.86
Pmid 04-172	Concrete pyramids	30°01.396	87°42.217	15.78	20.49
Pmid 04-173	Concrete pyramids	30°01.162	87°42.933	16.29	20.06
Pmid 04-174	Concrete pyramids	30°00.570	87°44.553	17.54	19.18
Pmid 04-175	Concrete pyramids	30°00.285	87°45.320	18.15	18.79
Pmid 04-176	Concrete pyramids	29°59.973	87°46.133	18.81	18.41
Pmid 04-177	Concrete pyramids	29°59.637	87°46.884	19.47	18.11

Reef Name	Material	Latitude	Longitude	Distance/ Nautical Miles Perdido Pass	Distance/ Nautical Miles Sand Island Light
Pmid 04-178	Concrete pyramids	30°02.741	88°00.111	26.32	8.81
Pmid 04-179	Concrete pyramids	30°01.800	87°59.659	26.44	9.82
Pmid 04-180	Concrete pyramids	30°00.821	87°59.372	26.74	10.83
Pmid 04-181	Concrete pyramids	29°59.869	87°58.926	26.94	11.85
Pmid 04-182	Concrete pyramids	29°58.974	87°58.396	27.09	12.84
Pmid 04-183	Concrete pyramids	29°58.001	87°58.204	27.55	13.82
Pmid 04-184	Concrete pyramids	29°57.008	87°57.825	29.93	14.86
Pmid 04-185	Concrete pyramids	29°56.148	87°57.205	28.10	15.84
Pmid 04-186	Concrete pyramids	29°55.283	87°56.650	28.34	16.81
Pmid 04-187	Concrete pyramids	29°54.431	87°56.037	28.58	17.79
Pmid 04-188	Concrete pyramids	29°53.525	87°55.542	28.95	18.78
Pmid 04-189	Concrete pyramids	29°52.641	87°54.852	29.23	19.82
Pmid 04-198	Concrete pyramids	29°52.209	87°57.261	30.93	19.59
Pmid 04-199	Concrete pyramids	29°53.067	87°57.841	30.64	18.64
Pmid 04-200	Concrete pyramids	29°54.056	87°58.207	29.15	18.00
Pmid 04-201	Concrete pyramids	29°54.987	87°58.690	28.90	16.94
Pmid 04-202	Concrete pyramids	29°55.539	87°59.121	28.82	16.29
Pmid 04-203	Concrete pyramids	29°56.791	87°59.730	28.55	14.89
Pmid 04-204	Concrete pyramids	29°57.737	88°00.213	28.09	13.92
Pmid 04-205	Concrete pyramids	29°58.686	88°00.760	27.96	12.87
Pmid 04-206	Concrete pyramids	29°59.600	88°00.900	28.19	11.80
Pmid 04-207	Concrete pyramids	30°00.729	88°01.057	27.93	10.61
Pmid 04-208	Concrete pyramids	30°01.662	88°01.628	28.00	9.60
Pmid 04-209	Concrete pyramids	30°02.623	88°02.175	28.10	8.59
Pmid 04-210	Concrete pyramids	30°02.572	88°03.491	28.72	8.62
Pmid 04-211	Concrete pyramids	30°01.617	88°03.463	28.85	9.58
Pmid 04-212	Concrete pyramids	30°00.482	88°03.460	29.06	10.72
Pmid 04-213	Concrete pyramids	29°59.494	88°03.451	29.12	11.76
Pmid 04-214	Concrete pyramids	29°58.345	88°03.510	29.40	12.95
Pmid 04-215	Concrete pyramids	29°56.907	88°03.315	29.58	14.46
Pmid 05-114	Concrete pyramids	30°03.380	87°32.471	12.15	27.57
Pmid 05-115	Concrete pyramids	30°02.507	87°32.467	13.01	27.83
Pmid 05-116	Concrete pyramids	30°01.657	87°32.475	13.86	28.10
Pmid 05-117	Concrete pyramids	30°00.805	87°32.467	14.71	28.41
Pmid 05-118	Concrete pyramids	29°99.961	87°32.496	15.53	28.78
Pmid 05-119	Concrete pyramids	29°59.083	87°32.479	16.42	29.08
Pmid 05-120	Concrete pyramids	29°58.226	87°32.490	17.28	29.44
Pmid 05-121	Concrete pyramids	29°57.530	87°33.129	17.96	29.26
Pmid 05-122	Concrete pyramids	29°58.375	87°33.440	17.11	28.63
Pmid 05-123	Concrete pyramids	30°00.452	87°33.450	15.04	27.76
Pmid 05-124	Concrete pyramids	30°01.307	87°33.435	14.19	27.45
Pmid 05-125	Concrete pyramids	30°02.591	87°33.806	12.92	26.71
Pmid 05-126	Concrete pyramids	30°03.412	87°33.777	12.10	26.48
Pmid 05-127	Concrete pyramids	30°03.122	87°35.414	12.51	25.22
Pmid 05-128	Concrete pyramids	30°02.277	87°35.373	13.34	25.53
Pmid 05-129	Concrete pyramids	30°01.616	87°36.005	14.07	25.26
Pmid 05-130	Concrete pyramids	30°00.784	87°36.204	14.92	25.43

The Ultimate Guide to Alabama Fishing

Reef Name	Material	Latitude	Longitude	Distance/ Nautical Miles Perdido Pass	Distance/ Nautical Miles Sand Island Light
Pmid 05-131	Concrete pyramids	29°59.947	87°35.931	15.71	26.00
Pmid 05-132	Concrete pyramids	29°59.052	87°35.930	16.59	26.40
Pmid 05-133	Concrete pyramids	29°58.227	87°35.921	17.41	26.80
Pmid 05-134	Concrete pyramids	29°57.343	87°35.946	18.28	27.22
Pmid 05-135	Concrete pyramids	29°56.483	87°35.929	19.13	27.68
Pmid 05-136	Concrete pyramids	29°55.645	87°35.956	19.96	28.11
Pmid 05-153	Concrete pyramids	29°55.959	87°38.293	19.99	26.26
Pmid 05-154	Concrete pyramids	29°56.757	87°80.340	19.12	25.93
Pmid 05-155	Concrete pyramids	29°57.627	87°37.891	18.29	25.63
Pmid 05-156	Concrete pyramids	29°47.482	87°37.897	28.25	32.15
Pmid 05-157	Concrete pyramids	29°59.363	87°37.912	16.61	24.74
Pmid 05-158	Concrete pyramids	30°00.101	87°38.420	16.01	24.00
Pmid 05-159	Concrete pyramids	30°00.927	87°38.534	15.25	23.54
Pmid 05-160	Concrete pyramids	30°01.82	87°38.211	14.32	23.42
Pmid 05-161	Concrete pyramids	30°02.621	87°38.107	13.53	23.20
Pmid 05-162	Concrete pyramids	30°02.542	87°39.256	13.94	22.31
Pmid 05-163	Concrete pyramids	30°02.422	87°40.206	14.37	21.60
Pmid 05-164	Concrete pyramids	30°02.876	87°41.002	14.27	20.78
Pmid 05-165	Concrete pyramids	30°02.788	87°41.993	14.76	20.04
Pmid 05-166	Concrete pyramids	30°02.707	87°42.940	15.27	19.33
Pmid 05-167	Concrete pyramids	30°02.652	87°44.021	15.84	18.52
Pmid 05-168	Concrete pyramids	30°02.565	87°45.036	16.44	17.78
Pmid 05-169	Concrete pyramids	30°01.686	87°45.007	17.12	18.25
Pmid 05-170	Concrete pyramids	30°01.759	87°44.041	16.58	18.93
Pmid 05-171	Concrete pyramids	30°01.837	87°43.064	16.06	19.63
Pmid 05-172	Concrete pyramids	30°02.089	87°42.101	15.42	20.26
Sand Island Light	Lighthouse at the entrance to Mobile Bay	30°11.227	88°02.980	26.03	0.00
Scott Barge	Wooden Hopper Barge	29°59.180	88°01.740	29.80	12.09
Southeast Banks	Natural Rock Live Bottom	30°01.425	87°56.900	25.07	11.12
Southwest Rock	Natural Rock Live Bottom	30°06.414	88°12.275	34.97	9.37
Sparkman	WW II Liberty Ship	29°59.676	87°43.050	18.43	20.77
Swingle Reef	Rig to Reef Oil Platform	29°25.160	87°35.600	60.00	51.87
Tank 100	M-60 Army Tank	29°58.325	88°02.460	30.80	12.91
Tank 101	M-60 Army Tank	29°58.474	88°00.568	29.39	12.92
Tank 102 A	M-60 Army Tank	30°00.518	88°00.147	27.65	10.95
Tank 102 B	M-60 Army Tank	30°00.535	88°00.148	27.64	10.93
Tank 102 C	M-60 Army Tank	30°00.530	87°59.965	27.51	10.97
Tank 38	M-60 Army Tank	30°00.691	87°43.911	17.88	19.58
Tank 39	M-60 Army Tank	29°59.715	87°43.791	18.69	20.22
Tank 40	M-60 Army Tank	29°58.652	87°43.714	19.60	20.90
Tank 41	M-60 Army Tank	29°57.670	87°43.645	20.45	21.55
Tank 42	M-60 Army Tank	29°56.674	87°43.534	21.32	22.27
Tank 43	M-60 Army Tank	29°55.640	87°43.409	22.22	23.04
Tank 44	M-60 Army Tank	29°54.637	87°43.330	23.13	23.78

Reef Name	Material	Latitude	Longitude	Distance/ Nautical Miles Perdido Pass	Distance/ Nautical Miles Sand Island Light
Tank 45	M-60 Army Tank	29°53.673	87°43.244	24.00	24.51
Tank 46	M-60 Army Tank	29°52.541	87°43.149	25.03	25.39
Tank 47	M-60 Army Tank	29°52.610	87°44.105	25.26	24.79
Tank 48	M-60 Army Tank	29°53.565	87°44.245	24.42	24.00
Tank 49	M-60 Army Tank	29°54.619	87°44.335	23.48	23.18
Tank 50	M-60 Army Tank	29°55.560	87°44.428	22.65	22.45
Tank 51	M-60 Army Tank	29°56.612	87°44.510	21.74	21.67
Tank 52	M-60 Army Tank	29°57.590	87°44.600	20.90	20.96
Tank 53	M-60 Army Tank	29°58.650	87°44.700	20.00	20.22
Tank 54	M-60 Army Tank	29°59.650	87°44.800	19.18	19.54
Tank 55	M-60 Army Tank	29°59.642	87°45.733	19.61	18.90
Tank 56	M-60 Army Tank	29°58.443	87°46.433	21.00	19.25
Tank 57	M-60 Army Tank	29°57.836	87°46.599	21.55	19.51
Tank 58	M-60 Army Tank	29°56.814	87°46.505	22.38	20.29
Tank 59	M-60 Army Tank	29°55.743	87°46.422	23.27	21.11
Tank 60	M-60 Army Tank	29°54.715	87°46.333	24.15	21.93
Tank 61	M-60 Army Tank	29°53.744	87°46.246	24.97	22.72
Tank 62	M-60 Army Tank	29°52.710	87°46.170	25.88	23.56
Tank 63	M-60 Army Tank	29°51.775	87°46.083	26.70	24.35
Tank 64	M-60 Army Tank	29°49.820	87°44.540	28.00	26.72
Tank 65	M-60 Army Tank	29°50.438	87°44.809	27.41	26.09
Tank 66	M-60 Army Tank	29°51.375	87°44.984	26.69	25.25
Tank 67	M-60 Army Tank	29°52.347	87°45.102	25.83	24.43
Tank 68	M-60 Army Tank	29°53.343	87°45.220	24.96	23.60
Tank 69	M-60 Army Tank	29°54.350	87°45.330	24.08	22.77
Tank 70 A	M-60 Army Tank	29°55.381	87°45.387	23.19	22.01
Tank 70 B	M-60 Army Tank	29°56.379	87°45.478	23.19	21.39
Tank 71	M-60 Army Tank	29°56.208	87°45.406	22.45	21.39
Tank 72	M-60 Army Tank	29°56.668	87°45.511	21.58	20.97
Tank 73	M-60 Army Tank	29°57.376	87°45.576	21.49	20.48
Tank 74 B	M-60 Army Tank	29°58.689	87°46.596	20.83	18.27
Tank 74 A	M-60 Army Tank	29°58.718	87°46.595	20.80	18.27
Tank 75	M-60 Army Tank	29°59.736	87°42.790	18.28	20.92
Tank 76 A	M-60 Army Tank	29°55.830	87°41.355	21.41	24.25
Tank 76 B	M-60 Army Tank	29°55.820	87°41.375	21.43	24.25
Tank 77 A	M-60 Army Tank	29°55.973	87°39.885	20.90	25.17
Tank 77 B	M-60 Army Tank	29°55.915	87°39.878	20.90	25.17
Tank 77 C	M-60 Army Tank	29°55.915	87°39.878	20.96	25.21
Tank 78 A	M-60 Army Tank	29°56.025	87°37.425	20.39	26.86
Tank 78 B	M-60 Army Tank	29°56.030	87°37.390	20.37	26.88
Tank 78 C	M-60 Army Tank	29°56.030	87°37.340	20.36	26.92
Tank 79	M-60 Army Tank	29°56.085	87°34.910	20.07	28.65
Tank 80	M-60 Army Tank	29°51.995	87°37.027	24.32	29.61
Tank 81 A	M-60 Army Tank	29°51.145	87°36.945	25.15	30.22
Tank 81 B	M-60 Army Tank	29°51.135	87°36.970	25.17	30.21
Tank 81 C	M-60 Army Tank	29°51.125	87°36.937	25.17	30.24
Tank 82 A	M-60 Army Tank	29°51.230	87°34.340	24.89	31.88

Reef Name	Material	Latitude	Longitude	Distance/ Nautical Miles Perdido Pass	Distance/ Nautical Miles Sand Island Light
Tank 82 B	M-60 Army Tank	29°51.305	87°34.345	24.82	31.84
Tank 83	M-60 Army Tank	30°00.704	87°42.796	16.91	20.40
Tank 85 A	M-60 Army Tank	30°00.845	87°40.185	16.35	22.31
Tank 85 B	M-60 Army Tank	30°00.820	87°40.235	16.38	22.28
Tank 86 A	M-60 Army Tank	30°00.965	87°37.830	15.62	24.07
Tank 86 B	M-60 Army Tank	30°00.950	87°37.785	15.62	24.12
Tank 87	M-60 Army Tank	30°00.063	87°35.279	16.13	26.47
Tank 88 A	M-60 Army Tank	30°00.998	87°35.260	15.20	26.09
Tank 88 B	M-60 Army Tank	30°00.985	87°35.285	15.21	26.08
Tank 89 A	M-60 Army Tank	30°01.963	87°33.975	14.16	26.77
Tank 89 B	M-60 Army Tank	30°01.977	87°33.950	14.15	26.78
Tank 89 C	M-60 Army Tank	30°01.990	87°33.945	14.13	26.78
Tank 90 A	M-60 Army Tank	30°03.925	87°34.185	12.21	25.97
Tank 90 B	M-60 Army Tank	30°03.890	87°34.190	12.24	25.95
Tank 90 C	M-60 Army Tank	30°03.858	87°34.223	12.25	25.95
Tank 91	M-60 Army Tank	29°59.953	87°37.681	16.57	24.64
Tank 93 A	M-60 Army Tank	29°59.913	87°41.961	17.80	21.43
Tank 93 B	M-60 Army Tank	29°59.928	87°41.964	17.78	21.40
Tank 93 C	M-60 Army Tank	29°59.950	87°41.980	17.78	21.40
Tank 94 A	M-60 Army Tank	29°58.316	88°04.590	32.33	12.99
Tank 94 B	M-60 Army Tank	29°58.286	88°04.485	32.27	13.00
Tank 94 C	M-60 Army Tank	29°58.300	88°04.420	32.22	12.98
Tank 95 A/B	M-60 Army Tank	29°59.358	88°06.236	32.98	12.20
Tank 96 A	M-60 Army Tank	30°00.642	88°06.818	32.78	11.09
Tank 96 B	M-60 Army Tank	30°00.643	88°06.816	32.78	11.09
Tank 97 A	M-60 Army Tank	29°58.345	88°05.884	33.26	13.12
Tank 97 B	M-60 Army Tank	29°58.355	88°05.608	33.05	13.07
Tank 98	M-60 Army Tank	29°57.339	88°05.985	33.88	14.13
Tank 99 A	M-60 Army Tank	29°59.405	88°06.956	33.51	12.33
Tank 99 B	M-60 Army Tank	29°59.421	88°06.905	33.45	12.30
Tank Broz A/B	M-60 Army Tank	30°00.308	88°01.803	28.97	7.99
Tank Sims A	M-60 Army Tank	30°00.296	88°02.106	29.20	11.01
Tank Sims B/C	M-60 Army Tank	30°00.319	88°02.112	29.20	11.01
Tensaw Bridge	Iron Tressel bridge spans	29°56.300	88°02.439	31.80	14.84
Tensaw Bridge	Iron Tressel bridge spans	29°56.543	88°02.440	31.80	14.93
Tensaw Bridge	Iron Tressel bridge spans	29°57.329	88°02.327	31.20	13.91
Tensaw Bridge	Iron Tressel bridge spans	29°57.753	88°02.691	32.20	13.47
Tensaw Bridge	Iron Tressel bridge spans	29°58.277	88°06.562	33.52	13.20
Tensaw Bridge	Iron Tressel bridge spans	29°58.357	88°06.566	33.55	13.27
Three Mile Barge	Flat Deck Steel Barge	30°13.709	87°32.953	2.44	26.06
Tulsa	Wood Hull Tug Boat	30°01.068	88°06.462	32.32	10.00
Unocal Rig	Oil Platform and Jacket	29°21.705	87°48.395	55.88	51.09
Wallace	WWII Liberty Ship	30°05.235	87°34.636	10.93	25.25

Mobile Bay

History-rich Mobile Bay is an extension of the Gulf of Mexico that stretches approximately 35 miles from the mouth of the Mobile River to the Gulf. It varies in width from eight to 18 miles.

The Mobile Delta to the north is an estuary that serves as a incredible nursery for numerous species of ocean life including shrimp. These juvenile shrimp grow and eventually migrate to the Gulf of Mexico, but first they must pass through Mobile Bay. There they must navigate a gauntlet of hungry species including speckled trout and redfish. Those species, as well as others in the bay, are intent on making sure the shrimp have short lives.

That being known, almost all fishermen fish Mobile Bay with live shrimp, dead shrimp or some plastic shrimp imitation.

There is no question that Mobile Bay offers its best fishing in the fall. There are numerous bait camps along the Causeway that sell live shrimp that time of the year and bait availability is rarely a problem.

For those intent on using artificials the D.O.A Shrimp is a no-brainer. A clear D.O.A. with a chartreuse tail works best in water that has little stain and the full chartreuse version works best in heavily stained water. Most are fished under a popping cork. Cocahoe Minnows and Sparkles Beetles are other favorite baits.

In addition to the shrimp, Mobile Bay also has a large concentration of menhaden in the fall. Chrome Rat-L-Traps and similar non-lipped crankbaits work well when casted over the bay's grass beds. These lures also work well around bridge pilings, piers and drop-offs.

Northern end of bay

There is a tremendous amount of one- to three-foot-depth water in the northern end, punctuated by narrow winding channels that eventually drop to the 10 and 12 foot depth that is generally uniform for the length of the bay. Fishermen work bridge pilings, snags and brush for fresh water bream in spring and fall, and go for speckled trout in deeper holes in winter.

This end of the bay plays host to a large number of saltwater species that migrate to the upper bay to forage on the huge numbers of shrimp that are available. A key area there is the "rock pile" where ships once unloaded their rock ballast. This is a good area for drift fishing while casting grubs. Numbers of speckled trout gather in this area.

Eastern Shore

Dozens of piers jutting into the bay are worked for flounder in spring, summer and fall. Live minnows, live or dead shrimp or a D.O.A. plastic bait tipped with a bit of squid is cast around the piers and slowly retrieved, bumping the bait along the bottom. The flounder lay on the bottom waiting for an easy meal to happen by and when it does, they pounce.

This same area can be worked successfully for speckled trout year around. They can be caught on live shrimp or minnow-shaped lures such as the blue and silver Rebel.

Each of the rivers flowing into the river can be worked for flounder, speckled trout and redfish, using the same type of lures.

The three species are the dominant gamefish of the lower bay and can be caught working channels, cuts, holes, piers, pipes, pilings and other structure—as sparse as it may be—for the length of the eastern shore, Bon Secour Bay and Fort Morgan peninsula.

Central Bay

Many fishermen avoid the Central Bay because of its lack of structure. The gas rigs are changing that somewhat. Found throughout the lower bay, these rigs are major vertical structures to provide cover and draw baitfish for the game species. Frequenting the rigs are speckled trout, flounder, redfish and some open gulf species.

In the middle section of the bay many top fishermen have success with speckled trout by "jump" fishing. They watch for gulls attacking shrimp and when that feeding frenzy is on, the fish aren't going to be far behind. A live shrimp or root beer sparkle Cocahoe minnow suspended two feet below a foam popping cork produces a lot of specks.

Southern Bay

On the southern end of the bay the area adjacent to the Mobile Ship Channel near the mouth of Mobile River is a key area. Speckled trout and redfish gather in numbers here in fall.

Again, live shrimp, artificial grubs, cut bait, white jigs and spoons will take these fish, with action being brisk at times.

Otherwise, the ship channel provides some action for Gafftopsail Catfish, redfish and an occasional Black Drum. The Gafftopsail Catfish can be distinguished by an extraordinarily long dorsal fin. Unlike the common saltwater catfish, considered a pest and inedible, the Gafftopsail is considered good table fare. It can be caught on cut bait, squid or dead shrimp. A depthfinder is essential if you plan to catch redfish in the channel.

Western Shore

From Dog River southward, the rivers and creeks flowing into Mobile Bay are prime locations for redfish, flounder, white trout and speckled trout.

Also, work the piers and the sparse structure for croakers, flounder, speckled trout and occasional redfish.

At the southern end of the bay—to either side of the Dauphin Island Bridge where Mississippi Sound joins the bay—work the bridge, the channel area and oyster reef area for redfish. Specks, Black Drum, Jack Crevalle, croakers and sharks.

Near the eastern tip of Dauphin Island, try for croaker, sharks and ground mullet (whiting).

Mike Bolton

Mobile Bay's best fishing spots

1. Choctaw Pass. This pass is just northeast of the mouth of Mobile River where the Tensaw River empties into Mobile Bay. The pass has a depth range ranging from about two to 18 feet. There is plenty structure there including submerged wood pilings. There is also an artificial reef there. Redfish, speckled trout, sheepshead, white trout, croaker, flounder and ground mullet can be caught at this unique location.

2. Upper bay flats and channels. Good bream in spring and fall. Fish holes for speckled trout in winter.

3. Daphne Pilings. Pilings are located (see map in this section) on the eastern shore of Mobile Bay near the town of Daphne. Numerous private piers here were destroyed by a hurricane. Some of the debris has been recovered while other debris floated away. There is enough debris left, however, to offer good fishing around what remains. The barnacle encrusted pilings attract huge numbers of sheepshead in the fall. Specks and redfish are also attracted to the area as well as Black Drum. There is also an artificial reef at the end of one of the old piers.

4. Middle Bay lighthouse. Built in 1885, this octagon-shaped lighthouse has been saved as a Historic Landmark and many fishermen in the area are glad that it was. The lighthouse, which once had a cow on its surrounding porch to provide milk for the light keeper's baby, is supported by seven steel legs. The lighthouse provides shade to the area underneath and the steel legs provide structure for baitfish. Speckled trout, flounder, white trout and redfish can be caught around the structure in October.

5. Weeks Bay and Fish River. Fall and winter for redfish, trout and flounder. Fish deep holes after cold snaps for speckled trout.

6. Magnolia River. Fall and winter for redfish, trout and flounder.

7. Eastern Shore piers. Countless piers jut into the bay from the northern end of the bay to Mullet Point where Bon Secour Bay begins, a distance of more than 10 miles. Fish piers and pilings from spring to January for speckled trout, summer trout and flounder.

8. Bon Secour Bay. Shallows extend many yards from shore, then drop quickly to between 7 and 10 feet. Fish the dropoff for reds, flounder and trout. Work the shallows at night with a light and gig for flounder.

9. Dixey Bar. Incredible fishing for redfish here. Lots of sharks so use a braided leader.

10. Oyster reef obstruction. Flounder, trout, drum, sometimes croaker. Depth rises from 8 to 6 feet.

11. Bon Secour River. Reds, flounder and trout. Work deep holes in fall and winter.

12. Flats behind Fort Morgan peninsula. Flounder follow this shoreline on their fall migration from the bay to the open gulf. Extremely good for flounder November through January.

13. Obstruction in bay. 12 feet. Trout, flounder, sometimes bluefish, blue runner, Spanish and northern mackerel.

14. Intercoastal Waterway channel. Use fishfinder to locate reds, trout, croaker schools along channel. Work channel for flounder until early winter.

15. Drilling rigs. Trout, reds, bluefish, flounder, mackerel in spring and summer, drum, croaker. Circle rig and find side where baitfish are holding.

16. Ship channel. Bay drops abruptly from 10 to near 50 feet, and runs from mouth of bay to docks at Mobile on northwest corner of bay. Not a tremendous amount of fishing in channel, but it can be worked successfully for Gafftopsail Catfish, redfish and black drum.

17. Eastern tip of Dauphin Island. Channels and flats can be worked for ground mullet, sharks and croaker. Specked trout on both inside and outside of island tip.

18. Back of Dauphin Island. This large area of channels and flats encompasses the Intracoastal Waterway, Dauphin Island bridge and utility line structures, oyster beds, and part of Mississippi Sound. Try for reds, speckled trout, croaker, flounder, Jack Crevalle, drum and sharks. Use bottom finder to locate structure and fish. Use bottom alarm, if equipped, as depth can change abruptly.

19. Cedar Point Pier. Bait, some tackle, fishing information, snacks available. Parking. All bay species.

20. Western shore has wide flats before dropping to between 7 and 10 feet. Try flats in warm weather at night for gigging flounder. Otherwise work area for croaker, flounder and speckled trout. Dropoff is primary area to check. A few reds are caught here, but it is not the bay's best area for reds.

21. Gaillard Island. This man-made island was built from the waste from dredging the ship channel. The island is surrounded by huge rocks designed to lessen erosion. There are several deep holes around the island made from dredging in the area so barges could get their heavy loads to the island. Flounder is the main species caught around the island while speckled trout and redfish can be caught in the deeper holes.

22. Fowl River and East Fowl River. Croaker, flounder, trout and reds, particularly fall and into winter.

23. Dog River. Same as 17.

24. - 39. Artificial reefs.

The Ultimate Guide to Alabama Fishing

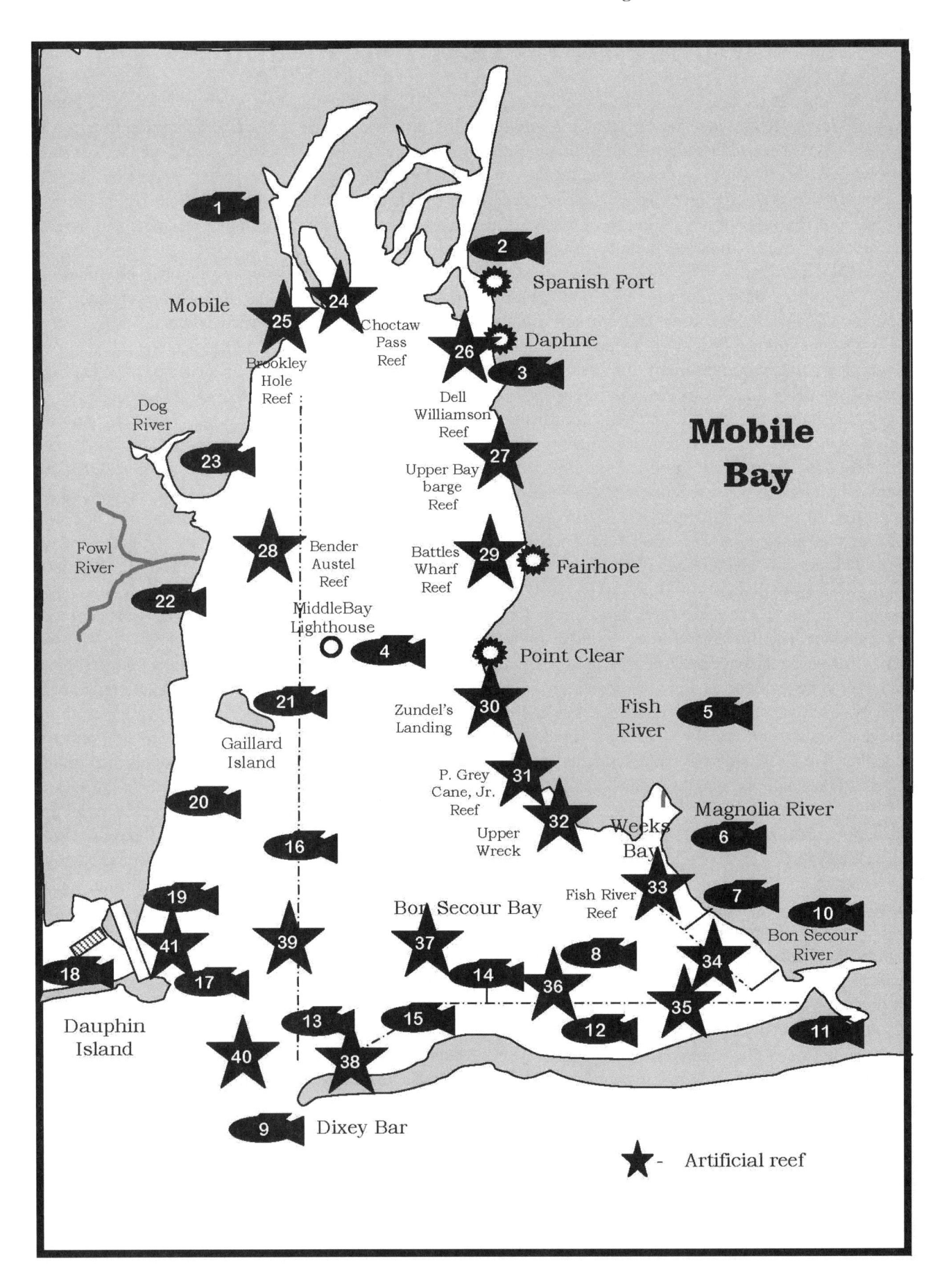

Mobile Bay

- Spanish Fort
- Mobile
- Choctaw Pass Reef
- Brookley Hole Reef
- Dog River
- Daphne
- Dell Williamson Reef
- Upper Bay barge Reef
- Fowl River
- Bender Austel Reef
- Battles Wharf Reef
- Fairhope
- MiddleBay Lighthouse
- Gaillard Island
- Point Clear
- Zundel's Landing
- Fish River
- P. Grey Cane, Jr. Reef
- Upper Wreck
- Magnolia River
- Weeks Bay
- Bon Secour Bay
- Fish River Reef
- Bon Secour River
- Dauphin Island
- Dixey Bar

★ - Artificial reef

278

Mike Bolton

Perdido Bay

Visiting this dividing line between Alabama and Florida by boat can be somewhat surprising. A narrow opening in the Gulf of Mexico suddenly turns into a massive body of water that stretches more than 20 miles to the northeast.

It's layout with its high dunes at the small entrance is so deceiving that the Spanish who discovered it actually lost it for a while. It is so well hidden from the Gulf that pirate Jean Lafitte actually used it for his base of operations in the 19th century.

The jetties on the east and west sides of Perdido Bay's mouth are the primary fishing locations.

Divers and fishermen share the territory, particularly at high tide when gulf water rushing into the bay clears the water sufficiently for diver's to have some visibility. Fishermen with boats anchor off the jetties and cast toward them with artificial, cut or live bait. Those without boats stand on the rocks and cast outward. Those on the rocks have a much higher risk of losing fish and tackle in the sharp rocks.

Flounder lie in the sand just off the jetties, while Bluefish, Spanish Mackerel and Mullet cruise in schools just off the rocks. The Bluefish can be caught on 0- or 00-size spoons or blue and silver Rebel-style lures, fished with a rapid retrieve. Spanish are attracted to any of dozens of small lures—ranging from silver spoons 0 or 00 size to jigs and multi-hook rigs, almost all of which are manufactured specifically for Spanish Mackerel and are available at area tackle shops.

The Mullet are caught in cast nets, but occasionally can be caught on tiny dough balls. Few fishermen go for Mullet with a hook and line, however. Live shrimp and bass-type, minnow-shaped crankbaits are used to cast the jetties for speckled trout. Live shrimp also are deadly for flounder, grouper and redfish. Grouper also are caught on live bait such as pinfish or Mullet minnows.

Dead shrimp or cut bait are effective for redfish, flounder, Sheepshead and grouper.

Redfish and flounder also can be caught on Sting Ray Grubs, which seem to be most attractive when tipped with a bit of cut bait or squid. They are cast, then slowly retrieved, allowing the grub to bounce bottom.

On the west side of the channel a concrete sea wall extends for several hundred yards. Many of the same species caught along the jetties can be found running along the wall. The same baits and fishing techniques can be used here as on the jetties. Except at slack tide, current is swift in the channel. While smaller hooks may do best for general fishing, pyramid sinkers of an ounce, sometimes more, work best to hold bait on bottom.

On the east side of the channel, perhaps a hundred yards north of the jetty, the sand is scalloped out, forming a calmer, deeper pool. Bits of debris (food to fish) settle in the pool, and small baitfish, seeking refuge from the current, move in and out of the pool. It is an excellent spot to try for flounder, using small pinfish, croaker or Mullet minnows that can be netted along the shore where you are fishing.

This pool can be cast into from shore or by wading out toward it.

Turn left into Cotton Bayou just after you cross under the bridge and you will move across shallows, then will find a deep hole. This hole also is worth trying for flounder, white and speckled trout. Mullet can be netted in this area.

Trout are occasionally taken on live bait and by casting small minnow-like lures, spoons, jigs or Sting Ray Grubs in Cotton Bayou, particularly around piers and in deep holes. Use your depth finder.

Due north of the Alabama Point Bridge is Terry Cove, where many charter boats dock. Water in Terry Cove drops from flats along the shore to about 10 feet. Fish the drop-offs for white and speckled trout.

The open water between Terry Cove and the tip of Ono Island sometimes holds Bluefish. This can be an iffy proposition, but a quick scan with the fish finder or a quick troll with small spoons, jigs or minnow-shaped lures will reveal their presence.

The channel leading to Wolf and Perdido Bays is a winding one, pocked with holes and flanked by sand bars. In winter, the deep holes hold speckled trout that can be caught on live shrimp fished on bottom. In summer, the trout are scattered, often foraging along the channel edges for food. Work lures along these cuts.

Another narrow channel leads immediately eastward from the bridge to Old River. Redfish and trout can be caught in the mouth of Old River, particularly where flats drop into deep water near the pilings.

The Ono Island bridge is a good spot for Sheepshead and speckled trout, while Mullet can be found running the river's shallower depths. On the eastern end of Ono Island, work the cuts for specked trout.

Just beyond the eastern end of the island, the Intracoastal Canal narrows and goes under the highway bridge that connects Perdido Key to Pensacola, Fla. Under the bridge is some hard rubble that holds trout, and is worth trying.

In Perdido Bay proper, Innerarity Point should be worked for trout and flounder. Much of the bay's shoreline is marshy, with shallow flats extending outward for many yards. Watch the water and your depth finder carefully or you can find yourself grounded. The bottom drops off abruptly from one or two feet to 10 to 12 feet. White trout, flounder, speckled trout and redfish should be sought along the drop-offs. While the entire shore can be worked, particular attention should be shown to the mouths of creeks, inlets and bayous.

On the northwest side of the bay, Red Bluff point and adjoining Soldier Creek should be worked. There are several piers jutting into the bay at Red Bluff and there are submerged pilings that tend to draw fish.

Soldier Creek as well as adjoining Palmetto Creek have channels of 8 to 10 feet that hold trout and redfish. Take care navigating into the creeks since shallows abound.

From the opening at Rose Point that leads to Wolf Bay to Sapling Point where Wolf Bay opens, the mouths of Roberts Bayou and Ingram Bayou should be worked for redfish and trout. Sapling Point also should be worked for specks.

In the upper end of Wolf Bay, Hammock Creek and Graham Bayou should be worked for redfish, white and speckled trout. At mid-bay, Mulberry Point juts outward from the west. Work the drop-offs around the point for specks and redfish.

Bay species such as redfish, trout, bluefish and flounder are much better at certain times of the year than others.

For example, remember that specks are extremely sensitive to drops in temperature. Thus, after a cold snap, look for them ganged up in any little hole of deep water they can find. Sometimes a hole only two or three feet deeper than surrounding water will hold a huge concentration of fish. These fish may be sluggish, but live shrimp fished right in their midst can bring strikes.

Croaker and whiting are almost year-around residents, however sometimes are hard to find in January and February, the coldest months.

Flounder also are found most frequently in spring, summer and fall. They tend to migrate into the open gulf in fall, and sometimes can be caught in great numbers during that migration, especially in the channel approaches to the mouth of the bay, which they all must pass through to get to the open gulf.

The spring, summer and early fall redfish tend to be small, with the larger Redfish, called bull reds, moving in mid-fall and remaining through January.

Sheepshead can be found around bridge abutments, piers or pilings—anywhere that crustaceans attach to structure. These fish feed on the crustaceans. They can be found year around.

Mike Bolton

Perdido Bay's best fishing spots

1. Jetties flank each side of the mouth of Perdido Bay. These stone structures can be reached by land and by boat. Exact depth and channel configuration changes from time to time due to erosion, silting, channel dredging, storms, etc. Species include grouper, mullet, sheepshead, Spanish Mackerel, flounder, redfish, bluefish, croaker, whiting, Jack Crevalle, a few sand sharks and trout.

2. A concrete seawall runs for many yards along western side of channel into Perdido Bay. Many of same species as found on the jetties can be found here, although incidence of grouper is diminished. Current is swift, requiring significant weight to keep bait on bottom.

3. Unless altered by recent dredging, there is a significant dip in the bottom behind the jetty on the east side of the channel. Good flounder fishing in this hole in spring, summer and fall.

4. Entrance to Old River has pilings adjacent to narrow channel. Check area in fall and winter for trout.

5. Ono Island Bridge. Fish around structure for sheepshead, trout, reds. Also check Old River channel for depth finder for same species, plus flounder and white trout.

6. A hole surrounded by flats just to the north and west of Perdido Pass Marina is frequently fished for trout, flounder and occasional reds. The shallower areas are worked by net for mullet.

7. Cotton Bayou. There is a free public boat ramp at the end of the bayou. The bayou can be worked for croaker, speckled and white trout, flounder and redfish. Canals cut into subdivisions are worked for trout and croaker. From Orange Beach Marine to Robinson Island, good area for white and speckled trout, ground mullet and croaker.

8. Canals cut into residential subdivisions are worked for trout and croaker.

9. From Orange Beach Marine to Robinson Island, good area for white and speckled trout. Fish dropoffs from deep side, casting to shallow and working to depths in warm weather. Check deep holes in winter for trout ganged up.

10. Bluefish roam cuts in area just inside mouth of bay, along with occasional blue runner and small mackerel in spring and summer.

11. Narrow and winding cuts and channel leading to main bay can be worked for speckled trout. Fish dropoffs from deep side, casting to shallow and working to depths in warm weather. Check deep holes in winter for trout ganged up.

12. Intracoastal canal. Check flats to either side for flounder and mullet, particularly in summer and fall. Work edges of channel in warmer months for trout, and deep holes for them in winter.

13. Bridge connecting Perdido Key to Innerarity Point has hard bottom around it. Particularly good in cool weather for trout.

14. Flats at Innerarity Point drop into channel of 20 feet. Check edges of channel for trout. In warm months use light and gig on flats to take flounder.

15. Tarkiln Bay and Bayou. Flounder, speckled and white trout.

16. Soldier Creek and Red Bluff and northward along shore. Work shallows in warm months for flounder at night. Fish drop-offs for speckled and white trout.

17. Perdido Beach area, Palmetto Creek and Spring Branch. Speckled and white trout, redfish

18. Rose Point area. Fish point for flounder and trout.

19. Roberts Bayou. Redfish and speckled trout.

20. Ingram Bayou. Redfish and speckled trout.

21. Sapling Point where Wolf Bay opens to the north. Redfish and speckled trout.

22. Hammock Creek and other small streams opening into Wolf Bay near its northeastern end. Speckled trout, white trout and redfish.

23. Northwestern end of Wolf Bay, including Graham Bayou. Redfish, white and Sspeckled trout, particularly in cooler months.

24. Mulberry Point. Redfish and speckled trout.

25. Intracoastal Waterway. Use depth finder to find fish roaming edge of channel. Good, especially in cooler months, for reds, trout, flounder and (with net) mullet. This is a particularly good area for fresh-water crappie fishermen in the fall since fish in the channel can be caught on live shrimp with the same bottom-fish-and-drift pattern used in Alabama. The waterway runs all the way to Mobile Bay and is spanned by the Alabama 59 bridge leading into Gulf Shores. Fishing is good along the length of the waterway. Particular attention should be given to fishing the several creeks flowing into the waterway, bridge structure, riprap and old tree structure remaining along the edge of the channel from when the waterway was cut.

26. Mill Point and Bear Point. Gig flounders in the shallows in warm months.

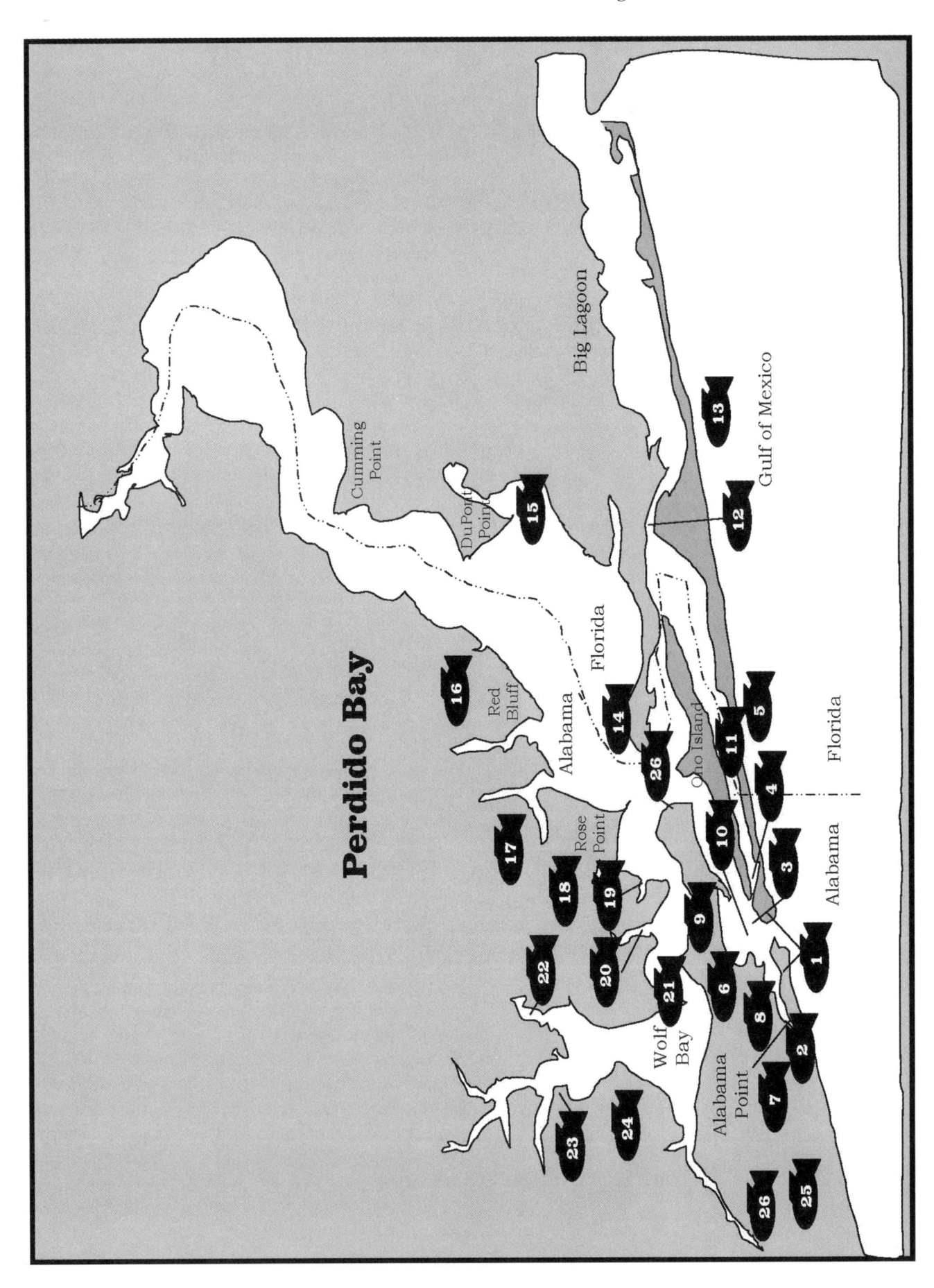

The Ultimate Guide to Alabama Fishing

Perdido Bay

Alabama's State Lakes

Alabama is blessed with incredible waterways that provide more than 20 major reservoirs to its fishermen. Unfortunately for many state residents, those lakes are not distributed evenly across Alabama. Several areas, including much of the southern third of the state and much of upper mid-western portion, lack the large reservoirs that so many Alabama fishermen enjoy.

But that doesn't mean Alabamians who live there don't have the opportunity to enjoy good fishing.

Wildlife and Freshwaters Fisheries saw this deficiency decades ago and created public fishing lakes in 20 counties concentrated in areas lacking sufficient natural waters to satisfy the fishing needs of the public. These lakes range from 13 acres to 184 acres.

These lakes are not lip service to silence whining anglers who might complain that they have nowhere to fish. The lakes are well-maintained, including weekly fertilization. These lakes gave up a tremendous average of 113 pounds of fish per acre, according to one study.

Studies have shown these lakes are undoubtedly the angler's best opportunity to catch a fish five pounds or larger in Alabama. Washington County lake did even better than that back in the early 1980s when it gave up what was then the Alabama state record—13 pounds.

All of these lakes are stocked with largemouth bass, bluegill and shellcracker and some of the larger ones have established crappie populations. Some also have hybrid bass, channel catfish and rainbow trout.

Many of these lakes are a haven for those who like to fish for catfish. Those lakes which have channel catfish have been stocked 50 to 200 catfish to the acre depending on fish populations.

A feeding program for catfish has been started on every lake stocked with them in an effort to concentrate them for fishermen. An average of 10 to 20 pounds of catfish food per day is thrown on every lake from May through October. Sixteen of the lakes are equipped with automatic feeders which throw food to the catfish on two-hour intervals throughout the day.

The lakes are fertilized weekly during warm months to increase the production of fish and prevent the growth of obnoxious aquatic vegetation. In addition, each lake is checked for lime requirement every three years.

Each lake is sampled every spring by electro-fishing to analyze fish development. In addition, each lake is sampled with seine nets once a month from May to September to determine the success of the bass spawn and to determine fish balance in the lake.

The excellent maintenance and resulting good fishing is only one of the plusses these lakes offer, many anglers believe. If you've ever fished one of Alabama's reservoirs, you've undoubtedly had your fishing interrupted at one time or another by water skiers, jet skiers, pleasure riders or swimmers.

These state lakes are strictly fishing lakes and other water activities are banned. Only five of the 20 lakes allow outboard motors to be used and then at trolling speed only.

Another management tool these lakes use is catch records. All fishermen are required to weigh every fish that is kept. These harvest records are computerized and the total harvests of all species on every lake is monitored. Harvests are regulated so no more than 30 pounds of bass per acre are removed each year.

For the convenience of anglers, each lake is provided with a concession building and an individual from the community is contracted to operate the concession.

The contractors are required to keep accurate records, weigh fish, fertilize the lakes, and keep the lake and grounds attractive. Everyone 12 years of age or older is required to purchase a $3 fishing permit for every trip to the lake and this is required along with a state fishing license. The concessionaires sell state fishing licenses.

Boat rentals are $5 per day and there is a $3 launch fee for those who bring their own boats. Paddles, cushions, life jackets, fishing supplies and food and drink are also available at concession stands.

The state lakes

1. Barbour County Lake
Location: Six miles north of Clayton on Barbour County 49.
Phone No: (334) 775-1054.
Lake acres: 75.
Fish species: Bass, bream, catfish, crappie and hybrids.
Additional facilities: Picnic areas, launching ramp, rental boats, camping available.

2. Bibb County Lake
Location: Five miles north of Centreville on Alabama 5.
Phone No: (205) 938-2124.
Lake acres: 100.
Fish species: Bass, bream, catfish and crappie.
Additional facilities: Launching ramp, fishing pier, rental boats, self-contained campsites.

3. Chambers County Lake
Location: Five miles southeast of Lafayette on Chambers County 55.
Phone No.: (334) 864-8145.
Lake acres: 183
Fish species: Bass, bream, catfish and crappie.
Additional facilities: Launching ramp, picnic tables, fishing pier, rental boats, barbecue pavillion.

4. Clay County Lakes (three lakes)
Location: One mile west of Delta on Alabama 47.
Phone No.: (256) 488-0038.
Lake acres: 13, 23 and 38.
Fish species: Bluegill, shellcracker, Florida largemouth, channel catfish and hybrids.
Additional facilities: Campsites, launching ramp, picnic table, fishing pier and rental boats. 343 land acres.

5. Coffee County Lake
Location: Four miles northwest of Elba on Coffee County 54.
Phone No. (334) 897-6833.
Lake acres: 80.
Fish species: Bass, bream, catfish, crappie and hybrids.
Additional facilities: Picnic tables, fishing pier, launching ramp, rental boats.

6. Crenshaw County Lake
Location: Five miles south of Luverne on on U.S. 331.
Phone No.: (334) 335-2572.
Lake acres: 53 acres.
Fish species: Bass, bream, catfish, crappie and hybrids.
Additional facilities: Self-contained campsites, picnic tables, fishing pier, launching ramp and rental boats.

7. Dale County Lake
Location: One mile north of Roy Parker Road (Dale County 36) in Ozark.
Phone No.: (334) 774-0588.
Lake acres: 92.
Fish species: Bass, bream, catfish, crappie and hybrids.
Additional facilities: Self-contained campsites, picnic tables, fishing pier, launching ramp, rental boats.

8. Dallas County Lake
Location: 11 miles south of Selma on Alabama 41.
Phone No.: (334) 874-8804.
Lake acres: 100.
Fish species: bass, bream, catfish, crappie and hybrids.
Additional facilities: Picnic tables and grills, fish-

ing pier, launching ramp and rental boats.

9. DeKalb County Lake
Location: One mile north of Sylvania off of DeKalb County 47.
Phone No.: (256) 657-1300.
Lake acres: 120.
Fish species: Bass, bream, catfish, crappie and hybrids.
Additional facilities: Kiddie playground, picnic pavilion, barbecue grills, self-contained campsites with restrooms, launching ramp, rental boats.

10. Escambia County Lake
(Leon Brooks Hines Lake)
Location: 23 miles east of Brewton off of Escambia County 11.
Phone: (251) 809-0068.
Lake acres: 184.
Fish species: Florida largemouth bass, bluegill, coppernose bluegill, shellcracker, channel catfish.
Additional facilities: Picnic tables, launch ramp, fishing pier, rental boats.

11. Fayette County Lake
Location: Six miles southeast of Fayette on Fayette County 26.
Phone No: (205) 932-6548.
Lake acres: 60.
Fish species: Bass, bream, catfish and crappie.
Additional facilities: Picnic tables and grills, launching ramp, rental boats.

12. Geneva County Lakes (Two lakes)
Location: 10 miles south of Enterprise on Alabama 27, then right on Coffee County 65 for seven miles. Turn right on Geneva County 64 for two miles, then right on Geneva County 63.
Phone No: (334) 684-0202.
Lake acres: 32 and 33.
Fish species: Bass, bream, catfish and crappie.
Additional facilities: Picnic tables and grills, launching ramp.

13. Lamar County Lake
Location: Eight miles west of Vemon on Alabama 18, then five miles north on Lamar County 21.
Phone No.: (205) 695-8283.
Lake acres: 68.
Fish species: Bass, bream, catfish and crappie.
Additional facilities: Fishing pier, launching ramp, rental boats, picnic tables and grills.

14. Lee County Lake
Location: Six miles southeast of Opelika on Alabama 169, then one mile west on Alabama 29.
Phone No.: (334) 749-1275.
Lake acres: 130.
Fish species: Bass, bream, catfish, crappie and hybrids.
Additional facilities: Picnic tables, launching ramp, rental boats.

15. Madison County Lake
Location: Seven miles east of Huntsville on U.S. 72, then left on Ryland Pike Road to Maysville.
Phone No.: (256) 776-4905
Lake acres: 105.
Fish species: Bass, bream, catfish, crappie, rainbow trout and hybrids.
Additional facilities: Launching ramp, fishing pier, rental boats, picnic tables and grills, self contained campsites.

16. Marion County Lake
Location: Six miles north of Guin on U.S. 43.
Phone No.: (205) 921-7856.
Fish species: bass, bream, catfish, crappie and carp.
Additional facilities: Rental boats, picnic tables and grills, launching ramp and fishing pier.

17. Monroe County Lake
Location: 12 miles north of Monroeville on Alabama 21 at Beatrice, then five miles west on Monroe County 50.
Phone No.: (205) 921-7856.
Lake acres: 94.
Fish species: Bass, bream, catfish, crappie and hybrids.
Additional facilities: Rental boats, picnic tables, launching ramp, fishing pier.

18. Pike County Lake
Location: Three miles south of Troy on Alabama 87, then right on Pike County 39.
Phone No.: (334) 242-3471.
Lake acres: 45.
Fish species: Bass, bream, catfish, crappie and hybrids.
Additional facilities: Picnic tables, rental boats, launching ramp, fishing pier.

19. Walker County Lake
Location: Three miles southeast of Jasper off old U.S. 78.
Phone No.: (205) 221-1801.
Lake acres: 163.
Fish species: Bass, bream, catfish, crappie, hybrids.
Additional facilities: Fishing pier, launching ramp, picnic tables, rental boats, self-contained campsites.

20. Washington County Lake
Location: Two miles west of Millry off of Washington County 34.
Phone No.: (334) 242-3471 or (251) 626-5153.
Lake acres: 84.
Fish species: Bass, bream, catfish, and hybrids.
Additional facilities: Launching ramp, rental boats, picnic tables, grills, overnight camping.

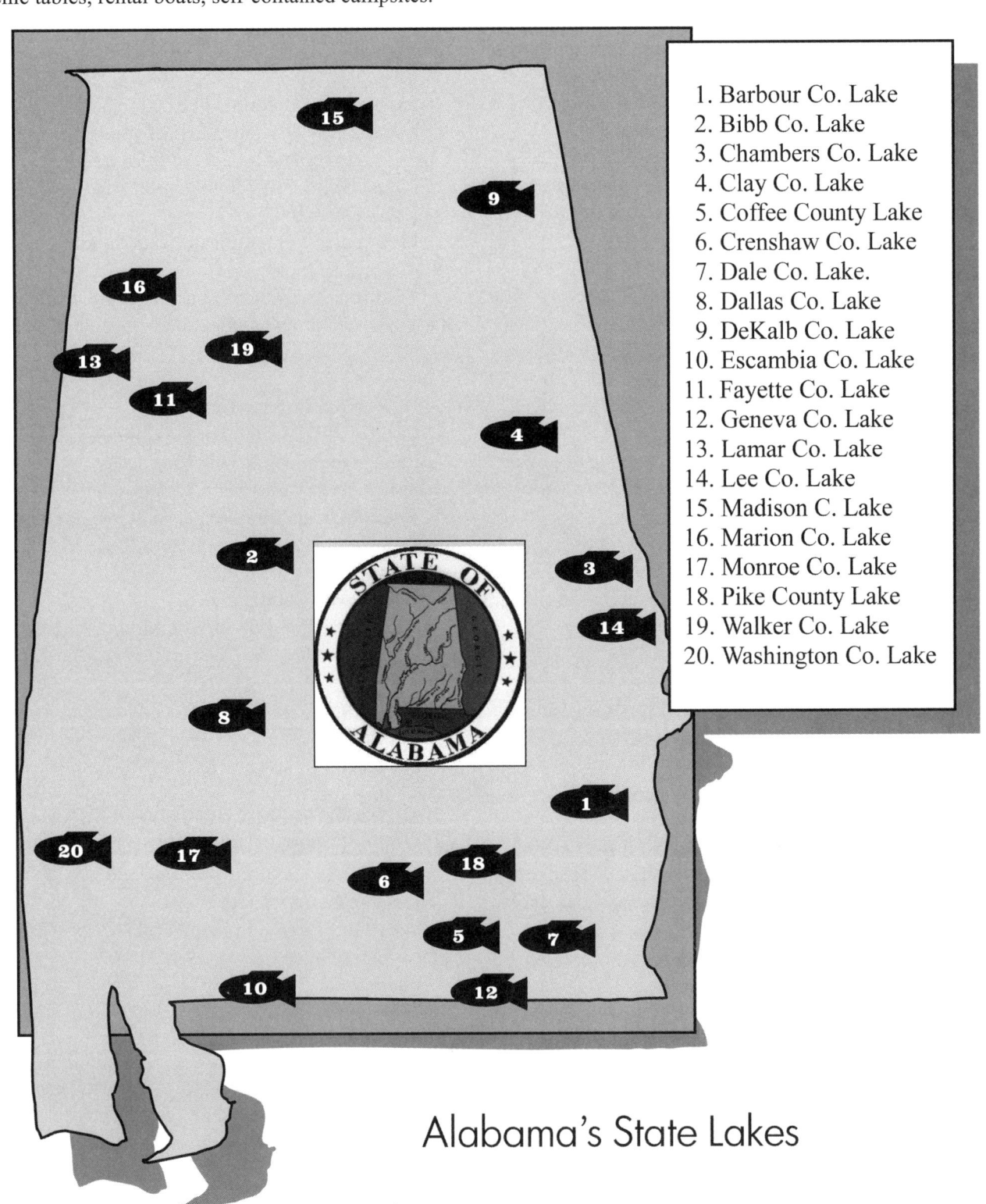

1. Barbour Co. Lake
2. Bibb Co. Lake
3. Chambers Co. Lake
4. Clay Co. Lake
5. Coffee County Lake
6. Crenshaw Co. Lake
7. Dale Co. Lake.
8. Dallas Co. Lake
9. DeKalb Co. Lake
10. Escambia Co. Lake
11. Fayette Co. Lake
12. Geneva Co. Lake
13. Lamar Co. Lake
14. Lee Co. Lake
15. Madison C. Lake
16. Marion Co. Lake
17. Monroe Co. Lake
18. Pike County Lake
19. Walker Co. Lake
20. Washington Co. Lake

Alabama's State Lakes

Mike Bolton

Pond Fishing

More than 40,000 farm ponds and private lakes dot the Alabama landscape and a great number offer wonderful fishing for those fortunate enough to have access to them. An incredibly large percentage of the bass larger 10 pounds and bream larger than one pound caught in Alabama these days come from those waters.

These farm ponds' and small lakes' better-than-average fishing is a result of two important factors.

First, fishing pressure is significantly less on farm ponds and private lakes than on public reservoirs. This lack of fishing pressure offers a calming effect to its inhabitants. Unlike public reservoirs where fish become "educated" by sometimes seeing five or six spinnerbaits or 10 or 15 plastic worms a weekend pass by their face, fish from these waters are less wary and strike more readily.

Secondly, the size of farms ponds and private lakes lend themselves to being better maintained than public reservoirs. Fertilization of these smaller bodies of water is common, whereas no public reservoirs receive fertilization.

Fertilization, it should be explained, accomplishes several things. One, properly fertilized water blocks out sunlight and prevents unwanted vegetation, such as lily pads, from growing from the bottom and choking out a lake.

Two, fertilization provides a natural food chain for fish. Fertilized waters cause the growth of plankton which provides food for insects, as well as small minnows and other baitfish. These nutrient-rich insects and small fish provide food for bream, which in turn provide food for bass and catfish. The result is healthier, faster-growing fish than you'll find in reservoirs.

Even in farms ponds and private lakes not fertilized by man, they often receive the watershed from farms and gardens and as a result are fertilized much in the same way.

Today, several pond management companies operate in Alabama. For those with the money, these lakes are limed to a correct pH level then fertilized. Many are stocked with shad and fathead minnows for baitfish.

Unlike many stock ponds across the state that were constructed to water cattle and fish were added as an afterthought, many ponds today are built with fishing in mind. Structure such as trees and boulders were added rather than removed.

There's no better proof of what well-maintained bodies of water can produce than the two consecutive state record largemouth bass which came from private bodies of water no less than 57 days apart in 1987.

The second, the 16-pound, 8-ounce largemouth which is still the record today, was caught by the late T.M. Burgin. It came from a 22-acre private lake in Shelby County.

That lake received periodic fertilization, lime applications and 100 pounds of floating catfish food per day. The lake also routinely gave up bream in the two-pound range and catfish over 25 pounds.

Today pay-to-fish "Super lakes" can be found all across the state. For a hefty fee anglers can experience a little of how the rich folks get to fish.

Fishing methods

One of the most exciting things about fishing farm ponds and private lakes is knowing that few are ever well-fished.

Unlike public waters where excellent fishermen chart every inch of every piece of structure with depthfinders, these smaller bodies of water are rarely ever attacked by anything stronger than a small jonboat and wooden paddle.

Fishing these farms ponds and small lakes is much like fishing reservoirs in that fish will relate to structure. Structure in these small bodies of water will range from fallen trees to ledges, humps, rocks, deep holes and standing drainpipes.

One of the most effective procedures in fishing these

waters is, if possible, locating the owner or person who built the lake and has seen it without water in it. Then ask questions.

How deep is the deepest portion? Is that deep area around the dam or elsewhere? Were trees cut when the pond or lake was built? If so, do the stumps remain? How about creek channels cutting through? Any ditches or ledges? Was debris piled when the lake was built? If so, where? Does shallow water offer sand or pea gravel? Are there incoming creeks? How about underwater springs that may offer cooler or warmer water than the rest of the lake?

Having a few of those questions answered can immediately put the lake's underwater image in your head and give you an idea where fish are likely to hide.

If that isn't possible, a portable depthfinder will work wonders. Slowly travel the lake charting its depth. Using a pen and pencil, draw the lake's outline as you imagine it. Fill in depths on the map and record any irregularities such as ditches, humps and deep holes.

If you don't have access to a portable depthfinder, a string and lead weight works well, albeit slower. Mark the string off in five foot increments and drop it often as you tour the lake. Hopefully, you'll sometimes hang in stumps, brush and limbs and break off the string. The time spent is well worth the knowledge gained.

This deliberate study of the lake may very well take up an entire fishing day, but it will result in better fishing. It is estimated that as much as 75 percent of the water in a lake is unproductive. Eliminating this water will mean you'll spend 75 percent more time fishing where the fish are going to be.

Bank fishing

Fishing from the bank of small lakes is an age-old fishing method that works well. But just as fishing from a boat, eliminating unproductive water is the key to successful fishing.

Bass fishing from the bank is sometimes more efficient than bream or catfishing because the bass angler is usually on the move and is more likely to stumble upon productive water.

Bream and catfish anglers are more likely to sit and wait in one spot for long periods of time before it dawns on them the area just isn't holding fish.

Locating structure from the bank isn't that difficult of a chore, however. A cheap pair of Polarized sunglasses will work wonders and many times when circling the lake on foot, the angler can see into the water and see fallen trees, stumps and other cover.

A small weight (no hooks) attached to the fishing rod's line makes for an excellent depthfinder. A cast and a glance at the line's angle will quickly tell you the waters depth. Throw often and look for ledges, humps and brush. A slow retrieve will help you feel for structure as the weight bounces along the bottom.

Mike Bolton

Alabama's most popular freshwater game fish

Bass, largemouth - (black bass, northern bass, bigmouth bass). The largemouth bass is the most common bass found in Alabama waters. The stumpy largemouth is easily identifiable by its coloration, which begins as a white belly, switching to moss green and eventually to dark green as you go up the body to the dorsal fin. A dark horizontal line runs the length of the fish's body and separates the white underbody from its darker upper half. Other characteristics that separate the largemouth from its cousin, the spotted bass, is a jaw that extends past the eye, and a toothless tongue.

The largemouth bass can be found in all reservoirs, rivers and streams in Alabama, as well as in private lakes and ponds. Two distinct strains of largemouth, the northern and Florida strain, are present in Alabama. The difference in these two species, however, is not obvious to the untrained eye. Scale counts and tissue samples are the only true methods of identification.

The state record for largemouth bass is 16 pounds, 8 ounces.

Bass, smallmouth - (smallie, brownie, bronzeback). The small-mouth bass has the almost identical shape and fin placement of its cousin, the spotted bass, but its coloration makes it impossible to confuse the two. The smallmouth has a dusky bronze color with distinct, dark vertical bars. It also lacks the dark horizontal bar that runs the length of the bodies on largemouth and spotted bass. The smallmouth has red eyes.

The range of the smallmouth is limited in Alabama. Almost all are confined to the Tennessee River and its drainage system. Some smallmouth have been stocked in Smith Lake, but catches are rare. Some smallmouth also have been stocked in Yates and Thurlow reservoirs.

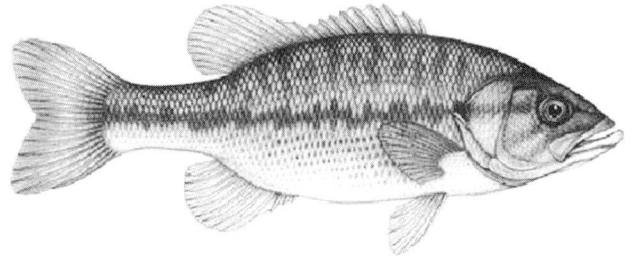

Bass, spotted - (spot, Kentucky spot). The spotted bass is often confused with its cousin, the largemouth, but distinguishing between the two isn't difficult. The spotted bass has more of a torpedo shape than the stocky-looking largemouth. It also has several rows of spots below the horizontal dark line that runs the length of its body. These spots may be faint or distinct.

There are other identifying characteristics. Approximately 95 percent of all spotted bass have teeth on their tongue. These teeth appear as a raspy patch. A light blue/green coloring also can be found on the jaw of most spotted bass. The spotted bass' jaw does not extend past its eye. Spotted bass can be found in all Alabama reservoirs but become less common south of Montgomery.

Bass, striped - (stripe, striper, saltwater stripe, rockfish, line-sides). The striped bass is a saltwater fish that has adapted to fresh water. This fish commonly swam from the Gulf of Mexico into Alabama waters early this century, but locks and dams have halted most of this migration. Striped bass are now stocked in most Alabama waters.

Having a slimmer body than the white bass or hybrid bass, the fish is easily identified by its silver appearance and its seven distinct, olive-gray stripes. These stripes are rarely broken, and the first stripe below the lateral line is complete to the tail. The striped bass' tongue also has two tongue patches.

Striped bass commonly grow to 20 pounds in Ala-

The Ultimate Guide to Alabama Fishing

bama. The state record is 55 pounds. The striped bass can be found statewide, but stocking has been discontinued in the Tennessee River.

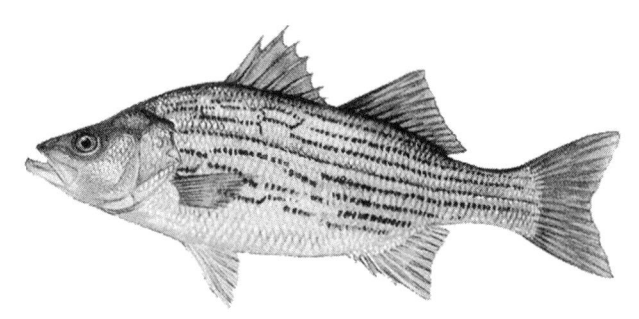

Bass, hybrid - (stripe, striper, hybrid bass). The hybrid bass in Alabama is a fish literally built in a laboratory. Biologists strip the eggs from the female striped bass and fertilize them with the milt of the male white bass. The resulting fry is technically a hybrid striped X white bass. The small fish are raised in laboratory jars, then holding ponds, before being stocked in Alabama waters.

This fish boasts several characteristics of both parents. It has the shorter, chunkier appearance of the white bass, as well as the white bass' frenzied feeding style. It has two tooth patches on its tongue, just as the striped bass, and the first stripe below the lateral line is distinct and complete to the tail. The remainder of its stripes are distinct and definitely broken, however.

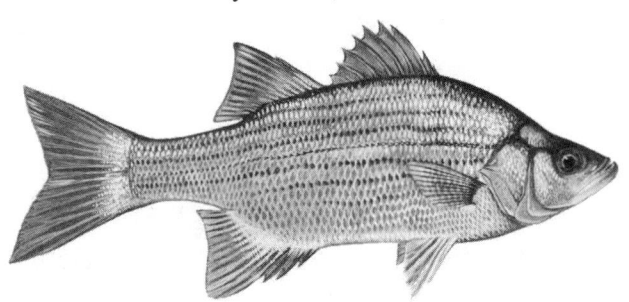

Bass, white - (stripe, white laker). The white bass is much shorter and chunkier in appearance than the striped bass, and rarely do these fish grow to more than 3 pounds. White bass are often confused with the striped bass and hybrid bass but need not be.

It has the silver color and seven stripes just as the stripe and hybrid, but its stripes are faint, and the tongue has only a single tooth patch. The first stripe below the lateral line is not distinct, nor complete to the tail. The white bass can be found statewide.

Bass, yellow - The yellow bass can be found in the lower Coosa drinage area and in the Mobile Delta, as well as Inland Lake. It resembles its close relative, the white bass, but its golden yellow sides and irregular stripes set it apart. It is smaller than most large bluegills, but has a more bass-like shape. The flesh is white, flaky and better tasting than the white bass.

Catfish, bullhead -There are several species of this catfish in Alabama. The black bullhead can be found statewide except for the Chattahoochee River. The brown bullhead and the yellow bullhead be found statewide. The snail bullhead and the spotted bullhead be found in the Chattahoochee River only. All are similarly shaped and their colors vary so greatly that colors can't distinguish. All have double dorsal fins, barbels, large heads and lack scales.

Catfish, blue - The blue cat often is confused with the channel catfish, but its deeply forked tail sets it apart. The blue catfish also has a straight margin on its anal fin as compared to the rounded fin of the channel cat. The blue catfish is found in all Alabama waters except the Chattahoochee River.

Catfish, channel - (spotted cat). Adults of this species are silver-gray to almost black in their upper body, while juveniles may have scattered black spots. It may be distinguished from its close cousin, the bullhead, by its slender tail base and its widely forked tail fin. The channel cat is found statewide.

290

Catfish, flathead - (Appaloosa cat, yellow cat). This is the largest catfish found in North America, and 100-pounders have been recorded. Both the body and head are flattened, and the protruding lower jaw further distinguishes this fish from other members of the catfish family. This fish is normally a dirty brown color and peppered with black sashes. This fish is found everywhere in Alabama but the Chattahoochee River system.

Catfish, white - (Albino cat) The white catfish is not white at all. The white cat is shaped like a bullhead, but has the coloration of the channel catfish.

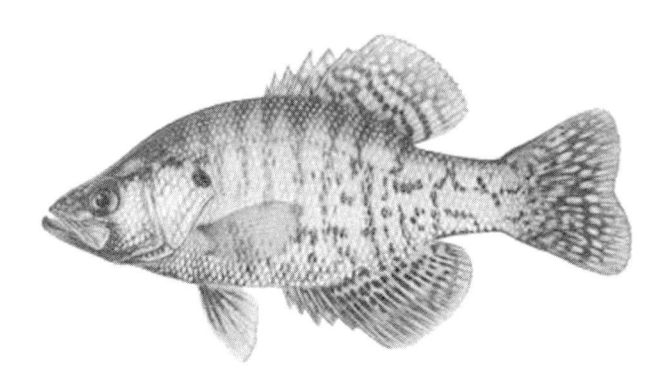

Crappie, white - (papermouth). Many anglers believe there is only one species of crappie, with the white crappie being the female and the black crappie being the male. That is not the case. The white crappie is a species that boasts a white/silver body with scattered dark spots that tend to form vertical bars on the body. The white crappie can tolerate much warmer water temperatures than the black crappie.

The state record for white crappie is 4 pounds, 8 ounces. It can be found statewide except for the Chattahoochee River system.

Crappie, black - (papermouth). The body of the black crappie is basically silver in color and often boasts a purplish sheen. The body also is heavily peppered in black spots. The black crappie has 7 to 8 dorsal fin spines.

The state record for black crappie is 4 pounds, 4 ounces. The blackcrappie can be found statewide.

Perch, yellow - (striped perch, lake perch) This fish, rarely found in Alabama, resembles a bass, but is a true member he perch family. Its most striking characteristic is its golden-yellow body with 6 to 8 dark bars which extend from the top of the back to the belly. It is also "sway-backed" in appearance. It can be found in the southern half of the state.

Pickerel, chain - The pickerel is often confused with the pikes and muskies because of its similar torpedo-like body shape, but it can be distinguished with its body markings. The sides are yellowish to greenish and overlaid with a chain-like pattern of black lines. This fish can be found statewide.

Sauger - The sauger is closely related to the walleye and the two are almost identical in appearance. The sauger has distinct dark spots he dorsal fin. The sauger has 17 to 20 soft rays in the second dorsal fin vs. 19-20 in the walleye. The flesh is firm, white and delicious. This fish can be found in the Tennessee River drainage system.

Sunfish, bluegill (bream) - The bluegill is found statewide in Alabama waters, including private lakes and ponds. A common forage for the largemouth bass, it is recognizable from other sunfish species by its solid black gill tab and the dark vertical bands that run the length of its body. Breeding males, which have a dark yellow or rust red breast, are well known for guarding the eggs in their beds during the spring spawn.

The bluegill is a noted hybridizer, often breeding with other members of the sunfish family to produce some unusual combinations.

Bluegill larger than 1-1/2 pounds are extremely rare, but the state record is 4 pounds, 12 ounces. That fish also is the world record. The bluegill is found statewide.

Sunfish, green (green perch, sand bass) - The green sunfish has a larger mouth and thicker, deeper body than most fish in the sunfish family. The fish has an extended gill flap with is bordered in black around light red, pink or yellow. The body is usually brown to olive green with a bronze to emerald green sheen, paling to yellow-green on the lower sides and yellow or white on the belly. There are emerald or bluish spots on the head. The fish has excellent, white, flaky flesh. This fish is common statewide.

Sunfish, longear (bream) - Although this fish rarely grows larger than 1/4 pound, its coloration arguably makes it one of the most beautiful fish in Alabama waters.

Its white-bordered, jet-black gill tab, which gets considerably large in the adult male, is the fish's most distinctive characteristic. The fish's back and sides are blue/green and broken by dark vertical stripes. The body also is peppered with yellow and emerald spots. The belly and sides of the head of breeding males are often bright orange-red.

The longear sunfish is found statewide.

Sunfish, redbreast (bream) - The redbreast sunfish's less-than distinctive shape causes it often to be referred to only as a bream, but the fish can be identified easily by two distinct characteristics. Easily noticeable is the flaming, burnt-orange breast of the adult and the greatly elongated black strap-like gill tab, which gets longer with age.

The redbreast sunfish is found only in the Coosa and Tallapoosa rivers and Lake Guntersville.

Sunfish, redear - (shellcracker, bream). The redear sunfish is found statewide in Alabama, but rarely is it called by that name. The well developed grinding teeth in its throat are used to crush snails, mollusks and other shellfish, hence its more popular Alabama name "shellcracker."

The redear sunfish is shaped much like its cousin, the bluegill, but several characteristics separate the two. The shellcracker is much lighter in color and lacks the light blue to dark blue gills. It also has a distinctive scarlet border on its gill tab, as compared to the solid black gill tab of the bluegill, and lacks the bluegill's vertical black bands.

The redear sunfish prefers clear, still water with abundant vegetation and finds its home around submerged logs and stumps. The state record for redear sunfish is 4 pounds, 4 ounces.

Warmouth - (bream) The rust and orange-colored warmouth has a more robust shape than its daintier cousins in the sunfish family. Several rust-colored streaks radiating from the fish's red eyes across the gill covers also help to identify the fish. Found statewide.

Trout, rainbow - The rainbow trout isn't native to Alabama waters, but has been stocked in a few locations. Colorization of this fish varies greatly, but it is noted for its broad pink stripe along the middle of the sides, interlaced with black spots. The fish is normally caught fly fishing, but can be caught on whole kernel corn and spinners. The flesh of this fish ranges from white to pink and is delicious. This fish has been stocked in Smith Lake, in the Smith Lake tailrace, at Tannehill State Park near Birmingham and Madison County Public Lake. Individuals have stocked the fish in several of the clear, cold streams near Springville in St. Clair County.

Walleye - The walleye is another fish imported into the state and stocked in small numbers. It is the largest member of the perch family. It has large glassy, opaque eyes. The walleye, often confused with the sauger, differs from that fish by its dorsal fin has a prominent dark blotch. The lower edge of the tail is white-tipped. The flesh is white to pink and one of the best meats of freshwater fish. The walleye was stocked years ago and can be found in state's upper half and in the Conecuh River.

Alabama's most popular saltwater game fish

Amberjack – An elongated fish easily identified by a dark slash mark across the eye. They are greenish gray on the back and lighter on the belly. The fish averages from five to 30 pounds and is frequently caught by charter boats which target them. Amberjack exceeding 100 pounds are occasionally taken from northern gulf waters. Inshore Amberjack are found over reefs and wrecks, seeming to prefer structure that offers a significant profile. The further offshore you fish, the larger the Amberjack seem to be.

These fish cruise over the reefs, preying on smaller reef fishes. They are caught by fishing live bait (chofers, white snapper or other baitfish) or jigging with 3/4- to 2-ounce jigs. Occasionally one is caught on cut bait. Thirty years ago, fishermen spurned Amberjack. However as other game species became scarce and restaurants popularized the firm, flavorful flesh, its popularity grew.

Always one of the toughest fighters in the sea, in the three decades the Amberjack has risen high in the rankings as a most sought after species. In fact, fishing pressure on Amberjack has led to stringent creel limits.

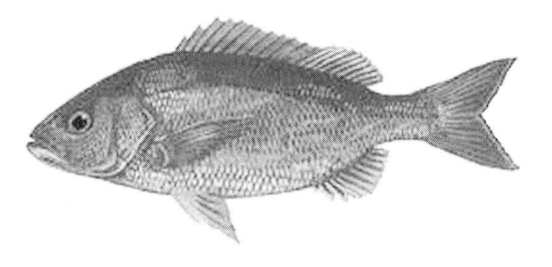

Beeliner – (Vermillion Snapper/Mingo Snapper) - This smallish red-colored snapper goes by many names and is hauled from the gulf by fishermen in tremendous numbers. These fish average a pound or less, but in areas of light fishing pressure, particularly offshore, they are caught ranging to seven or eight pounds. These fish are common around reefs and wrecks year around. They are caught by bottom fishing with cut fish or squid on small hooks.

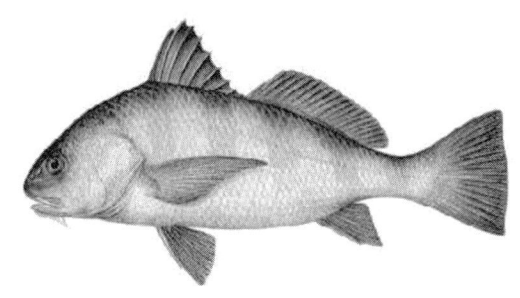

Black Drum - This heavy fish is distinguished by flesh-like chin whiskers. Younger ones are lighter with dark bars, becoming dark all over as they get older. While a record Black Drum will be more than four feet long and weigh more than 150 pounds, they average less than 10 pounds. These fish are found in bays, particularly in deep holes, along ship channels and around oyster beds. The fish is usually caught coincidentally on cut bait by bottom fishermen.

While the flesh is edible, the larger ones tend to be wormy. And, they have such tough hides many fishermen don't want to go to the trouble of cleaning them.

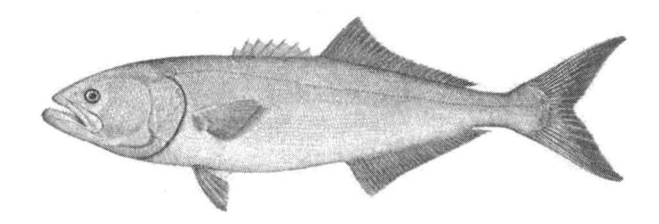

Bluefish - Shading from a blue-green back to silver sides, this longish fish commonly ranges from one to five pounds and one to two feet long. Larger ones as found along the Atlantic coast are not common in the northern gulf. They are found year-around, with several good runs coming in winter. This is a frenzy feeding fish, with cigar minnows the best natural bait and Rebel or Rapala lures the best artificials. A bucktail jig or bucktail spoon also is effective.

Lures are cast, must as a bass fisherman would cast and retrieve a crank bait. These fish are found around jetties, near reefs and wrecks and schooling into bays. The fish have strong, sharp teeth which should be avoided when extracting a hook. While the Bluefish reputation as a fighting game fish is legend, its reputation as a food fish is lacking. The flesh is considered only average tablefare, soft and with a strong taste. When fileting, take care to cut out the red streak running down each side. That improves the flavor dramatically.

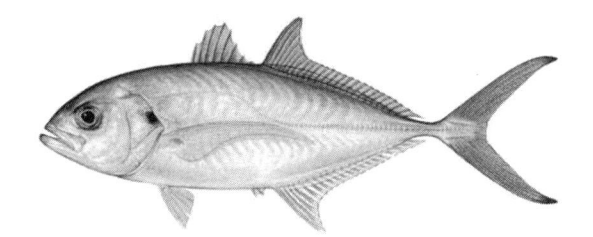

Blue Runner (Hardtail) - This smallish shallow-water fish is greenish on the back and upper sides, shading to a yellowish-silver, and with translucent fins. It runs about a foot long on average, peaking at about 20 inches. They average about a pound, and almost never reach five pounds. The Blue Runner is found in greatest abundance from spring to mid-summer. The fast, hard-hitting fish is important not so much for its food value, but because it is easily caught on small silver spoons or white jigs close to shore by surf fishermen, and because it makes such good live bait for larger game species.

They are caught along the beaches, and particularly around jetties, cuts and channels. They also range into the bays. They can be caught with light spinning tackle. Food value is only fair, and usually prepared grilled. Be sure to cut out the dark lateral line when cleaning. Removal will improve taste of the flesh.

Cobia (Ling) - This large fish ranges in color from a dark yellow-brown to a dark bluish-slate color on top and shading to white on its belly. Often a loner, this fish averages 12 pounds or more. Fourty pounders are not uncommon, with a record fish going over 100 pounds. It is caught in the spring off the beach just outside the surf by casting feather jigs, eels or live small fish to them. They also are caught the rest of the year over wrecks and artificial reefs with jigs and live bait. They are found both on the bottom and at mid-water over wrecks. Small Cobia are caught in bays in summer, but usually are under the minimum size limit. These fish are excellent food fare.

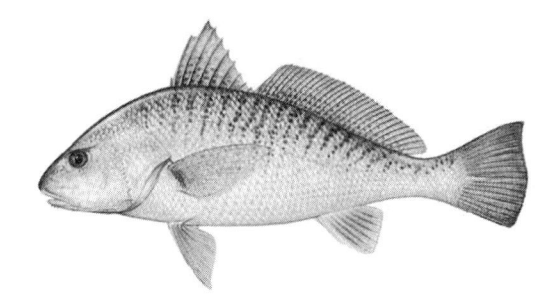

Croaker - This fish can be identified by a mouth slung just underneath a rounded nose, dark vertical bars running from their backs down their sides and their small chin barbels. This panfish ranges upwards to four or five pounds, but is commonly found to weigh a pound or so. They are found all year long in the northern gulf in every location—offshore reefs, bays, channels, the surf. The fish is an important one because it can be easily caught from shore with light tackle by the casual angler. Once located, they are easily caught with small hooks on bits of cut squid, shrimp or fish. They are among the best tasting of the common bottom fish.

Dolphin - Don't confuse this excellent game and food fish with the mammal called by the same name. The fish looks like a rainbow of blues, greens and yellows, with the stripes running the length of its long sleek body. The male differs from the female in that he has an extremely high forehead, giving him a squared-off look in the front. The Dolphin is among the fastest of all fishes, striking and fighting viciously. Size runs from small school Dolphin of two to five pounds to tackle-busters of more than 50 pounds.

They are found both inshore and offshore, but rarely are caught from land or piers. Most are caught trolling with spoons, jigs or bait; however, bottom fishermen catch them on cut bait from time to time. Like Amberjack, Dolphin are easier to catch if the first one hooked remains in the water drawing others into a feeding rally. Dolphin are frequently found around floating objects and floating mats of grass or weeds. The fish is common in the northern gulf from spring through fall. Food value, excellent.

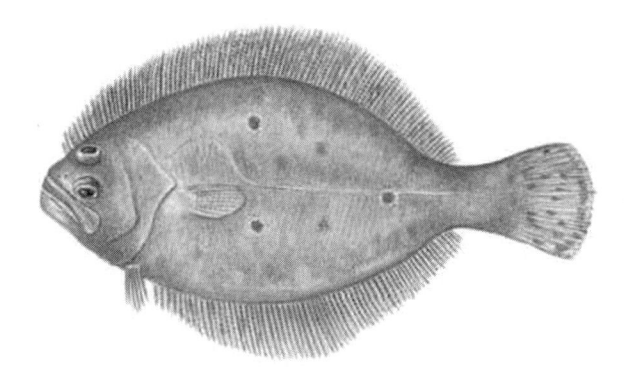

Flounder - This flatfish comes from a large family of fishes whose distinguishing characteristics are that they swim and lie on their sides, and often spend much of their time almost covered in the sand. A Flounder's top side is a mottled brown and it can change color to match that of its surrounding. Its other side (the one that sits on the sand) is white. This important game and food fish is a year-around inhabitant of the northern gulf, moving into shallow water, sometimes right up to the beach in warmer weather, and into deeper water offshore in cooler months.

It averages a pound or two and can range to well over 15 pounds, although that is rare. It can be found in many types of water and terrain—depending on the season.

They are caught in the gulf around reefs and wrecks, on sandy bottoms in bays, channels and the surf, and grass beds. Light casting or spinning tackle or light surf rigs can be used. Fish the bottom with live or dead shrimp or cut bait, live minnows or by casting a spoon or jig and retrieving it slowly along the bottom. Flounder also are gigged at night by shining a strong light into shallow water as you wade along or from a boat. If you move quietly and slowly, a Flounder will not move so long as the light is on it and can be easily gigged.

Food value, excellent.

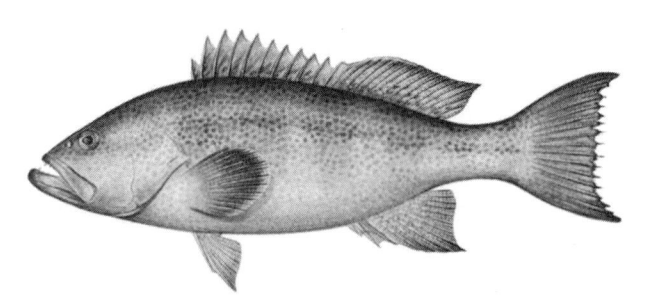

Grouper - There are 40 or more species of grouper, and several live in northern gulf waters—Jewfish, Gag (Black), Red, Kitty Mitchell, Yellowfin, to name a few. But more than 90 percent of those caught are the Red Grouper and Gag Grouper. Groupers as a whole are large fish and prolific ones. From the giant Jewfish (Now called Goliath Grouper) that can weigh more than 700 pounds to several sea basses that average less than a pound. The average Red or Gag grouper ranges from five to 25 pounds.

Most have the general shape of a freshwater largemouth bass. These fish are found in bays, offshore and inshore, year around. Virtually all groupers are bottom feeders and will take cut bait on occasion, however live bait such as small white snappers or pinfish are much preferred. These fish live around reefs, wrecks, rocks and other underwater obstructions. These are excellent food fish. Size, season and bag limits are under discussion by government authorities; check with local tackle shops for new legislation.

Jack Crevalle - It is a light olive-green color on the back, shading to gray, gold and yellow along the sides to the belly. It has a black spot on the gill cover and a deeply forked tail. It has a heftier body and blunter head than other members of the jack family. A savage striker and fighter, these are constantly moving fish. Due to their poor food quality, fishermen normally don't seek out Jack Crevalle.

However, once a fisherman catches one he likely will catch more since they travel in schools. On light salt water tackle a 20- to 30-pound Jack Crevalle can fight for a half hour or more before being landed.

They are fairly common in the gulf in spring and summer, moving into the bays in late summer and fall. They can be caught trolling just off the beaches, inshore and in channels along jetties. They also can be caught on cut bait. This fish ranges from three or four pounds to 40 pounds.

Food value, poor.

King Mackerel - This is one of the most important sport fish of the northern gulf. Anglers by the thousand troll inshore waters in the late spring, summer and fall for these sleek, silvery battlers, with whole fleets of charter boats dedicated to their capture. This pelagic, or migratory, fish makes its way up the Florida peninsula, arriving in northern gulf waters usually in April. Their arrival depends on the warming of inshore waters. The fish is black to navy blue on the back, shading to silver on the sides and white on the belly. Its tail is forked and its mouth has an array of pointed teeth. Migrating northward in the spring and then back southward in late September or October when the cool fall winds blow, the King Mackerel travel in schools, sometimes several miles off the shore, sometimes just outside the breakers. Some years they arrive in vast numbers; other years they are hard to find. Their size and where they swim seems to have little to do with each other. The pier fisherman casting with live minnows has a chance to catch a record-size king as does the charter boat trolling miles off shore. However, odds do greatly favor the charter boat because of its ability to move to the fish instead of waiting for the fish to come to it, as the pier fisherman must do.

These fish average five to 10 pounds, and rarely range to more than 50 pounds. Light to medium surf or boat rods are needed. Fishermen cast for king with live bait, jigs or spoons from piers or from boats anchored or tied to buoys. More often, fishermen troll for kings around buoys and over reefs and wrecks. In the warmest months, schools of kings can be seen cutting through baitfish on the surface and are trolled for with dusters trailing cigar minnows.

At other times they drop beneath the surface and the most effective trolled bait is natural or artificial squid. Trolling for king should be on the slow side. Due to fishing pressure, a limit has been imposed. Also there may be a closed season. Check with local authorities.

Food value, good. The flesh is a bit soft and to the oily side, making it a bit stronger than some.

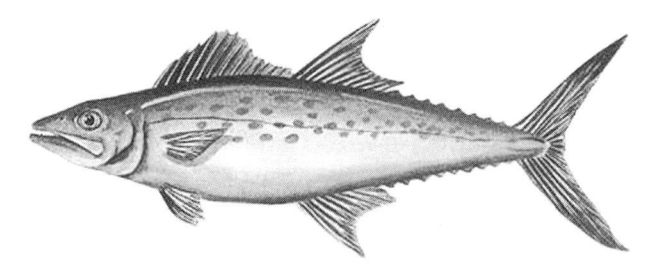

Spanish Mackerel- This mackerel, next to his big brother King Mackerel, is one of the most important and most sought after fish along the northern gulf. Many feel that it tastes better than the King, and he certainly can be found in the area earlier in the year than the King.

The Spanish is a long, slender fish of dark bluc to silvery white from back to belly, and with gold spots along its sides. It has a mouth of serrated teeth and a deeply forked tail. This is a smallish mackerel, ranging from an average of about two pounds up to five pounds or maybe a little more.

The heaviest run of Spanish along the northern gulf is in March and April, dying down as May wears on. There is another run in early August, usually of fish larger than those of the spring run. Stragglers can be caught throughout the warm-water months. These fish school about in the near shore gulf, into channels and the mouths of the bays, feeding on baitfish. They usually are caught by trolling small spoons or jigs or by casting from piers.

Marlin, Blue - These giants are blue on top, shading to white underneath. They average between 200 and 300 pounds, with the potential of reaching between 500 and 1,000 pounds. These billfish are found strictly offshore. While they never come inshore, they are found closest to land in the hottest part of summer. They can be found in northern gulf waters until the onset of winter. They are known for their speed, leaping ability and strong fight. Fishermen troll large jigs and natural bait at high speed in the ocean currents for the Blue Marlin. While fishermen may spot several marlin in a day and even have three or four nose the bait, a single fish in a day is a success story. Blue Marlin is growing in popularity as a food fish, and is frequently grilled like fresh tuna

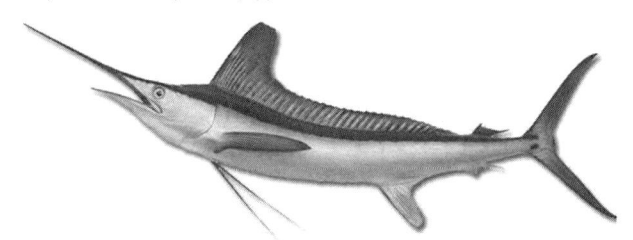

Marlin, White - This marlin, much smaller than the Blue Marlin, is caught in the same area as its larger cousin. The White Marlin averages 50 to 60 pounds, and usually does not exceed 150 pounds. It is bluish-black on top, shading to white towards its belly. It has colored lines or hues of purple along its back. It is fished for in the same manner and areas as the Blue Marlin, and at the same times of year. Again, it is not caught in great numbers, but rather is sought for its fighting ability.

While its food value is considered only fair, it is highly sought in certain parts of the orient.

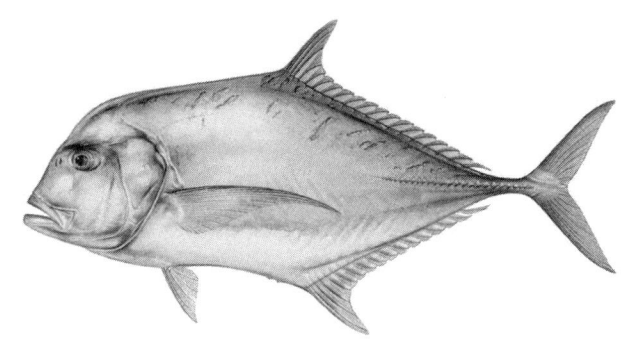

Pompano - It is a roundish fish, silvery to yellow-gold in color, with a deep forked tail. It ranges upward to an average of just under two pounds. One over five pounds is considered trophy size. They are most commonly found in the northern gulf in the early spring and again in late September and October.

The Pompano is caught in the surf on light tackle. Favorite baits are fiddler crabs, sand fleas and small jigs. Just cast the bait into the surf, then ever-so-slowly bump it along the bottom back to shore. Some believe this is one of the finest tasting fish found in northern gulf waters.

Porgy (White Snapper, Key West Porgy) - There are several species of the white snapper as this fish is widely known, but they all are such close cousins few outside the marine biology community appreciate the differences. Basically they are a whitish fish with pink or blue or green tinting, particularly around the head. One large species, sometimes called the Key West Porgy, is mottled light brown over white. They can easily be identified by the strong small jaws and the molar-like teeth. The average one runs a pound or two, but they can range to 10 pounds or more. They can be found in bays, channels, around piers and pilings, on occasion. But mostly they are caught bottom fishing in the open gulf around wrecks and reefs. They are primarily bottom feeders, living on shellfish and crustaceans.

Food value: good.

Redfish (Red Drum) - This is perhaps the most famous and most sought after of the croaking fishes. It is a goldish, red fish, with a heavy body, under-slung mouth and a dark spot at the tail, just before the tailfin begins. They are the largest and noisiest of the croakers. School reds range from three to eight pounds, while bull reds commonly run 15 to 20 pounds or more.

They are found year-around in the northern gulf, with bull reds arriving in large numbers in the late fall. In recent years, season on Redfish in some areas has been closed because of the toll taken by heavy harvesting. When they can be caught, they are found in the gulf and in the bay, in deep holes and along channels. They are caught on live bait such as pinfish or shrimp or on large spoons trolled slow and deep. Medium to heavy surf or boat rods are necessary.

Food value: good.

Sailfish - This is the smallest of the gulf billfish, and most commonly caught. Thanks to its huge dorsal fin—its sail—the Sailfish also is a unique game fish. The Sailfish's fin and upper back are a deep blue. It has a green stripe along its side, and a greenish white underside. It commonly reaches 40 to 60 pounds and, unlike other billfish such as the marlin and Swordfish, the Sailfish ranges into inshore waters. It rarely is caught within a mile of shore, although juveniles have been caught in the bays. Mostly it is caught more than a mile off shore by trolling, usually with the same baits used for marlin or King Mackerel. Trolling tide and weed lines, much as one would for Dolphin, seems to produce good results. It roams the northern gulf from April through September.

The pink flesh has excellent food value.

Sharks - The Hammerhead, Sandbar, Dusky, Bull, Nurse, Lemon, Tiger and Black Tip all are common in northern gulf waters.

All have the distinctive dorsal and tail fins, and most have the distinctive pointed head—except for the easily identified hammerhead and the more flat, rounded head of the Nurse. The Hammerhead is a brownish shark, weighing to 600 pounds. The Bull, Sandbar and Dusky sharks also are grayish to brownish, with the Dusky and Bull ranging to 600 pounds, while the Sandbar ranges to 250 pounds. The Black Tip, Nurse and Lemon sharks are somewhat smaller, ranging up toward 300 pounds maximum. The largest shark common to the northern gulf is the Tiger, a grayish shark with light stripes, that will go over 1,000 pounds.

They are year-around inhabitants, but summer months seem to bring them in larger numbers. Except for the Nurse Shark all have prominent teeth and can cause a nasty wound. All the sharks are found in the open gulf, with several such as the Black Tip and Sandbar also common in the bays.

They can be caught trolling mackerel-type lures or by fishing cut bait (the stinkier the better) in mid water, preferably near structure such as a reef or wreck. These sharks are primarily nocturnal feeders, but can be caught day or night. Sharks are curious fish; if you anchor and fish for them with cut bait such as Bonito or Skate, make lots of noise and, if you are fishing at night, put out lots of light. Sharks will come around to see what's going on.

Food value, excellent. However, if you intend to eat a Shark, gut and head it as soon as it is caught. The Shark stores urea in its liver. When caught the fish releases the urea into its body, tainting the meat. Dressing the fish immediately prevents the spread of the urea.

Sheepshead - This large member of the Porgy family is of typical panfish shape, and is distinguished by its dark vertical stripes on a lighter background. It has teeth like a sheep. The fish is common in the two to six pound range, and is frequently found in bays, channels and inshore.

It is a favorite target for spear fishermen and for bridge, jetty, bay and pier fishermen. The fish is found nibbling on the crustaceans attached to pilings and rocks and is usually caught on shrimp or fiddler crabs.

They are an excellent food fish, but difficult to clean due to heavy scales and big bones.

Red Snapper *Black Snapper*

Snapper, Black and Red - These prized food fishes are of similar size and shape, however are easily distinguished from each other by the Red Snapper's bright red color and the Black Snapper's gray with just a hint of red tint.

They are found in northern gulf waters year round. Heavy fishing pressure reduced their numbers but limits and season have resulted in dramatic increases and populations, especially off the Alabama coast.

Ranging from keeper-size of just over a pound or so to trophy weight in excess of 40 pounds (the Black Snapper ranges to about 10 pounds), the Red Snapper averages two to 10 pounds. These snappers are commonly found in depths from 50 to more than 300 feet, hanging around reefs, wrecks or artificial reefs. While most are caught in the open gulf, more than one surprised angler has found them in the bays. They are notorious nibblers, and are caught bottom fishing on cut bait (squid or fish), dead minnows such as cigar minnows or live baitfish.

These fish will work their way up the water column and larger ones are known to hang out at the edge of a wreck; thus, larger snapper are sometimes caught on an unweighted line with a live baitfish. These snappers are considered among the world's finest food fish.

Spadefish - This member of a tropical fish family is shaped similar to the Triggerfish and the Angel Fish. It is a silver fish with prominent vertical stripes. The bigger the fish gets, the lighter the stripes. The Spadefish averages just over a pound and can grow to more than five pounds. It is a schooling fish, lazily cruising back and forth over reefs and wrecks; sometimes moving near shore to be around jetties, piers or seawalls—anywhere it can find an abundance of jellyfish, a staple of Spadefish diet. They are easily located in warm months and can be caught on small hooks baited with bits of jellyfish.

It is considered to have good food value.

Speckled Trout - This member of the drum family is shaped something like a freshwater trout, thus its name. It is darker on the back, changing to a silver color on the belly. It is spotted on the upper half of the body. Speckled Trout average from one to eight or nine pounds.

They are found in channels, bays and brackish rivers emptying into salt water year around. However, cooler months are considered best for them because the fish frequently gang up in deep holes in bays and rivers when the weather turns cold. Otherwise they scatter across the wide bays, foraging for food. In winter months the best bait is live shrimp. In summer, however, bays are flush with smaller "trash" fish that will steal the shrimp before a trout ever has a chance at it. A live bait minnow, spoon or minnow-like plug that you would use for bass fishing are preferred lures. In summer concentrate fishing around grass flats where they roll into deeper water; fish from the deep-water side.

They are considered good table fare, expecially when prepared fresh. The Speckled Trout is a soft-meat fish and does not freeze well.

Swordfish - This offshore gamefish is easily identified by the prominent sword-like bill, much broader than that of the marlins.

The fish is purplish on the back and lighter on the belly. An eight-foot Swordfish is not unusual and it has been known to grow to 15 feet. It is a loner and lives in the deep waters of the offshore currents, but is sometimes seen basking near the surface or cutting through surface baitfish. It is caught by trolling natural bait, usually squid, mostly during the warmer months.

It is a prized food fish.

Trlggerflsh - This rounded fish is easy to identify by its solid to mottled gray rough thick skin and by its three dorsal spines. One of these spines "unlocks" the others so they can be depressed and folded into a notch on the fish's back—thus, the trigger of the Triggerfish.

It averages about a pound, and can range to more than 10 pounds. Triggerfish are found in bays, channels, wrecks and reefs inshore and offshore. They are extremely common on any inshore obstruction. They will take a bite out of anything that looks like meat, and are notorious bait stealers with their big squirrel-like teeth and tough lips

A note here: This can be a foul-tempered creature that won't hesitate to take a plug out of you if you are so careless as to get within biting proximity. Use needle-nosed pliers to take out a hook!

They are found year around, caught bottom fishing with cut bait, small hooks and light to medium tackle.

For many years the Triggerfish was considered a trash fish, perhaps because its tough skin made it a little troublesome to clean. During the past several decades, however, fishermen have come to prize the excellent firm, white meat.

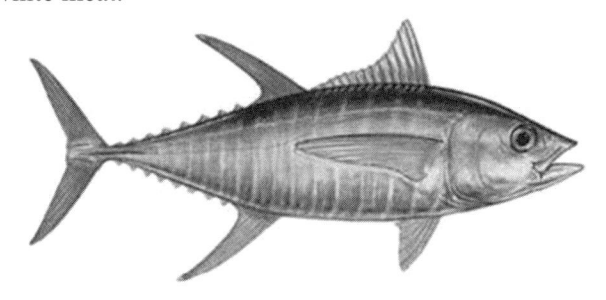

Tuna - These fishes are shaped like torpedoes. They are swift, muscular, with pointed heads and forked tails. They vary in color but generally have a bluish, blackish to white cast, sometimes tipped with some other color as a distinguishing mark.

The three most commonly found in northern gulf waters are the Bluefin, Yellowfin and Blackfin Tuna.

The Bluefin is found offshore year around and is the largest of the three, reaching 1,000 pounds or more in gulf waters. It feeds almost constantly on fish and squid, and has been the target of heavy commercial fishing.

The Yellowfin is another year-round offshore species. It is caught as close as 40 miles from shore in summer months, while is is most common more than 150 miles out in the winter. The Yellowfin ranges to 250 pounds or more, and is caught trolling as it feeds near the surface.

The Blackfin is the tuna most frequently caught by northern gulf fishermen because it is the one found closest to shore, sometimes within just a few miles of shore. Once in a while one is even caught from a pier. It also is found offshore. The fish averages from 25 to 30 pounds, but can range to more than 60 pounds. The fish is found mostly in the summer as an incidental catch to mackerel fishing, and on rare occasions is brought up by bottom fishermen.

Wahoo - This fish has the same general shape of the King Mackerel, and may be confused with that fish. However the Wahoo can be identified by the longer nose and dark bars that run vertically on the body. They also are generally larger than King Mackerel, ranging to more than 100 pounds. The fish is found mostly in the summer inshore, but offshore year round. It is caught by trolling with mackerel-type and small versions of marlin-type lures.

The Wahoo is an excellent food fish.

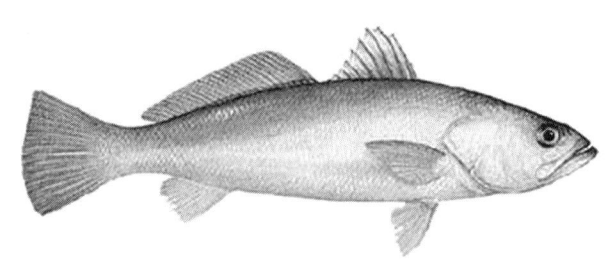

White Trout (Silver Trout) - This fish looks a bit like a small version of the Speckled Trout, but without the coloring or spots. It is white or silvery colored and generally runs from slightly less than a pound to about three pounds. It is common in bays and channels year around and can be caught in much the same way as Speckled Trout. Live or dead shrimp, sand fleas and fiddler crabs are favorite baits, as is a Sting Ray Grub artificial bait.

The trout's soft meat is good when cooked fresh. It does not freeze well.

Whiting - Summer waders see the smallest of Whiting—little ghost-like minnows darting in and out of the surf. They grow to about three pounds, while they average a pound or less.

The Whiting is a long, thin fish, mottled to silvery white, with an under-slung mouth and a barbel on its chin. They are found in the surf and around channels leading into the gulf. They grub around the sand for bits of food. They are common and caught anywhere in the surf from the shoreline to beyond the breakers almost all year long. Light spinning tackle and smallish hooks baited with bits of shrimp, squid, cut bait or sand fleas is fine for Whiting fishing.

Food value is excellent.

Rods and reels

Fishermen, who at times have been known to stretch the truth, have for centuries told about days when the fish literally jump into the boat. Personally, I've never seen anything but a carp fish jump in a boat. The fish that matter to me are more prone to jump out of the boat. That's why I own fishing tackle.

Fishing tackle is the bread-and-butter of fishing. An angler's rod and reel, or even a cane pole, is the single-most important piece of fishing equipment he owns. Fishing success is directly tied to that piece of equipment and how the angler makes it work. A wide array of fishing tackle is used in Alabama. From a cane pole to a $100 graphite-composition rod tipped with a $100 reel, anglers have many tackle choices. The type fishing you plan to do and your budget should figure into the purchase of this type of equipment.

Here's the most commonly used tackle in Alabama:

"Cane" poles

Fishing with a cane pole is fishing in its most simple form. A cane pole is simply a long piece of cane which has been shellacked or lacquered to make it waterproof. A cane pole consists of the pole, a piece of line approximately as long as the pole, and the terminal tackle—usually a hook, float and a split shot.

An angler is severely limited in what he can do with a cane pole. Although it possible to outfit the pole with a small, inexpensive plastic reel; the line on a cane pole cannot be cast. The line rather must be swung out to the desired location or lifted and dropped into a spot. Naturally, that location cannot be further away than twice the length of the pole.

Cane poles have other drawbacks, too. They are cumbersome and difficult to transport, even though some have ferrules that allow them to be broken into sections. Cane poles also are heavy. They can be tiresome during a long day of fishing.

Cane poles are not as common as they used to be because spincast outfits have gotten so inexpensive they are now within the budget of almost all anglers.

Also, many anglers who still find the need of a pole have switched to an ultra-lightweight, fiberglass, telescoping pole.

These inexpensive poles still have a draw-back in

that it is not possible to cast the bait. A plus is that a bait can be dropped precisely into tight cover where otherwise it might get entangled by other casting methods.

Spincast tackle

Spincast reels, sometimes called a closed-faced reel, is the easiest to fish with, least expensive and most popular fishing reel on the market today.

Spincast tackle is almost always used in freshwater because its working parts are rarely stainless steel and they rust easily in saltwater. These parts also are contained within the reel housing and they are not easily accessible to be cleaned.

The inexpensive spin cast outfit is perfect for panfish such as small bass, small catfish, crappie and bream. A spincast outfit is definitely the best outfit for a beginning or occasional freshwater angler because of its price and simplicity.

Spincast tackle is basic in that the line is held on a narrow, flat spool and is wound onto that spool with a pickup pin attached to a revolving disc. The entire retrieval apparatus is contained within a housing, with the line entering and exiting through the reel's nose cone. A push-button release that sets the line free and allows the lure to be cast can be found on the back of the reel.

Spincast reels are usually equipped with a simple drag system that allows the angler to fight and play bigger fish with the lessened chance of snapping the line. These drags usually consist of a star drag adjacent to the retrieval handle or a dial above the push button release.

Most of these reels are equipped with a anti-reverse button which prevents the handle from turning backward. This forces the drag system to absorb the pressure of an outgoing line when a fish makes a run.

Another plus of the spincast outfit is that, unlike bait casting tackle, it's possible to fish a wide variety of weighted lures. The spincast is so versatile that either a tiny 1/4-ounce jig or a heavy topwater lure such as a Jitterbug can be cast equally well.

Baits fished with a spincast outfit usually consist of live baits such as worms, crickets, minnows or grubs, or artificial baits such as jigs, plastic worms or smaller crankbaits and topwater baits.

Spincast outfits are often neglected by the anglers which own them, but they take punishment remarkably well. A once-a-year cleaning (more often if the reel is ever submerged in water) and a twice-a-year line replacement is often all that is needed to keep the reel in top shape.

It should be noted that using too-heavy line (anything above 10- pound test is too heavy, 6-pound test to 8-pound test is perfect), allowing line to grow old and failure to keep the reel spool full are the most common maladies which make spincast reels inoperable.

Spinning tackle

Because of the availability of widely varying sizes of spinning tackle, it is the most versatile fishing tackle on the market. Ultra- light spinning tackle can be used to make a 10-ounce shellcracker feel like a fighting tarpon, or heavy-duty saltwater spinning tackle can be used to manhandle a feisty 35-pound king mackerel.

A spinning reel works on much the same principle as a spincast reel in that the weight of the lure is used to pull the coiled line from the reel. There are several major differences in these two reels, however. First, the spool isn't contained in a housing, hence it's common name of "open- faced" reel. Second, instead of sitting atop the rod as spincast and bait casting reels do, the spinning reels hangs below the rod. This makes the rod and reel combo well balanced.

The line is placed back onto the spool by being thread through a revolving bail, and this bail has a roller (to prevent line wear) on which the line rolls when it is retrieved. This bail flips open to disengage the line so the lure may be cast.

Unlike spincasting and bait casting reels which usually have handles permanently mounted on the right side, spinning reels usually have a handle that is interchangeable right or left. A spinning outfit is usually cast with the right hand and reeled with the left, but many anglers have a problem adjusting to that and choose to reel right-handed.

Spinning rods are usually limber, but their tell-tale giveaway is their abnormally large guides. These oversized guides allow the coiled monofilament to pass through them with ease.

As said before, spinning tackle comes in all sizes and can be used for almost any type fishing that is found in Alabama.

Ultra-light spinning tackle is a common sight when freshwater panfish such as bream and crappie are the quarry and the angler wants more action than simply winching the fish in with heavy tackle.

Medium-weight spinning tackle is common in the hot summer months when bass are suspended on ledges and a lightweight lure such as a grub or Sassy Shad is needed as a bait. Medium-weight spinning tackle is also common below Wheeler, Wilson and Pickwick dams in the fall when smallmouth make their move in the turbulent waters. Many anglers also opt for medium-weight spinning tackle when light line, a small bullet weight and a small plastic worm is needed for bass in clear water situations.

Heavy-duty spinning tackle is common along the Alabama Gulf Coast. This heavy tackle can usually be found being used by anglers fishing from Gulf Shores Pier, in the surf or sometimes trolling.

Bait casting tackle

Bait casting tackle is the tackle of choice for many bass fishermen. The lightweight construction, smooth retrieval and excellent drag system found on today's bait casting reel makes it a natural for this type of fishing.

The bait casting reels differs vastly from spincast and spinning reels in that the spool in this reel revolves as the lure is cast. This sometimes results in a spool over-run at the end of the cast. These overruns are called "backlashes" or "bird nests." The difficulty in fishing with this type of reel has turned many anglers against it.

Many modem bait casting reels have integral devices designed to lessen the chance of these backlashes. A series of magnets inside the reel's left side plate slow the spool's spin at the end of the cast. This "centrifugal" brake has made it possible for more anglers to use bait casting tackle.

The bait casting reel, which is nothing more than a winch, is equipped with a level wind system. The level wind system is a traveling guide for the line which puts the line onto the spool evenly.

This level wind system disengages when the lure is cast. The reels that are commonly used for bottom fishing and trolling in the Gulf are technically bait casting reels, but never is the bait cast with these reels. In bottom fishing, the spool is placed into the free-spool position and the bait is simply lowered to the bottom.

In trolling, the spool is placed in the free-spool position and the bait is released to the proper distance behind the boat. In both cases, a thumb is usually placed on the revolving spool to prevent backlash. These heavy-duty reels differ from their smaller counterparts in that most do not contain a level-wind system (you must use your thumb to direct the line onto the spool evenly), nor do they have, or need, magnetic cast control.

Fly fishing tackle

Fly fishing is considered by many modern day anglers as a type fishing best suited for pipe-smoking old stuffies and other elitists, but learning the art of "throwing line" can add an exciting new dimension to anyone's fishing.

Because the artificial flies used in fly fishing weigh virtually nothing, it is impossible to cast such a lure using its weight. That is why in fly fishing, the line, rather than the lure, is what is cast.

This makes fly casting dramatically different from other casting methods. Unlike in spin casting, spinning and bait casting tackle, the reel is a relatively unimportant part of the system. The fly reel is nothing more than a spool where extra line is stored.

To understand fly casting, you first must understand the basics of fly line. Modern fly line consists of a braided Dacron or cotton core covered with a PVC outer sheath. This line is extremely strong (30- to 40-pound test), waterproof and lightweight. Most fly lines are intended to float because the angler is attempting to lure a fish to the top of the water to catch an insect that has supposedly fallen there.

Floating lines come in many styles, including level lines, double tapers, weight forwards and "shooting heads."

Level lines are the same diameter throughout the length of the line. These lines are inexpensive, yet good enough for the beginning fly fisherman.

Double taper lines are standard thickness throughout their length, except for a taper at each end. This taper drops the line size down gradually except for the final feet, where the line is a small, standard diameter to the end. This configuration enables the fly to be dropped into the water with the most gentle of splashes and is designed as to not spook fish.

Weight forward tapers have a thick bulge near the end of the line, with a pronounced taper with the final few feet being a standard diameter. This configuration enables the angler to throw heavier lures such as bass popping bugs.

Another type of floating fly line is best known as "shooting head." Like weight forward lines, it is designed for long distance casting. Shooting heads are attached to long, low test-weight monofilament leaders.

Choosing fly line is simple. A universal code can be found on all fly line packaging to explain what is inside. It looks like this:

DT - double taper.
L - level.
ST - shooting taper
WF- weight forward
Numbers 3-12 - weight of line.
F - floating.
S - sinking.
F/S - floating/sinking.

A typical code may look like this:
L5F - level line, 5-weight, floating.
DT1F - double taper line, 7-weight, floating.

Although fly line is light, it is not small in diameter. It is for that reason that a length of monofilament leader (5 to 12 feet) must be attached to the fly line so that a fly may be tied to it. This leader may be attached to the fly line by either a knot or a special eyelet which slides up into the line and will not back out.

Choosing line and the right fly rod and reel is easy. Learning to cast a tiny fly to a desired spot often proves to be much more difficult for some. Practice makes perfect and practice is needed to become a good fly fisherman. Start in an open field (water isn't needed at first) with a flyrod, line and a fly with the hook cut off.

Lay 10 to 15 yards of line directly out in front of the rod and grasp the rod in your right hand. Take the fly line in the left hand. Snap the line up and back over your right shoulder, stopping the rod at the 1 o'clock position. Pause to allow the line to pull tight, then snap the rod back to the front pulling the line with it.

Practice this until the line snaps smartly in both directions and the fly rolls out to the ground in a gentle touchdown. Once that is accomplished, you may begin to increase the length of the cast by pulling line out of the reel with your left hand.

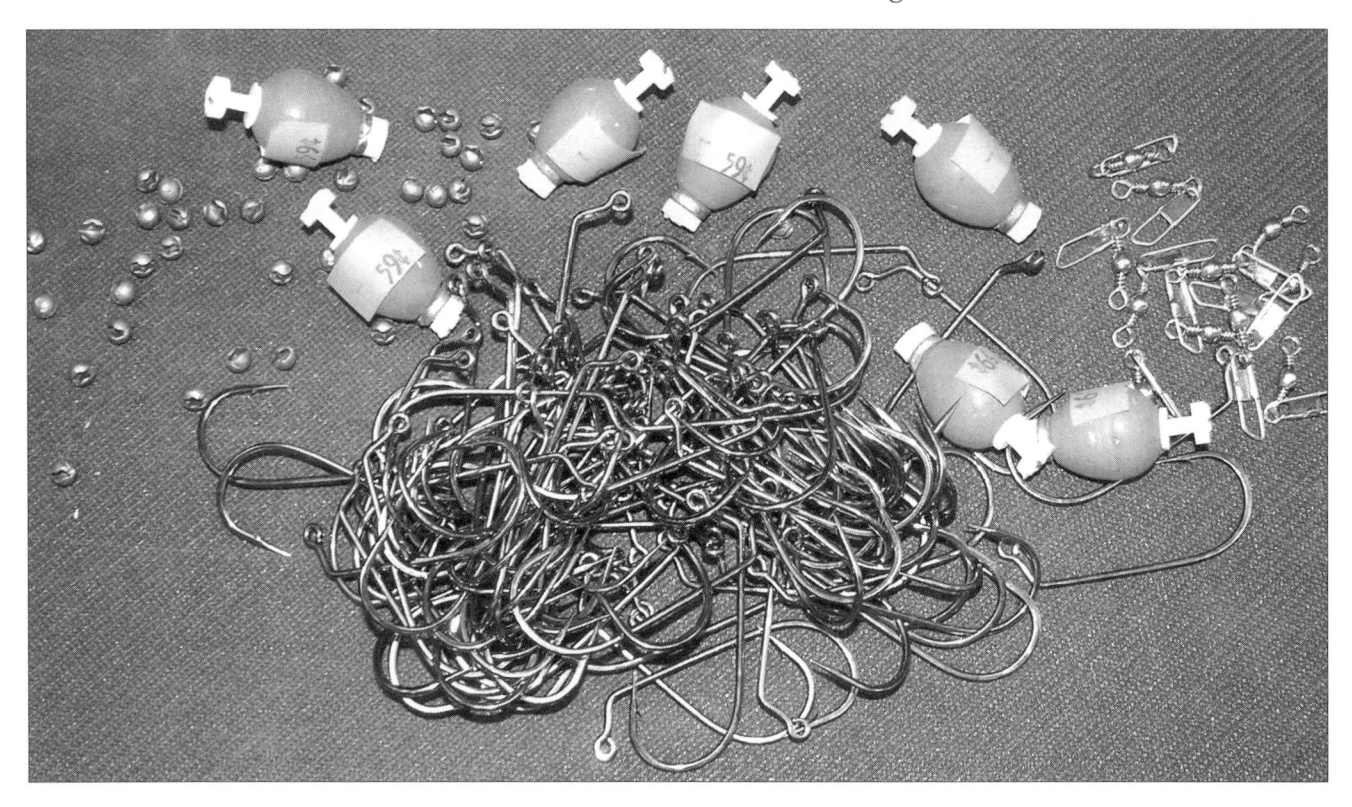

Terminal tackle

It's possible to own the most expensive rod and reel known to man, but as a fishing outfit, that expensive rig is no better than the terminal tackle that is affixed to it.

Terminal tackle is simply any of the tackle—hooks, weights, floats and swivels—that you attach to your line to aid in fishing.

Understanding the total terminal tackle picture can be a confusing nightmare. Literally thousands of options and variations in terminal tackle can be found as you circle the globe in search of bonefish, peacock bass, tuna, halibut, salmon, permit and the like. Fortunately for Alabamians, the options needed to catch the species of fish found in this state aren't nearly as complicated or numerous.

Hooks

Let's back up a bit. Hooks, a main component in fishing tackle, can be terribly complicated. If you were to make a point of trying to understand the seemingly endless variations in sizes, thicknesses, shank shapes and lengths, types of eyes and hook finishes, not to mention styles such as Aberdeen, Sprout, Kirby, Perfect, O'Shaughnessy and Carlisle, you'd probably never have time to go fishing.

Understanding hook sizes is among the most confusing aspect of it all. Hook sizes range from size 22 to 1 (the larger number being the smallest hook), then once size 1 is reached, sizes run 1/0 to 16/0 (the largest number being the largest hook). Sizes 22 through 1 come in even numbers (22, 20, 18, 16, etc.), except, for some unexplainable reason, one odd number—size 1. To make matters more complicated, when hooks become larger than size 1, they are known as 1/0, 2/0, 3/0, etc., all the way to 16/0. These hooks sizes come in odd and even numbers.

Confused yet?

There are numerous styles of hooks on the market, each having a particular purpose. Each also has its own shank length, shank profile and shank thickness, not to mention point style.

There's no reason to cloud your mind with a complete understanding of hooks, however. Make a point of noticing what hook better anglers are using for your particular type of fishing. Apply that knowledge to your fishing.

A few rules of thumb for hooks:

Largemouth, spotted bass, smallmouth – Hooks used to rig plastic worms usually range from 2/0 to 6/0, with 2/0 and 4/0 being the most common. These hooks may have an offset shank that allows the hook to be imbedded in the plastic worm without crimping the worm.

A worm that doesn't hang straight on the hook will rotate when retrieved and will twist the line. Some worm hooks may also have barbs along the shaft to make it difficult for the worm to slip off. Another option sometimes found on a worm hook includes a small wire loop that makes the worm weedless.

For live bait fishing for bass, a No. 2 to 2/0 wire hook causes little damage to live bait such as minnows and crawfish, but the smaller hooks could straighten under the weight of an extremely large bass.

For larger live bait such as shad, a 3/0 or 4/0 hook is needed.

Hybrids, striped bass – Hybrid and striped bass are, on the average, larger than largemouth, spotted and smallmouth bass. Live bait in the form of shad is normally used in Alabama for these species. If fishing exclusively for hybrids and small stripe, a 3/0 or 4/0 wire hook will suffice. For large stripe, a forged 5/0 hook may be needed. Striped bass that top 20 pounds are common in Alabama.

Crappie – Minnows are the most common bait for crappie in Alabama. Gold-plated, extra fine wire hooks, No. 2 to No. 6, work well because they cause less damage to minnows than thicker shank hooks.

Bream – The sunfish family has special hook needs. The mouths of these fish are extremely small, yet they have the ability to completely swallow most small hooks. The answer is a small throated hook with an extra-long shank. This extra-long shank makes hook removal easy. A No. 10 to No. 18 long-shanked, extra fine wire hook works well when crickets, worms and meal worms are used as bait.

Catfish – Catfish in Alabama range from the half-pound squealers found in farm ponds to the 50-pound-plus monsters found on the Tennessee River. Naturally, a wide range of hooks can be used for these fish.

Most cats you will fish for in Alabama will range from 1 to 5 pounds. A No. 2 to 2/0 wire hook works well on catfish this size.

Catfish have extremely tough, spongy skin around their mouths and the thin wire hook penetrates easily. These hooks also work well with livers, gizzards, cut baits or other common catfish baits.

If custom-made catfish bait such as doughballs or stink baits are used, many anglers opt for treble hooks. No. 6 to No. 10 treble hooks work well, especially the hook is of the doughball-type hook

Floats

Floats have been used for centuries for two purposes. They alert anglers when they are getting a bite, and to keep bait off the bottom and to present it at a level that fish can find it easily. Floats come in many forms and are made from a wide variety of materials including cork, plastic, foam, balsa wood, balloons and porcupine quill. Floats can be used for any type of fishing that is done in Alabama.

Alabamians traditionally use corks too large for their needs. A cork should be chosen so it is large enough to ride upright in the water, but small enough that it can be pulled under easily and offer no real resistance to the fish. Fish that mouth a bait rather than swallowing it often spit it out when the feel the resistance of the cork. An easily sinkable float allows the fish to pull it under and swim away without detecting something is amiss.

Here are some float options available in Alabama:

Plastic floats – The most common float used in Alabama is the red/white plastic bobber. This traditional float is inexpensive and its use is wide-spread. Bobbers are not the only float option available, however, and other type floats are often more effective in certain situations. The bobber's non-aerodynamic style makes it offer a lot of resistance to being pulled underwater.

Cylindrical-shaped plastic floats are much more effective than bobbers.

Cork floats – Cork is naturally buoyant and thus makes for an excellent float. There are several drawbacks to cork floats, however. Cork is a lightweight material and the result is a lightweight float. These cork floats are too lightweight to cast with most tackle. Most light cork floats may be cast on spinning tackle only.

Cork, in its natural color, is not very visible. Many cork floats are painted a bright color to increase their visibility.

Porcupine quills - Porcupine quills, although a dying breed of float, make excellent floats for panfish. These floats stand on end easily, are slim enough to be easily pulled underwater and cast well.

Balloons – Small, round party balloons make excellent floats for stripe when live baits need to be drifted away from the boat in clear, shallow water in locations such as Smith Lake or Lake Martin.

Foam Floats – Foam floats are probably the most popular floats used in Alabama. They are lightweight, but heavy enough to cast. They are also bright and inexpensive. Foam floats come in many shapes and sizes, including one that pops when retrieved. This popping action attracts some species of fish like schooling hybrids.

Some foam floats are equipped with a lead ring attached around the float at the bottom. This allows for long casts.

Balsa floats – Balsa floats are durable and are used most often by the angler who takes his fishing for panfish seriously. Balsa floats are the most expensive type float found in Alabama, but they last longer and are the most sensitive to a fish's bite.

Weights

Weights, which are traditionally made of lead, have several purposes. Most importantly, they get the bait down where you want it and keep it there. They also, in some circumstances, provide weight to aid in casting.

A wide array of weights are used in Alabama. Some of the most popular include:

Bullet weights – These bullet-shaped weights are used when fishing for bass with plastic worms. The weight not only drops the worm to the bottom quickly, the tapered head guides the worm through underwater obstacles and helps avoid snags. These weights come in a variety of sizes.

Split shot – This is the most commonly used weight in Alabama. This easy-to-use weight simply clamps on to the line by squeezing it with pliers.

One version, removable split shot, has ears that may be squeezed so that the split may be reopened and the weight removed from the line. Split shot is used in all types of fishing in Alabama, but is most commonly used in bream, crappie and catfish angling.

Egg sinker – Eggs sinkers are commonly used when a large, heavy sinker is needed, as in catfish fishing. The egg shape allows the weight to be retrieved through rocks and other obstacles while minimizing the chance of snagging.

Bank sinker - Teardrop shaped bank sinkers are most commonly used in Alabama as weights for trotlines, and also by anglers fishing rock piles and jetties. This weight's shape allows it to be pulled through rocks easily. This weight should not be used in current for its shape allows it to slide easily.

Casting sinker – A casting sinker is basically a bank sinker, but has a brass wire hanger rather than a cast-lead eye as the bank sinker. This weight casts better than the bank sinker because it doesn't wrap up in the line as easily as a bank sinker.

Clincher sinker – This long, thin sinker is basically an elongated split shot. The line is placed in the weight's slot and ears on each end are pressed down to keep the weight at the proper place on the line. Some versions have a rubber insert in the slot to keep the line from being damaged.

This weight should be used when a heavy split shot is needed, or in trolling situations.

Pyramid sinkers - Pyramid sinkers have sharp edges that dig into sand and mud, thus allowing the weight to hold to the bottom in current. They are commonly used below dams when fishing for catfish, drum and hybrids.

Swivels

Swivels, like weights, come in many shapes and sizes. Most consist of two or three round metal eyes that rotate independently of each other to perform such functions as preventing line twist, attaching several components such as a bait and a sinker and to make for ease in switching lures.

The most commonly used swivels in Alabama are:

Barrell swivels – Barrel swivels are used to connect fishing line to a leader. They are often used to keep a weight away from the bait. This is done by placing the weight on the line, attaching the barrel swivel and adding a length of leader.

Ball bearing swivel – These swivels are used when preventing line twist is a must, such as trolling with a lure such as a spoon that twists and rolls.

Snap swivel – The most common swivel used in Alabama. The safety pin-like snap on this swivel opens and closes easily to allow lures to be changed without having to cut line and re-tie knots. Inexpensive.

Interlock swivel – Similar to the snap swivel, this swivel is used for large fish when a stronger swivel is needed.

Lock swivel – The end of this wire snap hooks around itself and uses spring tension to keep the snap closed. This swivel is used in freshwater for big catfish and striped bass, but it's more common application is for bottom fishing in the Gulf.

Three-way swivel - This swivel is used to connect fishing line to two components. A common rig using this swivel consists of the line tied to one loop, a leader to one loop and a leader and weight attached to the other.

This rig can allow a weight to sit on bottom while a live bait or cut bait is held off the bottom. This rig is used for crappie fishing with minnows, catfishing in current with live or cut bait and bottom fishing in the Gulf with either live or cut bait.

Mike Bolton

Lines and knots

Line is one of the most easily damaged components in your tackle system. As excellent as present-day lines may be, all eventually wear out or need replacement for other reasons.

Relying on it means you need to understand its strengths and vulnerabilities. No line is perfect for every situation. Knowing what line to use in various situations requires an understanding of breaking strength, knot strength, UV resistance, castability, limpness, stretch, abrasion resistance, and visibility.

Modern fishing line

Fishing lines today are far better than those of 20 years ago. Almost all of today's lines were created in a laboratory and their components are too complicated to discuss here. They were created from artificial substances, including nylon, fluorocarbon, polyethylene, Dacron and Dyneema.

The most common type of line is monofilament made of a single strand of polymer. It is popular with fishermen because of its buoyant characteristics and its ability to stretch under load. Alternatives to standard nylon monofilament lines today are copolymers or fluorocarbon, or a combination of the two materials. Fluorocarbon line is valued for its ability to be less visible to fish. Fluorocarbon is also a more dense material, and therefore, is not nearly as buoyant as monofilament. Anglers often utilize fluorocarbon when they need their baits to stay closer to the bottom without the use of heavy sinkers.

There are also braided fishing lines, co-filament and thermally fused lines, also known as "superlines" for their small diameter, lack of stretch, and great strength relative to standard nylon monofilament lines.

Synthetic lines have gained acceptance because of their superiorities which include greater strength relative to diameter, durability, freedom from rot and mildew and, in some cases, less visibility to the fish.

Selecting the correct line

Line selection should be based on the type and size fish being sought, the fishing conditions and the type of rod and reel being used. For example, you need a tough, abrasion-resistant line for fishing around brush and underwater objects that might damage your line's surface. For ultra-light spinning, you don't want a mono lacking a little "give"; line lacking any stretch won't withstand the shock of hook-set in really light-line tests. But these are only two qualities line manufacturers formulate to balance performance and durability. To varying degrees, these and other engineered qualities tend to affect how often you need to change line.

Line characteristics affecting your ability to cast, set the hook and see the line can be looked upon as control qualities. In this regard, it is important to consider limpness, stretch and visibility when choosing a line for a given kind of fishing.

For example, dead limp lines do not cast well from spinning reels because they do not easily spring free from underlying line coils on the spool. Resistance from this clinging shortens casting distance.

Similarly, lines that are too springy do not cast properly from revolving-spool reels, like baitcasting reels. When cast, such lines tend to fluff-up loosely over underlying line coils. Backlashes often result.

Some monos, cofilament and braided lines, by way of contrast, lie down smoothly on the spool. This helps them cast well from revolving-spool reels. But this very quality works against their best casting performance with spinning or closed-face spincast reels.

Line stretch – The most popular nylon monofilament lines stretch enough to withstand the shocks and stresses of hook-set and long, sustained runs. For bass, some anglers prefer lines in which stretch is minimal, making hook-setting easier in such hard-mouthed species.

Braided and co-filament lines have relatively little stretch compared with monos, making them great for hook-setting. They are also very strong. But one should respect their breakload ratings when battling heavy, fast-moving fish due to this lack of stretch.

Visibility – Visibility refers to the color characteristics of line that allow you to see it while casting, retrieving and playing fish. Opinions vary among anglers and line scientists as to whether clear, color-dyed or fluorescent lines are the least visible to the fish. One respected chemist feels there is little reason to believe a line's color frightens fish. He cites more trophy-class fish being caught on a colored fluorescent line than on any other during a given year.

Many experienced anglers, on the other hand, believe there are justifiable reasons for using clear line in extremely clear water and fluorescent line when poor light conditions or turbid water make watching clear lines difficult.

Durability – How often you fish, the fishing conditions, and the strength and durability of your line influence how often you need to spool on fresh line. Most lines are damaged by coming into contact with fish teeth, gill plates, scaly skins and bills (in the case of sailfish, swordfish and marlin). They also are damaged when scuffed over underwater objects, like rocks and submerged stumps and trees. Severe line damage may occur when lines are cast through cracked or scored line guides. Additionally, the shock of hook-setting, pull of heavy fish and heat generated by the line moving rapidly under stress across guide surfaces also take their toll on a line's strength.

Strength –Various kinds of strengths are formulated into lines to prevent breakage and prolong their life. Knot strength prevents a line from breaking when setting the hook or fighting fish heavy handedly.

The knots you use also are critical. So is how you tie them. Some knots are stronger than others. That's because they are designed to cushion the tough, synthetic lines to prevent their cutting into themselves.

Correct knot-tying techniques are needed to assure optimum knot strength. The twists or coils comprising a knot should be drawn up carefully. Lubricate monos with a little water to prevent their heating up while being drawn tight. Braided lines need no lubrication while knotting. The knot should be firmly and smoothly cinched up with a steady pull—never a quick jerk.

Breakload – Breakload is the pound-test rating of the line printed on the package label. Tensile strength often is confused with breakload. Tensile strength actually is a calculated value, rather than a lab-tested evaluation of break load. Tensile strength expresses the mathematical relationship between break-load and line diameter.

Abrasion strength – This refers to a line's ability to withstand friction when coming into contact with objects above and below the surface. Much improvement has been made in nylon abrasion strength. Despite this, all nylon lines (braid and cofilament too) eventually become worn enough to require replacement. Knots formed inadvertently in the line while fishing should be snipped off and the lure retied.

Age – With synthetic lines, age is not the problem it was with lines made from natural materials. If synthetic line is protected from excess heat, sunlight, ozone and chemicals, it should retain its best qualities for at least a year, often longer. Nylon, for example, is virtually ageless when stored properly, say the chemists.

The real enemies of monofilament line

Prolonged exposure to sunlight breaks down the chemical composition of nylon line. A clue to this is a bleached-out color. Excess heat literally cooks out nylon line strength. Troubleshoot this by looking for a chalky surface, excess stiffness and curl. Certain chemicals (like battery acid), chemical fumes and ozone either destroy or seriously weaken lines.

Prevent this by storing them away from chemicals and appliances with electric motors (like freezers and refrigerators) that generate ozone. Some anglers seal their bulk spools of line in large, zip-top plastic bags for fume and ozone protection.

Fishing damage - To detect line damage from fishing, inspect the line by sight and touch at least once an hour and after landing every fish. Check the line from the lure's knot up through where it flexes against the rod tip while casting (two to three feet). If it feels rough or looks kinky and curly, snip off the suspect line and re-tie your lure.

When to change line – Line-wise anglers usually change line on all their reels well before the start of each new fishing season. Select lines with qualities that fit with your tackle and methods, as well as the fishing conditions. Choose abrasion-resistant lines for fishing around brush and underwater objects. If you fish extremely light tackle, make sure your line has a little "give" to prevent break-offs when hook-setting.

Use flexible lines for revolving-spool reels; those with slightly more springiness for closed-face spincast and spinning reels.

If you fish only a few days a year, then it is possible that changing line once a year will suffice. But if you fish frequently, such as one or more times a week, you may actually need to change line before every trip. Frequent anglers often examine the full casting length of the line after each day of fishing, changing line immediately if it appears the worse for wear.

Don't forget to store your reels and bulk spools of line as far as possible from heat, direct sunlight and chemical. As one of your most important partners in fishing, your line deserves some mighty thoughtful attention.

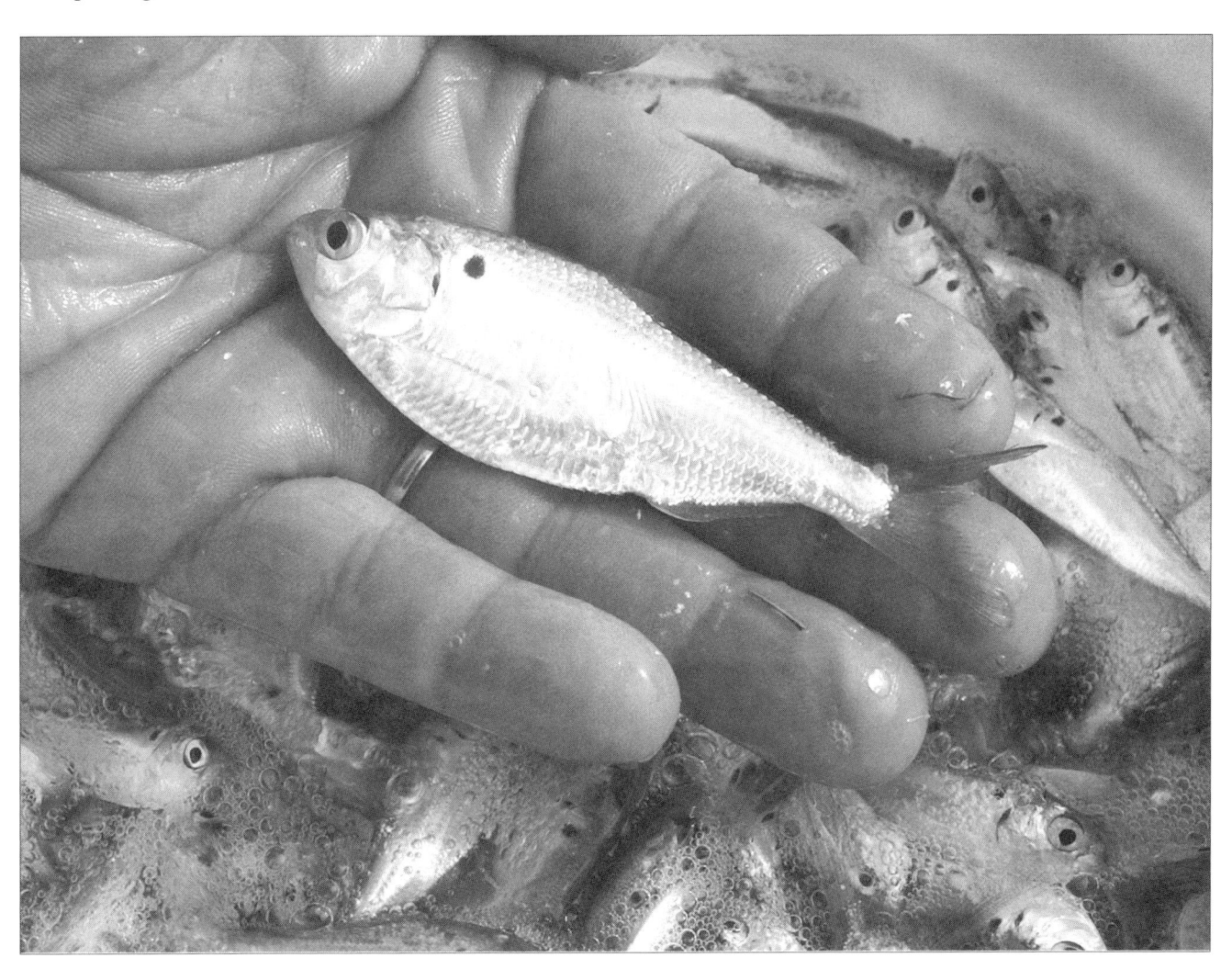

Live Bait

There's no question that there's something special about fooling a trophy fish into taking an artificial bait. The challenge undoubtedly holds the fascination of many of Alabama's 500,000-plus licensed fishermen.

But never doubt that natural bait fishermen outnumber artificial bait fishermen. Crappie fishermen with their minnows, bream fishermen with their worms and crickets, stripe fishermen with their shad, catfishermen with their liver and other smelly baits and trotliners and jug fishermen with their cutbaits are in the majority in Alabama.

Fishing natural bait has its pros and cons, but mostly pros. There's no question it's easier to fool a fish with the real thing, and fish aren't as likely to spit out a natural taste such as worm or minnow as they are a painted piece of plastic or steel.

Also, most natural baits are easily available and inexpensive, often free. Red worms, wigglers, night crawlers, crickets, grasshoppers, grubs, wasp larvae, caterpillars, and catalpa worms often can be found right in the back yard.

Shad and minnows can be netted or trapped in many lakes, and crawfish can be found in many small streams and water-filled ditches by anyone willing to turn over a few rocks. Mealy worms can be raised in a box at home.

For the angler in a hurry or the angler unwilling to hunt his own bait, most natural baits can be purchased rather inexpensively at tackle stores, bait shops or marinas.

Natural baits do have a few drawbacks. They tend to be messy, and some carry odors. Natural baits also require special care and storage, and many require special containers. Unlike artificial baits, they can't be tossed into the tackle box and forgotten until the next fishing trip.

Here's a look at the most popular natural baits used in Alabama:

Red worms, wigglers and night crawlers – Without question, worms are the most popular bait of any kind used in Alabama. Primarily used as a bait for bream and catfish, worms offer fish a natural-tasting food they love.

There are three types of worms primarily used in Alabama. Red worms are small, thin worms normally used for bream. Wigglers are slightly larger worms that have a lot of action. Wigglers are a versatile bait. They are used for bream, catfish and even bass.

Night crawlers are large, skinny worms with a flat head that are imported to Alabama from Canada, North Carolina and several northern states. They are usually 6 inches long or longer and have a tough body. They are expensive and usually are packed 16 to a box. Night crawlers are primarily used by commercial fishermen on trot lines.

Red worms and wigglers can be purchased from most lakeside tackle supply houses, or they can usually be dug wherever the angler can find damp, rich, loose soil. Shady hillsides with undergrowth and hardwoods are prime worm habitat, as are damp, shady agriculture areas left bare by animal traffic.

Many anglers construct their own worm beds or attractors by building up mounds of loose, black soil mixed with sawdust and peanut hulls and covering them with carpet or other heavy cloth.

Periodic dousing of the carpet or cloth with a water hose insures the worms will stay close to the top of the bed where they can easily be dug.

The worm you choose should be based on what you plan to fish for. If bream are small enough to steal the worm and avoid the hook, go with red worms and stick the hook point into one of the worm's pointed ends and thread the worm onto the hook. This prohibits smaller fish from stealing bait and forces larger fish to inhale the hook when they take the bait.

For larger fish such as big bream, catfish and bass, a wiggler, or several wigglers, should be laced onto the hook by running the hook in and out of the worm along the length of its body. Unlike threading a worm on a hook, this juicy offering allows the worms to wiggle and further attract fish.

Fiddle worms – Fiddle worms were once a staple for our grandfathers' fishing, but the inability to raise them in captivity and the difficulty in keeping the delicate worm alive has just about ended their use.

These worms, sometimes more than a foot long, derive their name from the age-old practice of "fiddling" them out of the ground. This is done by driving a stake into the ground and sawing through the stake with a hand saw. The resulting vibrations drive the worms from the ground. Fiddle worms should be rigged like night crawlers.

Crickets, grasshoppers – Crickets are a wonderful bream bait, especially in spring when fish are bedding. A wiggling cricket floated over a bream bed is the safest bet in fishing that you'll get a strike.

Crickets are inexpensive to purchase—about four cents apiece. At that price, purchasing them makes more sense than chasing them around the backyard.

Only one method should be used to hook a cricket. Using a thin wire hook, enter the cricket at the rectum, pushing the hook up through the anus and out directly behind the cricket's head. This allows the cricket to sit upright in its natural position. Crickets are hardy and can be kept alive for up to three weeks by placing them in a plastic gallon milk jug and punching the top full of air holes. Crickets need to be fed and watered. A piece of raw potato and a damp paper towel dropped into the jug will provide the crickets with both food and water.

One good trick better bream fishermen use when bream are on bed is freezing crickets. Freezing will kill crickets, but will keep them fresh. Grasshoppers also make a good bait for bream, but grasshoppers are not sold commercially. If you want to save a few dollars, catching a few crickets out of the backyard beforehand isn't a bad way to go.

Caterpillars, grubs and Catalpa worms – All three of these creatures are easy to find and make excellent bait.

Both caterpillars and catalpa (often pronounced ka-taw'-ba in Alabama) worms are the larva stage of butterfly-like insects and can be found in trees. Some build nests that are easily identifiable by a white, silky, opaque appearance.

These insects can be kept for long periods of time by placing them in glass jars (air holes should be punched in the lids) along with some of the leaves off of the tree from which they were taken. They can be kept in the refrigerator or in a dark, cool basement.

They can also be frozen and used all times of the year. The insects are messy, but worth the trouble. Both bream and catfish devour them.

The larger catalpa worms often are split down the middle with a knife or razor blade and turned wrong-side out before being placed on the hook. This release their juices and gives fish a taste they won't let go of.

Grubs also make excellent bait, especially for bream. The white insects can be found in most backyards under rocks and at the base of shrubbery.

Wasp larvae - Were it possible to poll bream, wasp larvae might be their favorite snack, even more than the cricket. The tiny, white morsels stolen from wasp nests make a delectable bait just the right size for hand-size bream.

Collecting wasp larvae can be a problem, however, and a painful one at that. Wasp larvae are present during the summer months, and because they are the wasps' young, they wasps protect the nests with vigor.

The easiest, but most expensive solution is to use wasp spray, the kind that shoots a 20-foot stream of immediate death.

Another solution is to use a cane pole. The bait gatherer can quietly slip up on the nest, give it a long-distance whack and run like hell. A well-placed whack will knock the nest loose and let it fall to the ground. The nest can be retrieved later when the wasps decide to leave in search of a better neighborhood.

The nests can be kept in re-closable plastic bag until time for the fishing trip. The individual larva can be placed onto small bream hooks much like a worm or grub.

Wasp nests can be placed in re-sealable plastic bags and kept in the refrigerator until fishing time.

Minnows – Minnows have a natural place in the food chain of bass, bream, crappie, catfish, white bass and many other species found in Alabama. It's only natural that the small fish make excellent bait.

Two types of minnows are primarily used in Alabama—toughies and shiners. Toughies are a small, slick-skinned minnow sometimes called fatheads.

These minnows are hardy and take changes in temperature and rough handling well. Shiners are a larger, but softer and more delicate minnow. Sudden changes and temperature and rough handling will kill them.

A large variety of minnows can be found swimming across the state and all make good bait. They can be trapped in a minnow trap. The easy-to-use traps are available at many larger fishing supply houses.

Minnows may be hooked in one of two ways. If the minnow is going to be cast and retrieved, or trolled, it should be hooked through the lips. A thin wire hook should be forced through the lower lip and out the upper lip.

This allows the bait to be pulled through the water naturally. If tight-line fishing, the minnow should be hooked through the fleshy portion of the back directly behind the dorsal fin, high enough to miss the backbone. Piercing the backbone with a hook will result in immediate paralysis of the minnow and it will soon die.

Crawfish – Crawfish are an often overlooked bait, although they are naturally found in the food chain of largemouth bass, spotted bass, smallmouth bass, catfish, hybrid bass and striped bass.

As a bait, the crawfish's main drawback is availability. Rarely are they sold commercially, and collecting them can be time consuming. The smart angler sends the neighborhood kids on a mission into nearby creeks, streams and ditches with a 25-cent reward for every crawfish found.

Crawfish should be hooked through the fleshy tail and allowed to sink to the bottom. The pinching crustaceans will do your work for you as they crawl around the bottom and alert fish to their presence. Crawfish make an excellent bait on stumpy points and around underwater brushpiles.

Shad – Shad, like minnows, are a natural part of the food chain for many Alabama fish, especially largemouth bass, spotted bass, smallmouth . bass, striped bass and hybrid bass.

Two species of shad are predominant in Alabama: threadfin shad and gizzard shad.

Shad have some drawbacks because they are less-than-hardy fish. Reduced oxygen, rough handling and sudden changes in water temperature can kill shad quickly.

A shad tank such as stripe fishermen use is a necessity for anyone fishing shad regularly. The tank should be white to reflect heat, round to prevent the shad from piling up in a corner and killing themselves and insulated to prevent sudden temperature changes.

The water should be kept cool (65-72 degrees) and should be aerated with salt. Salt acts as a medicine and keeps the shad healthy and frisky. A good rule of thumb is to add 32-ounces of salt per 30 gallons of water.

Shad can be hooked in either of two ways. Many anglers hook shad through the lower lip and out the upper lip. This method is okay, but there is a better way.

The stripe guides on both Smith Lake and Lake Martin hook the shad through the nostrils. This location is free of vital organs and is tough. The shad can breath through its mouth and it stays alive and frisky.

Shad are plentiful and easy to catch on most bodies of water in the state. Shad school around dams, concrete walls near dams, in many sloughs and pockets, and along sea walls. A cast net will easily take shad schooling in shallow open water, while a shad dip net will take shad along concrete structures.

Frogs – Frogs are a natural bait often overlooked by fishermen. In lily-pad or grass infested lakes and ponds where frogs naturally appear, a kicking frog on top of the water in spring or fall is an explosion waiting to happen.

Frogs should be lip-hooked and cast gingerly around shallow-water structure when bass are hitting topwater.

Frogs aren't available commercially, so they have to be caught. Driveways and back porches near water tend to attract frogs at night During the summer. In early summer months, when frogs hatch, it is not uncommon to find hundreds of small hoppers.

The night before a fishing trip, a flashlight and a foam cooler is about all you need. Shine the light in the frog's eyes to stun it, then capture by hand and place in the cooler.

Shrimp – In the Mobile Delta, where brackish water results in the intermingling of both freshwater and saltwater species, shrimp are an excellent bait. Shrimp should be hooked halfway up the fleshy portion of the tail, allowing the shrimp to swim freely. Shrimp may be Carolina-rigged and bottom fished, or hung from a bobber and floated off the bottom. Shrimp make an excellent bait for largemouth bass, as well as saltwater species such as redfish, speckled trout and sheepshead.

Other tackle needs

Unfortunately for the ol' paycheck, there's more to fishing than rods and reels, terminal tackle and bait. Just as the accessories make the fashion-conscious female, they also make the fisherman.

Accessories can make or break a fishing trip. This author once went on a three-day deep sea fishing trip and left his polarized sunglasses hanging on the rearview mirror of his truck in the parking lot. When he returned, he was suffering snow blindness from looking at the bright white boat for three days.

Other trips have been hampered by spending time fashioning fish stringers out of shoelaces, building raincoats from Hefty bags and trying to tighten reel screws with dimes and the eyes of fish hooks.

One especially memorable trip featured this author and two companions spraying down their arms with WD-40 to repel biting gnats in the Mobile Delta. A new can of bug spray sat in the front seat of the car in the parking lot.

Before assembling a wide array of fishing accessories, sit down with a piece of paper and make a list of priorities. Having an electric hook sharpener or a set of electronic, digital, hand-held scales is nice, but not at the expense of not having a pair of needle-nosed pliers or a life jacket. The tendency is to go shopping and come away with neat gadgets, but that tendency can lead to some long, miserable days of fishing.

While putting together a list of accessories, these items should be at the top of your list:

Essential Items

Personal safety items – Personal safety items such as needle-nosed pliers, wire cutters, sunglasses, sunscreen, mosquito repellant, goggles (if you ride in a boat) and weather radio are covered in the section on personal fishing safety. Don't skimp here. Having these items with you can make the difference between a situation being minor and being trip-ruining or even life-threatening.

Tacklebox - A tackle box is an angler's purse. It's simple as that. In theory, a tackle box allows the angler to take the equipment necessary for a successful trip. Lures, terminal tackle, monofilament, stringers and a wide variety of other goodies can be carried with ease in a tackle box.

Tackle boxes have made an incredible transformation through the years. Wicker baskets were replaced by steel boxes and then by aluminum boxes. Plastic boxes came along after that and better plastics made boxes plastic worm proof.

These days, soft-sided tackle boxes are the rage. These tackle boxes hold individual clear tackle boxes that can be swapped in and out depending on what is needed for that fishing trip. If you are going bass fishing on a freezing day, obviously there is no need to take a box of buzzbaits along. That box can be swapped for another box of jig-and-pigs or something else you might use that day. If you are going crappie fishing and don't plan on doing any bass fishing, you can take your crappie boxes, along and remove the bass tackle boxes.

These bags also have other benefits. Most offer front and side pockets that have room for sunglasses, goggles (for boat riding), a neatly-folded rainsuit, mosquito repellant, sunscreen, weather radio, hand-held scales, fish attractant, wire cutters (for cutting hooks out of humans), needle nose pliers (for removing hooks from fish), jeweler's screwdriver (for reel repair), an extra drain plug for the boat and even an extra baseball cap (to replace the one that always blows out of the boat).

These bags are also moldable to a certain degree and can be forced to fit in boat compartments or in your vehicle where hard tackle boxes couldn't fit.

What tackle box is right for you? You'll have to decide that. You may have $5 worth of tackle, or you may have enough to fill the back of a dump truck. Just be sure not to skimp and buy a box so small that it forces you to leave needed items—especially safety items—at home.

Life-jacket - If you plan to fish from a boat, or if you fish from the bank and cannot swim well, a life jacket is a must.

This is a place not to skimp, but not for the reason you probably believe. In this day and time, any life jacket that can be found on the market can save your life. The secret is to purchase a jacket comfortable enough that you'll wear it, and of quality workmanship so it will be in once piece when you need it.

You'll find the old, orange, collar-type vest inexpensive and tempting to buy, but this is a life jacket you're probably not going to wear. A SOSpenders type vest that wears like suspenders but inflates once it hits the water is the best bet. Spend the money to purchase a fishing vest. You'll be glad you did.

Stringer - If you fish from the bank, or from a boat that doesn't have a live well, you'll need a way to contain the fish you want to save and eat. A stringer is a must.

You can go two routes here, both of them inexpensive. A rope stringer usually sells for $2 or less, while a chain stringer costs only a little more.

If you have a choice, choose the chain stringer. On a rope stringer, fish are slid on top of one another, gill-to-gill, and they die easily.

With a chain stringer, fish are connected individually to safety pin-like snaps. Used sensibly, this stringer allows fish to stay alive for a long time.

Nail-clippers – Probably no inexpensive item is worth more to the angler than a pair of nail clippers. These clippers easily handle the difficult chore of cutting monofilament fishing line.

Many anglers are quite adept at biting line in two with their teeth, but not only is this bad on the teeth, it also leaves a flattened end on the line which doesn't go through hook eyes or weights easily.

Fishing towel – Of the fishing items that can be found free, nothing is more valuable than a fishing towel. Keeping hands free of worm and fish slime can prevent an expensive rod and reel from being ripped from your hands when a fish strikes suddenly and unexpectedly. On more than one occasion, this angler has caught a fish, unhooked it, returned it to the water and on the very next cast thrown the rod and reel overboard because of slippery hands.

A fishing towel also serves as savior for clothes, for the natural reaction to dirty hands while fishing is to wipe them on clothes.

Towels have other uses, too. I keep several in the boat and use one to dip into the water to wash my face. A towel dipped into ice water and wrapped around your neck on a hot day not only prevents "redneck," it also rejuvenates you.

A towel may be nothing more than and old kitchen towel, or a rag out of the rag bag at home. An inexpensive, but wise investment is a golfer's towel. A golfer's towel has a brass eyelet stamped into it and a snap ring which lets you snap the towel to a belt loop or tackle box handle.

Insulated ice chest – If you eat the fish you catch, an insulated ice chest is a must. Too many anglers catch fish, do a good job of keeping them alive on a stringer, then throw them in the trunk or in the back of the truck for the ride home.

Fish have fragile meat which spoils easily. Fish should be kept in top-notch condition from the second they are caught until they are placed on the table.

If a fish can't be kept alive on a stringer, it should immediately be placed on ice. Fish being transported should always be kept on ice.

A few non-essential items

As you get more involved with fishing, there will be items that you'll want to add to your fishing repertoire. Consider these:

Hook sharpener - Fish hooks become dull with age. A sharp hook catches more fish.

Rain suit – On a cold, rainy day, it can prevent the life-threatening situation of hypothermia.

Camera – There will be days you'll be proud of your catch and will like to show it off, but would also like to return that catch to the water. A camera is the only way that can happen.

Filet knife – If you plan to dress the fish you catch, a good filet knife makes the task so much easier.

Binoculars - Binoculars may seem like a strange item to have on a boat, but many good fishermen use them religiously. During the spring and fall months, they'll help you locate schools of fish hitting shad on top of the water.

Hand-held scales - Have you ever kept a fish you would normally put back just because you want to know what it weighs? Scales help promote catch-and-release.

Plug-knocker – Fishing lures are expensive. A plug knocker attaches to your line and slides down it to knock loose plugs that are hung up on stumps and other structure.

Mike Bolton

Catching fish

People may claim to go fishing only to clear their minds, but never doubt that catching fish doesn't help the mind clear more quickly.

Catching fish is dependent on many factors. Anglers must first fish where the fish are, present a bait in an appealing manner, successfully detect the strike, hook the fish, fight the fish and land it. To catch a fish, the angler's mind, reflexes and tackle all must be in good working order.

Detecting the strike

One of the most difficult tasks for the novice angler is detecting the strike—the moment in which the fish takes the bait in its mouth. On rare occasions, a fish may inhale a bait and detecting the strike is unimportant. More often, however, a fish will carefully mouth a bait and if anything feels unnatural—hard plastic, steel, a strange taste or the feel of line pressure—it will spit it out. It's for that reason that detecting the strike is so important.

Reaction to a strike depends on the bait and type of fish. Live bait feels and tastes natural to fish so they are less likely to spit it out.

Different species of fish attack a bait differently, too. Catfish, buffalo, carp, drum and redhorse suckers often will lightly mouth a bait, spit it out and mouth it again. These fish will often swim off with the bait, feel line pressure, and spit out the bait again.

Small bream, meanwhile, will attack a bait and run, trying to keep other fish from getting it. Slab-size bream will often give an undetectable strike and the angler's first inkling a fish is on is seeing the line move.

Crappie also will often lightly inhale a jig or minnow, and the strike may be almost undetectable. Bass literally suck in a bait, giving an electric-feeling tap-tap on the line. They'll just as quickly spit out the lure if it doesn't feel natural.

Striped bass and hybrid bass hit a lure on the run, never slower than 6 miles per hour, biologists say.

Setting the hook

The action of imbedding the hook in a fish's mouth is known as setting the hook. In trolling situations, or situations of a quickly retrieved lure, fish normally hook themselves and setting the hook isn't necessary. But in most situations, the angler must set the hook himself.

The type of fish, the type of hook and the strength of the line determine how the hook should be set.

In fishing for bream or crappie, the angler normally has to only watch a cork disappear, then lightly lift the rod tip to set the hook. The thin wire hooks used in this type of fishing penetrate the skin in the fish's mouth easily. The mouth of these fish tear easily and a strong hook set will pull the hook free.

In fishing for non-game species such as catfish and carp, the angler usually must watch his line and wait for the fish to swim off before setting the hook. To set the hook on these fish, the angler must give the rod tip a good pull from the 10 o'clock position to the noon position. Heavy wire hooks and treble hooks don't penetrate the rough mouths of the fish easily and a hard hook-set is needed.

317

In bass fishing with a worm, the angler must wait for the electric-feeling tap-tap before using a forceful snap of the rod, pulling up with both hands from the waist toward the face.

When bass fishing with crankbaits and other lures with multiple hooks, a lighter hook set is needed.

When tight-lining live bait for stripes and hybrids with rods mounted in holders on a boat, the fish swim away with a bait. These fish will mouth the bait from the rear, then turn the bait around head first before swallowing it. It is for this reason that a delay is needed before setting the hook.

Playing the fish

When it dawns on a fish that a hook is in its mouth and it is suddenly and unexplainably being pulled away from the direction it wants to go, its first reaction is understandably to escape.

The method of escape varies. Smaller largemouth and spotted bass may dive into cover, wrapping the line around stumps, pilings, brushtops or anything else that is handy. If that doesn't work, dazzling aerial displays often follow with the fish erupting from the water and trying to sling the lure by shaking their heads back and forth. Large bass rarely jump, choosing instead to break off the line in structure.

Smallmouth bass often try to sling a bait or lure by dramatic dives, followed by jumps of sometimes five and six feet into the air.

Catfish, drum, carp, striped bass and hybrid bass attempt to escape by trying to put enough pressure on the line to break it. Those species, when caught below dams in swift water, often run with the current and are successful in their escapes.

Smaller fish, such as bream and crappie, usually find today's modem tackle no match and can be reeled directly to the boat.

Some fish-fighting tips

1. Allow the action of the rod and the reel's drag system to absorb sudden runs by the fish. Never point the rod directly at the fish. This allows the line to absorb the shock, instead of the rod.

2. Adjust the drag on a reel every time you go fishing. Once a big fish is on, it's often too late.

3. It's best to set the drag too loose than too tight. When a big bass or striped bass makes a long run against the drag, the drag heats and expands, which tightens down on the drag.

4. Jumping fish are fun to watch, but the chances of losing a fish increase dramatically any time a fish leaves the water.

If the fish leaves the water, drop the rod tip and give it slack. This will make it more difficult for it to sling a lure by shaking his head back and forth. Better yet, discourage the fish from jumping by putting the rod tip down, even into the water, if necessary. This pulls down on the fish rather than pulling up on it.

4. Learn the technique known as "pumping." This age-old technique involves reeling the line tight, lifting the rod above the head to lift the fish up, then dropping the rod tip quickly as you crank in line. The rod tip is then lifted above the head again and the process begins again.

This procedure allows the angler to pull in a big fish when the big fish is putting up an extremely strong fight.

5. Never get a fish to the boat before it's ready. The shorter line it's on, the better its chance of escape. Wear a fish down, especially a big fish.

6. Once the fish is worn down and beside the boat, back off the drag considerably. Should the fish make an unexpected run on six inches of line, the loose drag will possibly contain it.

Landing fish

Which method to use in landing a fish is the choice of an angler. Smaller fish may be swung into the boat or onto the bank or dock, but larger fish need special landing procedures.

Fishing nets – The net is the most common landing device, but landing a fish with a net isn't as simple as it looks. Many a well-meaning angler has lost his fishing buddy's fish-of-a-lifetime thanks to sloppy net work.

Never attempt to land a fish from the rear with a net, or make a swipe at a still-feisty fish as it makes a run at the boat. Wait until the fish is worn out and calm and lying on its side beside the boat.

Place the net into the water vertically, 3/4 of the way in. Use the fishing rod to slide the subdued fish into the net. Any sudden lunge the fish might make will drive it into the net.

Fish learn from birth to escape anything chasing it. A net coming at the fish from behind triggers that instinct. It's rare that you'll be able to net a fish from the rear.

Lipping - Many anglers, and especially tournament bass anglers, prefer to land bass without a net.

How is that possible? Smaller bass are often simply swung into the boat to conserve time. Larger bass are "lipped."

Lipping requires a thumb to be placed inside the bass's mouth and an index finger to be placed under the lower lip. The fish can then be lifted out of the water.

When a fish is lifted in this manner, it is usually temporarily paralyzed and the hooks may be removed easily.

The drawback to this method, besides the fact the fish must really be worn down, is that the angler must handle the fish in close proximity to the hooks. A pair of needle-nose pliers should be used to remove the hooks to avoid a hook injury.

Safe release – If the fish is going to be released, it is often not necessary to put the fish into the boat or onto the bank or dock for hook removal. With the fish still in the water, a pair of needle-nose pliers can be used to take the hook out.

When using this method, wet your hands before handling the fish.

Beaching – On occasion, the angler fishing from the bank will catch a large fish and find himself without a net.

The best method is play the fish in deeper water allowing the fish to tire itself out. The fish can then be led on its side into shallow water where it may be lipped. On medium-sized fish, it is often possible to lead the fish into shallow water and actually drag it onto the bank.

Bridge gaffs – Anglers fishing from platforms below dams in Alabama, or anglers fishing from the piers on the Alabama Gulf Coast, are often faced with the predicament of having to land a big fish by pulling it up a great many feet. Monofilament fishing lines are often not up to the task.

These anglers often choose to use a bridge gaff. This device is made from either three or four large hooks welded together and weighted. This device hangs from a single rope.

The bridge gaff is lowered into the water, usually by a bystander, and the large fish is led over it. The bridge gaff is then lifted and the fish is impaled and raised to the platform.

Gaffs – The gaff is a standard feature on saltwater boats. A gaff is a barbless hook lashed to a long stick or pole.

Once a fish is brought beside the boat, the angler, or deckhand, uses the long-handled hook to impale the fish and lift it aboard.

Catch and release

One of the most important advances in sport fishing in Alabama is the advent of catch and release. Catch and release is a conservation tool in which fish are released back into the water unharmed rather than eaten or given away.

Unless the angler understands the art of catch and release, he may be defeating his purpose. Fish that are handled roughly are often returned to the water only to die later.

The largemouth bass is the most commonly released fish, and for good reason—it receives the most fishing pressure.

To release a bass, it's best to never remove the fish from the water. Handling the fish sometimes removes its protective coating of slime. While holding the fish in the water, use a pair of needle-nose pliers to remove the hook.

On occasion, it is necessary to bring a fish aboard the boat or onto the bank or dock to remove the hook. If that's the case, handle the fish by touching it only in the mouth area. Be careful not to allow the fish's body to brush against your clothing, the surface of the boat or dock, or the fishing rod or line.

Largemouth bass, spotted bass and smallmouth bass lend themselves to catch-and-release. Other species, such as bream, crappie, catfish, drum and carp, also can be released, but their release isn't as critical to an ongoing fishery as the release of the three bass species.

Striped bass and hybrid bass do not lend themselves well to catch-and-release, especially during Alabama's warmer months. Studies have shown these fish become heavily stressed during the long battle that is usually needed to subdue them. They often die after a long battle.

Preserving tournament fish

Tournament fishing presents a different catch-and-release problem. Bass caught during a tournament cannot be released immediately. They must be kept alive until the end of the tournament so they can be weighed.

Fish can be kept alive in your boat with clean, cool water, oxygen, and chemical treatment to calm the fish and help them recover from injuries, including the touch of your hands.

How you handle the fish greatly increases or decreases the fish's chance of survival. If the fish is destined for release, use heavy enough tackle to bring the fish in without exhausting it. Then be careful handling the fish when you take the hook out. The use of nets isn't recommended because they can severely damage a fish.

The Bass Anglers Sportsman Society once ran an experiment at major BASS tournaments using chemicals and ice in livewells and not a single fish was lost, BASS officials said.

Livewells in boats were treated with Catch and Release, made by Jungle Laboratories, and the livewell water was cooled with ice.

Catch and Release is a tranquilizer and water conditioner. It also adds electrolytes to water to restore them to the fish. The water treatment also stimulates slime protection by fish. When a fish is stressed by being caught and handled, the cells which produce slime shut down. Taking off the slime has an effect similar to taking the skin off a human. There is a great likelihood of infection.

When adding ice to the livewell, try to lower the water temperature about 15 degrees from lake water temperature. Block ice works best, but bagged crushed ice will suffice. Loose crushed ice will not. A good rule of thumb is five to eight pounds of ice per livewell.

The best way to monitor water temperature is to put an aquarium thermometer in the livewell. The experiments with BASS disproved the old tale that you can't drop temperature on bass or it will kill them. Finally, make sure the water in the livewell is clean. If you get your boat in shallow water and the pump sucks mud into the live well, change the water, even if it means you have to dip it out. It's less harmful to the fish to change the water and leave them out of water for a minute or two than to leave them in dirty water.

New boats have aeration and re-circulating systems in livewells. Older livewells circulate water from below the boat into the livewell.

By keeping your fish alive, you can return to the water those you don't need for a tournament or don't want for dinner. They can grow and be caught another time.